World Sustainability Series

Series Editor

Due to its scope and nature, sustainable development is a matter which is very interdisciplinary, and draws from knowledge and inputs from the social sciences and environmental sciences on the one hand, but also from physical sciences and arts on the other. As such, there is a perceived need to foster integrative approaches, whereby the combination of inputs from various fields may contribute to a better understanding of what sustainability is, and means to people. But despite the need for and the relevance of integrative approaches towards sustainable development, there is a paucity of literature which address matters related to sustainability in an integrated way.

Notes on the quality assurance and peer review of this publication

Prior to publication, the quality of the works published in this series is double blind reviewed by external referees appointed by the editor. The referees are not aware of the author's name when performing the review; the referees' names are not disclosed.

More information about this series at http://www.springer.com/series/13384

Walter Leal Filho · Amanda Lange Salvia ·
Luciana Brandli · Ulisses M. Azeiteiro ·
Rudi Pretorius
Editors

Universities, Sustainability and Society: Supporting the Implementation of the Sustainable Development Goals

Springer

Editors
Walter Leal Filho
HAW Hamburg
Hamburg, Germany

Manchester Metropolitan University
Manchester, UK

Luciana Brandli
Faculdade de Engenharia e Arquitetura
University of Passo Fundo
Passo Fundo, Rio Grande do Sul, Brazil

Rudi Pretorius
Department of Geography
Unisa
Pretoria, Gauteng, South Africa

Amanda Lange Salvia
Faculdade de Engenharia e Arquitetura
University of Passo Fundo
Passo Fundo, Rio Grande do Sul, Brazil

Ulisses M. Azeiteiro
Department of Biology and CESAM
University of Aveiro
Aveiro, Portugal

ISSN 2199-7373 ISSN 2199-7381 (electronic)
World Sustainability Series
ISBN 978-3-030-63398-1 ISBN 978-3-030-63399-8 (eBook)
https://doi.org/10.1007/978-3-030-63399-8

This Springer imprint is published by the registered company Springer Nature Switzerland AG
The registered company address is: Gewerbestrasse 11, 6330 Cham, Switzerland

Preface

Sustainable development is a matter of wide societal relevance, for two main reasons. Firstly, the pillar upon which societies are built, based on principles such as the rule of law, democracy, social inclusion, cohesion and justice, need to be complemented by elements such as well-being and the economic prosperity of its citizens. Secondly, in order to allow them to prosper, societies need to be based on the concept of environmental sustainability.

As institutions with a mandate to educate, perform research and outreach, universities are well placed to explore learning opportunities, and to expand the competences of students, as change leaders, allowing them to make a contribution towards the sustainability of the societies they are part of. No matter if in Europe, North America, Latin America, Africa, Middle East or Australasia universities should use their influence to not only educate students, but also encourage them to seek solutions and develop a new thinking that may promote and enable their societies to become more sustainable.

It is against this background that this book has been produced. It is a truly interdisciplinary publication, useful to scholars, social movements, practitioners and members of governmental agencies, with an interest on the connections between universities, sustainability and societies.

The aims of this book are as follows:

i. to present initiatives (i.e. curriculum innovation, research, activities, practical projects) as they relate to sustainable development at university level and links with society at large;
ii. to foster the exchange of information, ideas and experiences acquired in the execution of projects, from successful initiatives and good practice;
iii. to introduce methodological approaches and projects which aim to integrate the topic of sustainable development in the curriculum of universities.

Last but not least, a further aim of the book is to document and disseminate the wealth of experiences available today. This book is structured around two parts:

Sustainability-Based Approach and Methods
Case Studies on Sustainable Development at Universities.

We thank the authors and reviewers for their contribution. We hope that the contributions on this volume will provide a timely support towards the implementation of sustainability initiatives and will foster the global efforts towards promoting sustainable development works which link universities and societies. By doing so, a significant contribution towards the implementation of the UN Sustainable Development Goals may be provided.

Hamburg, Germany — Walter Leal Filho
Passo Fundo, Brazil — Amanda Lange Salvia
Passo Fundo, Brazil — Luciana Brandli
Aveiro, Portugal — Ulisses M. Azeiteiro
Pretoria, South Africa — Rudi Pretorius
Summer 2021

Contents

Sustainability-Based Approach and Methods

Moving Towards Sustainable Agriculture Through Higher Education

Jessica Zahra and Mark C. Mifsud

1 Introduction

Agriculture, as defined by Lehman, Clark and Weise (1993) consists of "activities which foster biological process involving growth and reproduction to provide resources of value. Typically, the resources provided are plants and animals to be used for food and fibre, although agricultural products are used for many other purposes also" (p. 127). Agriculture has been evolving tremendously throughout the years, with notions of sustainable agricultural practices becoming more popularised, where economic viability is being given equal importance to environmental, habitat and resource protection, along with the well-being of farmers and consumers of the produce (Gold 1999).

The main aim of this study was to qualitatively research the local full-time arable farmers' perceptions on issues pertaining to sustainable agricultural practices. The main research questions being:

- Are farmers aware of the negative impacts of commercial arable farming practices on the environment?
- Through their life experiences and knowledge on the field, what are their views of moving towards more sustainable farming practices?
- Are farmers willing to seek means of improving their practice while tackling daily agricultural issues?

J. Zahra (✉) · M. C. Mifsud
Centre for Environmental Education and Research, University of Malta, Msida MSD 2080, Malta
e-mail: jessica.zahra.4@ilearn.edu.mt

M. C. Mifsud
e-mail: mark.c.mifsud@um.edu.mt

W. Leal Filho et al. (eds.), *Universities, Sustainability and Society: Supporting the Implementation of the Sustainable Development Goals*, World Sustainability Series,
https://doi.org/10.1007/978-3-030-63399-8_1

The study will also delve into how universities and higher educational institutions may come about tackling issues of sustainability within society, in this particular case seeking the role of the SDGs in sustainable agricultural practices.

The major findings of this study focus on the current local, Maltese farming situation, delving into issues and possibilities of targeting more sustainable farming practices, considering the environmental, economic and social aspects of this local primary industry. This study aims to bring to light farmers' concerns and daily issues faced in their practice, helping identify their perceptions and attitudes towards sustainable agriculture, helping to find the most viable means of adapting such sustainable strategies through higher education input.

2 Literature Review

2.1 Sustainable Agriculture

> Sustainable agriculture is not agriculture without risk, which would be unrealistic to expect. It is agriculture that is able to manage risk and maintain its resilience in the face of change and inevitable surprise.
>
> (Pinter 2007, p. 22).

Agriculture has a major role to play in environmental protection, social wellbeing and economic growth. Alonge and Martin (1995) discuss the various changes taking place within the agricultural sector, all directed to minimise the negative effects this primary industry has, especially on the environment.

The Food and Agricultural Organization (FAO 2016) expects that by 2050, 60% more food production is needed to feed an anticipated 9 billion global population. The increase in global food production is expected to impact the natural resources, mainly water used for irrigation purposes, the land and its nutrients, such as phosphates (GIZ 2015). The 2030 Agenda (FAO 2016), recognizes that agriculture today faces more challenges than ever before, needing to feed more people, with a lack of water, reduced land productivity and a decreasing agricultural workforce, enforcing the need for global agricultural sustainability.

Scherer, Verburg and Schulp (2018) discuss the role of sustainable intensification and global food security, stating that "Ideally, sustainable intensification implies more production on the same land area while reducing environmental impacts and maintaining ecosystem functioning" (p. 43).

Farmers need to be open to adopt innovative techniques and technologies in order to cater for more sustainable practices. However, it is not only the job of farmers to guarantee sustainable practices, but consumers also play a crucial role. The more industrialised agriculture has become, the more people have begun to realise "that agriculture does more than simply produce food, animal feed and energy. It also has impacts on the climate, human health, and global ecosystems" (GIZ 2015, p. 5).

2.2 Agriculture and the 17 Sustainable Development Goals

Agriculture has a major role to play in all of the seventeen Sustainable Development Goals (SDGs). The SDGs are a continuation of the eight Millennium Development Goals (MDGs) adopted for fifteen years till 2015 (UN 2015), interlinking the three main pillars of sustainable development: "economic growth, social inclusion and environmental protection" (FAO 2017). These MDGs were replaced by the new seventeen SDGs, which have been in place since 2016 and which will remain in place till the year 2030 in view of the 2030 Agenda for Sustainable Development (FAO 2017). This Agenda was adopted by all 193 members of the United Nations, including Malta.

The Food and Agriculture Organization of the United Nations (FAO), has the task of targeting sustainable agriculture in view of all the SDGs, focusing on the importance of cooperation and partnerships between the relevant stakeholders across the globe. The main five strategic priorities of the FAO (2016 p. 21) in the 2030 Agenda are:

- To help eliminate hunger, food insecurity and malnutrition
- To make agriculture, forestry and fisheries more productive and sustainable
- To reduce rural poverty
- To enable inclusive and efficient agricultural and food systems
- To increase the resilience of livelihoods to threats and crises.

The FAO (2017) states that:

> As the fundamental connection between people and the planet, sustainable food and agriculture are at the heart of the 2030 Agenda, without proper nourishment, children cannot learn, people cannot lead healthy and productive lives, and societies cannot prosper. Without nurturing our land and adopting climate-resilient agriculture, future generations will struggle to feed a growing population (p. 5).

2.3 Education for Sustainable Development and Adult Education: Role in Agriculture

Environmental adult education seeks to bring changes in people's attitudes and behaviours. It aims at bringing about a sense of transformation, in the way people act and view the environment around them. Adult education plays a crucial role in allowing participants to see the link between the social, political and the environment, exploring each one and seeing the links between them as more of a holistic approach (UNESCO 1997). Through this study, the researchers aimed to gather knowledge on sustainable practices within the local agricultural community, seeking means of understanding the farmer's thoughts on sustainable practices.

UNISA (2015) discuss Paulo Freire's belief of the ability of adult education to bring about a sense of consciousness and the ability to critically become conscious of one's own situation while being able to act on a social and political level to bring about the necessary changes, overcoming silence. Freire believed that everyone has knowledge to offer and contribute towards learning. Adult education should offer the opportunity for the participants and adult educators to work on the same level, sharing their power, feeling comfortable to share and communicate their ideas effectively. "Through what Freire calls 'a pedagogy of question', adult learners are given an opportunity to reflect more critically upon their experiences and begin to re-shape their ideas, experiences and visions of the world" (Mayo, as cited in Clover 2002).

Looking for means of raising awareness amongst farmers, the questions asked during interviews for this study, sought to encourage reflection about one's practice, questioning whether there are more sustainable means or not. Engel and Engel (2012) discuss how the ultimate agents who ought to bring about change are the members of the community, in this case, a community of farmers who are to adopt new practices and implement other measures to cope with global issues, such as the threats of global warming. Farmers are the key stakeholders whose decisions will directly impact "either the environmental conservation or degradation" thus adequate training on developing and meeting their needs sustainably is essential (Freitas et al. 2015, p. 324).

University research may help minimise the gap between the higher institutions and public entities, in this case farmers. Researching means and ways of effectively communicating and implementing sustainable farming strategies, is essential to guarantee a sustainable future. "The preservation of the environment depends on an ecological awareness, which depends on education," (Gadotti 2009, p. 108) and through such training courses the participants are encouraged to awaken their relationship with nature, feeling the need to protect and preserve its resources, becoming conscious of their own practices (Gadotti 2009).

Caruana (2015) through his four case studies set in the Mediterranean context, states that one of the roles of Environmental Education is to help encourage the shift from the dominant paradigm of thinking, which predominantly deals with issues of unsustainable farming practices where profit is given priority, to an emerging paradigm where one's behaviour is more in line with their values, encouraging commitment through community solutions, motivating one another. ESD allows for community-dialogues, whereas the emerging paradigms, which are more sustainable, are not only discussed, but also implemented at a community level. When it comes to adult education, "passion and motivation seem to be crucial" (Caruana 2015, p. 671) in encouraging a shift in mind-sets towards more sustainable measures.

2.4 Agriculture in Malta

Agricultural land in Malta takes up around 110 km^2, with 9370 total holdings, characterised by small scale holdings (average of 1.2 ha). It contributes to 1.7% of Malta's economy and 2.9% of total employment (EU 2016).

The latest official statistics collected with regards to the farming population was through the Census of Agriculture and was last carried out in 2010 (NSO 2012), while a Farm Structure Survey was conducted by the NSO in 2013, having a total of 1372 full-time farmers, including both arable and pastoral farmers. This research project focused on arable farmers, totalling 725 according to NSO (2016) statistics. Since the agricultural census is held every 10 years, the most recent data (JobsPlus 2018) further indicates that registered full-time mixed crop growers, amount to 162 farmers while farmers having both fields and animals total 136 farmers. This indicates an overall total of 298 farmers. Most of the local farmers practice mixed farming, meaning they do not specialise in one specific cash crop, but grow a variety of seasonal crops (NAP 2018).

Maltese farmers have to face a variety of constraints some of which were outlined in the National Agricultural Policy (NAP) paper (2018, p. 24):

- Scarcity of land related to the small size of the islands and dense population.
- Lack of natural resources, in particular, water scarcity.
- Urbanisation, land use pressures and opportunity cost of land.
- Dependence on imported fodder and other inputs that are costly in view of a limited bargaining power.
- Fragmentation of human and physical resources.
- Individualism and general inability to exploit economies of scale.

Malta's membership within the EU has guaranteed farmers subsidies and aid, but has also introduced new regulations. The Common Agricultural Policy (CAP) aims to increase agricultural productivity, self-sufficiency, maintaining farming jobs by guaranteeing improved income for farmers. It also aims to stabilise markets and guarantee stable and reasonable consumer food prices (Waugh 2002). The Maltese government has also negotiated a Special Market Policy Programme, aiming to transform this small-scale industry into a self-sustainable one, rehabilitating, encouraging and reviving this traditional activity (Ministry for Rural affairs and the Environment, n.d.).

3 Methodology

The gap in knowledge, which this research project aimed to fill, is the lack of data on the views of full-time arable farmers on sustainable development. This research aimed to gather farmers' perceptions and attitudes towards a sustainable farming future, and most importantly, to give a voice to the few remaining full-time conventional, arable farmers on the Maltese Islands. In this section of the paper, the qualitative approach to this research will be outlined.

The aim of qualitative research is to collect data which is as in depth as possible (Lichtman 2006), having the opportunity to discuss matters with the farmers, seek their knowledge and perceptions on the issue at hand. Qualitative research "aims to achieve depth rather than breadth" (Blaxter et al. 2001, p. 64). This may be considered as a critical element of qualitative research as it "involves looking deeply at a few things rather than looking at the surface of many things" (Lichtman 2006, p. 13). Ary et al. (2006) discuss how qualitative researchers seek to make sense of reality and the uniqueness of each individual, how they think and how they feel in face of particular issues. It is also important to seek understanding through an insider's perspective, an "emic" perspective.

When doing qualitative research, the researcher must be aware of the risk of bias, being aware that his/her interpretation of things is influenced by their own view of the topic, thus having an influence on the interpretation of the results (Lichtman 2006). Interpretative studies are the basis of this part of the research inquiry. "They provide rich descriptive accounts targeted to understanding a phenomenon, a process, or a particular point of view from the perspective of those involved" (Ary et al. 2006, p. 463). Qualitative data, on the other hand lacks generalisation, because it is not its aim to collect data from a large number of participants, focusing on quality rather than quantity (Creswell and Clark 2011).

The method deemed to be the most adapt to gather qualitative data from the farming sample, was through individual interviews. Interviews allow the data collected to reflect the participants' opinions, beliefs and feelings better (Ary et al. 2006). Neuman (2006) defines interviews as "a short-term, secondary social interaction between two strangers with the explicit purpose of one person's obtaining specific information from the other" (p. 305). The type of interview carried out is a semi-structured interview. Semi-structured interviews minimise deviations by asking open-ended questions, but yet allow for follow-up questions and probing questions to clarify answers and go further in depth (Newby 2014).

The disadvantage of using interviews is that it provides the researcher with a large volume of in-depth data in a relative short period of time. It is time-consuming to analyse and classify the data collected through interviews (Ary et al. 2006). There is also the risk of the interviewees not willing to provide certain information, or rather providing false information (Ary et al. 2006). Cohen, Manion and Morrison (2007) discuss how voice recording the interview allows the researcher to retain eye-contact with the interviewee, although short note taking is also recommended. All of the interviews conducted with farmers were audio-recorded, with their consent, and this allowed for better communication while allowing the farmers to feel more at ease as it was more conversational.

The interviews were carried out with 10 full-time arable farmers around Malta and Gozo, one of which was a hydroponics farmer. In order to include the views of organic farmers, another 2 interviews were conducted, totalling 12 interviews in all. The participation in interviews was on a voluntarily bases. The participants were selected through purposive sampling.

The aim of conducting interviews with full-time farmers was to seek their truthful opinion on specific issues which are related to farming. During the interview, farmers expressed their opinion and worldview on current issues related to the sustainability of the agricultural industry on the Maltese Islands. The time spent answering questions and discussing issues ranged from twenty minutes to an hour and forty-five minutes. The interviews proved to be of high interest and very fruitful as all farmers showed great knowledge in their area of specialisation. Due to the very limited

Table 1 List of interviewed farmers

Interview	Date	Location	Interview length
Farmer 1	28-01-2019	Mġarr	18 min 22 s
Farmer 2	05-02-2019	Rabat	56 min 44 s
Farmer 3	19-02-2019	Mġarr	1 h 21 min 25 s
Farmer 4	22-02-2019	Żebbiegħ	1 h 40 min 29 s
Farmer 5	01-03-2019	Rabat	50 min 08 s
Farmer 6	08-03-2019	Mġarr	43 min 54 s
Farmer 7	10-03-2019	Organic	13 min 29 s
Farmer 8	15-03-2019	Organic	28 min 56 s
Farmer 9	30-03-2019	Kerċem, Gozo	30 min 46 s
Farmer 10	30-03-2019	Kerċem, Gozo	38 min 51 s
Farmer 11	31-03-2019	Nadur, Gozo	47 min 31 s
Farmer 12	06-04-2019	Hydroponics	16 min

number of organic farmers locally, the location of the organic farms is not being disclosed to safeguard the identity of the farmers (Table 1).

4 Results

Thematic analysis was used to encode all the qualitative data available through all of the twelve interviews (Boyatzis 1998). The twelve farmers interviewed came from different districts from across Malta and Gozo; Six from the Northern Region, two from the Western Region, one from the South-Eastern Region and three from Gozo (Fig. 1).

Through the use of open coding, a variety of themes began to emerge from the transcribed data. This was followed by axial coding leading to the identification of sub-categories, and finally selective coding led to five major themes (Strauss and Corbin 2008; Neuman 2006). Through these themes the data was organised using colour coding and highlighting.

The main themes:

- Challenges in farming
- Impacts of farming on the environment
- Indigenous farming knowledge
- A sustainable farming future
- ESD and farming.

Fig. 1 Map of Malta showing regions (NSO 2017)

4.1 Challenges in Farming

From all the interviews carried out with full-time farmers, various issues emerged which farmers considered as challenging. The most prevalent challenges faced by farmers which were mentioned by practically all of the farmers interviewed included: long hours of work without rest on Sundays; changing weather patterns and storms which destroy their crops; small land parcels; competition from abroad and the quantity of imported produce; the ever increasing diseases and pests coming along with imported produce; and, the local system of auctioning local produce at the Pitkali markets. Evidence of such challenges are such statements:

Farmer 1 Farmers do not have specific working hours, it is hard work, and you risk having all your produce taken away by storms. One day you have crops in your field, and the next everything may be taken away by storms.

Farmer 2 Produce imported from abroad impacted us very negatively because many pests and diseases came along with it. For example the white fly and the red spider did not exist in Malta before. All of these came because of the produce from abroad.

One of the worst things, which I noticed this year is the importation of carrots and oranges with leaves. This is not supposed to happen because it is a way to distinguish between local and foreign produce, and to reduce the risk of diseases being imported.

4.2 Impacts of Farming on the Environment

Human activity, in this case the use of land for agricultural purposes, inevitably brings along both positive and negative changes in the natural state of the environment. Most of the farmers expressed their opinion that without farming, the Maltese landscape would take on a totally different scenario. Farming helps maintain the fields in a good condition, at least visually. Some of the farmers expressed that it is of utmost importance that farmers understand the environment around them and work to maintain it in the best possible manner through proper care.

Farmer 7 Farming helps the environment. First of all, it takes care of the environment. Whether it is rubble walls and fields which are being worked and kept clean at least there is life rather than everything left on its own. There is a sense of organisation, when there are farmers working the fields.

One of the most prominent aspects of the positive impacts of farming on the environment which was mentioned by most of the conventional farmers, were rubble walls. Rubble walls may act for the purpose of dividing one field from another, or otherwise help in reducing soil erosion due to runoff. Rubble walls also act as an ecological habitat within themselves. Farmers understand the importance of keeping rubble walls intact.

Agricultural progress along the years has seen the development of new machinery and equipment to aid in easing the work of the farmer. Over the past years, conventional farmers had to take on a more of a commercial role, increasing the quantities of their produce in order to be able to compete with the imported produce. This has inevitably led to more pressure on the available natural resources of the Islands, especially soil and natural water resources, that is groundwater. Added production has also led to an increase in use of artificial chemicals and nitrates.

Farmer 5 It is impossible to work without using pesticides! Can a doctor work with natural remedies and not give us any medicines? He could and he tries to as much as possible and so dothe farmers! I can assure you, and it is important to understand, that the majority of farmers know what they are doing and they do not use pesticides for nothing. Pesticides are very expensive, very, and if one could not use pesticides, he would not!

Ground water is the only natural water resource available all year round in the Maltese Islands. Another seasonal water resource available, which farmers may utilise, is the water available during the rain period in valleys across Malta and Gozo.

Due to the dependence on mainly groundwater, the Maltese Islands experience water scarcity, especially in harsh Summer months, where the demand on water increases drastically. Some farmers also lamented about the fact that overconstruction on the Islands is also leading to a decrease in groundwater.

Farmer 2 Buildings and roads are both a threat. These large concrete roads. As water is not percolating through the rocks, it runs off. When we talk about table water, there are many who say that farmers are using the table water. But before, rain water had permeable surfaces to pass through, for example in valleys where a lot of water collects, water seeps slowly through. But nowadays if you have roads, or ceilings, water runs off and ends up being lost at sea, as we are not collecting such water. So, that is one thing against agriculture, as you must water the fields, just as we have to eat and drink.

Another prominent issue related to the agricultural industry and farming, is the level of nitrates present in groundwater. To limit the level of nitrates, present in water, solid manure must be stored between the 15th of October until the 15th of March. During the remaining period, the dry season, farmers can apply organic matter and fertilizers to their fields. All the farmers interviewed were aware of this Nitrates Action Programme and the time frame they are allowed to fertilize their fields.

In order to minimise the pressure on groundwater by the agricultural sector, the treatment of waste water in Malta is being implemented. This has not only led to an improvement in sea water quality, but also led to the production of Class A Secondary Water. This New Water is being offered to farmers in various localities across Malta and can be used for irrigation purposes. This is aimed to reduce the uptake of groundwater through boreholes.

4.3 Indigenous Farming Knowledge

A few of the farmers, especially the elderly farmers, remember a lot of different farming practices which were used in the past but are no longer in use nowadays. Most of the methods used in the past have now been replaced by machinery or other methods which make life easier for the farmers. Farmers spoke about how not all fields were irrigated, as some of the fields had no access to groundwater.

There are two types of fields, which in Maltese are called *bagħli* and *saqwi*. *Bagħli* is dry-land, meaning it is not irrigated, is only watered when it rains. *Saqwi* on the other hand means irrigated, through groundwater resources. As one may expect, nowadays the number of non-irrigated fields have decreased drastically and basically the majority of fields are irrigated through different means, increasing the production of crops. Farmers also spoke about changing irrigation practices throughout the past years. The improvements in irrigation practices led to less water being lost as farmers use the appropriate amount necessary for the plant.

Farmer 7 Before, they used to water using a water mill powered by a mule, going around with buckets, of water, filled up from the borehole. Nowadays, we use pumps. There used to be water channels, made of stone around fields to transport water.

Speaking to conventional farmers, almost all of them were sceptical on organic farming, especially about local organic farms. Some of the reasons why they are unconvinced is the fact that given that the Maltese Islands are so small, it is rare to find a field which is enough distance away from another field which may be sprayed. The buffer zone was the most discussed issue of why they do not believe that it is possible in Malta. Another issue which was raised is the loss of produce when one converts to organic, hindering farmers from opting to venture into organic farming. Even terraced fields may hinder farmers from considering organic farming as they discussed how the chemicals in the soil may percolate down to other fields.

4.4 A Sustainable Farming Future

Farmers had a lot to say about a sustainable farming future. When asked what they understand by sustainable farming, most delved into aspects of the farmers' wellbeing, financial stability and also enhancement of the environment, which they depend on.

Farmer 6 Sustainable agriculture means that one grows produce which is viable and provides a living for the future and the present. It is important that one does not grow any crop, thinking that it may or may not produce. Nowadays, you cannot do that. The expenses are huge, so one has to be careful which produce one chooses to grow and how.

Most of the farmers were not so optimistic about farming in Malta, especially as a full-time job. The number of full-time farmers has been on the decline, along with the number of young farmers. Farmers commented about the fact that lately a number of fields are being bought or used for recreational purposes rather than for farming as such. It could be that due to an increase in part-time farmers and a decrease in full-time farmers, farming is being seen as more of a pastime rather than a full-time job.

Farmer 3 The future of agriculture in Malta, if I take you around in fields, you could see fields with rooms in it, not store rooms, but with a BBQ, that is the future I am seeing. Maybe I am being pessimistic, but…

Encouraging young farmers is the key for guaranteeing a future for this industry in Malta. The government is helping by providing subsidies and helping farmers invest in this industry through European Union funding. The question is: Is it enough to encourage the younger generation to take on farming as a full-time job? University courses and training workshops may be a step in the right direction, attracting the younger generations to specialise in farming practices. Promoting training courses and offering initiatives for the middle aged and older generation of farmers, may also help boost the local agricultural sector and help them understand the importance of adopting new sustainable farming techniques, showing them, through first-hand experiences, that such methods are affective.

Through these interviews conducted with farmers, one could not help but notice that each of them mentioned various eco-friendly sustainable methods which they use in order to make conventional and organic farming more sustainable. Such methods mentioned by farmers include the use of flowers to detect any diseases, use of bees for pollination, use of mulching, crop rotation, natural sprays used on plants as a prevention and the most common one, the manual pick up of weeds rather than using herbicides.

One of the common things between conventional farmers and organic farmers is the use of sulphur and copper sulphate, which are both natural chemicals used as fungicides. These have been used for years and are still commonly used by both sectors of farmers.

Farmer 8 We use a lot of methods which were used by our ancestors. We use the same methods as them. We make use of sulphur (*kubrit*) and copper sulphate (*kupru*), the method used before chemical pesticides were used in Malta.

Farmers had a lot to offer with regards to suggestions in order to improve their practice or even make farming more sustainable in the long run. They suggested for better rain-water cathcement projects in valleys, just competition, more marketing of local produce, the use of organic matter collected in households which may be used in fields and most importantly, better labelling, packaging, grading and traceability of local produce.

4.5 *Education for Sustainable Development and Farming*

Most of the farmers acknowledged the fact that various courses are provided. Using the principles of ESD during such training courses, would help farmers embrace the principles of sustainable farming practices, along with their current good practices. Engaging farmers in decision making, will also help them in embracing such decisions and implementing them within their line of work. A sustainable farming future may guarantee the continuation of this primary industry, seeking means of improving it, making it more profitable, seeking means of reducing the impact on the environment and guaranteeing the well-being of farmers.

Farmers feel that often, within society in general, they are looked down at. They feel their work is not appreciated enough. Sustainable Development is not only about promoting the care of the environment, but also helping in the well-being and social aspects, in this case of farmers. Empowering farmers is a need. To feel appreciated and trusted in your line of work will encourage one to work even harder.

Farmer 6 We have to be careful, as they say agriculture is not important, but agriculture is of utmost importance to our country! Certain people may not appreciate as they are served, but since you have a local Maltese product, a healthy product and a very good one, in my opinion, if you do not take care of them and help them, one day in the future it is going to be a shame as then you cannot bring them back.

The need for more dialogue between the government, different stakeholders and farmers themselves was mentioned by a number of farmers. They feel the need to be consulted and voice their opinions on decisions taken in the name of farmers. ESD is built on giving a voice to all relevant stakeholders and giving a voice to the sectors of society which may not always be heard.

Another relevant issue raised by one of the farmers to guarantee a better future for farmers, was the need to work with school children. This would enable them to get to know more about the farmer's work, appreciating their work and understanding where their food comes from. Seeing the produce grow, one would become aware of the whole process involved, knowing how it is produced and finally ending at home on their plates.

Farmer 3 The thing I like most about farming, is the need to carry a sense of responsibility, to produce your crops in the most nutritional means and cultivate it, as at the end we are producing food for our consumers.

4.6 Discussion on Main Findings

This research project elected to provide a clearer picture of the local farmers' attitudes and perceptions towards sustainability issues in farming. It is evident that farmers are familiar with the concepts of sustainability, not necessary the word 'sustainability' itself, but rather the concepts which form an integral part of sustainable agricultural practices. Through interviews, farmers integrated the social, economic and environmental aspects of their practice.

It is apparent that farmers do practice some sustainable measures, which were mostly passed from previous generations, such as crop-rotation, use of natural fertilizers and herbicides, the importance of upkeep of rubble walls, maintaining their soils' nutrients, etc. Meanwhile, it was also evident that the majority of farmers feel that they cannot totally do away with using artificial chemicals, such as pesticides, pointing out the fact that lately they have been witnessing an increase in invasive pests and diseases damaging their crops.

In view of Malta's natural water resource, farmers acknowledge that it is a scarce resource, pointing out the fact that in recent years the majority of farmers have been using more efficient water irrigation methods, such as drip-irrigation. New Water, having been lately introduced in new areas around the Islands, has great potential of helping reduce ground water extraction. Some of the farmers are still uncertain about using it, fearing that it will be of a detriment to the quality and taste of their produce.

More distribution points are also needed. The future of the farming industry in Malta, according to the farmers is in a dire state, noticing the decrease in the number of younger full-time farmers.

5 Conclusion

This research project, carried out through the University of Malta highlighted sustainability issues pertaining to the local farming community that require clarification, better understanding and increased engagement from the farming community. This was the first phase of the research project. In the next phase the university will design outreach activities in order to address these lacunae in the form of training workshops and short courses in order to help farmers

i. better integrate sustainable farming strategies in their work
ii. improve farmers' social and economic status
iii. develop the skills necessary that lead to increased empowerment.

The main thrust will be to support local communities, helping reduce the costs and carbon footprint related to transportation of goods. Further studies should focus on how consumers can aid local farmers and how the local economy may benefit farmers, the community and the environment through increased sustainability. The university should also research ways in which it can increase its engagement with schools and the younger generations. Schools may be a potential platform through which farmers and the public together with the younger generations engage together, encouraging dialogue and build up trust and understanding.

References

Alonge AJ, Martin RA (1995) Assessment of the adoption of sustainable agriculture practices: implications for agricultural education. J Agric Edu 36(3):34–42. https://doi.org/10.5032/jae.1995.03034

Ary D, Jacobs LC, Razavieh A, Sorensen C (2006) Introduction to research in education, 7th edn. Thomson Wadsworth, Belmont, USA

Blaxter L, Hughes C, Tight M (2001) How to research, 2nd edn. Open University Press, Buckingham, United Kingdom

Boyatzis RE (1998) Transforming qualitative information: thematic analysis and code development. SAGE Publications, London, United Kingdom

Caruana C (2015) Civic action for sustainable futures: what role for adult environmental education? In: Leal Filho W et al (eds) Intergrative approaches to sustainable development at university level. World Sustainability Series, pp 663–674. https://doi.org/10.1007/978-3-319-10690-8_45. Retrieved from https://www.um.edu.mt/library/oar/handle/123456789/41926

Clover DE (2002) Environmental adult education. Retrieved from http://journals.sagepub.com/doi/abs/10.1177/104515950201300201

Cohen L, Manion L, Morrison K (2007) Research methods in education, 6th edn. Routledge, Oxon, United Kingdom

Creswell JW, Plano Clark VL (2011) Designing and conducting mixed methods research, 2nd edn. SAGE Publications, London, United Kingdom

Engel KE, Engel PH (2012) Building resilient communities: where disaster management and facilitating innovation meet. In: Wals AE, Corcoran PB (eds) Learning for sustainability in times of accelerating change. Wageningen Academic Publishers, Wageningen, Netherlands, pp 133–147

EU (2016, March) CAP in your country: Malta. Retrieved from http://ec.europa.eu/agriculture/sites/agriculture/files/cap-in-your-country/pdf/mt_en.pdf

FAO (2016) Food and agriculture: key to achieving the 2030 agenda for sustainable development. Food and Agricultural Organization of the United Nations. Retrieved from http://www.fao.org/3/a-i5499e.pdf

FAO (2017) Food and Agriculture: Driving Action across the 2030 agenda for sustainable development. Food and Agriculture Organization of the United Nations. Sustainable Development Goals. Retrieved from http://www.fao.org/3/a-i7454e.pdf

Freitas MR, Matias SV, Macedo RL, Freitas MP, Venturin N (2015) Nonformal environmental education: a case study with farmers in a town of the South Minas Gerais state, Brazil. Chin J Popul Resour Environ 14(4):324–331. https://doi.org/10.1080/10042857.2015.1078491

Gadotti M (2009) Eco-pedagogy: extending the educational theory of Paulo Freire to sustainability. In: Corcoran PB, Osano PM (eds) Young people, education and sustainable development: exploring principles, perspectives, and praxis. Wageningen, Netherlands, pp 107–112

GIZ (2015) What is sustainable agriculture?. GIZ, Bonn and Eschborn, Germany

Gold MV (Sept, 1999) Sustainable agriculture: definitions and terms. Alternative Farming Systems. USDA. Retrieved from https://books.google.com.mt/books?id=zwAUAAAAYAAJ&printsec=frontcover&dq=Sustainable+Agriculture:+Definitions+and+Terms+WithMary+V.+Gold&hl=en&sa=X&ved=0ahUKEwjci7j92pzmAhXRYcAKHaHKBvgQ6AEIOzAC#v=onepage&q&f=false

Lehman H, Clark A, Weise SF (1993) Clarifying the definition of sustainable agriculture. J Agric Environ Ethics 6(2):127–143

Lichtman M (2006) Qualitative research in education: a user's guide. Sage Publications, London, United Kingdom

Ministry for rural affairs and the environment (n.d.) Malta fact file: agriculture, food, fisheries. Retrieved from http://cap.europe.bg/upload/docs/2012-05/loadfile.ashx.pdf

NAP (2018) Agricultural policy for the Maltese Islands: 2018–2028. Parliamentary secretary for agriculture, fisheries and animal rights. Retrieved from https://agriculture.gov.mt/en/agricultural_directorate/Documents/nationalAgriculturalPolicy/napFinal.pdf

Neuman WL (2006) Social research methods: qualitative and quantitative approaches, 6th edn. Pearson International Edition, Boston, USA

Newby P (2014) Research methods for education, 2nd edn. Routledge, Oxon, United Kingdom

NSO (2012) Census of agriculture 2010. National Statistics Office, Valletta, Malta

NSO (2016) Agriculture and fisheries (2014) National Statistics Office, Valletta, Malta. Retrieved from https://nso.gov.mt/en/publicatons/Publications_by_Unit/Documents/B3_Environment_Energy_Transport_Agriculture_Statistics/Agriculture_and_Fisheries_2014.pdf

NSO (2017) Regional statistics Malta: 2017 Edition. National Statistics Office, Valletta, Malta. Retrieved from https://nso.gov.mt/en/publicatons/Publications_by_Unit/Documents/02_Regional_Statistics_(Gozo_Office)/Regional%20Statistics%20MALTA%202017%20Edition.pdf

Pinter L (2007) A strategic approach to influencing agricultral policy and practice through measurement. In: Hani FJ, Pinter L, Herren HR (eds) Sustainable agriculture: from common principles to common practice. International Institute for Sustainable Development and Swiss College of Agriculture, Manitoba, Canada, pp 19–24

Scherer LA, Verburg PH, Schulp CJE (2018) Opportunities for sustainable intensification in European Agriculture. Global Environ Change 48:43–55. https://doi.org/10.1016/j.gloenvcha.2017.11.009

Strauss A, Corbin J (2008) Basics of qualitative research: techniques and procedures for developing grounded theory, 3rd edn. Sage Publications Inc, London, United Kingdom

UN (2015) The Millenium development goals report 2015. New York. Retrieved from https://www.un.org/millenniumgoals/2015_MDG_Report/pdf/MDG%202015%20rev%20(July%201).pdf
UNESCO (1997) 6a Adult environmental education: awareness and environmental action. In: UNESCO, A series of 29 booklets documenting workshops held at the fifth international conference on adult education. UNESCO Institute for Education, Hamburg, Germany, pp 273–286. Retrieved from http://unesdoc.unesco.org/images/0011/001159/115957eo.pdf
UNISA (2015) Principles and theories of adult education tutorial letter 501/3/2015. Department of Adult Basic Education and Youth Development, UNISA. Retrieved from http://unesdoc.unesco.org/images/0024/002451/245104E.pdf
Waugh D (2002) Geography: an integrated approach, 3rd edn. Nelson Thornes, Cheltenham, United Kingdom

Jessica Zahra is a Geography teacher at a Middle school in Malta. She has graduated with a Bachelor of Education (Honours) at the University of Malta, specialising in the teaching of Geography and History. Having both education and environment at heart, Jessica has recently embarked in studies and research in the field of Education for Sustainable Development. She is currently finalising her Masters course in Education for Sustainable Development at the University of Malta under the Centre for Environmental Education and Research (CEER).

Mark C. Mifsud is a Chartered Biologist, a Chartered Environmental Scientist, a Chartered Science Teacher, a Certified Wildlife Photographer and a Prince II Certified Practitioner with a wide range of experience in Environmental Education and Sustainable Development. Mark has drafted management plans for nature parks, reviewed EIAs and advised on ODZ applications as an independent member on the NHAC board within MEPA. He has more than 20 years teaching experience (from primary to tertiary levels), and is currently involved in environmental communication, teaching, research, academic refereeing and in a number of international projects. Mark has been entrusted as a member on various national and international boards. He is presently co-ordinating the masters in education for sustainable development at the University of Malta and holds adjunct status within the faculty of Integrated Science and Technology at James Madison University, Harrisonburg, Virginia in the US. He also held teaching posts at the University of Ca Foscari Venice, Italy, at Southern New Hampshire University in the US and at the University of Yaounde, Cameroon. Mark has published in excess of 40 publications in the area of education for sustainable development.

Balancing Sustainable Livelihoods and Conservation in the Marine Environment

Mark C. Mifsud

1 Introduction to MPA Management and People's Livelihoods

Marine protected areas (MPAs) are now a widely used tool for marine conservation and fisheries management (Charles and Westlund 2016). In coastal areas, it has become clear that the success of MPAs, and the achievement of sustainable fishery production, requires a combination of effective management and conservation frameworks, maintenance of decent fisheries livelihoods, and a governance system that allows for effective participation of coastal communities, fishing people, and other users in considering, designing and implementing MPAs (Charles and Westlund 2016).

MPAs are an important management tool that can either help or hurt local livelihoods, depending on how they are developed, designed, and implemented. It is not unusual that fishermen resist the establishment or expansion of MPAs. This is usually done because fishermen may fear a reduction in their livelihood and also a decrease in their ability to access to their fisheries. The establishment of MPAs may therefore result in conflicts between fishers and state agencies. A further issue is that there may be increased conflicts between fishers and the tourism sector. This stems from the fact that the objectives and programmes of MPAs often in favour of tourism at the expense of other sectors (CANARI 2005).

Research generally points towards the effective importance of incorporating stakeholders in meaningful participation for effective marine conservation planning and management (Pomeroy and Douvere 2008; Voyer et al. 2012). In actual fact, some

M. C. Mifsud (✉)
CEER, University of Malta, Valletta, Malta
e-mail: mark.c.mifsud@um.edu.mt

W. Leal Filho et al. (eds.), *Universities, Sustainability and Society: Supporting the Implementation of the Sustainable Development Goals*, World Sustainability Series,
https://doi.org/10.1007/978-3-030-63399-8_2

researchers state MPAs are more likely to meet their biological and social goals if the human dimensions is integrated into the MPA design and evaluation process (Pollnaca et al. 2010). MPA management strategies that find the "middle-ground" between government-led and community-based approaches may be most effective (Jones 2002).

The aim of this research project was to understand the perceptions and attitudes of stakeholders involved in the fisheries sector towards the implementation of new MPAs especially with regard to the impact these MPAs would have on their livelihood. Although there have been some studies carried out in other countries, this is one of the first such qualitative studies to be undertaken in the Maltese Islands. The study also attempts to highlight the role of higher education institutions and other various authorities in helping to achieve a balance between the fisheries sector and conservation.

2 Livelihoods and the Local Context

Livelihoods are usually defined as

> "…the assets (natural, physical, human, financial and social capital), the activities, and the access to these (mediated by institutions and social relations) that together determine the living gained by the individual or household" (Allison and Ellis 2001). A livelihood is sustainable when it can cope with and recover from stresses and shocks and maintain its capabilities and assets both now and in the future, while not undermining the natural resource base (Chambers and Conway 1991). The term 'livelihoods', can have different meaning ranging from 'enhanced' livelihoods, 'complimentary' livelihoods, or 'alternative' livelihoods on a continuous variable depending on the actual amount and type of change occurring in the livelihood of the economic stakeholder (Fig. 1).

Although heavily advocated as a solution to the many problems facing fishermen, the provision of supplemental and alternative livelihoods has had only limited success

Enhanced Livelihoods

Primarily focus on adding value to ongoing traditional or historical activities through improved management. Basically keeping current livelihood with better effective management practices. Fishermen should not be forced from the sea because of bad management or management that favours other sectors.

Complimentary Livelihoods

A diversification strategy that includes elements of adopting "supplemental" strategies (making current practices more sustainable). This strategy is less risky than alternative livelihoods

Alternative Livelihoods

Change in livelihood and enhanced management. Usually the least desirable option of the three. When changes in livelihood strategies are required, they should draw on and respect people's interests and capacities.

Fig. 1 Different types of livelihood. Adapted from Pomeroy

in most cases. It is assumed that people only fish as an occupation of last resort, and that people would give up fishing if a more economically viable option as available. These assumptions rarely hold true because they often fail to consider people's strong attachment to the fishing lifestyle (cultural and personal identity).

Therefore to ensure effective participation from economic stakeholders (especially fishers), reduce conflict and sustain long term sustainability, MPA management should include provisions that help fishers achieve enhanced livelihood as much as possible. This mainly involves active and meaningful participation throughout the process through which management regulations and provisions are taken.

Maltese fisheries represent a typically Mediterranean artisanal type. They are not species selective and are frequently described as multi-species and multi-gear fisheries. This means that fishers switch from one gear to another several times throughout the year (NSO 2016). The social and cultural importance of the Maltese fishing industry far outweighs its negligible economic contribution, which is equivalent to about 0.1% of the national Gross Domestic Product (NSO 2016). The most active fishing ports in the Maltese islands are located in the Southwest area with more than 1047 boats registered while the most common size of boats utilised are boats up to 5 m in length or from 5 to 10 m in length (NSO 2016). The main fishing port in Malta is Marsaxlokk Harbour, in the southeast of the island while the most important fishing port on the island of Gozo is Mgarr Harbour, where over 70% of the island's fleet berth. There are four fishers' cooperatives in Malta and all professional fishermen are affiliated to one of them.

3 Methodology

The methodology for this study included extended interviews with a number of stakeholders employing constructivist/interpretivist approaches. The study included the participation of 26 separate stakeholders (entities, co-operatives or associations) of which 12 were economic stakeholders. This included both economic stakeholders and those whose livelihood was not necessarily dependant on the marine environment such as government agencies and regulatory authorities, along with NGOs as seen in Fig. 2. These stakeholders were deemed to be of importance in order to obtain a more comprehensive view of the current situation.

The planning of the research process started off with a 'desk study' process in order to set up a plan of action with regard to research. An introductory meeting with stakeholders was first set up during January (2018). During this meeting this study was introduced and the scope of the livelihood analysis explained. Having already considered the importance to use qualitative research, a number of interview questions were drafted. The interviews took the structure of a focus group or group interview as more than one representative from the same association was present. The author personally acted as the moderator or interviewer in all the interviews and tried to 'create a permissive and nurturing environment that encourages honest

Fig. 2 Main consulted stakeholder groups

perceptions and points of view, without pressuring participants to vote, plan or reach consensus'.

The interview process included the following structure/procedures:

- The site selection of the interviews were according to the wishes of the stakeholders. Most interviews were held in their actual official premises.
- The interviews were digitally recorded and transcribed by the researcher.
- The language used was according to the wishes of the participants. Most interviews were held in Maltese. Although English featured more prominently in the diver and NGO interviews.
- Interview questions were sent (usually digitally) to the stakeholders at least three days prior to the actual meeting.
- The interviews were performed from January to April 2018.

The interviews were then transcribed in the original language and the verbatim used eventually translated into English. The scripts were then examined thematically.

4 Results

The next part of this study takes the form of a narrative report. It reviews the main themes, and provides extensive quotations and paraphrases from individual participant responses, meant to illustrate and support each theme. In some instances sections of dialogue within the interview were used to better illustrate participant thoughts and opinions about various issues, in addition to individual participant responses. The data was analysed according to an adaptation of the approach identified by Vaughn et al. (1996):

- *identifying the big ideas*
- *unitising the data*
- *categorising the units*
- *identifying themes and use of theory.*

4.1 Main Themes

From the interview study a number of categories started to emerge, and after further analyses, the general stakeholder responses were categorised in the following main themes:

- Perceived Shortcomings in Current MPAs
- Blaming Game—Need to build bridges
- Enforcement and Livelihoods
- Visions of Sustainability
- New MPAs and possible effects on livelihoods
- Suggestions for Improvement.

4.1.1 Perceived Shortcomings in Current MPAs

All the stakeholders to varying degrees consider that the current MPAs are not functioning. They state that rules are non-existent as is enforcement. Ten stakeholders did not want to comment on the proposed MPAs due to current issues surrounding the existing ones which are not considered acceptable by them.

> *Nothing has ever been done to protect marine species in Malta. Nothing.*
>
> *We were voted best marine diving in Europe…but there are no fish.*
>
> *(Dive Club Association 1)*

> *No one cares about the MPAs. You can fish everywhere – Fishermen should understand that when MPAs work, fishing will become better for them. How can we propose another eight if we are not even taking care of the current ones we have. We do not need reports and books so everything remains on paper.*
>
> *(NGO 1)*

Others stated that the existing rules and regulations have not been communicated clearly.

> *We started with Marine protected areas some years back, and I am speaking on the existing ones. What can I do, can I throw a net? I do not know what can be done. None of the MPAs are working.*
>
> *(Recreational Fishermen Association 1)*

4.1.2 Blaming Game—Need to Build Bridges

The majority of the stakeholders attributed the reduction in fish to other users. Most stakeholders either put the onus on recreational fishermen and on spear diving.

So we have created a desert by fishermen and you are worried about the little corals on the cave that can be effected by air bubbles.

(Dive Centre 1)

There is enforcement for the Fishermen but no one talks about the frogman – they capture a lot of fish. There are many zones in which we cannot fish why only us.

(Professional Fishermen Association 1)

The marine environment has now turned to the worst – There are divers that go down at night for groupers on wrecks and at night and it is illegal. Therefore, the top predators have been reduced. Caves have been denuded from the presence of Sertella due to the bubbles from the divers. We put underwater signs but they were removed after three days. Dive centres do talk about conservation etc. *but when it hits them financially, they ignore these regulations.*

(Diving Club 2)

Nonetheless, a number of diving centres did single out spearfishing as one of the main threats to the sustainability of the sea, while other diving centres mentioned both recreational and commercial fishing as the main activities to blame for the current situation in the sea. An NGO did not make a difference between recreational and commercial fishing and just mentioned that the important principle is sustainability.

This is San Lawrenz (shows pics)-look at the nets, no fish. At Crocodile rock we are diving and they are throwing the nets. The main problem are the recreational Fishermen not the professional as they do not follow any rules.

(Diving Centre 2)

Coastal fishing is going to go down – I agree that it should be restricted. In the sense that an S or an MFC – but with proper regulations. The professional Fishermen has to compete with these and this is unfair competition as they do not pay tax. What about a bag limit for the recreational Fishermen? And they sell at a reduced price. Recreational fishing as it is harming us a lot. How can you be recreational if you catch 80 kg of pixxi San Pietru. There should be a lot of control on them.

(Professional Fishermen Association 1)

4.1.3 Enforcement and Livelihoods

Most stakeholders were extremely critical of the current levels of enforcement in the MPAs. This fact alone attaches a very negative perception of MPAs in the stakeholders and it is one of the major reasons why ten stakeholders did not want to comment on the impact the proposed MPAs could have on their livelihoods. Some stakeholders mentioned that the rules are not well known to most of them.

There is no enforcement at all for the existing MPAs. There are no rules. If there are rules we would know about them. The rules are not defined as they cannot enforce them. The rules have to be crystal clear and enforced.

(NGO 1)

Who is responsible for enforcement at sea. There are no guidelines and rules on existing MPAs, only some rules on certain wrecks. Unless we create a green water wardens – we need a gang of these to control Malta and Gozo – following rules which are set and clear. We are not against fishing but they have to understand that there are no fish and that something has to be done. They cannot continue to fish everywhere.

(Diving Club 1)

If there are any rules in the current MPAs we would have heard about them and if they are enforced, someone gets fined, we would get to know.

(Professional Fishermen Association 2)

There has to be enforcement – as the areas will not work. Currently enforcement is not enough.

(NGO 2)

The Police department appear to be hindered in carrying out proper enforcement because of a lack of both human and physical resources.

We have a maximum patrolling capacity of one dinghy in the south and one in the north and usually do not venture out more than 5miles. Gozo and Comino are not within our remit. Most illegalities are mooring boats in swim zone, non registration Recreational fishing and commercial fishing not really within our remit. Report comes in from fisheries department. If there will be so many protected areas, enforcement will be too difficult. We have four dinghies but not enough personnel to use them all at same time. I need at least double the number of people

(Government Entity 1)

The Civil Protection Department state that they try to assist the police and AFM through reporting, as they cannot enforce the rules.

We assist both the police and AFM. We can report but do not need to enforce – we usually help out ERA as needed in research but we do not need any increase in people.

(Government Entity 2)

4.1.4 Visions of Sustainability

When asked to state their main concerns in various open ended questions the majority of stakeholders mentioned ideals of sustainability in a way in which they can continue with their activities—but also mentioned the importance of conservation in order to keep alive what we have.

Following numerous meetings and seminars held during the past years, which we attended, Birds, caves, sand banks and reefs have always been the main issues discussed. However despite our repeated questions about what are the parameters involved to protect these environmental issues, we never received an adequate answer. Our main activity is fishing. We wish to know what is allowed and what is not allowed. For this reason one cannot answer what impact these and other MPAs have on fishing activities, whether as a livelihood or as recreational. On the other hand we wish protection of the marine flora especially fish and allied creatures from the destructive and non selective fishing gear. Unfortunately this issue is never upheld by the Environment Authorities and they refer this issue to other entities.

(Recreational Fishermen Association 1*)*

Our interest in the sea is not money or income – but we encourage people to use the sea. We want the MPA to be set up in a way that enhances the wildlife in those particular areas – that is what divers pay to come and see. In these last 20 years since we have started diving here, the sea has been denuded of fish. The divers bring in a significant amount of income – there are 70,000 divers who come year very year. This does not cost anything to the government – they only built some toilets at Cirkewwa.

(NGO 3)

Some stakeholders mentioned the issue of bunkering and that it has to be stopped in the way it is happening. Some also argued that if the regulations are enforced similarly to the ones on Filfla in the future they might as well stop fishing.

No one knows how fishing is restricted, ERA or Fisheries, if they restrict bunkering much better as the sea bed is being destroyed. The amount of fish has decreased a lot throughout the ages. Perhaps only tuna was not so affected. We are all monitored through GPRS. But the S are not monitored like us full time Fishermen. If the rules are the same as those there are on Filfla and they are enforced the coastal Fishermen can just stop fishing.

(Professional Fishermen Association 3)

Most stakeholders discussed the positive aspect of their activity and how with proper management the fish stocks would be replenished.

Divers are environmentalists in their nature. The long term sustainability of the industry depends on us protecting/safeguarding this god sent natural resource. The diving industry generates much more for the country in the way of revenue than the other stakeholders using these dive sites. Fish numbers have deteriorated to an extent that dive sites are EMPTY especially when the water is still cold. We know what it was like 20 years ago and that is why we are angry. The positive side of this story is that the sea is generous and it will not take too long to repopulate our dive sites.

(Dive Centre 1)

4.1.5 New MPAs and Possible Effects on Livelihoods

Different stakeholders reacted differently to the different proposed MPAs. While all of the stakeholders were of the opinion that they should have been given an indication of the rules that would prevail in the MPAs. The majority (fifteen) did not want to commit themselves, while one Commercial Fishermen association stated that they will not accept new reserves if situation remains as is. Nonetheless some general trends do appear. The stakeholders who are involved in fishing feel that their livelihood is under threat. Dive centres seem to feel less of a threat to their livelihood.

Reserves will not make a difference. Enforcement of the law is non-existent.

We will not accept more protected areas if the situation remains as is. We have accepted trawlers within the 25-mile zone to compete with the small Fishermen. A lot of Fishermen are now employing Indonesian individuals that work nonstop. These industrial fishermen are negatively effecting the small Fishermen. Small Fishermen usually fish on one boat and do not employ anyone else. Cannot compete with Fishermen that employ Indonesians – and these should not be working within the 25-mile radius. How can I compete with them?

(Professional Fishermen Association 3)

The inshore sites do not affect us at all. But our trawlers work on the other side (offshore) so we will not be affected. But the ones that trawl for shrimp are affected. Numbers 3, 2 and 4 offshore may mainly affect us negatively as we do shrimp trawling there

(Professional Fishermen Association 4)

Divers usually appear more resilient to the proposed MPAs and in general are of the opinion that alternative sites can be found.

Commercial fishers are not fishing coastal, the recreational fish coastal and create issues. Site 1 is very good place for diving – nice caves, but is very prone to weather – it would be good to be protected…if this is an exercise as before on putting these on a piece of paper it will not change any of my livelihood at all. If it is enforced and diving not allowed – yes it will affect my livelihood. Site 3 appears to have left out Xlendi And Mgar ix xini and both have caves. The site that includes Fomm ir rih – is an inaccessible part of the island. Least frequented- mostly cliff.

(Diving Club 1)

The stakeholders that are not dependent on the sea for their livelihood questioned the idea of livelihood and the fact that we should be talking about quality of life at this stage rather than livelihoods.

No real impact on livelihood but on our quality of life.

MPAs should not remain only on paper – there is also a great need for education.

(Dive Centre 1)

4.1.6 Suggestions for Improvement/Alternatives

A number of different suggestions or alternatives were proposed by the different stakeholders. A common proposal was to install permanent moorings.

We need permanent moorings so as not to damage the seabed continuously. Ok it will destroy one metre by one metre, but that is much better than having anchorings in different areas daily destroying. Dripping anchors really destroys the seabed.

(NGO 1)

We swim over the posidonia – we do not damage it – anchoring damages it. We already proposed fixed moorings to ERA. Divers do not want to come back because there are no fish.

(Dive Centre 4)

One Dive Centre mentioned the idea of permanent moorings but also a large number of other proposals that aim to make the sea more sustainable.

There should be fixed moorings in special areas such as Mgarr ix-Xini and Comino Clear signage indicating cave features, diver experience grade requirements, travelling lines & potential environmental/personal danger spots Limitations to trammel nets in breeding areas, appropriate mesh sizes outside of them - No fishing/no catch areas in clearly defined strategic locations. Consider whether catch & release programs or artificial reef projects could benefit the underwater environment. Educational campaigns for both local and visiting divers Dedicated marine enforcement units covering both surface and underwater activities with assistance of land based units & clear channels of communication with stakeholders.

(Dive Centre 1)

A large number of proposals that aim to make the sea more sustainable were also mentioned by a Professional Fishermen Association.

> *A task force should be set up with individuals selected from Fishermen so any illegal activities in relation to fishing are reported. Only traditional fishing methods should be allowed in Marine Protected areas. S registered boats should not be allowed to do any type of fishing. Trawling should be made illegal within the 25mile zone as is causing total devastation to marine life and traditional Fishermen. This happened after Malta entered the EU. This is a main reason why fish decreased. Another reason for the decrease is that the amount of recreational Fishermen has increased. They still sell fish to restaurants.*
>
> (Professional Fishermen Association 3)

Both one Professional Fishermen Association and one Dive Club association are of the opinion that a ban should be introduced to take stock of the situation.

> *All the inshore sites are targeting Fishermen. If need be we ban fishing for 2 years. But then we need to have studies done.*
>
> (Professional Fishermen Association 3)

> *There have to be areas that are pro life. We should put a 5 year moratorium. Then after the moratorium the Fishermen are allowed in some zones and not in others. We proposed this but when I went to discuss with politicians (name) he told me that you are going to make me lose votes. MTA endorses our idea.*
>
> (Dive Club 3)

Government entities mention different logistical issues in order to be able to cope with the increased numbers of MPAs

MPAs should not be implemented all at the same time, but there should be priorities, gps updates until and gradually introduced, including an educational campaign to all people not only stakeholders. Even at schools.

(Government Entity 3)

> *Fisheries department know where the Fishermen boats are through satellite. A normal GPS shows when you are approaching Filfla that it is protected and you cannot put down anchor* etc. *All MPAs should be plotted in this way on GPS. It would make everyone more aware of where they are even for enforcement reasons.*
>
> (Government Entity 1)

One NGO mentioned the importance of regulation and awareness raising, while the Maritime Civil Protection Department mentioned the importance of media to reach out.

> *All the marine area around Malta should be protected. They should have been extended. There should be some more protection inside ports and creeks. Fishing should be restricted in ports especially when they catch juvenile fish. We want to protect all the Maltese coast. Divers should be aware and understand when they can cause damage to caves and refrain from entering. We agree with maximum sustainable yield.*
>
> (NGO 3)

> *Notice to Mariners are checked by the commercial Fishermen but not by the recreational Fishermen – media should be used more.*
>
> (Government Agency 1)

It should be regulated, especially if non-certified divers are not allowed to dive there. Only divers with cave diving experience should be allowed to dive here (MPAs with caves)

(Diving Centre 4)

The issue of better communication between stakeholders was put forward by an NGO.

Stakeholders that carry out economic activity - they have to agree on the rules and give alternatives and they have to agree between themselves to reduce conflicting activities – divers-tourists- Fishermen. Our feeling with the fishing communities is that they are not heard by the authorities. The fishing communities really fear what the rules will be. We want fishermen to remain fishermen to continue fishing in a sustainable way.

(NGO 2)

The importance of having few MPAs that actually work, as opposed to a lot that do not, was highlighted by one Diving Club.

MPAs should be there as nurseries to restock other fishing areas. Better not make more protected areas – and it is business as usual…what is the point. Better have fewer sites – and you will have opposition - but then there will be long term benefits. Hopefully, we will not only care about photo opportunities.

(Diving Club 3)

Two fishing associations have issues of trust both with the people who do studies on the marine environment and also with certain authorities. This may hinder honest communication.

We know that Posidonia is very important because it creates life and we try not to destroy it. We have serious doubts on a number of studies that have been made – either because they choose wrong people or because of other issues.

(Professional Fishermen Association 2)

4.2 Discussion and from the Thematic Analysis of Qualitative Data

This research project mainly focused to understand the attitudes and perceptions of stakeholders on the impact of new MPA introductions.

It is evident that the "stakeholders" of the sea—fishermen, diving centres, free divers, tourist operators and people of the island—who developed an idea of protection based on the needs of their own category need to be listened to and given a 'voice' by the authorities. It is also apparent that there are issues of trust between stakeholders and the authorities and these have to be addressed in order for progress in sustainability to be registered. It is understandable that participatory process is fundamental to rebuild local identities through the protection of traditions and environment, strengthen confidence with institutions and develop a sustainable local maritime economy. The following more specific results are considered the main outcomes of the thematic analysis:

- Some stakeholders did not want to discuss the effect of the new MPAs on their livelihoods—mostly because the current ones are considered to not be working at all and because they wanted to know the regulations which will be in effect beforehand.
- Most stakeholders are concerned about the importance of sustainability of the marine environment—they understand the importance of rules to achieve this.
- Spearfishing is an issue that has to be properly regulated and enforced—It has been identified as a problem by both fishermen and divers.
- Perception is that enforcement is not done at all. The Administrative Law Enforcement is understaffed and requires more resources.
- Regulations are at best unclear or worse inexistent to most stakeholders.
- Recreational fishermen also appear to be perceived as problematic to the sustainability of the sea.
- There are communication and trust issues between some stakeholders and the authorities.
- There are communication and trust issues between some stakeholders.
- There is mistrust in a number of marine studies especially from the fishing community.
- Bunkering is frequently mentioned as destructive to marine life—permanent moorings mentioned as a mitigating measure.
- Stakeholders want to be more involved in the decision making level not only at the consultee level.

5 Recommendations

Most recommendations involve management approaches to be taken in order to adopt a participatory attitude and involvement from the stakeholders concerned. The involvement of Higher Education Institutions is important as neutral representatives of goodwill and objective expertise and to be able to train the trainers in communication skills and to clarify objectives. The following table represents the main recommendations and the authority/higher education institution that should take the lead in the initiative (Fig. 3).

6 Conclusion

This research project, highlighted sustainability concerns pertaining to the local stakeholders involved in the sea. These concerns require clarification, better understanding and increased engagement from the fishing community. In this phase of the research the concerns of the stakeholders with regard to the introduction of new MPAs were mapped. In the next phase the university together with other authorities will need to design outreach activities in order to address these lacunae in the form

Recommendation	Responsible authority/institution
Meetings to be held from November to March. This will help to accommodate fishers in attending. Divers, water sport operators and boat trip operators will also be more accessible during these months.	Environment and Resources Authority
The links of biodiversity protection and their livelihood should be made apparent to stakeholders.	University of Malta
Key messages should be tailored for different target audiences. A proper communication and education strategy should be drawn up by an ESD expert in higher education.	University of Malta
The use of fishermen, divers and other stakeholders to endorse the planning process and for awareness raising should be given high priority. The idea of having fishermen presenting to fishermen should be actively explored.	Environment and Resources Authority/ University of Malta
The use of high profile national celebrities for awareness raising campaigns should be explored but given less importance than the actual stakeholder 'champions'.	University of Malta
Exchange visits to other MPAs to train general stakeholders and economic stakeholders on MPA sustainable management.	Department of Fisheries
Training workshops should be provided and should be structured into relevant various areas as the implementation of the MPA progress.	University of Malta

Fig. 3 Main recommendations and responsible authorities

of training workshops and short courses in order to help the maritime stakeholders in the following:

i. better integrate sustainable practices in their work.
ii. help stakeholders develop the skills necessary to explore enhanced or complimentary livelihoods.
iii. Increase effective discussions between the various stakeholders.

From the stakeholder interviews and the findings of the present research it is suggested that intrinsic motivation is utilised as a long term solution to local MPA management vis a vis stakeholders. Higher education institutions are of paramount importance in order to achieve effective communication to build bridges, give a 'voice' to the stakeholders and to increase awareness on sustainability.

MPA management can be successful if it promotes the appropriate behaviour of users of the marine area in question. Such approaches should recognize the values, concerns, knowledge and customary practices of stakeholders through providing for active participation in MPA management processes from the beginning Fiske (1992), Kelleher (1999). A Management Option Continuum is being presented in which the possible extremes of management to change behaviour are presented: on the one side management that involves extrinsic approach (e.g. carrot and stick) with

Management Option Continuum		
Emphasis on Enforcement		Emphasis on stakeholder participation and cooperation
Executive cross sectoral authority		Community-based partnership
Education to justify restrictions and promote compliance		Education to promote support and participation
Reliance on comprehensive scientific information		Science used for guidance where appropriate and available

Fig. 4 Management option continuum. Adapted from Jones 2002

stakeholders, while on the opposite side of the spectrum is a management option which in principle is more based on intrinsic motivation (Fig. 4).

Further studies should focus on how the stakeholders can be directed to various alternative livelihoods (such as maritime tourism) and how the local economy may benefit stakeholders, the community and the marine protected areas through increased sustainability. The university should also research ways in which it can increase its engagement with the maritime community in order to act as a mediator between the stakeholder organisations and the government authorities.

Acknowledgements This research study was undertaken as part of the studies of the Project LifeBahar for N2K (Life12NAT/MT 000845) financed through the EU life + programme.

References

Allison E, Ellis F (2001) The livelihoods approach and management of small-scale fisheries. Mar Policy 25:377–388

CANARI Policy Brief No. 5 January 2005. Marine Protected Areas and Sustainable Coastal Livelihoods, Fernandes Industrial Centre, Administration Building Eastern Main Road, Laventille Trinidad and Tobago

Chambers R, Conway G (1991) Sustainable rural livelihoods. IDS Discussion Paper 296. London

Charles A, Westlund L (2016) Fishing livelihoods as key to marine protected areas: insights from the World Parks Congress. Aquatic Conserv Mar Freshw Ecosyst 26(Suppl. 2):165–184

Fiske SJ (1992) Sociocultural aspects of establishing marine protected areas. Ocean Coast Manage 18:25–46

Jones P (2002) Marine protected area strategies: issues, divergences and the search for the middle ground. Rev Fish Biol Fish 11:197–216. Netherlands

NSO (2016) Agriculture and fisheries 2014. National Statistics Office, Valletta. ISBN 978-99957-29-50-9

Kelleher G (1999) Guidelines for marine protected areas. IUCN, Gland, Switzerland and Cambridge, UK

Pollnaca R, Christieb P, Cinner J, Daltona T, Dawd T, Forrestere G, Grahamc N, McClanahanfa T (2010) 1, Marine reserves as linked social–ecological systems. Proc Nat Acad Sci 107(43):18262–5. Available from https://www.researchgate.net/publication/41532358_Marine_reserves_as_linked_social-ecological_systems. Accessed 20 January 2020

Pomeroy R, Douvere F (2008) The engagement of stakeholders in the marine spatial planning process. Mar Policy 32:816–822

Vaughn S, Schumm SJ, Sinagub J (1996) Focus group interviews in education and psychology. Sage Publications, London

Voyer M, Gladstone W, Goodall H (2012) Methods of social assessment in marine protected area planning: is public participation enough? Mar Pol 36:432–439

Mark C. Mifsud is a Chartered Biologist, a Chartered Environmental Scientist, a Chartered Science Teacher, a Certified Wildlife Photographer and a Prince II Certified Practitioner with a wide range of experience in Environmental Education and Sustainable Development. Mark has drafted management plans for nature parks, reviewed EIAs and advised on ODZ applications as an independent member on the NHAC board within MEPA. He has more than 20 years teaching experience (from primary to tertiary levels), and is currently involved in environmental communication, teaching, research, academic refereeing and in a number of international projects. Mark has been entrusted as a member on various national and international boards. He is presently co-ordinating the Masters in Education for Sustainable Development at the University of Malta and holds adjunct status within the faculty of Integrated Science and Technology at James Madison University, Harrisonburg, Virginia in the US. He also held teaching posts at the University of Ca Foscari Venice, Italy, at Southern New Hampshire University in the US and at the University of Yaounde, Cameroon. Mark has published in excess of 40 publications in the area of Education for Sustainable Development.

Linking Biodiversity and Human Health to Achieve Sustainability

Deborah A. Gatt and Mark C. Mifsud

1 Introduction

In Malta, a migration towards academia and urban posts, at the expense of lower-profit rural occupations, has lent weight to the loss of appreciation of the local environment and knowledge regarding biodiversity and its benefits in terms of health and well-bring. The trend towards urban environments and technological lifestyles (for example, computers, television, tablets, mobile phones) has further promoted the disconnection from nature among the "modern and civilised" man.

Education of the local population is required to increase awareness of biodiversity as not simply an aesthetic benefit but an essential part of life with significant psychological and physiological benefits, including the proper and robust functioning of the various systems within the human body.

Higher Education Institutions are highly respected as sources of information, and educators in such institutions are well-placed to impart knowledge and skills regarding sustainable development to students, and ultimately the community (Chinnasamy and Daniels 2019). Unfortunately, topics and skills relating to sustainable development may be regarded as non-essential components of courses, and are not necessarily incorporated (Waas et al. 2010).

In addition, Higher Education Institutions are strategically sited to develop research centres that encourage the active participation of members of the local communities (Granados-Sánchez et al. 2011).

D. A. Gatt (✉) · M. C. Mifsud
Centre for Environmental Education and Research, University of Malta, Msida MSD 2080, Malta
e-mail: drobeah@gmail.com

M. C. Mifsud
e-mail: mark.c.mifsud@um.edu.mt

W. Leal Filho et al. (eds.), *Universities, Sustainability and Society: Supporting the Implementation of the Sustainable Development Goals*, World Sustainability Series,
https://doi.org/10.1007/978-3-030-63399-8_3

However, to implement effective outreach, the current situation must first be understood. The literature reviewed revealed a lack of studies on Maltese adults. This study addresses this gap through the surveying of Maltese adults in the age bracket of 30 to 45 years old, with a view to contributing towards establishing a baseline for current understanding of local biodiversity and its effects on human health.

2 The Maltese Context

The Maltese Archipelago, situated in the Mediterranean Sea, consists of small, low islands. The islands host a population of over 460,000 people, and more than 2.4 million tourists visit the islands annually. Slightly less than a third of the country is classified as "urban" or built, with just over half of the country utilised for crop production. Less than a fifth of the islands are undeveloped, with *garigue* habitat being the dominant component of the remaining natural areas on the islands. Stevens et al. (1995) observed that the detrimental impact of humans on the environment was significant.

2.1 The Maltese Environment and Biodiversity

The Maltese Islands are home to a wide variety of species of fauna and flora, yet the local adult population is not necessarily aware of this diversity. This lack of awareness may be partially attributed to a highly competitive educational system that used foreign textbooks and included little local environmental education (Mifsud 2010; Pace 2009).

Several studies have assessed the knowledge, attitudes and behaviour of schoolchildren, youths and young adults towards the local environment. Similar results were obtained in Malta, Cyprus, and Tokyo, to name but a few, where a general awareness of environmental issues and a positive attitude towards the environment are present, yet pro-environmental behaviour is not highly evident (Mifsud 2010; Nisiforou and Charalambides 2012; Soga et al. 2016). Such studies have generated recommendations for educational programmes targeted towards the younger generations, using education to shape environmentally beneficial behaviour.

Similar studies on older members of the Maltese population are not readily available but national campaigns that focus on sustainable activities are conducted, promoting specific topics such as reduction of food waste and avoidance of single-use plastics. While not directly related to biodiversity, such campaigns promote behaviours that benefit the environment.

2.2 *Pollution and Health*

In 2016, 210 premature deaths in Malta were attributable to the concentration of airborne particles of less than 2.5 microns in size ($PM_{2.5}$) (European Environmental Agency, 2019, p. 68).

The global issues of land and water pollution also apply locally, with improper disposal of chemicals, sewage, pesticides and fertilisers, plastics, and plastic bags comprising some of the threats faced. Besides water-borne diseases, polluted water poses risks to foetal health and development following ingestion by the mother (Currie et al. 2013), and diversity of aquatic species is affected by the physical blockage of watercourses, eventually resulting in eutrophication (Dodds 2007).

The loss of flora on land results in increased erosion and dust storms, thus reducing air quality (Aenab et al. 2013), followed by a subsequent decrease in soil quantity. The consequential alterations to soil biodiversity may directly affect human health through the reduction or elimination of specific components, such as certain microorganisms that lessen the prevalence of allergies and allergic disease following exposure (Wall et al. 2015).

2.3 *Effects of Biodiversity on Human Health and Well-Being*

Exposure to biodiversity in the natural environment is known to ease a variety of ailments (Kuo 2015), and increased physical activity reduces the risk of obesity and cardiovascular disease. Using this premise, the "Green Prescription" programme implemented in New Zealand allows for General Practitioners to prescribe countryside walks instead of medication. The programmes generated mixed reviews, yet improvements in physical and mental health were reported (Patel et al. 2012).

Although further research is required to strengthen knowledge regarding the links between biodiversity and health, Sandifer et al. (2015) find that biodiversity benefits humans on psychological and physiological levels through aesthetic, cultural and socio-economic pathways, as well as a result of the materials obtained from ecosystems.

2.4 *Biodiversity, Stress and Mental Health*

Studies show that contact with biodiversity promotes relaxation and assists in the recovery from stress, with measurements of stress indicators, such as salivary and hair cortisol, quantitatively proving these effects (Gidlow et al. 2016).

The psychological benefits of biodiversity are directly related to the richness of the biodiversity present (Fuller et al. 2007), and significantly improved benefits were noted upon exposure to the physical natural environment as opposed to a simulated environment (Kjellgren and Buhrkall 2010), including in the treatment of depression and anxiety (Barton and Pretty 2010).

2.5 Biodiversity and Physical Health

Several possible mechanisms through which humans benefit from nature have been proposed, including immune system benefits, reduction of blood glucose levels, and improved quality of sleep (Kuo 2015). Two of these mechanisms are described below.

Phytoncides are volatile organic compounds (VOCs) secreted by trees and plants, that repel or combat bacteria, fungi, insects and animals. Research indicates that phytoncides may assist in reducing the incidence of cardiovascular diseases, boost immune function, increase natural killer (NK) cell activity, and reduce anxiety (Li 2010; Tsao et al. 2014).

Microbiota—People with allergies who live in areas of low environmental diversity have significantly lower generic diversity of gammaproteobacteria on their skin when compared to individuals who do not suffer from allergies (Hanski et al. 2012). These bacteria help protect against allergies and inflammatory disorders, and may eventually be utilised in the prevention and treatment of allergies (Kaesler et al. 2018).

2.6 Ecosystem Services

The Millennium Ecosystem Assessment Board (MA) (2005) states that ecosystem services are "the benefits people obtain from ecosystems" (preface, page V). Studies on human perceptions show that many factors influence the value placed on ecosystem services, with differences evident between residents in rural and urban areas, and between countries. The differences could be a result of personal backgrounds and the ecosystem services required (Lapointe et al. 2019).

The uptake of pollutants and deposition of particulate matter by plants is an essential ecosystem service (Mori et al. 2018). In the urban environment, pollution-tolerant trees can sequester several tonnes of carbon annually (Nowak et al. 2018).

Temperature extremes may be mitigated by trees and green roofs, provided correct planning is utilised (Calcerano and Martinelli 2016). Buildings are thus shaded during the summer and shielded from cold winds in winter (Laaidi et al. 2014). This, in turn, reduces energy consumption (Kuronuma et al. 2018).

High Water Quality can be achieved using aquatic species for the cleaning of water bodies (McLaughlan and Aldridge 2013). Some wastewater treatment systems and urban waterway management programmes incorporate such species into their designs (Mounir et al. 2017; Wang et al. 2017).

2.7 Loss of Human-Nature Connection

The manner in which people perceive nature is positively related to the amount of time spent in nature throughout their lives (Soga et al. 2016).

Many people tend to spend most of their time indoors or away from nature. This relationship between man and nature (often referred to as the "Human-Nature Connection", HNC) has deteriorated and this disconnection impacts human health and well-being (Ives et al. 2017; Soga and Gaston 2016).

Soga and Gaston (2016) report several data sets that show a decline in outdoor activity across several generations and propose that the consequences of this decline include changes to health and well-being, as well as emotional, attitudinal and behavioural changes. These consequences in turn may further aggravate and intensify the disconnection with nature.

2.8 Sustainable Exploitation

The likelihood of sustainable exploitation is higher if the local biodiversity is used for personal needs, rather than as a source of income with no personal impact (Pardo-de-Santayana and Macia 2015).

A local case study involves a parasitic plant (*Cynomorium coccineum*) believed to have medicinal properties and found primarily in one site in the Maltese Islands. During the rule of the Knights of the Order of St. John, the perceived value of the plant was so high that samples were gifted to visiting dignitaries and the site was placed under military surveillance. In this case, protection was imposed for the personal gain of the Grandmaster (Deidun 2010).

2.9 Humans and Their Perceptions of Nature

Possession of knowledge, awareness and a positive attitude towards nature and the benefits of biodiversity does not necessarily translate into environmentally friendly behaviour. Whether knowledge influences behaviour, or whether it influences attitudes, which in turn influence behaviour, is an ongoing debate with support for both options (Eilam and Trop 2012). Other potential pathways exist, and Smith and Louis (2009) postulate that group norms are essential in influencing the relationship between attitude and behaviour. When a practice becomes accepted by society or the referent group to which an individual belongs, that practice is adopted.

2.10 Research Gaps Identified

A high rate of construction activity is reducing the percentage cover of remaining natural land in Malta, thus reducing the amount of biodiversity present (Deidun 2008). Ecosystems provide services that benefit humans, yet studies show that awareness and understanding of the term "biodiversity" is low (Lindemann-Matthies and Bose 2008; State of NSW and Office of Environment and Heritage 2017). Also, possessing knowledge about the environment does not necessarily promote environmentally responsible behaviour (Mifsud 2010; Nisiforou and Charalambides 2012). Nowadays, humans spend little time engaging in outdoor activities, thus preventing the formation of a bond or relationship with nature (Soga and Gaston 2016).

Few studies are available that focus on adults and their awareness of the effect of human activity on biodiversity, and the impact of biodiversity on humans. This study intends to address this gap and seeks to query the knowledge held by the local population with regard to biodiversity and its effects, and the extent to which biodiversity is considered as integral to their health and well-being.

3 Methodology

Data collection involved semi-structured interviews performed with at least two participants from each of the six districts (according to LAU1/NUTS 3 Classification) in the Maltese Islands. The districts are Northern, Northern Harbour, Southern Harbour, South Eastern, Western, and Gozo and Comino. The interviews were balanced between persons who grew up in rural or semi-rural and urban environments.

The target population was Maltese adults aged between 30 and 45 years old ($\text{Maltese}_{30\text{–}45}$), as few studies have been performed on this age group and this target population is:

1. Able to make informed decisions, and implement new behaviours (workplace and home) (Worthy et al. 2011);
2. Likely to have an influence on children (Pearson et al. 2009).
3. Expected to have a strong connection with nature as they grew up when open spaces were more accessible, tablets and mobile phones were non-existent, and playing outdoors was an everyday activity (Soga and Gaston 2016).
4. Easily accessible, as the researcher belongs to the same age group.

3.1 The Interviews

The interviews were of a semi-structured format and the development of the questions followed the Interview Protocol Refinement (IPR) framework proposed by Castillo-Montoya (2016). Four types of questions were included.

1. *Introductory*: neutral, eliciting general information.
2. *Transition*: linking introductory and key questions.
3. *Key*: related to the research questions.
4. *Closing*: opportunity to close.

The Interview Protocol Matrix (IPM) for this study, as recommended by Castillo-Montoya (2016) for the evaluation of the interview questions, is depicted in Table 1. The classification of the questions is according to the four types described above.

3.2 Data Analysis

Fifteen interviews were conducted and the audio was recorded (with the interviewee's consent), transcribed and analysed using the modified version of the method proposed by Vaughn et al. (1996), where the main ideas were identified, units created and categorised, and themes formed following examination of the main ideas and categories.

The following table describes some demographics of the interview participants (Table 2).

Theme 1: Biodiversity affects Physical and Mental Health

When referring to biodiversity, interviewees mentioned trees almost exclusively, and the stated effects of biodiversity were limited to air quality and effects on mental health and stress reduction.

> *Plants and trees clean the air, so without them, there would be even more pollution, and then this would be linked to more of certain types of cancers.*
>
> (Interviewee 11)

> *Even coming home to a small front garden, for example, or being able to breathe clean air ... all of these things would improve my overall well-being and mental health.*
>
> (Interviewee 5)

One interviewee, although familiar with the role of plants in relation to air quality, was unaware of the fact that species diversity was preferable to a single species.

> *As regards animals and plants affecting my health, I imagine that mostly I need enough trees around to change carbon dioxide into oxygen, but I don't know if they need to be diverse or if they can be one kind of tree.*
>
> (Interviewee 7)

Differing levels of education did not affect the benefits of biodiversity mentioned, although one (post-graduate) was aware of the detrimental effects to human health when ecosystem imbalances were present.

> *... certain rodents, or the pigeons for example, they're not doing good for people who suffer from asthma or allergies ... Biodiversity should affect people in a positive way, but we have*

Table 1 IPM evaluation of the interview questions. I = Introductory, T = Transition, K = Key, and C = Closing

No.	Question	Socio-demographics	Are Maltese$_{30–45}$ aware of the concept of biodiversity?	Are Maltese$_{30–45}$ aware of how their actions affect biodiversity?	Are Maltese$_{30–45}$ aware of how biodiversity impacts their health and quality of life?
1–11	Socio-demographics	**I**			
12	What or who were your biggest influencers in learning about the environment?		**T**		
14	Have you ever heard of the word "biodiversity" before? Can you explain biodiversity briefly?		**K**		
15	Do you think that pollution (air, land or water) affects biodiversity?			**K**	
16	Do you think that biodiversity impacts your physical health?				**K**
17	With respect to the natural environment, countryside and urban environment in Malta, what could be done to improve your surroundings and your quality of life?				**K**

(continued)

Table 1 (continued)

No.	Question	Socio-demographics	Are $Maltese_{30–45}$ aware of the concept of biodiversity?	Are $Maltese_{30–45}$ aware of how their actions affect biodiversity?	Are $Maltese_{30–45}$ aware of how biodiversity impacts their health and quality of life?
19	Do you think that your education has played a part in your attitude and behaviour towards the environment? And in your perception of the local natural environment and how you perceive your surroundings?		**C**	**C**	

Table 2 Demographics of interview participants

District of residence	Interviewee No.	Sex	Age	Area in which Interviewee grew up	Role in industry	Level of education
Northern	4	Male	33	Urban, house/apartment with plants	Management	Undergraduate
	11	Female	30	Urban, house/apartment with plants	Homemaker/Professional	Undergraduate
Northern Harbour	13	Female	45	Semi-Rural	Administrative Staff/Homemaker	College
	15	Male	44	Semi-Rural	Senior Management	Post-graduate
	5	Female	36	Urban, house/apartment with plants	Self-Employed	Undergraduate
	10	Female	38	Urban, house/apartment with plants	Administrative Staff	Diploma
Southern Harbour	2	Male	43	Rural	Sales Executive	College
	7	Female	38	Urban, house/apartment without plants	Junior Management	Undergraduate
South Eastern	1	Male	45	Semi-Rural	Management	Undergraduate
	3	Male	30	Semi-Rural	Operator	Diploma
	6	Male	38	Semi-Rural	Senior Management	Doctorate
Western	8	Female	31	Urban, house with plants	Management	Diploma
	12	Female	40	Semi-Rural	Senior Management	Undergraduate
Gozo and Comino	9	Female	44	Semi-Rural	Professional	Post-graduate
	14	Male	40	Urban, house with plants	Self-employed	College

damaged the biodiversity as well in certain circumstances, and that is the effect we are going through because of the damage man has caused in the first place.

(Interviewee 9)

Theme 2: Formal Education and Experience

Although formal education directly impacts the theoretical aspect of learning about nature, most interviewees claimed their attitude and perception of nature were most affected by personal experiences.

Yes, I think it [education] played an important role. Although … most of my thinking about the environment has developed rather late in the education process … my biggest influence came from friends, and this happened mostly when I was 18 to 20.

(Interviewee 6)

Yes, I think so. And it's not just a matter of education in the sense of schooling; it's also education, the nurture part, which is due to family culture, friendships, relationships, community. I think they all have an important role ....

(Interviewee 9)

Although no longer necessarily participating in a formal education environment, schools can reach parents through the children, who return home eager to practise what they learned.

... when my eldest was in school ... they started separating the waste, and it was 12 years ago, so we didn't use to separate the waste yet, the recycling. But my son insisted that we should start at home as well.

(Interviewee 13)

Formal education is instrumental in the development of critical thinking skills. Such skills promote an inquisitive nature, encouraging the search for further information.

Yes, I think the more you educate yourself, the more you learn, and the more your ideas and perceptions change. If you don't make an effort to be educated, then all of your perceptions and all of your ideas are just your own and are not necessarily based on fact or science. So you could basically believe anything.

(Interviewee 8)

Theme 3: Infrastructure and Nature

The main suggestions for improving quality of life included increasing the number of pedestrian areas and accessible green areas within the urban setting.

Plant more trees. Leave more open spaces.

(Interviewee 2)

The current approach to construction was mentioned, with suggestions of possibly building upwards rather than outwards and developing abandoned properties instead of new ground.

... I would minimise the number of permits that are being issued. At least ... build upwards on the already developed plots, because if you expand, you're always reducing the environment and creating more buildings.

(Interviewee 3)

... first don't grant permits for the little ODZ [Out of Development Zone] areas we have left. Obviously, our space is really limited, and we need to increase our open spaces rather than decrease them.

(Interviewee 12)

Although environmental laws are in place, these laws are not respected or enforced.

Well, enforcement of the law. Basically, there was a law passed in Malta whereby ... the coastline should be considered to be public. I mean, if you enforced that ... it would make it more difficult for developers to continue developing on the same sites.

(Interviewee 1)

Theme 4: The Meaning of Biodiversity
All interviewees had previously heard of the word "biodiversity", but almost half were unsure of the definition. Although unsure, most were able to propose a partial definition.

> *I don't know, maybe there is something diverse in nature?*
> (Interviewee 10)

> *I don't know. Everything that makes up the earth, humans and everything that is alive, in my opinion.*
> (Interviewee 12)

4 Discussion

Ecosystems provide a variety of services (Nowak et al. 2018), yet the main service mentioned in the data collected relates to air quality.

Unfortunately, the Maltese culture favours the use of motorised vehicles as the primary form of transportation. Although a minority exists that prefers alternative forms of transport, such as bicycles, neither the culture nor the country infrastructure caters for these practices.

The reliance on motorised vehicles has led to a situation of extreme congestion and an ageing infrastructure that is unable to cope with the volume of traffic.

The resultant decrease in air quality causes respiratory issues (Balzan et al. 2012; Drago et al. 2013) and affects the quality of life of those subjected to acute situations of pollution, potentially negating attempts at healthier lifestyle choices.

> *One thing that I try to do, even for the benefit of my children, is that I always take them to school on foot ... However, by the time I get to the top of the hill, often there are so many trucks and cars that I almost choke.*
> (Interviewee 11)

Several projects are currently under way—namely the Marsa Junction Project and the Central Link Project—that aim to reduce congestion and vehicular emissions through the improvement of road infrastructure. However, given the current rate of acquisition of motorised vehicles (National Statistics Office 2019), this may only be a temporary solution.

Roadside vegetation assists in reducing air pollution from vehicular emissions (Mori et al. 2018). However, urban vegetation is lacking in Malta. At least 33% of the interviewees supported this view.

> *... it really annoys me that there's nowhere in walking distance that I can go and just, you know, have some countryside for my kids.*
> (Interviewee 7)

In a study conducted by Mifsud (2010), Maltese youth postulated that the Government avoids enforcing environmental laws due to fear of losing votes. This study linked the lack of enforcement with a high rate of construction, including on ODZ land (Out of Development Zone). Although the total government expenditure in addressing environmental challenges has increased in recent years (Environment and Resources Authority 2018), the enforcement and execution of the policies, management plans and legislation is not apparent to the Maltese$_{30–45}$ population.

> *There are certain laws which are there on paper, which haven't been respected for quite some time. And yet, we close an eye on them, no enforcement.*
>
> (Interviewee 15)

Biodiversity positively affects human physical health and well-being, with long-term exposure reducing the risk of obesity through increased physical activity, lowering mortality associated with cancer (Li 2010) and respiratory and cardiac diseases (Tsao et al. 2014), and reducing the effects of stress (Barton and Pretty 2010). Exposure to the natural environment, both during gestation and in the first few years of life, may have positive effects on the mental health of children (Engemann et al. 2018) and reduces atopic sensitisation and the prevalence of chronic inflammatory diseases such as asthma (Hanski et al. 2012).

Even among university-educated interviewees, the data collected indicates that these benefits are not widely known. Apart from the impact on air quality and mental well-being, some interviewees were unaware that biodiversity had any impact on human quality of life. The interviewees appear to believe the effects of biodiversity to be indirect, rather than direct, and excursions into the countryside may be perceived as "hassle", possibly due to a cultural mentality where such ventures require the inclusion of commodities, such as food and beverages.

> *For example, nowadays you say "let me take them for a day out to Buskett", and you have to stay carrying the sausages, for example,* etc. *It's easier to go to Sliema and find everything ready and you just pay for it.*
>
> (Interviewee 3)

The value placed on the natural environment may stem from the extent of time spent outdoors as a child (Soga and Gaston 2016); several interviewees attributed their positive environmental perspectives to their personal experiences outdoors. It is possible that, rather than the *quantity* of time spent outdoors, it is the *quality* of time spent outdoors that has the most significant impact on adult perspective and behaviour towards the environment (Ives et al. 2017).

Examination of the data collected showed that the understanding of the impact of biodiversity on human health and quality of life was limited to mental and emotional well-being, the purification of air and the production of oxygen. None of the interviewees mentioned other benefits and services that ecosystems provide, such as regulation of climate, carbon dioxide sinks, and direct health benefits. However, all interviewees were aware that pollution and some human activities negatively affect biodiversity, which in turn impacts humans.

The main outcomes of the study are:

1. Maltese$_{30–45}$ understand that biodiversity improves the quality of air, positively affects well-being and mental health, and that more biodiversity would improve

their quality of life. However, awareness of the direct benefits of biodiversity on physical health is not present.
2. Periods of high-quality time in the natural environment are beneficial in shaping attitudes through the formation of significant experiences. This supports the Human-Nature Connection, which emphasises quality of experience.
3. Roadside vegetation and urban green spaces are scarce, yet desired.
4. Malta entertains a "car culture" which does not promote alternative forms of transportation, and excessive vehicular emissions counter attempts at having a healthy lifestyle.
5. Environmental legislation is present but enforcement is not apparent.

5 Assessment of Study Findings

The findings of the study were grouped into three categories, critically assessed, and a visual model was created to better understand the links and processes involved.

Source of Influence—how awareness and attitude may be influenced.

- Education (formal, informal and non-formal), media, campaigns and enforced or promoted behaviour by local authorities.
- Development of critical thinking skills, often through higher levels of education (for example, university level).

Process—the current state, or processing, of knowledge that may affect subsequent actions.

- Understanding of the effects (direct and indirect) of biodiversity on humans. This understanding may be more extensive in persons with higher levels of education.
- Awareness of alternative and sustainable behaviour, with or without feasible solutions to obstacles in implementation. Educational bodies, particularly Higher Education Institutions, can play a significant role in generating awareness of alternative behaviours.

Action—formation of environmentally beneficial behaviour.

- Imposition of specific behaviours by external entities, such as school regulations demanding reusable containers for lunches, or separation of waste for recycling and composting.
- Voluntary behaviours inspired through significant experiences and repetitive excursions in the natural environment to increase the Human-Nature Connection.
- Widely disseminated information regarding alternative forms of behaviour, encouraging the adoption of these behaviours by society in general and creating new *social norms.*

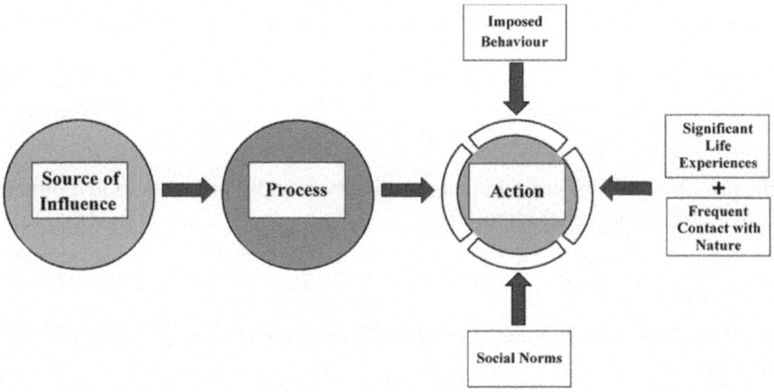

Fig. 1 IPA model illustrating the links between influence, process and action

6 The Influence-Process-Action Model

The model above (Fig. 1) illustrates the links between the three categories mentioned above—Influence, Process, and Action (IPA). The "state" of a person (*processing*), with regard to their knowledge and attitudes, is influenced by one or more *sources*, including media, family, and exposure to education. A person's attitude and level of knowledge will impact their behaviour (*action*) but may not be enough to fully shape environmentally responsible behaviours. Other factors, including imposed behaviours by government and schools, social norms and expectations, together with significant experiences and frequent contact with nature, may be required to inspire a shift towards behaviours that are more favourable to the environment.

7 Recommendations

The following are recommendations to address shortcomings identified in the course of this study. While the University of Malta is suggested as a coordinating body for several actions, this role may be adopted by any suitable Higher Education Institution.

Action	Coordinating body
Perform Community Outreach and Education	
• Provide information about biodiversity, associated benefits, and ecosystem services through media	University of Malta
• Distribute information leaflets about health benefits of biodiversity to households on seed-impregnated or easily compostable paper	University of Malta
• Promote changes in mentality, encouraging brief excursions into the natural environment without the "hassle" of food and drinks	University of Malta
• Organise workshops regarding the traditional use of plants and the harmonisation with modern pharmaceuticals	University of Malta

(continued)

(continued)

Action	Coordinating body
• Offer activities for urban gardening	University of Malta
Increase Availability of Products	
• Discuss the possibility of importing environmentally friendly products at competitive prices with local importers	Government of Malta
• Support local suppliers to increase their presence on the market	Government of Malta
Environmental Support	
• Embellish urban areas with greenery to increase pressure on creating new social norms	Local Councils
• Increase patrols and law enforcement for environmental infringements	Government of Malta

8 Conclusion

This study allowed for a brief foray into the complexities of the current situation regarding the relationship of the Maltese$_{30–45}$ adult population with biodiversity. The results obtained were interesting in that in most cases they corroborated studies performed in other countries and supported local studies on similar topics. It is clear that this university research has contributed to increasing understanding of the various linkages between the knowledge and behavioural domains. Higher Education Institutions have been identified as being key instruments in performing community outreach. As the main stakeholder, these institutions should take the lead in applying and contextualising the recommendations presented by working in partnership with both the Government of Malta and the Local Councils. The implementation of the recommendations listed above will initiate the process of empowerment of the population through understanding their environment using Education for Sustainable Development.

Upon considering the results obtained through this research, and the concluding remarks, the following ideas for further research are being proposed:

- A study on the effectiveness of Higher Education Institutions in achieving increased biodiversity awareness through community outreach.
- Research on how ESD practices can bring forth, among the locals, a sense of social responsibility that contributes towards protection, appreciation, and ownership of biodiversity.
- A study focusing on the local population that attempts to analyse the applicability of the IPA model with different age groups.

The research proposed above would increase the literature available on the local population, particularly if performed among older age groups, thus further contributing to the closure of the gap in literature identified earlier within this paper. Additionally, such studies would enable the monitoring of the effectiveness of any implemented community outreach programmes and allow for improvements in the programme strategies.

References

Aenab AM, Singh SK, Lafta AJ (2013) Critical assessment of air pollution by ANOVA test and human health effects. Atmos Environ 71:84–91. https://doi.org/10.1016/j.atmosenv.2013.01.039

Balzan M, Cibella F, Ruggieri S, Bilocca D, Drago C, Zammit C, Viegi G et al (2012) Prevalence of allergic respiratory diseases in school children (age 11–14) in Malta and Sicily. In: Operational CROSS-BORDER PROGRAM Italia-Malta 2007–2013, p 1

Barton J, Pretty J (2010) What is the best dose of nature and green exercise for improving mental health? A multi-study analysis. Environ Sci Technol 44(10):3947–3955. https://doi.org/10.1021/es903183r

Calcerano F, Martinelli L (2016) Numerical optimisation through dynamic simulation of the position of trees around a stand-alone building to reduce cooling energy consumption. Energ Build 112:234–243. https://doi.org/10.1016/j.enbuild.2015.12.023

Castillo-Montoya M (2016) Preparing for interview research: the interview protocol refinement framework. Qual Rep 21(5):811–831. Retrieved from https://nsuworks.nova.edu/tqr/vol21/iss5/2

Chinnasamy J, Daniels J (2019) The role of universities and educators in developing and implementing sustainable developmental goals. Andragoška Spoznanja 25(3):47–60. https://doi.org/10.4312/as.25.3.47-60

Currie J, Graff Zivin J, Meckel K, Neidell M, Schlenker W (2013) Something in the water: contaminated drinking water and infant health. Can J Econ 46(3):791–810. https://doi.org/10.1111/caje.12039

Deidun A (2008, September 7) The best use for "disused land." In: The times of Malta. Retrieved from https://www.um.edu.mt/library/oar/handle/123456789/36498

Deidun A (2010) Challenges to the conservation of biodiversity on Small Islands: the case of the Maltese Islands. Int J Arts Sci 3(February 2004):175–187. http://malta.academia.edu/ALANDEIDUN/Papers/303618/Biodiversity_conservation_challenges_on_small_islands

Dodds WK (2007) Trophic state, eutrophication and nutrient criteria in streams. Trends Ecol Evol 22(12):669–676. https://doi.org/10.1016/j.tree.2007.07.010

Drago G, Balzan M, Ruggieri S, Bilocca D, Zammit C, Colombo P, Cibella F et al (2013) 10–15 years old children living in Malta are at higher risk for developing allergic respiratory diseases than those living in Southern Sicily. In: Operational cross-border program Italia-Malta 2007–2013, p 1

Eilam E, Trop T (2012) Environmental attitudes and environmental behavior-which is the horse and which is the cart? Sustainability 4(9):2210–2246. https://doi.org/10.3390/su4092210

Engemann K, Pedersen C, Arge L, Tsirogiannis C, Mortensen P, Svenning J-C (2018) Childhood exposure to green space—a novel risk-decreasing mechanism for schizophrenia? Schizophr Res 199:142–148. https://doi.org/10.1016/j.schres.2018.03.026

Environment and Resources Authority (2018) State of the environment report 2018—summary report. Marsa. https://doi.org/10.1007/s13272-011-0027-3

European Environmental Agency (2019) Air quality in Europe—2019 report. No. 10/2019. https://doi.org/10.2800/62459

Fuller RA, Irvine KN, Devine-Wright P, Warren PH, Gaston KJ (2007) Psychological benefits of greenspace increase with biodiversity. Biol Let 3(4):390–394. https://doi.org/10.1098/rsbl.2007.0149

Gidlow CJ, Randall J, Gillman J, Smith GR, Jones MV (2016) Natural environments and chronic stress measured by hair cortisol. Landscape Urban Plan 148:61–67. https://doi.org/10.1016/j.landurbplan.2015.12.009

Granados-Sánchez J, Wals AEJ, Ferrer-Balas D, Waas T, Imaz M, Nortier S, Au T et al. (2011) Sustainability in higher education: moving from understanding to action, breaking barriers for transformation. In: G. U. N. for I. (GUNI) (ed) Higher education in the World 4. Higher education's commitment to sustainability: from understanding to action. Palgrave Macmillan UK, pp 193–207

Hanski I, von Hertzen L, Fyhrquist N, Koskinen K, Torppa K, Laatikainen T, Haahtela T et al (2012) Environmental biodiversity, human microbiota, and allergy are interrelated. Proc Natl Acad Sci. https://doi.org/10.1073/pnas.1205624109

Ives CD, Giusti M, Fischer J, Abson DJ, Klaniecki K, Dorninger C, von Wehrden H et al (2017) Human–nature connection: a multidisciplinary review. Curr Opin Environ Sustain. https://doi.org/10.1016/j.cosust.2017.05.005

Kaesler S, Skabytska Y, Volz T, Biedermann T (2018) The biodiversity hypothesis and immunotolerance in allergy. Allergo J Int 27(5):140–146. https://doi.org/10.1007/s40629-018-0072-0

Kjellgren A, Buhrkall H (2010) A comparison of the restorative effect of a natural environment with that of a simulated natural environment. J Environ Psychol 30(4):464–472. https://doi.org/10.1016/j.jenvp.2010.01.011

Kuo M (2015) How might contact with nature promote human health? Promising mechanisms and a possible central pathway. Frontiers Psychol 6(August):1–8. https://doi.org/10.3389/fpsyg.2015.01093

Kuronuma T, Watanabe H, Ishihara T, Kou D, Toushima K, Ando M, Shindo S (2018) CO_2 payoff of extensive green roofs with different vegetation species. Sustain (Switzerland) 10(7):1–12. https://doi.org/10.3390/su10072256

Laaidi K, Zeghnoun A, Dousset B, Bretin P, Giraudet E, Beaudeau P, Vandentorren S (2014) Brogan and Partners the impact of heat islands on mortality in Paris during the August 2003 heat wave. 120(2):254–259

Lapointe M, Cumming GS, Gurney GG (2019) Comparing ecosystem service preferences between urban and rural dwellers. Bioscience 69(2):108–116. https://doi.org/10.1093/biosci/biy151

Li Q (2010) Effect of forest bathing trips on human immune function. Environ Health Prev Med 15(1):9–17. https://doi.org/10.1007/s12199-008-0068-3

Lindemann-Matthies P, Bose E (2008) How many species are there? Public understanding and awareness of biodiversity in Switzerland. Human Ecol 36(5):731–742. https://doi.org/10.1007/s10745-008-9194-1

MA (Millennium Ecosystem Assessment) (2005) Ecosystems and human well-being. Synthesis report. Island Press, Washington, DC

McLaughlan C, Aldridge DC (2013) Cultivation of zebra mussels (Dreissena polymorpha) within their invaded range to improve water quality in reservoirs. Water Res 47:4357–4369. https://doi.org/10.1016/j.watres.2013.04.043

Mifsud M (2010) Maltese youth and the environment: a qualitative study. J Teacher Edu Sustain 12(2):110–128. https://doi.org/10.2478/v10099-009-0058-6

Mori J, Fini A, Galimberti M, Ginepro M, Burchi G, Massa D, Ferrini F (2018) Air pollution deposition on a roadside vegetation barrier in a mediterranean environment: combined effect of evergreen shrub species and planting density. Sci Total Environ 643:725–737. https://doi.org/10.1016/j.scitotenv.2018.06.217

Mounir M, Ainouche A, Nelson M, Cattin F, El A (2017) Long-term investigation of constructed wetland wastewater treatment and reuse: selection of adapted plant species for metaremediation. 201:120–128

National Statistics Office (2019, October 28) Motor Vehicles: Q3/2019 [News Release], pp 1–8

Nisiforou O, Charalambides AG (2012) Assessing undergraduate university students' level of knowledge, attitudes and behaviour towards biodiversity: a case study in Cyprus. Int J Sci Edu 34(7):1027–1051. https://doi.org/10.1080/09500693.2011.637991

Nowak DJ, Hirabayashi S, Doyle M, McGovern M, Pasher J (2018) Air pollution removal by urban forests in Canada and its effect on air quality and human health. Urban Forestry Urban Greening 29:40–48. https://doi.org/10.1016/j.ufug.2017.10.019

Pace P (2009) Emerging from Limbo: environmental education in Malta. In: Taylor N, Littledyke M, Eames C, Coll RK (eds) Environmental education in context: an international perspective on the development of environmental education. Sense Publishers, Rotterdam, pp 73–81
Pardo-de-Santayana M, Macía MJ (2015) The benefits of traditional knowledge. Nature 518(7540):487–488
Patel A, Kolt GS, Keogh JWL, Schofield GM (2012) The green prescription and older adults: what do general practitioners see as barriers? J Primary Health Care 4(4):320–327. https://doi.org/10.1071/hc12320
Pearson N, Timperio A, Salmon J, Crawford D, Biddle SJH (2009) Family influences on children's physical activity and fruit and vegetable consumption. Int J Behav Nutr Phys Act 6(34). https://doi.org/10.1186/1479-5868-6-34
Sandifer PA, Sutton-Grier AE, Ward BP (2015) Exploring connections among nature, biodiversity, ecosystem services, and human health and well-being: opportunities to enhance health and biodiversity conservation. Ecosyst Serv 12:1–15. https://doi.org/10.1016/j.ecoser.2014.12.007
Smith JR, Louis WR (2009) Group norms and the attitude-behaviour relationship. Soc Pers Psychol Compass 3(1):19–35. https://doi.org/10.1111/j.1751-9004.2008.00161.x
Soga M, Gaston KJ (2016) Extinction of experience: the loss of human-nature interactions. Front Ecol Environ 14(2):94–101. https://doi.org/10.1002/fee.1225
Soga M, Gaston KJ, Koyanagi TF, Kurisu K, Hanaki K (2016a) Urban residents' perceptions of neighbourhood nature: does the extinction of experience matter? Biol Cons 203:143–150. https://doi.org/10.1016/j.biocon.2016.09.020
Soga M, Gaston KJ, Yamaura Y, Kurisu K, Hanaki K (2016) Both direct and vicarious experiences of nature affect children's willingness to conserve biodiversity. Int J Environ Res Public Health 13(6). https://doi.org/10.3390/ijerph13060529
State of NSW and Office of Environment and Heritage (2017) Who cares about the environment?
Stevens DT, Lanfranco E, Mallia A, Schembri PJ (1995) Biodiversity conservation and utilisation in the Maltese Islands. Identifying and monitoring biodiversity and its utilizations in commonwealth Small Island developing states, November. https://www.um.edu.mt/library/oar/handle/123456789/21112
Tsao TM, Tsai MJ, Wang YN, Lin HL, Wu CF, Hwang JS, Su TC et al (2014) The health effects of a forest environment on subclinical cardiovascular disease and heath-related quality of life. PLoS ONE 9(7):11–13. https://doi.org/10.1371/journal.pone.0103231
Vaughn S, Schumm JS, Sinagub J (1996) Focus group interviews in education and psychology. Focus group interviews in education and psychology. SAGE Publications, Inc, Thousand Oaks. https://doi.org/10.4135/9781452243641
Waas T, Verbruggen A, Wright T (2010) University research for sustainable development: definition and characteristics explored. J Clean Prod 18:629–636. https://doi.org/10.1016/j.jclepro.2009.09.017
Wall DH, Nielsen UN, Six J (2015) Soil biodiversity and human health. Nature 528(7580):69–76. https://doi.org/10.1038/nature15744
Wang W, Wang Y, Li Z, Wei C, Zhao J, Sun L (2017) Effect of a strengthened ecological floating bed on the purification of urban landscape water supplied with reclaimed water. Sci Total Environ 623:1630–1639. https://doi.org/10.1016/j.scitotenv.2017.10.035
Worthy DA, Gorlick MA, Pacheco JL, Schnyer DM, Maddox WT (2011) With age comes wisdom: decision-making in younger and older adults. Psychol Sci 22(11). https://doi.org/10.1177/0956797611420301

Deborah A. Gatt recently entered the field of Education for Sustainable Development and has conducted this study under the umbrella of the Centre for Education and Environmental Research (CEER) at the University of Malta. She holds a B.Sc. (Hons) in Biology and Chemistry and is finalising her Masters in Education for Sustainable Development. Deborah is professionally involved in Quality Control and Assurance for pharmaceutical manufacturing and has a personal interest in herbal medications and the effects of the natural environment on human health.

Mark C. Mifsud is a Chartered Biologist, a Chartered Environmental Scientist, a Chartered Science Teacher, a Certified Wildlife Photographer and a Prince II Certified Practitioner with extensive experience in Environmental Education and Sustainable Development. Mark has drafted management plans for nature parks, reviewed EIAs and advised on ODZ applications as an independent member on the NHAC board within MEPA. He has more than 20 years teaching experience (from primary to tertiary levels) and is currently involved in environmental communication, teaching, research, academic refereeing as well as in several international projects. Mark is a member of various national and international boards. He is presently coordinating the Masters in Education for Sustainable Development at the University of Malta and holds adjunct status within the Faculty of Integrated Science and Technology at James Madison University, Harrisonburg, Virginia, in the US. He also held teaching posts at the University of Ca Foscari Venice, Italy, at Southern New Hampshire University in the US and at the University of Yaoundé, Cameroon. Mark has published more than 40 publications in the area of Education for Sustainable Development.

Understanding Scale in Wicked Problems of Sustainable Development: Who Needs Dedicated Courses in Higher Education?

Sven Linow

1 Introduction

Are we teaching the right skills, competences and knowledge at university? This is an age-old question, which constantly needs re-evaluating: Today's graduates will live in a world which is in serious danger of losing essential parts of its biosphere (Briggs 2017), with a fast changing climate triggering all kinds of secondary responses in the system-earth (Wallace-Wells 2019), strong political reactions to slowing economic growth in rich countries, as well as the siren-song of digital wonders and more. Our society needs students to acquire robust problem solving skills, to understand large-scale problems in the hope to find a viable path into a desirable or at least acceptable future (Levrini et al. 2019), and to somehow survive and prosper in this evolving mess by managing our large, urgent and actively developing problems. Our future holds many wicked problems (Rittel and Webber 1973), which are by definition complex and not constrained by discipline. They start with vague, unclear and conflicting goals. Viable paths forward will include innovations and exnovations in technology and society (e.g. Jonassen et al. 2006).

Learning and teaching to approach these challenges (often wicked problems, Rittel and Webber 1973) of the real world must draw the real world into the classroom (Biggs and Tang 2011) and to learn the necessary skills demands knowledge about real world contexts and the workings of the system-earth. Such knowledge is not acquired in an aseptic, pure and abstracted learning environment relying solely on well-structured problems. Learning to tackle real-world problems needs real-world problems in the classroom. Such problems are ill structured and present conflicting goals that depend on the viewpoint of affected parties. Simplifying them in order to

S. Linow (✉)
Department of Mechanical and Plastics Engineering, Hochschule Darmstadt,
University of Applied Sciences, Schöfferstraße 3, 64295 Darmstadt, Germany
e-mail: sven.linow@h-da.de

W. Leal Filho et al. (eds.), *Universities, Sustainability and Society: Supporting the Implementation of the Sustainable Development Goals*, World Sustainability Series,
https://doi.org/10.1007/978-3-030-63399-8_4

make them more accessible for teaching and learning is often only possible by setting specific assumptions (and being open about these assumptions), but their scale is one of the fundamental characteristics that must be understood and assessed.

The question addressed here is which discipline would gain from dedicated courses in higher education that teach how to systematically grasp and assess scale of urgent problems as a dedicated learning outcome (Biggs and Tang 2011). The question arose in a workshop attended by teaching staff and students at Hochschule Darmstadt during the 4th international climate action week in 2019.

Hochschule Darmstadt has a dedicated and well established Sozial-und Kulturwissenschaftliches Begleitstudium for engineers and other non-social sciences, with the purpose of teaching about social impacts of technology. Thus, the question could be framed differently as to who gains from a dedicated course teaching the basics of technology, natural sciences and basic methods for quantifying scale in the (super-wicked, Levin et al. 2012) problems of our time: only the non-technical disciplines, or all students? The argument is structured by first trying to understand the context of competences for problem solving into which assessing scale belongs. A second perspective comes from real student's assignments from different disciplines.

2 The Meaning of Scale

Scale refers to the physical characteristics of a system in the wider sense; and is measurable and comparable. The scale of a problem can be assessed in many different ways, where physical measures (e.g. size, volume, mass, energy, power, temperature, composition) are one relevant set of indicators, the dynamic of the system (e.g. growth rate, velocity, emission rate, diffusion rate, reaction rate) another. Both are typically used in an engineering or natural-sciences perspective. Indicators from social sciences (e.g. well-being, migration rate, resilience, vulnerability) or health science (e.g. toxicology, exposure, emission, prevalence, body weight) provide a different perspective about scale that may or may not match with other approaches. Economic indicators (e.g. income, wealth distribution, monetary benefit, cost) instruct about scale as well.

Activities like weighting the relative importance of problems or assessing possible solutions needs the identification and understanding of relevant scales. From a different perspective, it is the inability of most of us to grasp scale, that leads to successful greenwashing, or the implementation of non-evidence based measures in politics.

As an example of scale: it is hard to grasp that some 100 m of glass tubes filled with algae and today painstakingly mended by scientists will in the near future be able to provide approximately 6 km^3 or 6.000.000.000.000 l of gasoline the global economy combusts in a year.

3 A Classification of Work Assignments

Teaching new or complex subjects in higher education involves using many different approaches. Lectures are only a part of the learning experience; all kinds of additional formats add to learning (seminars, laboratories or tutorials). A relevant part of teaching is assignments, where the students do some kind of reflection or application of learned content with the intent to advance their understanding, enable their development of questions, hone their art and craft, acquire skills and become fluent in using them.

The solution of exercises happens early in courses, often as a first level of assignments which grow more complex over time. Here, exercise is understood as an "*objective which can be achieved using known means*" (DE "*Aufgabe*", VDI 2221-1). In an exercise, there is exactly one solution, there is one or a small number of possible paths to reach the solution. All necessary information to reach the solution is given with the description of the exercise; it is fully defined. The possible paths for the solution are usually identical from a purely mathematical perspective. Exercises need to be very abstract, as this is necessary to enable their simple structure. Thus, context including scale of real-world problems is seldom a part of an exercise. Numbers, while often relevant for the discipline, are often chosen to help students understand some typical numbers in this discipline, but can also be totally arbitrary.

Example Calculate the mass of 5 m^3 dry air at 2 bar using the ideal gas assumption.

The didactic value of exercises stems from the experience that learners will need to start from simple well-defined tasks, to acquire the basic competences, skills and tools of their discipline. Thus exercises are of high value in teaching.

On the other hand, no one expects that graduates will solve exercises in their jobs. Quite the contrary, they will work on real world problems where a problem is understood as an "*issue whose solution is not obvious and cannot be directly specified using familiar methods*" (DE "*Problem*" VDI 2221-1).

Example Define a stationary gas turbine for pumping natural gas through a pipeline. The rated flow of the pipeline is …

This second example includes many tasks that could have been framed as an exercise as well, but completing them would require a clear understanding of the problem, its context and a definition of boundary conditions (not fully given in the problem description).

There is a very clear distinction between an exercise and a problem, as Table 1 illustrates. This is not a gradual difference of scale (more or less of the same). On the contrary, these are fundamentally different concepts, which need separate terms. Using additive adjectives, as is often seen in the literature (e.g. *well-structured* vs. *ill-structured*), is descriptive but does not illuminate the difference sufficiently.

Learning and teaching switches to problems at a later stage: when the basic concepts are understood and have been trained through exercises, and when the students have established sufficient contextual knowledge. In the framework of

Table 1 Aspects of exercises versus problems

Aspect	Exercise	Problem
Synonym	*Well-structured problem, Story problem, Word problem*	*Ill-structured problem*
Specification of the task	Complete	Incomplete, often contradictory Defining it is part of the solution
Degree of complexity	Low, constrained	High/usually unconstrained/often infinite
System boundary	Complete, well defined, closed	Incomplete, contradictory, open Defining it is part of the solution
Context	None or irrelevant for the exercise	Real world, incomplete, contradictory Conflicting aims Defining it is part of the solution
Number of solutions	Exactly one	Zero or any number
Solution method	One, obvious for the expert If there are more than one, they are mathematically equivalent	Non-obvious or non-existent Choosing one is part of the solution
Criteria for the solution	Obvious, part of the assignment	Non obvious, not known, incomplete, contradictory Defining them is part of the solution
Quality of the solution	Deviation from exact solution is measurable or can be calculated	Unknown Defining criteria is part of the solution
Disciplines involved	Exactly one	Many (all, depending on perspective)
Interactions	None	Feedbacks into other systems Feedbacks from evolving other systems will change and influence
Explanation	Obvious It is part of the assignment	Many and sometimes contradictory explanations are possible Providing one is part of the solution

competence-oriented teaching, exercises are part of acquiring the two lower levels of Bloom's taxonomy, namely (context) knowledge and comprehension (Biggs and Tang 2011), thus problems are a necessary part in learning the levels of application and beyond.

There exists a relevant hurdle both for student and teacher in going from exercises to problems, caused by the large systemic gap between these concepts as Jonassen et al. (2006) argues:

- From the student's perspective, exercises come first in a specific course or during learning. Students get used to this structure and may start to believe that doing

abstract exercises is the aim of the course or its learning outcome. Their impression—often gained in earlier phases of education—can be that learning at university is connected with abstract, dull exercises only. The transition to problems during a course, as well as during their studies, can thus be hindered by their expectations, it may even not be noticed by them. Context-rich real world problems may still be approached like overcomplicated exercises. Students still seek for the simple formula or basic recipe that will match with some of the words provided in the problem description; they get lost without understanding why. The teacher needs to be clear about the transition from exercises to problems.
- From a teacher's perspective, exercises are a simple and reliable tool. It is easy to generate them in large numbers; students request them for their exam preparation and textbooks are full of good or old examples. Teachers had their own learning experience using exercises and they may see this as the usual approach to teaching. Teachers that never actively worked on real-world problems during their career may lack experience or feel insecure about creating and working with complex context laden problems. They may consider that the messiness and resource demand for the introduction of the extra context for real-world problems overburdens precious teaching time. Teaching with real-world problems also leads to aspects not all disciplines are used to. Ethical questions arise and political implications may cross with the belief systems of course attendants, including the teacher. The preparation of problems is different from the preparation of exercises, and may need substantially more effort not all universities are prepared to support. Students may feel unprepared or are unwilling to discuss non-disciplinary issues, and courses can fail even when the teacher is well prepared and motivated.

If the intended learning outcome of a course or a degree programme is the ability to approach real-world problems, reach a deper understanding of the system earth, to understand the messiness and trans-disciplinary interactions or feedbacks between subsystems, and the scale of real-world problems, then exercise-based teaching is clearly insufficient and must give way to problems as relevant tools. Dedicated approaches for teaching ill-structured problems are, for example problem-based learning (Jonassen et al. 2006) or interdisciplinary projects, where understanding context and resolving group interaction becomes part of the assignment.

4 Learning to Solve a Complex Problem

In an exercise the rules are laid out, cookbook schemes are available and are usually memorised (and may therefore serve as a substitute for understanding). Addressing problems needs a systematic approach that covers all relevant aspects of the problem in a meaningful way. The existing number of methods for addressing real-world problems is large—many disciplines have developed their own set of specific approaches or tools. However, there is an underlying generic approach found in most of these (e.g. Bardwell 1991; Bizer and Führ 2014; ISO 14000, VDI 2221-1), which is summarised as a generic method in Table 2. The actual number of steps in the sources varies, Table 2 provides a detailed number of steps. The core to these approaches is always to create a clear understanding of the problem and defined criteria for a good solution before any systematic generation of solutions starts. Each step may make it necessary to go back to prior tasks and revise the results there. The steps are interconnected and interdependent (Rittel and Webber 1973). The ultimately worst approach is having

Table 2 A generic method for solving problems

Step	Aspect	Aim
1	Teamwork, diversity, skills	Address the problem in a team: if possible where the diversity of expertise and disciplines of the team-members matches the expected necessary skills
2	Evidence based	Get rid of dogma and beliefs, accept a scientific approach, agree on facts. These are fundamental steps for meaningful teamwork and for the intellectual flexibility necessary to address wicked problems
3	Problem understanding	Agree on a joint understanding of the problem: reach a consensus[a] Accept that the understanding of the problem is an ongoing process and part of the project itself Describing the scale of the problem helps with understanding, comparing and agreeing on the level of urgency
4	Criteria for good solution	What shall the project achieve on what scale and how is it described best? Look for measurable or quantifiable criteria, as in SMART goals[b] Criteria should include aspects of unintended consequences, where known or expected
5	Solution-finding	Generate possible solutions; generate many different possible solutions Some solutions will and need to challenge prior assumptions
6	Decide	Use the criteria developed in step 4 to choose a best solution
7	Implement	Implement your solution
8	Monitor	Carefully monitor the implementation: look for non-anticipated effects Steer or change where indicated, abort when necessary Monitoring can involve refining of goals and criteria
9	Document	Document what you have learned including your understanding of the cause of the problem

[a]Consensus is a "*general agreement, characterized by the absence of sustained opposition to substantial issues by any important part of the concerned interests and by a process that involves seeking to take into account the views of all parties concerned and to reconcile any conflicting arguments. NOTE Consensus need not imply unanimity.*" (IEC/ISO Dir 2)
[b]SMART describes goals that are for example Specific, Measurable, Achievable, Realistic, Timely. Such goals are already part of a specific solution as they imply relevant simplifications

a solution before steps 1 through 5 have been finished carefully (even though this is the typical approach in politics and current public discourse) (Ansar et al. 2014).

The amount of planning deemed necessary in projects has varied substantially over time. These are changes to the outer technical aspects of the approach to problem handling, but the core aspects never change and are extremely valuable skills:

- How to reach a consensus?
- How to generate sound but simple criteria, accessing the scale and urgency of a problem and its possible repercussions into the system?
- How to generate good and new possible solutions?

Managing problems is often a better frame than solving them. However, addressing our emerging problems needs more than just the problem solving competence; it still needs disciplinary expertise, the ability to visualize the invisible, hands on knowledge of the living earth, understanding numeracy and scales as well as interdisciplinary systems thinking, and imagination and self-knowledge (e.g. Buck 2019). This needs to be part of meaningful teaching for the future.

5 Creating a Hierarchy of Problems

Problems can be organised with respect to their scale and the complexity necessary to manage them. No two problems are identical, even similar looking problems may not be similar at all, or differences in the approach to manage them may lead to different understanding and outcomes. Still there are indications for the severity and complexity of specific problems, which can be used for creating a hierarchy. These include:

- The scale of the problem. Is it local and constrained (e.g. by a building or fence)? Is it affecting a larger area or community or is it global?
- The urgency of the problem. Is wait-and-see a good option to solve the problem? This is often the case when the scale is limited and there are no effects affecting relevant values like health or life. Or is action extremely urgent; would inaction cause unacceptable loss.
- The environmental impacts connected to the problem (ISO 14001). Are the environmental impacts beneficial or acceptable? Will the environmental impacts create losses? What is the scale of the losses and are they acceptable?
- The view from society. Who will benefit? Are impacts internalised or externalised from the system boundary of the problem? Who is taking the risk and who will experience losses? What is the scale of eventual losses and is this acceptable to all affected or not?
- The frame of the problem. Where is the system boundary set? What are my values? What is my willingness to accept losses elsewhere (other regions, the future)?
- The measures used for the assessment of solutions. As an example, in politics and international business cost-benefit analysis is often used. By using cost-benefit

analysis any problem will be framed as a discounted monetary value; this intends to discount the future. Energy and Entropy are non-discountable and are very robust leading measures as well (e.g. Hall and Klitgard 2018 argue that available energy, not money is the relevant measure of economic value). Human or societal benefit as well as justice, given in non-monetary terms are less simple but very valuable measures.

A generic hierarchy of problems becomes possible when this is taken into account (VDI 2221-1; Levin et al. 2012; Rittel and Webber 1973). Such a hierarchy is not objective in the sense that there is one single measure that allows for unambiguous comparison, it is rather an approach that will depend on the frames and values used and may mirror the interest of a specific party. Different criteria or views will lead to different arrangement of problems:

(a) A *tame problem* (Rittel and Webber 1973) can be solved individually and disciplinarily, as many disciplines have developed methods and standards that enable the inclusion of relevant non-disciplinary aspects in a simplified way. Tame problems are thus often approached as disciplinary even though they are not. Designing a product may be received as a typical example of a highly disciplinary tamed problem, even when constraints from other disciplines are part of the specification (e.g. safety, ergonomics, mode, brand identification, pricing, environmental impact). Tame problems are well suited for disciplinary teaching (but may be misinterpreted as overtly complex exercises).
(b) *Wicked problems* (Rittel and Webber 1973) overwhelm a disciplinary approach even when constraints from other disciplines are known and included in the process of problem understanding. Wicked problems have relevant repercussions in society. Here most possible solutions will generate effects that are unacceptable—at least in a sustainable development frame, as these solutions create new and larger problems elsewhere (externalisation).
(c) Super-wicked problems (Levin et al. 2012) tend to overwhelm governments, policy and states or the willingness and ability of elites to do what is necessary. All problems given in the introduction are super-wicked.

This hierarchy provides some relevant insight when approaching a problem. For example, it can be hard to get to a joint understanding or consensus of the severity of a specific problem, as any comparison depends on personal understanding, expertise, individually set systems boundaries, accepted externalisations, and one's specific belief system.

Teaching about real world problems needs to include teaching about scale in different directions. Without grasping scale by using suitable measures, students are unable to compare problems, or if working on a single problem, compare solutions in a meaningful way. Choosing an adequate scale is part of the problem-solving process itself (Table 2 line 4, Bardwell 1991) and is a separate competency in teaching. Choices for scales will depend on the students or expert's ability and expertise. Calculating a scale can be a complex undertaking in itself. Approaches range from using simple heuristics to complex methods like vulnerability assessment (GIZ 2017) or life cycle assessment (ISO 14040). They all have their merits but require in depth learning.

6 Using Scale in Assignments—Examples from Different Disciplines

Engineers are trained to measure or calculate numbers of all kinds, so should understanding scale come easily for them? To understand if there is a measurable difference between disciplines, results from assigned projects from two different study programmes are compared. The courses and projects were centred around ill-defined real-world problems that served as graded learning outcomes. Students were required to apply appropriate problem solving methods and to involve context and understand scale for their assignments. In view of the argument, outcomes from two different elective courses are compared with an external benchmark:

(a) A masters level course on environmental conscious design (ECD) mainly addressing mechanical engineers (it was however open to other degree programmes). The learning is arranged around student's projects, where small groups do their own design. The aim of the project was to make a new design for a product or to redesign a product that falls under the scope of the Energy Related Products Directive (2009/125/EU). Design is understood as a "*set of processes that transforms requirements into specified characteristics or into the specification of a product, process or system*" (ISO 14062).
(b) A masters' level course addressing sustainable development (H:NE) with a focus on specific aspect of sustainable development (Linow et al. 2018). This course consisted of a public lecture, where relevant stakeholders provided input, and an accompanying seminar. Students deepened their understanding of the input in the seminar and, as a learning outcome, generated own proposals for the improvement of specific problems. Some of the student's outcomes are accessible (https://www.youtube.com/channel/UC3BJ_sw6WFCsRrp8lMYkmTA). Students were heterogeneous with societal sciences and engineers attending and teams tended to be interdisciplinary.
(c) The Bundespreis Ecodesign defines itself as the highest ranking governmental award on eco-design in Germany (https://bundespreis-ecodesign.de/). This provides a good benchmark for student's assignments. It provides a different view with an implicit focus on design as aesthetics-centred (not to be confused with the generic design concept as defined in ISO 14062).

The relevant question is how scale was included in the projects. Consideration of scale needed to be part of the projects at different stages:

- The object or issue chosen: Is it a relevant part of technology or infrastructure? Does it bind relevant resources during its lifetime?

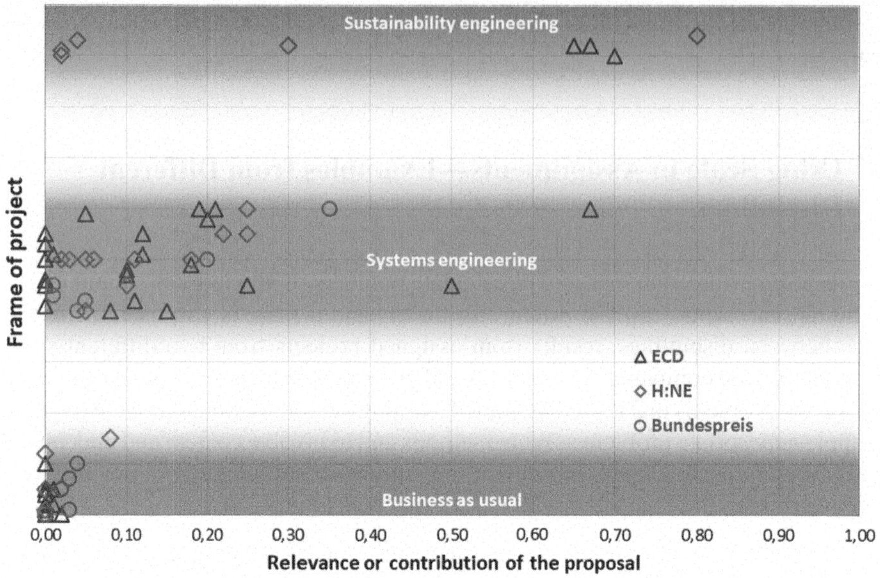

Fig. 1 Comparing outcome (x-axis) and problem framing (y-axis) for two master level courses (ECD, H:NE) and the Bundespreis Ecodesign

- The design or changes the team agrees on: Are major environmental impacts reduced by a relevant factor? Can the solution be scaled covering all regions or all comparable products?
- When new resources are considered: Are they available at the necessary scale? Is the environmental impact of these new resources sufficiently smaller than the gain made through using them?

Figure 1 provides an overview of the results, where all projects have been assessed in two dimensions.

The x-axis gives a rough quantitative estimate of the impact the proposed solution will have, when successfully applied in the intended scale. Student's projects provided a detailed discussion of gains and necessary resources throughout the course and in the individual reports, which were used for the assessment. Most of the projects addressed a single product or service or a regional issue; as the H:NE course specifically focused on the Darmstadt region. Therefore, scale can be regional or small with respect to the German economy, but still being substantial with respect to the systems boundary.

The y-axis essentially places the project in a taxonomy for its implicit or explicit understanding of sustainability that has been developed (Seager et al. 2012):

- *Business as usual* is here characterised by its obvious neglect of the question of resources and scale, a neglect of the underlying questions posed by the project itself especially with respect to environmental impacts. System boundaries tend to be blurred and optimistic assumptions about availability of cheap abundant energy and material are inherent, but often not explicitly considered. This frame seems to hinder a realistic view of the future or even a meaningful consideration (Levrini et al. 2019).

- *Systems engineering* is characterised by tight system boundaries, a good grasp of efficiency and an understanding of relevant environmental impact of the product. The core service of the product and its generic outline are usually not questioned, but assessed through a life-cycle perspective. An understanding of scale and of time is included, often presented through a definition of aims that deviate clearly from today's state. Discussed ideas for solutions tend to be pragmatic, cost effective and working. Groups consider aspects from other disciplines for the solutions.
- *Sustainable engineering science* starts from a deeper scepticism with respect to any technical solution and seeks to minimise the product itself. Problems are transferred from the technical to a societal level and aspects of sufficiency become a relevant part of the discussion. Scale and feedbacks as well as global aspects must be part of the considerations.

This classification is based on the reports as well as on discussions with the project teams during the course. This taxonomy is understood as essentially digital, but allowing for some variation in the classes.

Figure 1 shows as a generic trend, that a deeper understanding of and involvement with sustainability (y-axis) can lead to stronger effects (x-axis). On the other hand, some of the projects undertaken with deep insight resulted in negligible change to the systems performance. These were cases where the product itself was quite optimal or the service near impossible to achieve with a substantially altered product. The large cluster near the origin mirrors groups that had problems grasping the idea of the course itself or problems in applying a meaningful method for their work.

An interesting and not expected outcome is the low performance of the projects that received a Bundespreis Ecodesign award. The assessment of these contributions is a result of in depth discussion with both courses, where students provided relevant perspectives and interpretations. These discussions offered some possible explanations for the rather low performance, where neglect of scale is one of the more relevant aspects, as many proposals did not consider the scale of the problem addressed, or for the resources needed to roll out the solution at scale.

The data in Fig. 1 stems from 4 years of teaching both courses. Most students select these courses because of the subject and the project-approach. These students like challenging projects and started working interdisciplinary with only minor nudging, and grasp scale when methods to assess it are provided. Both courses include open discussion of approaches taken and time for peer-review. All teachers are clear on the intent to consider scale of problems and solutions as a relevant part—this is communicated as a learning outcome and thus as part of the grading (Biggs and Tang 2011). As a major outcome, there are no significant differences between students depending on their disciplines, i.e. between engineering and social science.

Part of the generic toolkit discussed in depth in both courses is a life-cycle perspective which "*considers the entire life cycle of a product, from raw material extraction and acquisition, through energy and material production and manufacturing, to use and end of life treatment and final disposal. Through such a systematic overview and perspective, the shifting of a potential environmental burden between life cycle stages or individual processes can be identified and possibly avoided*" (ISO 14040). By getting back to this perspective, both courses make disciplinary boundaries visible, and stress the importance of contributions from many different views. In the ECD-course, societal aspects as well as acceptance issues by potential users are more often highlighted by the teacher. In the H:NE course, students tend to need more help with technical aspects of complex products. Life-cycle perspective discussions in both courses include aspects of potential user's reluctance to accept relevant changes to a product, possible incentives, and nudges or regulations needed for substantial changes to product systems.

Other formats at Hochschule Darmstadt show that that grasp of scale is not a given, but is a skill that needs to be learned in a dedicated environment. Public discourse in terms of scale is much worse. Not only is scale regularly ignored, but scientists trying to bring scale as part of the scientific consensus (IPCC reports) into the discourse are either ignored or regularly attacked as being unreasonable or fear-mongers.

7 Conclusion—Whom to Teach About Scale

The core argument made here is that the competence of addressing problems includes skills in grasping scale, thus all students and society gain from understanding scale in wicked problems. Being able to address problems in a meaningful way would enable politics, organisations and society to manage humanities urgent, large and wicked problems. Assessing scale is an intricate part of analysing a problem as it helps to get to a common understanding and is a necessary part of the assessment of possible solutions in a meaningful way.

Neither a systematic generic approach to interdisciplinary problem solving, nor assessing scale are part of all curricula; especially not at the Bachelor level. Engineers learn some basics with the technical design methodology (e.g. VDI 2221-1) and they are able to act interdisciplinary within technical fields. For them, the full understanding of wicked problems based on the inclusion of non-technical (especially human and societal aspects) is hard, as this tends to be offloaded into dedicated courses or is actively discouraged by some as being "un-engineering behaviour". For social sciences, it is hard to overcome the barrier of scaling of technical aspects, as this is not an explicit part of their studies. Therefore, all would benefit from a dedicated teaching of scale in wicked problems This argument is supported by the data which shows that all are able to understand the challenge and include it in their work.

The work environment of most graduates and our common wicked problems demand that students are prepared; the accessed courses were electable, not mandatory. If a university wants to elicit problem-handling as a basic skill, it must introduce this as a mandatory element of their programmes. Our students actively requested a problem-based approach to teaching and learning in interdisciplinary projects for such a course. Some universities are at the beginning of the process (for example Hochschule Darmstadt is getting to a join understanding of the problem itself). Other universities understood this to be a major challenge for teaching and requirement for universities and developed their own approaches (e.g. Weiser et al. 2018).

Key aspects that enable understanding of scale and support the acquisition of the competence to assess scale independent of discipline are:

- Using real world problems in teaching and learning. These should be society's actual wicked problems at regional or global scale, not some abstract or historical case.
- Project based learning and grading based on the project outcome—constructive alignment helps to communicate clearly, as this demands that the learning outcome is aligned with the grading criteria (Biggs and Tang 2011).
- Teach and demand a methodical approach to addressing problems. Discourage early fixation on a single approach. Request views from different disciplines or perspectives, which should include a comparison of different possible solutions.
- Include in teaching and expect in assignments relevant generic concepts like life-cycle thinking, systems thinking, feedback-loops in interacting systems.
- Demand quantitative assessment and discussion of relevant scale-effects; provide basic methods.
- Teach underlying skills, like accepting interdisciplinary approaches and seeking input from beyond their own discipline; investigating and finding good sources; the ability for systematic scaffolding of their own knowledge.
- Broad contextual knowledge helps with fast erection of own knowledge scaffolds, a simple way to introduce this is by interdisciplinary co-teaching.

These are generic skills independent of a disciplinary view. One can specify this in more detail, if climate-crisis, biodiversity loss, or sustainable development with a focus on vulnerable communities is the context of a course. Relevant issues evolve from the content:

- The basics of the earth-system, especially energy- and mass flows, climate, harvestable energetic flows of the earth system, ecosystem dynamics (in depth Kleidon 2016, bachelor level Linow 2019).
- Energetic basics of human society, especially power and energy, energy density, energy returned on energy invested (Hall and Klitgard 2018), resource demand for (energetic) activities.
- Dissipation of energy and material in human activities.

After defining the problem, providing criteria for good solutions, and getting to a relevant possible solution, the next step is implementation, which will be the harder task in more reluctant organisations.

References

Ansar A, Flyvbjerg B, Budzier A, Lunn D (2014) Should we build more large dams? The actual cost of hydropower megaproject development. Energ Pol 69:43–56
Bardwell LV (1991) Problem-framing, a perspective on environmental problem-solving. Environ Manag 15:603–612
Biggs J, Tang C (2011) Teaching for quality learning at university. McGraw-Hill, Maidenhead
Bizer K, Führ M (2014) Praktisches Vorgehen in der interdisziplinären Institutionenanalyse. Ein Kompaktleitfaden. sofia-Diskussionsbeiträge 14-7. Darmstadt. https://www.sofia-darmstadt.de/fileadmin/Dokumente/Diskussion/2014/Netzversion_Stufenheuristik.pdf
Briggs JC (2017) Emergence of a sixth mass extinction? Biol J Linnean Soc 122:243–248
Buck HJ (2019) After Geoengineering. Climate Tragedy, Repair, and Restoration. Verso, London
Deutsche Gesellschaft für Internationale Zusammenarbeit (GIZ, 2014) The vulnerability sourcebook. Concept and guidelines for standardised vulnerability assessments. Eschborn. https://www.adaptationcommunity.net/?wpfb_dl=203
Deutsche Gesellschaft für Internationale Zusammenarbeit (GIZ, 2017) Risk Supplement to the vulnerability sourcebook: guidance on how to apply the vulnerability sourcebook's approach with the new IPCC AR5 concept of climate risk. Eschborn. https://www.adaptationcommunity.net/vulnerability-assessment/vulnerability-sourcebook/
Hall CAS, Klitgard K (2018) Energy and the wealth of nations. An introduction to biophysical economics. Springer, Cham
ISO/IEC Directives, Part 2 (2018) Principles and rules for the structure and drafting of ISO and IEC documents
ISO 14040 (2006) Environmental management—life cycle assessment—principles and framework
ISO/TR 14062 (2002) Environmental management—integrating environmental aspects into product design and development
Jonassen D, Strobel J, Lee Chwee Beng (2006) Everyday problem solving in engineering: lessons for engineering educators. J Eng Edu 95:139–151
Kleidon A (2016) Thermodynamic foundations of the Earth system. Cambridge University Press
Levin L, Cashore L, Bernstein S, Auld G (2012) Overcoming the tragedy of super wicked problems: constraining our future selves to ameliorate global climate change. Pol Sci 45:123–152
Levrini O, Tasquier G, Branchetti L, Barelli E (2019) Developing future-scaffolding skills through science education. Int J Sci Edu 41:2647–2674
Linow S (2019) Energie - Klima - Ressourcen. Quantitative Methoden zur Lösungsbewertung von Energiesystemen. Hanser, München
Linow S, Führ M, Kleihauer S (2018) Aktivierende Ringvorlesung mit begleitender Konzept-Werkstatt Herausforderung: Nachhaltige Entwicklung—Klimaschutz in und um Darmstadt. In: Leal W (ed) Nachhaltigkeit in der Lehre. Eine Herausforderung für Hochschulen. Springer, Berlin
Rittel HWJ, Webber MM (1973) Dilemmas in a general theory of planning. Pol Sci 4:155–169
Seager T, Selinger E, Wiek A (2012) Sustainable engineering science for resolving wicked problems. J Agric Environ Ethics 25:467–484
Wallace-Wells D (2019) The uninhabitable Earth. Life After Warming. Tim Duggan Books, Ney York
Weiser A, Hill M, Picht L et al (2018) Forschendes Lernen an der Leuphana Universität Lüneburg: Das Leuphana Semester. In: Reinmann G, Lübcke E, Heudorfer A (eds) Forschendes Lernen in der Studieneingangsphase. Empirische Befunde, Fallbeispiele und individuelle Perspektiven. Springer, Berlin

Presenting an Interdisciplinary Teaching Approach to Involve University Students in Issues of Sustainable Development

Franziska Körner, Henning Strubelt, and Hartwig Haase

1 Introduction

In 2015, all United Nations member states agreed on the 2030 Agenda for Sustainable Development to tackle the challenges our world is facing like the climate crisis, hunger, social inequality and loss of biodiversity which will harm living and future generations, if no action is taken (United Nations n. d.). At its core are the 17 Sustainable Development Goals (SDGs) as a call for action.

The fourth goal *Quality Education* aims to guarantee "inclusive and equitable quality education and promote lifelong learning opportunities for all". One of its subordinate targets (4.7.) is to "ensure that all learners acquire the knowledge and skills needed to promote sustainable development, including, among others, through Education for Sustainable Development" (United Nations n. d.).

Even before the Sustainable Development Goals were formulated and passed, the importance of Education for Sustainable Development was highlighted by the UNESCO Education for Sustainable Development World Decade 2005–2014 and the following Global Action Program. In its implementation scheme for the Education for Sustainable Development World Decade, UNESCO states that Education for Sustainable Development "is for everyone, at whatever stage of life they are. It takes place, therefore, within a perspective of lifelong learning, engaging all possible learning spaces, formal, non-formal and informal, from early childhood to adult life. [Education for Sustainable Development] calls for a re-orientation of educational approaches – curriculum and content, pedagogy and examinations" (UNESCO 2005). Since Education for Sustainable Development must be integrated at all levels

F. Körner (✉) · H. Strubelt · H. Haase
Institute of Logistics and Material Handling Systems, Otto von Guericke University Magdeburg, Universitätsplatz 2, 39106 Magdeburg, Germany
e-mail: franziska.koerner@ovgu.de

W. Leal Filho et al. (eds.), *Universities, Sustainability and Society: Supporting the Implementation of the Sustainable Development Goals*, World Sustainability Series,
https://doi.org/10.1007/978-3-030-63399-8_5

and institutions of education, universities and higher education institutions where future leaders of our societies are trained should play an important role. This should not only be about the acquisition or generation of knowledge, but students must be empowered to deal with the complexity of the issues at hand and to reflect on the impact of their decisions and behavior from a global perspective (Barth et al. 2007).

In this paper, an example is given for how Education for Sustainable Development can be implemented in higher level education. The introduced concept for a university course not only covers a topic of the field of sustainable development: climate change adaptation, but also fosters skills in students which the authors think are helpful to tackle the given challenges and to find solutions for our pressing issues, like creative and critical thinking, communication, collaboration, conflict management, decision-making and problem-solving (cf. UNESCO 2005). Additionally, students gain a deeper understanding of the political system in their federal state (in this case Saxony-Anhalt).

By addressing climate change adaptation, the course tackles a current real-world problem as climate change is not a remote possibility, it is changing our everyday lives now and in the future. It leads to diverse effects on nature, society, and economy. Rising average temperatures and sea levels as well as an increase in extreme weather events are direct consequences of climate change that we can already perceive. The fact that the climate is changing and that there are consequences associated with it has already been scientifically proven numerous times. It is safe to assume that the consequences of climate change will become even more severe in the coming years. The question now is: how do we deal with the advancing climate change and its consequences?

Essentially two policy responses to climate change exist: mitigation and adaptation. While mitigation addresses the causes by reducing greenhouse gas emissions, adaptation is about reducing the risks posed by the impacts of climate change. The subject area of climate adaptation deals specifically with the measures that can be taken now to counteract the current and future consequences of climate change and to minimize the damage and dangers that are likely to arise.

To ensure a world worth living in for future generations arguably both response strategies seem similarly important. However, climate change adaptation measures and their importance do not appear to be as widely known as climate change mitigation actions. The course presented in this paper focuses on climate change adaptation, but also incorporates the fact that adaptation and mitigation need to be considered as interdependent.

The course is carried out as a political simulation game. Students from different study programs work together in small groups and take on the role of ministers and government officials at state level (in this case Saxony-Anhalt) to identify relevant issues in their area of responsibility with regard to climate change and to develop bills that address these issues. Within their group, they are free to share the work actively shaping the learning process, thus strengthening their self-reliance (Barth et al. 2007).

The opportunity for students from different backgrounds to get together allows them to learn from each other and about their different perspectives on the problem the group decided to focus on. Moreover, it reflects that "sustainable development is a matter of negotiation" (Barth et al. 2007). This is special for teaching at universities which is often characterized by disciplinary settings and socialization (Barth et al. 2007). However, interdisciplinary appreciation and skills are needed to mirror the complex problems and demands addressed by Education for Sustainable Development (Dale and Newman 2005). To emphasize interdisciplinarity even more, this course not only relates to the group of students and the problems they are working on, but is also represented in the teaching staff being from different faculties of the university and the involved experts of various fields of climate change (adaptation) and political sciences. These experts are invited to the classroom before the students start to work in their small groups, to share their expertise and to provide insight into recent developments.

This more traditional way of education, i.e. experts giving talks, is used in the curriculum to provide background information for the exploratory interdisciplinary group work that follows and may also give opportunities for informal learning, not explicitly intended by the curriculum (Livingstone 2001), e.g. about how to organize as a group, which according to Schugurensky (2000) can also happen inside formal educational institutions individually or in a group situation. In their analysis of the development of key competencies for sustainable development in higher education, Barth and colleagues (2007) find indications that not only formal but also informal learning is relevant, referring to informal learning in extracurricular activities at the universities. The authors of this paper consider this also applicable to this course as they tried to recreate a similar setting within the curriculum by giving the student groups the utmost freedom. However, it is up to the students themselves, if they seize the opportunity for learning and acquiring new competencies (Barth et al. 2007).

By framing the course as a political simulation game and putting the students into a real-life scenario where they are in charge of mapping out legislation to avoid and reduce harmful consequences of climate change in Saxony-Anhalt, this teaching approach starts from real life problems, which Jucker (2001) considers as crucial basis for effective Education for Sustainable Development, based on his international survey and review of the international debate on Education for Sustainability. In his view, "starting from real problems and aiming at real solutions can also overcome one of the most persistent problems of the past 30 years of conventional environmental education, namely that raised awareness and increased knowledge do not automatically lead to more sustainable behavior".

2 Concept of the Course

The idea and the concept for this course were developed at Otto von Guericke University Magdeburg by teaching staff from the Departments of Logistics, Political Science with Focus on Sustainable Development and Business Didactics and Didactics of Economic Education based on previous successful collaborations. After receiving funding from the German Federal Ministry for the Environment, Nature Conservation and Nuclear Safety, the course was introduced for the first time in the summer semester of 2019. About 50 students from engineering, social sciences and teacher training study programs participated in the first run.

The sessions throughout the semester were split into the kick-off event, in which the students were introduced to the concept and the topic of the course, six workshops with experts, in two of which students were obliged to take part in, two parliamentary debates, in which students' bills were discussed and voted on, and the closing event, in which a press conference was simulated (see Fig. 1). To do justice to the applied character of the seminar, an attempt was made to find venues outside the university with practical relevance for the different sessions, especially the workshops.

The Scenario. During the kick-off event, students are introduced to the scenario in which the political simulation game is situated: After the ramifications of climate change became more obvious in recent years and were especially showing in the impressive heat of the last summer, voters in Saxony-Anhalt decided that something must be done and the Climate Adaptation Party 21 (see Fig. 2 for the logo) surprisingly managed to get an absolute majority from a standing start in the recent state-wide elections. Therefore, the party has come to power and is in charge of forming the new government. Due to this surprising success and a lack of personnel in the young party, students are now being asked to fill in and take over the opening ministerial offices to work on translating their election promises into actual laws and governmental actions. To visually support this scenario, photomontages with impressions from the election campaign, for example from television appearances or election posters, were shown (for an example, see Fig. 3).

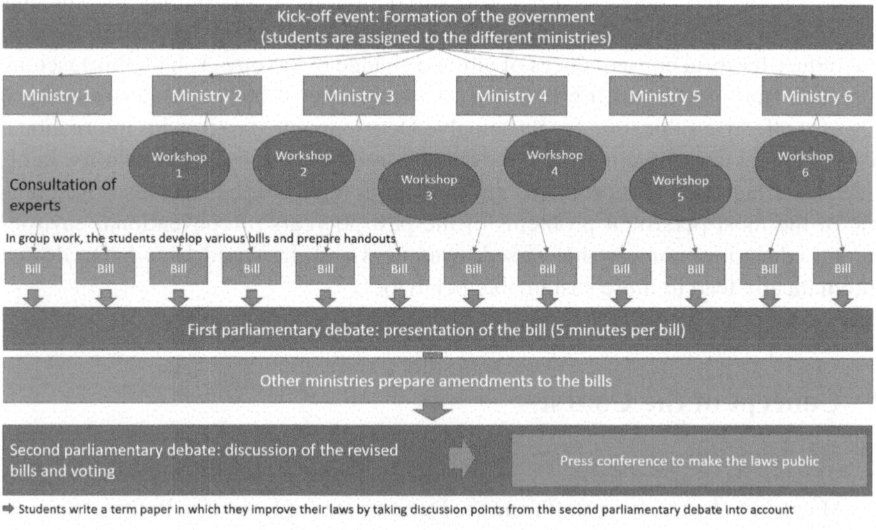

Fig. 1 Conceptual design of the political simulation game

Fig. 2 Logo of the climate adaptation party 21

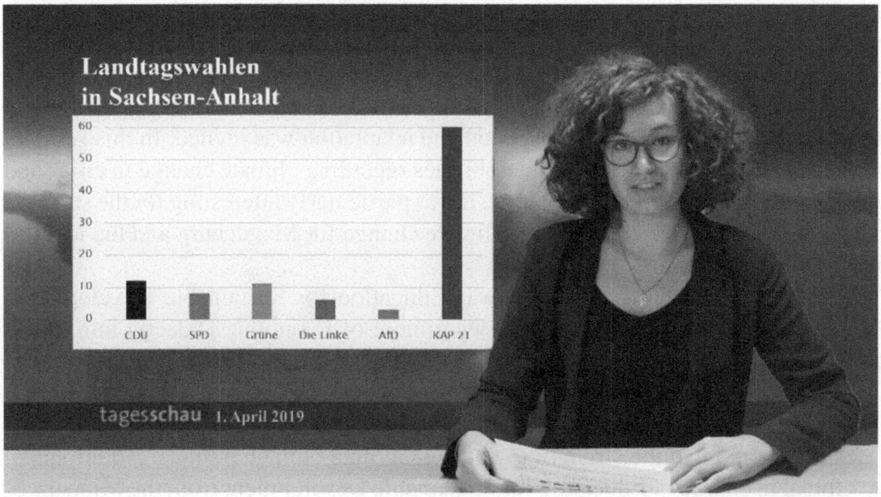

Fig. 3 Photomontage to introduce the scenario: KAP 21 wins the absolute majority in the state elections

The goal of the kick-off event is also to make the students familiar with the overall concept, the course of events and their tasks during the semester. To show differences and similarities between the simplified procedures in our political simulation game and the actual legislative process in the state parliament of Saxony-Anhalt, the lecturers from the political sciences carried out a reality check in which the state legislative processes were explained.

As an introduction to the context of the seminar, the State Office for Environmental Protection Saxony-Anhalt provided the exhibition "Climate Change Adaptation in Saxony-Anhalt" and a staff member gave an insight into the data of the long-term climate analysis in Saxony-Anhalt. Another important part of the first meeting was assigning the students to the small interdisciplinary groups, i.e. state ministries in which they were working together in the course of the semester to develop bills according to the responsibilities of their ministry. The ministries to be staffed were:

Labor and Social Affairs, Education and Science, Economy and Energy, Environment, Agriculture and Regional Development and Transport. The Ministry of Finance was filled with the teaching staff.

The Workshops. In the following weeks of the semester, workshops were held on relevant topics in the field of climate change, climate change adaptation and sustainability. Due to funding received from the Federal Ministry for the Environment, Nature Conservation and Nuclear Safety, it was possible to offer the speakers a refund for their travel costs and expenses.

The first workshop on the topic of politics took place in the state parliament of Saxony-Anhalt. Students were given a guided tour of the building by the Visitors' Service, before the political sciences teaching staff explained how the legislative process in Saxony-Anhalt works and which actors play a key role in it. In order to also offer a practical perspective, a speaker of the Green Party parliamentary group in the state parliament was invited. He talked about current issues of climate policy in Saxony-Anhalt and answered students' questions.

For the second workshop, an engineer from an institute which consults local authorities on climate change mitigation and adaptation was invited. In this session, students learned how to assess vulnerabilities regarding climate change in cities and which measures there are to counteract. It was particularly interesting for the students to hear about the expected effects of climate change for Magdeburg and the ideas to avert and reduce them.

The topic of the third workshop was Education for Sustainable Development. The teaching staff member from the department of Economic Didactic introduced the students to the ideas of Education for Sustainable Development, its goals and where it takes place. Based on current events, students had a discussion on the Fridays for Future movement, which was relatively new at this moment, and whether it would last and had the strength to change German climate politics. Additionally, the official in charge of Education for Sustainable Development from the Ministry of Environment of Saxony-Anhalt was invited for this session. In his presentation, he explained the historical background to Education for Sustainable Development and what activities in this area exist in Saxony-Anhalt. At the end students participated in a common dilemma game as an example for a teaching or training method to stimulate a conversation and discourse on the topic. In this game, players are assigned to different fishing boats on a lake with limited fish stock over several seasons. The aim is to catch as many fish as possible over the years. Each group secretly announces how many fish they would like to fish in each season. Fish that remain in the lake after each fishing season multiply by a certain factor to simulate the natural recovery of fish stocks. The group with the most fish in the end wins (cf. BMU 2017). During the game, the students experienced the difficulties involved in reconciling ecological and economic aspects in a sustainable manner.

For the workshop on climate change and health a member of staff from the Federal Environment Agency was invited. In his talk, he described the health challenges we are and we will face as a result of climate change. The focus was on the increase in heat extremes. Also, the development of Saxony-Anhalt's climate change adaptation strategy was presented by an official of the Ministry of Labor, Social Affairs and

Integration of Saxony-Anhalt before answering students' questions on this topic. Since education and risk communication play an important role in the field of health, the students finally developed an idea for an education campaign on a topic of their choice in the area of health risks from climate change in group work.

The fifth workshop was on energy and mobility. In his presentation "No energy turnaround without a turnaround in transport", a mobility researcher from the Social Science Research Centre Berlin, presented possibilities and examples for future sustainable mobility. In addition to electric mobility, active mobility and public transport were presented as the main solutions to prevent further climate change. Afterwards, a staff member of the Chair of Electrical Power Networks and Renewable Energy from the university spoke about the challenges of energy system transformation and the role of electric mobility. The question how electrical networks should be designed to meet future and changing requirements was discussed. Additionally, a member of a local initiative promoting better infrastructure for cyclists was invited as a representative of civil engagement. He encouraged voluntary work in this area and gave examples of how this is possible in Magdeburg.

The last workshop at Magdeburg City Hall was about climate friendly city planning. The head of the urban planning office in Magdeburg presented many examples of ways in which a city can be adapted to climate change and made sustainable. He focused on urban densification and the need to keep climate-compensating and risk areas free. A staff member from the Office for Environment then presented the climate change adaptation concept of the city of Magdeburg. The focus here was primarily on the different degrees to which the various districts are/will be affected.

The bills. Subsequently, the students worked in their groups on the drafts for their bills. Each group was asked to write two bills corresponding to the responsibility of the ministry they were in. To give them orientation on what needs to be included in their bills a template was provided. The template included the issues presentation of the problem, goal of the bill, measures to achieve this goal, target group and success indicators. Additionally, students were asked to assess the impact of their plans regarding sustainability, which is obliged for new laws in Germany.

The developed bills were discussed in two parliamentary debates. For the debates, students took on the roles of members of parliament (in order to be able to vote on the bills). To make the sessions as realistic as possible, the teaching staff adopted the role as state parliament presidium with one of them guiding the debate as the speaker of the parliament.

In the first debate, all bills were briefly presented in five minutes by one member of each ministry. Subsequently, the other ministries were asked to submit amendments to the proposals, which they had prepared in writing beforehand. With their bills, students addressed a wide range of topics, some narrower and more specific, others broader and more comprehensive. For example, they made a case for higher funding for research on climate change adaptation, pleaded for sustainability consultants at schools, developed plans to make cities greener and rewetting marshlands.

In the second parliamentary debate, which took place in a round-shaped auditorium to enhance the similarities to the parliament floor, the bills were briefly presented again with focus on the changes that had been incorporated into the bills following the proposals from the other ministries. Afterwards, the bills were discussed in plenary session, before a vote was held on each bill. Three of the proposed laws were not considered suitable and were rejected by parliament. In order to make the discussions of the government program (i.e. all the passed bills combined) somewhat more realistic—not every desired measure can be implemented due to financial restrictions—the Ministry of Finance set a limited (point) budget of 20 points. The costs of the implementation of each bill were estimated by the Ministry of Finance (= the teaching staff) in points from 0 to 5, also. Consequently, the students were asked to vote on the ranking of laws in a second ballot to decide which laws should actually be included in the government program.

To celebrate the success of the consultation process, a "press conference" was held in the last session, to which the speakers and supporters of the course were also invited. After a press release by the "government spokesman" (member of the teaching staff), in which the background of the seminar was briefly explained, students from each group presented the individual laws as spokespersons of their ministries. Afterwards, the audience had the opportunity to ask questions.

3 Feedback and Evaluation

Besides the press conference, the evaluation of the course was also part of the last session. Before the event ended informally in the local beer garden, the students were asked to give feedback on the course on prepared flipcharts: What did they like? What needs to be changed/improved? Do they have further comments on certain aspects of the course like organization, their motivation and what they learned? Overall, the course received a lot of positive feedback. Students were interested in climate change adaptation. They liked the workshops with the different experts taking place in changed venues, especially the workshop on climate change adaptation and energy and mobility, and the debates in the end of the course, which "were much more interesting than expected". The students reported the course had "good communication, open dialogue and a pleasant atmosphere". A number of students held that the expenditure, meaning the time and effort they put in the course, was too high and others complained that it was difficult to integrate the alternating dates for the sessions into their schedule. This is partly due to the desired interdisciplinarity, as the different study programs of the participants also result in different schedules. Here, the goal is for an even better coordination with the various courses of study in 2020.

In addition to the overall evaluation at the end of the course, students were asked to give feedback after each workshop, as well. For that, a really short questionnaire with questions regarding the topic of the workshop, the speakers, the overall comprehensibility, the gain of knowledge, the interaction, the overall assessment (grade) and further comments was prepared and given to the students at the end of each workshop. The individual evaluations of the various workshops were also positive, for example, the students reported back that they liked the guided tour through the state parliament. Sometimes, they wished for more breaks or even more opportunities to get involved.

Since the students taking part in the course came from different study programs, different requirements had to be met so that the course could be put on their record. Some only needed a certificate of attendance (fewer credit points), while others were required to hand in a term paper, in which they further elaborate the bill they were working on and evaluate the group working process. In the assessment that students made in those term papers, one complaint occurred more frequently, namely the varying motivation among the students to contribute to the group work due to the different requirements in their study programs. Students who did not require a graded record seemed to hold back and left the majority of the work to the other students who needed to prepare a term paper in the end.

4 Conclusions

In this paper, an example is given how Education for Sustainable Development can be implemented in higher education. The described course for university students from different study programs focuses on climate change adaptation. The special requirements of this multidimensional topic were taken into account by trying out an unconventional teaching format, precisely a political simulation game working on real problems, which emphasizes interdisciplinarity and also nourishes skills for sustainable development in students like creativity, cooperation and problem-solving.

Due to the realistic approach of the political simulation game, the learning locations outside the university (state parliament, ministry of social affairs, etc.), a humorous approach, and a collegial cooperation between teaching staff and students, (most) students were highly motivated and committed. They took their roles in the political simulation game seriously. This resulted in very concrete and realistic bills, which were also favored by the workshops that were very implementation-oriented and geared to the concrete tasks and current priorities of the ministries and the administration. Large, visionary drafts outside the box looking far into the future, which the authors were also hoping for, were not included.

Limitations. In general, it can be said that the concept for the course worked. Students taking part in the course were interested and gave positive feedback in the end. The authors were very happy with the outcome. However, to assess the students' learning progress more thoroughly, a scientific evaluation needs to follow.

It would be insightful to see exactly what learning achievements and knowledge gains have remained after the course has been completed, as the course covers a great deal of content within one semester. Simplifications to reduce the students' workload have already been implemented, especially in the political simulation game.

Most of the background knowledge was offered in the workshops. However, students were required to attend only two workshops, yet were free to do more. The quality of the workshops depends on the experts invited and their ability to present their topic. Another side effect of the dependence on experts is that funding is needed at least to pay the travel costs. Alternatively, it is quite conceivable that the workshops may be run by the local teaching staff.

Future prospects. So far, the course was held once at the university in Magdeburg. The authors' goal is to establish the course as a fixed part of the curriculum in the long run. Additionally, the authors promote the idea of the course in the context of Education for Sustainable Development for other educators. Even though the topic of the course is climate change adaptation, the authors agree that the flexible concept of the course can be transferred to a wide range of topics. It is particularly suitable for those topics and problems that require interdisciplinary solutions and have social relevance, which is true for issues regarding sustainable development.

In the workshops, pundits can represent different aspects and/or perspectives on the chosen topic. The group work and the discussions in the parliamentary debates encourage students to elaborate the relevant issues and form an opinion for themselves. The political simulation game in itself can be easily adjusted to different states or countries. Depending on the constitution of the respective administration level students might need to take on different roles "to have the right" to propose bills. No matter what roles students take on and how the legislation process is organized in the political simulation game it is important to have the reality check so that students learn about the real political system they are simulating.

As emphasized before the authors consider the course a success and would like to encourage others to try new unconventional approaches in their teaching as well.

Acknowledgements We would like to acknowledge our colleagues Lars Berker, Prof. Dr. Robert W. Jahn and Dr. Ulrike Zeigermann (from Otto von Guericke University Magdeburg) who collaborated with us in conceiving and holding this course successfully.

The project in which this course is developed received funding by the Federal Ministry for the Environment, Nature Conservation and Nuclear Safety based on a decision of the German Bundestag.

References

Barth M, Godemann J, Rieckmann M, Stoltenberg U (2007) Developing key competencies for sustainable development in higher education. Int J Sustain High Educ 8(4):416–430

BMU (2017) Umwelt im Unterricht. Prinzipien der nachhaltigen Fischerei. Retrieved from https://www.umwelt-im-unterricht.de/unterrichtsvorschlaege/prinzipien-der-nachhaltigen-fischerei/

Dale A, Newman L (2005) Sustainable development, education and literacy. Int J Sustain High Educ 6(4):351–362

Jucker R (2001) Sustainability? Never heard of it! Some basics we shouldn't ignore when engaging in education for sustainability (EfS). Int J Sustain High Educ 3(1):8–18

Livingstone DW (2001) Adults' informal learning: definitions, findings, gaps and future research. WALL Working Paper, 21

Schugurensky D (2000) The forms of informal learning: towards a conceptualization of the field. WALL Working Paper, 19

UNESCO (2005) United Nations decade of education for sustainable development (2005–2014): international implementation scheme [PDF]. Retrieved from https://unesdoc.unesco.org/ark:/48223/pf0000148654

United Nations (n. d.) Sustainable development goals. Knowledge Platform. Retrieved from https://sustainabledevelopment.un.org/sdgs

Methods of Transdisciplinary Collaboration Within Sustainable Research and Development Projects

Andrea Heilmann and Hardy Pundt

1 Introduction

In 2015, The United Nations Member States adopted the 17 Sustainable Development Goals (SDGs, see UN 2018). They provide a framework of aims and measures for a better and more sustainable future and should be achieved by 2030. The SDGs are interconnected and therefore partnerships of different groups and people, scientists, as well as non-scientists, on different levels are required. Research and development, however, require the cooperation of actors from various scientific disciplines and practitioners from a broad range of sectors. The involvement of stakeholders and practitioners has become additionally important due to the need for sustainable concepts on the one hand, and their concrete implementation on the other. The collaborative work of scientists and practitioners on real life problems is referred to as transdisciplinary research (TDR). TDR must be differentiated from interdisciplinary research, which is referred to as an interaction between two or more disciplines. TDR appears to be a pathway to sustainability because it integrates scientific expertise and practical needs, as well as legal regulations and rules. All these aspects must be considered, if decisions are envisaged that are accomplishable in societal frameworks. Establishing this as a basis, the collaboration between universities and other societal institutions and organizations, namely TDR, becomes a necessity to achieve sustainable solutions for a great variety of problems.

Schulte and Heilmann (2019) conducted a literature review and analyzed several definitions of TDR. Finally, she defined TDR as a combination of Pohl (2011) and Cronin (2008):

A. Heilmann (✉) · H. Pundt
Department of Automation and Computer Sciences, Harz University of Applied Sciences, Friedrichstr. 57-59, 38855 Wernigerode, Germany
e-mail: aheilmann@hs-harz.de

W. Leal Filho et al. (eds.), *Universities, Sustainability and Society: Supporting the Implementation of the Sustainable Development Goals*, World Sustainability Series,
https://doi.org/10.1007/978-3-030-63399-8_6

> In transdisciplinary research the research activities are defined as those that deal with a lifeworld problem within an interdisciplinary team. The researchers are working in cooperation with people who are affected by this lifeworld problem - the stakeholders - who provide the necessary practical knowledge and are involved in the research process. The intensity of cooperation depends on the type of project and can therefore be subject to fluctuations. In the course of this cooperative research, it is possible to create (transcend) an amount of knowledge and understanding that goes beyond the various disciplines, which is used to solve the problem or that can make a positive contribution to coping with future problems (knowledge platform or knowledge unit).

Bergmann et al. developed a guideline for TDR, in which they mention the following five criteria of transdisciplinary research projects (Bergmann et al. 2005, p. 15):

1. Reference to everyday problems and questions
2. Working in interdisciplinary research teams
3. Inclusion of relevant practical knowledge
4. Integration of the knowledge gained in the disciplines and practice
5. Formulation of new questions or solutions in order to bring the gained insights into practice and science

TDR is usually a long-lasting process during which researchers and stakeholders cooperate closely. A continuous evaluation of such processes is necessary to guarantee quality of results. The process can be divided into four phases: the input phase, the throughput, the outcomes and output and the outcome and impact (Schulte and Heilmann 2019). Every phase includes different criteria, although the criterion sustainability is an indispensable part of every phase. This illustrates the point that TDR is a relevant tool to implement the SDGs within the society.

TDR-related research and development can be carried out based on different methodological approaches such as Third Mission, Citizen Science and Living Labs. These approaches have been seen as independent methods so far. One outcome of our research carried out over the past few years has been that instead of focusing on only one procedure, a combination of the approaches can contribute significantly to a sustainable implementation of measures that were derived collaboratively. In order to develop a long-lasting decision support methodology, it is therefore necessary to determine the specific characteristics of the above-mentioned approaches. In the following sections these methods are presented, compared and an application example for the implementation of the SDG 13 "Climate action" is given.

2 Description of Approaches for Transdisciplinary Research and Development

First, the three approaches introduced before will be explained briefly aiming at laying a foundation for the following sections.

Third Mission

The term "third mission" has been defined by different authors covering several aspects. Schulte et al. present a definition of Third Mission, which is mainly influenced by the definition of Pasternack and Zierold (2015), but it is more specific:

> Third Mission is defined as a university's activity, which fulfils the following requirements:
>
> It is related to university's core processes - teaching and research - or their strategic aims.
>
> making use of the university's resources, e.g. professors, students or rooms.
>
> shaping the university's non-academic environment, e.g. by cooperative activities with non-academic agents.

This definition is accepted within the framework of this article. Universities have a variety of options for Third Mission activities. These include student projects, joint workshops or further training as well as specific (disciplinary or interdisciplinary) research projects. The projects are often of short duration, and locally linked to the university's region.

Citizen Science

Citizen Science is a form of participation which integrates people who do not work solely in science but participate in scientific processes. The European Citizen Science Association specifies this research method as a flexible approach that can be adapted to different circumstances and disciplines, but should take ten fundamental principles into account (ECSA 2015). Bach et al. (2017), classify citizen science projects depending on the type of participation into:

Co-design: Citizens are involved in the research process from the beginning, i.e. a joint project planning and implementation, which can lead to shared responsibilities.

Co-production: Citizens work under the guidance of scientists in the research process (e.g. data acquisition and evaluation).

Virtual participation: Large amounts of virtual data are gathered, for example, via crowdsourcing or distributed sensors. In addition to recording, citizens often take responsibility for the evaluation of digital data.

Educational projects with elements of original research: Beside new scientific results, the implementation of the knowledge takes place in close cooperation of researchers and citizens.

Autonomous research: This encompasses all activities of individuals or interest groups who generally conduct independent research without any institutional connection. Support can be given, for example, by providing scientific resources (e.g. laboratories).

Depending on the degree of participation, the amount of time and resources required is higher compared with Third Mission activities.

Living Lab

Living Labs are generally understood as an infrastructure that enables and favors a user-centered research methodology. According to a general understanding, the living lab approach (also often referred to as real-world laboratories) comprises a user-friendly, real-world research environment in which not only science, business and organizations carry out research and development together, but above all, users themselves take over an active role in the innovation processes.

Definitions of living labs are based on several publications. Specific requirements were developed as part of a pilot project by the European Commission named "European Network of Living Labs (ENoLL)". Based on the results of the network, a methodological manual containing five main elements was developed (Malmberg 2017):

Multi-method approaches	• Use of diverse methods, adapted to the respective question
User engagement	• Participation from the start
Multi-stakeholder participation	• Broad stakeholder participation, which includes representatives of the private and public sectors as well as academics and citizens
Real-life setting	• Implementation in a real-world environment
Co-creation	• Equal involvement of users in the process by not being the subject of the investigations

Living Labs represent the most comprehensive approach in relation to the implementation of transdisciplinary research. They are carried out over a long period of time, often in a relatively large area. Their focus is not only on the development but also on the testing of new solutions. This requires the provision of more extensive resources compared to the approaches described before.

Analysis of the approaches

Based on Bergmann's criteria (2005, p. 15) and on the afore-mentioned approaches, the following table was drafted. It can be used as a decision support tool in the sense that it enables users to select the appropriate approach for their specific projects. Based on the criteria mentioned in the rows it is possible to assess which criteria play more or less important roles.

Comparing the approaches, it becomes clear that the cooperation of actors from science, practice, as well as citizens, is the common ground for developing sustainable problem solutions (except for interdisciplinary research). In addition to the criteria in Table 1, the different requirements regarding the spatial, resource und time framework need to be taken into consideration. Aiming at supporting the implementation of the SDGs in a collaborative, TDR-based approach, the criteria, even if they are discussed separately in Table 1, have to be merged. This will be illustrated using a concrete example which is presented in the next section.

Table 1 Degree of fulfillment of TDR-related criteria using different methods

Criteria	Living lab	Citizen science	Third mission	Interdisciplinary research
Everyday problems (1)	+++	+++	++	+
Interdisciplinary research teams (2)	+++	+++	+	+++
Relevant practical knowledge (3)	+++	+++	+	o
Integration of the knowledge (4)	++	++	++	++
Formulate new questions (5)	+++	++	+	++

o … none, +… weak, ++ middle; +++… strong consideration of the criteria

3 Application of the TDR Approaches in the Field of Climate Adaptation

3.1 Climate Adaptation as a Case Study

The UNFCC stated that "Adaptation, in the simplest terms, refers to the actions that countries will need to take to respond to the impacts of climate change that are already happening, while at the same time preparing for future impacts. Successful adaptation activities also call for the effective engagement of stakeholders - including national, regional, multilateral and international organizations, the public and private sectors, and civil society and the management of knowledge for adaptation at each step" (UNFCCC 2017).

The quote clearly hints on the need of transdisciplinary research in this area thus aiming at strengthening resilience and adaptive capacity concerning climate-related hazards. This corresponds with the implementation of the SDG 13 ("Climate Action"), but climate adaptation is also relevant for other SDGs like SDG 14 and 15, or SDG 6.

The Harz University of Applied Sciences has carried out several transdisciplinary research projects in the field of climate adaptation in a specific pilot region. The pilot region is the administrative district of Mansfeld-Südharz situated in the south of the German federal state of Saxony-Anhalt covering an area of nearly 1450 km^2. Land use in its rural areas is dominated by agriculture and forestry. The following case study can serve as an illustrative example for the complex interaction between the different approaches mentioned in Table 1. The study shows that the application of only one approach is not sufficient for the decision on and implementation of measures that should be achieved according to the SDG 13, "Climate Action".

3.2 TDR Methods Applied in Three Projects Aimed at Climate Change Adaptation

The administrational authorities of Mansfeld-Südharz have been working on local climate change adaptation measures since 2011. The work has been carried out during three phases. The goal of the first phase (project "Klimpass", 2011–2012) included the compilation of a local adaptation strategy. Interdisciplinary research, Citizen Science as well as Third Mission Activities were applied. Within the second phase (project "Klimpass-Aktiv", 2013–2016), the advancement, implementation, and evaluation of specific adaptation measures were envisaged. This phase comprised several activities based on interdisciplinary research, Citizen Science as well as Third Mission Activities. The project "BebeR", running from 2017–2020, is looking especially at the improvement of the existing methodologies for soil erosion vulnerability assessment. BebeR fulfils the requirements of a Living Lab.

In all phases, the development of an interactive project website/platform consisting of vulnerability maps and an online discussion forum to enable information sharing were key elements (Pundt et al. 2017, 2020). The interactive platform is based on internet technologies such as Content Management Systems (CMS) and Web-GIS (Geographical Information Systems). The website enables all relevant actors to collaborate in different forms. In such a way, the website widens effectively the opportunities of actors to participate and collaborate due to the integration of the advantages of the different approaches mentioned before. To clarify this, some aspects of each approach, identified as being relevant for the projects, will be described briefly.

Third mission was applied when researchers and students of the Harz University and the Magdeburg University of applied sciences provided their specific knowledge in different areas (e.g. climate adaptation, GIS, CMS, and others). For instance, soil erosion rates were calculated using a simulation software. This was carried out by researchers from from the water management department of the university in Magdeburg. The results were included in the interactive website and thus formed an effective discussion basis for workshops with stakeholders from local water management, nature protection, regional planning, agriculture, and forestry. Students were involved with project works addressing specific issues, e.g. the production of videos about challenges of climate change in that particular region. Besides, members of the university acted as coordinators and moderators between the different stakeholders.

Citizen Science includes the knowledge and expertise of stakeholders of the region as well as local citizens aimed at involving these people in the development and implementation of climate adaptation measures. The interactive website and in particular the online maps played an important role in this respect. Contingently, data captured by citizens can complete the datasets and contribute to qualitatively better results. In various cases, different actors (classified into stakeholders, e.g. representatives from administrative authorities, forestry or agriculture, and citizens, e.g. members

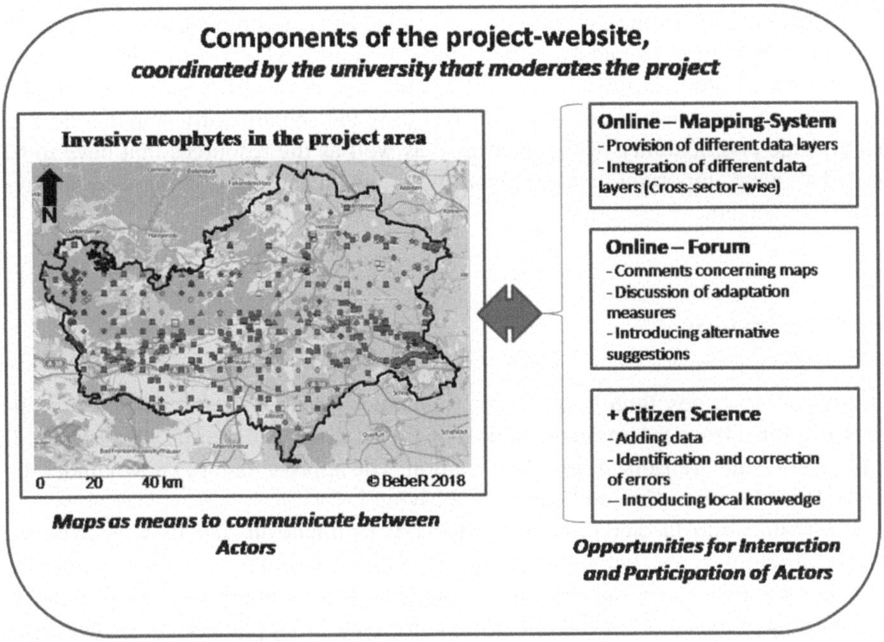

Fig. 1 Components of the project-website within the climate adaptation projects in the pilot region Mansfeld -Südharz (Pundt and Heilmann 2020)

of NGOs, inhabitants of villages) provided knowledge that would not have been considered without applying this approach. Figure 1 shows the overall collaboration structure within the complex and long-lasting process of climate adaption in the pilot region.

A further example of applying citizen science is the linkage of the website to an external project that deals with capturing data on invasive species. Such new species invade due to climate change. Data capture is carried out using a mobile, smartphone-based app which enables users to directly send the data to a specific website (e.g. www.korina.info). The results of this collaboration were included in leaflets and videos produced by students, aiming at improving information provision for local people. Referring to the definitions given above, these activities integrate Citizen Science and third mission activities.

Living Lab—A main goal of the BebeR-project was to develop and assess measures which reduce the vulnerability of the pilot region in terms of soil erosion. This had to be done under explicit consideration of the varying climate change-induced conditions. A multi-method approach, integrating the three TDR approaches mentioned before, was used in this project to achieve sustainable decisions. Therefore, the interactive website was expanded, for instance, a multi-criteria analysis tool was integrated. This multi-criteria analysis was utilized in order to prioritize possible options for the minimization of soil erosion. This was only possible through the active participation of the different actors from science and practice, thus assessing results of simulations and analyses in workshops combined with field trips. In such a way, the criteria for a living lab were finally met, e.g. the real-life setting and implementation in a real-world environment. Not only the scientifically sound concepts, but also the practicability of the measures, as well as the resources that have to be provided, were important needs defined by many stakeholders. They were always taken into account.

4 Conclusions

The paper describes three approaches suitable to deal with the complex problems resulting from the implementation of the SDGs. An essential element of all approaches is the transdisciplinary foundation. The approaches are known under the terms Third Mission, Citizen Science and Living Labs.

Due to the multi-faceted problems provoked by climate change, these approaches follow different goals, but each has specific advantages because they suggest to work cooperatively in a transdisciplinary manner. Within the framework of a three-phased project that was carried out at the University of Applied Sciences Harz, the different approaches were briefly discussed. The experiences resulting from these projects prove that TDR is an appropriate way to deal with the complex problems of climate adaptation. Particularly, TDR is suitable to make collaboration of scientists, practitioners, stakeholders and possibly citizens more effective and enables an all-embracing view of problems, thus looking on them from diverse perspectives.

Describing the various methods, Third Mission, Citizen Science and Living Labs should be investigated more deeply concerning their general suitability for the decision support of complex, multi-faceted problems. An important result of our projects is that different approaches can be used simultaneously in one project. This is illustrated in the case studies. However, in order to integrate the individual methods in one project, comprehensive coordination is required. This task can ideally be carried out by a university due to its independent state, and the exclusive interest in scientifically sound results. An important prerequisite for TDR is that the principles and methods are well known and appreciated by all participating actors, the researchers, the stakeholders, the citizens, and others.

References

Bach BN, Blanckenburg C, Göbel S, Jaeger-Erben M, John R, Dienel HL, Rückert-John J, Talmon-Gros L, Teichler T, Vohland K, Wiatr M, Ziegler D (2017) Konzept zur Anwendbarkeit von Citizen Science in der Ressortforschung des Umweltbundesamtes. In: Umweltforschungsplan des Bundesministeriums für Umwelt, Naturschutz, Bau und Reaktorsicherheit, vol. 2481. Institut für Sozialinnovation Consulting UG, Berlin

Bergmann M, Brohmann B, Hoffmann E, Loibl MC, Rehaag R, Schramm E, Voß J-P (2005) Qualitätskriterien Transdisziplinärer Forschung. In: Institut für Sozial-Ökolgische Forschung (ISOE), Frankfurt am Main, ISOE-Studientexte, No. 13

Cronin K (2008) Cronin, Karen: transdisciplinary research (TDR) and sustainability. In: Institute of Environmental Science and Research Limited (ESR). www.learningforsustainability.net/pubs/Transdisciplinary_Research_and_Sustainability.pdf

ECSA (2015) Ten principles of citizen science—Bürgerwissenschaften. https://ecsa.citizen-science.net/sites/default/files/ecsa_ten_principles_of_cs_german.pdf

Malmberg K (2017) Ines Vaittinen (European Network of Living Labs). Living Lab Methodology Handbook. https://u4iot.eu/pdf/U4IoT_LivingLabMethodology_Handbook.pdf

Pasternack P, Zierold S (2015) Strategieentwicklung trotz Hindernissen. Hochschul-aktivitäten und Bedarfslagen in schrumpfenden Regionen. In: Fritsch M, Pasternack P, Titze M (eds) Schrumpfende Regionen—dynamische Hochschulen, pp 255–278. Springer Fachmedien, Wiesbaden

Pohl C (2011) What is progress in transdisciplinary research? Futures 43(6):618–626

Pundt H, Heilmann A, Scheinert M (2017) Assessing vulnerabilities as a step toward climate change induced hazard prepardness. Int J Saf Secur Eng 7(2):137–146

Pundt H, Heilmann A (2020) Building collaborative partnerships: an example of a 3rd mission activity in the field of local climate change adaptation. In: Leal WF, Alves F, Azeiteiro U, Manolas E (eds) Universities as living labs for sustainable development. supporting the implementation of the sustainable development goals, vol II. Spinger Nature Switzerland, Cham, pp 621–636

Schulte R, Heilmann A (2019) Presentation and discussion of an evaluation model for transdiziplinary research. In: Castelli GP (ed) European Journal of Sustainable Development 8(3):1–10. ISSN 2239-5938 (print), ISSN 2239-6101 (online)

UN (2018) United Nations Sustainable goals development. Retrieved from: http://www.un.org/sustainabledevelopment/sustainable-development-goals/. Accessed January 27th, 2018

UNFCCC (2017) United Nations framework convention on climate change. Available at: http://bigpicture.unfccc.int/. Accessed 17th January 2017

Professor Dr. Andrea Heilmann holds a University Diploma in Water Engineering (Dipl.-Ing.) from Dresden University of Technology (1990). She took her doctoral degree in the field of waste management at Dresden University of Technology (1999). From January 1991 to September 2000 she worked as a consultant in waste management. Since 2000, she has been Professor of Environmental Management and Technology at Harz University of Applied Sciences. Her research focus is on sustainable development as well as on climate protection and adaptation. At Harz University she leads the Sustainable Management Group, which develops and implements sustainable projects. She is also the responsible person for the environmental management system according the EMAS III.

Professor Dr. Hardy Pundt studied Geography, Geoinformatics, Geology and Botany at the University of Münster, Germany and holds a University Diploma in Physical Geography (Dipl.-Geogr.). He worked as scientific employee at the University and the Chamber for Agriculture and spent scholarships in the UK, Brasil, and the US. He did his Ph.D. in Geoinformatics in 1995 and worked as a scientific assistant at the Institute of Geoinformatics, University of Münster between 1996 and 2002. Since 2002 he has been Professor ofGeographical Information Systems at the Harz University of Applied Sciences in Wernigerode, published around 100 scientific worksand cooperated with national and international partners in various projects. He was the editor of several books covering areas like Geoinformation and Image Processing, eGovernment, and Climate Change Adaptation. He was member of the Council of the Association of Geographic Information Laboratories in Europe (AGILE), Europes leading association that fosters teaching and research in all fields of spatial data collection, analysis and visualization, between 2009 and 2017.

Universities as 'Global Citizens': Reflections and Learning from a Project on Female Leadership for Sustainable Development of Fishing Communities in Colombia

Valeria Ruiz Vargas, Alicia Prowse, and Lina Maria Barrios

1 Introduction

Research funding for projects in countries that are on the list of Official Development Assistance (ODA)—a measure of flows of international aid—has increased in the United Kingdom in recent years (UKRI 2020; Centre for Global Development 2019). Universities have been working on projects through these funding streams that are increasingly focused on sustainable development and the sustainable development goals (SDGs). The SDGs have been developed by a consultative process spanning nations across the globe (UN 2015). Sustainable development is a model with policy relevance at international level (Estes 2010; Baker et al. 2005). However, sustainable development has been criticised for representing the views of some communities over the views of others (Andreotti 2016, 2014; Rajan and Thornhill 2019). For instance, indigenous people's worldviews or that of feminist political ecologists may not fit into the economic growth agenda that is part of the SDGs: for example SDG 8, Decent work and economic growth.

With the range of perspectives that different human communities may value, universities working on sustainable development projects in ODA countries, may face ethical dilemmas (Stein et al. 2019) that are not necessarily taken into account by

V. R. Vargas (✉) · L. M. Barrios
Department of Natural Sciences, Manchester Metropolitan University, Chester Street, Manchester M1 5GD, UK
e-mail: V.vargas@mmu.ac.uk

L. M. Barrios
e-mail: Lina.Barrios@mmu.ac.uk

A. Prowse
University Teaching Academy, Manchester Metropolitan University, Oxford Road, Manchester M15 6BG, UK
e-mail: a.prowse@mmu.ac.uk

W. Leal Filho et al. (eds.), *Universities, Sustainability and Society: Supporting the Implementation of the Sustainable Development Goals*, World Sustainability Series,
https://doi.org/10.1007/978-3-030-63399-8_7

traditional ethical processes. One of these dilemmas is that sustainable development research in the Global South funded by international aid can be seen as paternalistic and neocolonial (Andreotti 2016, 2014; Rajan and Thornhill 2019). Therefore, trying to avoid the pitfalls of neocolonialism requires an intentionally reflexive approach from those responsible for the initial framing and implementation of projects.

Global citizenship is a framework that looks at ways of addressing social issues. This is similar to the ODA related funding for universities' research. However global citizenship, especially its conceptualisation in recent years, brings together issues of social and environmental wellbeing whilst critically engaging with a range of 'political' communities such as indigenous communities or neoliberal western communities (Pashby et al. 2020). Therefore, global citizenship is an appropriate framework to explore the intentions of universities and researchers in ODA countries working in sustainable development projects.

Exploring the intentions of a university's research involves researchers questioning the funding and conception of this research. How should universities take steps to interrogate the framings of their international research projects in order to 'decolonise' the academy and to be congruent with a desire to do public good (Walker and McLean 2013)? How do universities see themselves as 'good global citizens' while also seeing themselves as competitive businesses? These are questions that have yet to be explored in the literature (Stein et al. 2019) and is the focus of this paper.

As universities find their place on the global stage there have been moves towards a more outward-looking perspective (Hudzik 2014). Although this could lead to a conception of the university as an agent for global citizenship, one material consequence has been an increasing tendency towards a competitive 'globalised outlook' (Engel and Sizcek 2018). This is often conceptualized as 'internationalisation' of the university but is limited in its extent appearing to be self-serving rather than focussed on public good (e.g. Friedman 2018), with national interest as the key driver (Engel and Sizcek 2018). These authors thus conclude that based on their exploration of university internationalisation strategies these "limit the possibilities of advancing central tenets of global citizenship" (Engel and Sizcek 2018: 749).

Consideration may thus be given to whether, as Engel and Sizcek (2018) suggest, global citizenship should be principally seen as an outcome of increased international co-operative efforts or whether it is a motivational driver of these efforts. Although UK higher education has been said to have an increasing audit culture (Erickson et al. 2020), which has helped to stifle engaged citizenship in its staff (Spooner 2017), it is still possible for individual staff to be motivated by a sense of service.

Universities characterise two main areas of activity: education and research. Funding of international research in the UK in recent years has been via the Global Challenges Research Fund (GCRF), as part of the UK's ODA commitment. However, there is little information on the direct benefit to local actors versus the return to the developed world (e.g. what proportion of funding returns to, or never leaves, the UK).

It has been well-documented that in some sectors, the funding of local efforts for self-help are routinely marginalised. For example, in humanitarian responses to emergencies only 0.2% of the overall funding response goes to local actors, even though these are the people who are inevitably the first on the scene (Roepstorff 2019).

A university acting as a 'global citizen' might be expected to model an ethical approach to funding of its activities in relation to its obligations to both ODA and to its core purpose for public good. This could translate into a critical view of not only what research is funded, but also on the ethics of the power relationships and the uncritical use of Western paradigms of knowledge that are inevitably self-reproducing (Dawson 2020). In practice that might mean that the university as an organisation, "promotes critical understanding of how we can think and act without prejudice in our diverse and interconnected world" (Mansouri et al. 2017: 5).

These issues frame the reflections of three researchers from a UK institution who devised and conducted a research project funded by the UK DfID via the Global Challenges Research Fund.

2 Project Background

2.1 Project on Sustainable Development and Women's Leadership

The research project was funded by GCRF and carried out in Colombia, one of the ODA countries. Women leaders belonging to fisher associations were selected by the regional federation (Federacion de Pescadores, acuicultores y Agricultores del Canal del Dique-FEPASACADI) via a meritocratic selection process to allocate 24 places for a residency, representing more than 30 associations. During the residency, the researchers gave a general background on concepts related to participatory research, biodiversity and sustainability (e.g. tragedy of the commons), sustainable development goals (SDG) and Aichi targets, followed by activities that allowed researchers to gain feedback on participants' perspectives on these topics. The women leaders and researchers explored together the meaning of female empowerment and networking. They also explored how they could use their strengths as women leaders in the area to promote innovative projects and entrepreneurship in their areas of interest. The women leaders acknowledged that they benefited from the project in the way they wanted to benefit, dependent on their level of participation. Researchers reflected on their teaching, learning, and research experience during the planning, execution and analysis of the project. The researchers also encouraged women leaders to reflect on their journey from the beginning to the end of the residency.

One key idea that the researchers kept in mind was that although 'development' is typically predicated on a one way flow of knowledge and values from the 'developed' to the 'less developed' nations, (Stein et al. 2019) this can lead to the reproduction of epistemic and economic dominance even within higher education partnerships designed to address this.

2.2 Community Based Participatory Research

Community based participatory research (CBPR) is a type of qualitative research (Kesby 2005) that provides structure and mechanisms for collaborative and rigorous research, using traditional or new methods, but with a community focus (Horowitz et al. 2009). In CBPR community members and researchers contribute equally and recognise each other's experience, there is trust, collaboration, shared decision making, and shared ownership of the findings, with acquired knowledge that benefits all partners. The partners commit to long-term research relationships, and aim for local capacity building, systems development, empowerment and sustainability (Horowitz et al. 2009). The community benefits by enhancing their empowerment, co-learning with scientists, informing the wider community of organizing efforts, and linking research to policy action (Balazs and Morello-Frosch 2013). There are several levels of CBPR, from a basic level where the community are study participants, the design is arranged by the researchers and results are socialised with the community, to a higher level in which the community are partners in the research and highly involved throughout the process, from conception of the project, data collection, fundraising, publication and ownership of results (Balazs and Morello-Frosch 2013). In all the levels the community is recognised as a unit of identity, but higher levels of collaboration imply longer times of execution and complexity, to ensure relevance of research, rigor, and maximum reach to stakeholders (Balazs and Morello-Frosch 2013), which may not always be possible to achieve. Examples of CBPR with native communities in Latin America have proven to be a very effective tool to collect, analyse and monitor environmental conditions and promote sustainable use of resources within the community (Oldekop et al. 2012; Saavedra et al. 2016).

Research question: what could the three UK based researchers learn about researcher motivations towards designing and implementing international sustainable development projects in the context of the public good/global citizenship mission of higher education from a project in the Colombian Caribbean?

3 Methods

The research design for this study was based on reflective practice in teaching and learning (Brookfield 2002) and research (Chiu 2006; Wilson 2007; Nicholls 2009). The practical element that led to the reflection was a participatory research project focused on women's leadership in fisher communities in the Dique Channel region of Colombia (Barrios et al. 2020). The whole research project was done over the course of eighteen months (August 2018 until December 2019). The 31 participants (including the researchers) spent 6 days/5 nights in residential fieldwork in a working farm in the Dique Channel, Colombia. During the design of the research project and preparation of the residential fieldwork, the researchers started to develop a reflective dialogue (Chiu 2006; Wilson 2007; Nicholls 2009).

Reflective methods are important in participatory research. This is because they help researchers to become aware and include a wide range of voices and perspectives (not only the most prominent) from the participants as well as from themselves (Chiu 2006; Wilson 2007; Nicholls 2009). Brookfield (2002) provides four lenses for critical reflection (originally designed for community college learning and teaching experiences), which are: autobiographical experiences of learning, the learners' eyes, our colleagues' experiences and the theoretical literature. The researchers used these lenses to help orientate reflections. In this reflection, the three researchers conceptualised themselves as learners.

This reflective process was highlighted by the discussions about ethical approval for the project. During the fieldwork, the three researchers met every night to reflect on the day and made any changes required for the subsequent days of data collection. During these meetings the three researchers developed their reflective dialogue, making notes of key aspects to be further discussed.

After analysing the data collected during the fieldwork, writing and submitting a paper to an academic journal (Barrios et al. 2020), the researchers met to record a final reflective dialogue. In preparation for this dialogue the researchers read their individual and collective reflective notes written during the research project. The dialogue was driven by the researchers' questions to each other. In this respect it was similar to an unstructured interview. However, the three researchers acted as interviewers and interviewees simultaneously. The whole dialogue (55'28" long) was audio recorded and divided in three. Each researcher did a verbatim transcription of about 18 min (i.e. a third of the whole audio recording).

The three researchers triangulated the results of the data analysis whilst engaging in a process of self-reflection to acknowledge the biases inherent in their own research personas. Thus, the three researchers coded the whole reflective dialogue independently. A framework was developed throughout the research project whilst having the reflective dialogues. The framework included four main aspects:

1. Researchers perceptions of power relations between the researchers and the Dique Channel participants.
2. Agency of the researchers as perceived by themselves
3. Identity of the researchers as perceived by themselves.
4. Interactions between 1, 2 and 3.

The first stage of coding was done assigning text to a unit of analysis for aspects 1, 2, and 3 of the analysis framework and to identify themes. The second stage focused on the 4th aspect to analyse the interactions between 1, 2 and 3. Each researcher worked in stages 1 and 2 of the data analysis independently. The results were only discussed once the two stages of coding were completed by all the researchers.

Researcher A: She defines herself as a multidisciplinary researcher. She started to work in academic research before starting or completing her PhD. Her background is in music, arts and sustainable development. Researcher A has 6 years' experience working in higher education. Her workload is focused on research with a small teaching load. Previously she was teaching music in further education (FE). Her approaches to teaching have shifted over the years depending on the subjects taught. In FE, whilst teaching Music Theory and Solfege, she perceives her approach as structured and teacher-led. When teaching contemporary music and sustainable development she perceives her approach as more student-led, less structured and holistic. Researcher A has a keen interest in power relations and the impact they have on people professionally, personally and within inter- and intra-organisational stakeholder networks in the education and research sector.

Researcher B: Researcher B is originally an ecologist, completing a PhD on the ecological impacts of an invasive non-native species. She has worked as a professional actor; teacher of English to speakers of other languages; woodland surveyor for a national park; and collaborated in funded art/science projects. She has taught undergraduate and postgraduate Biology and Research Methods. She now works in a centre for learning and teaching at a university in the UK, working closely with academic staff to enable them to develop their teaching practice. She has led projects at the institutional level in global citizenship, student transition and personal tutoring.

Researcher C: Researcher C is a marine biologist with a PhD in tropical marine ecology. Her work involves comparisons between tropical and temperate marine environments around the world. In the UK she has worked in four leading universities, teaching and supervising students (bachelor-Ph.D. levels) on topics related to marine biology and ecology. She coordinated the Newton Fund project Fishing for Life (www.pescandoparalavida.org) between 2016–2018, the base for the women's leadership project. In Colombia, she worked for the national marine research institute (coordinating research projects and a research line) and the Ministry of Defence (as advisor for the Colombian Maritime Director and the Colombian representative at UN-IMO). During the last 3 years she has coordinated several binational projects (UK-Colombia). Her participation in Newton and GCRF projects has increased her involvement in scientific-social research.

4 Results

The coding analysis showed 18 different themes identified as a result of the reflection on the research experience with the women leaders in Colombia (Fig. 1), from which power relations, agency and identity are the most relevant to the aim of this paper.

4.1 Analysis of Power Relations (Symmetrical and Asymmetrical)

The researchers' reflections were characterised strongly by a consideration of power relations. A basic open coding showed that the researchers considered both symmetric

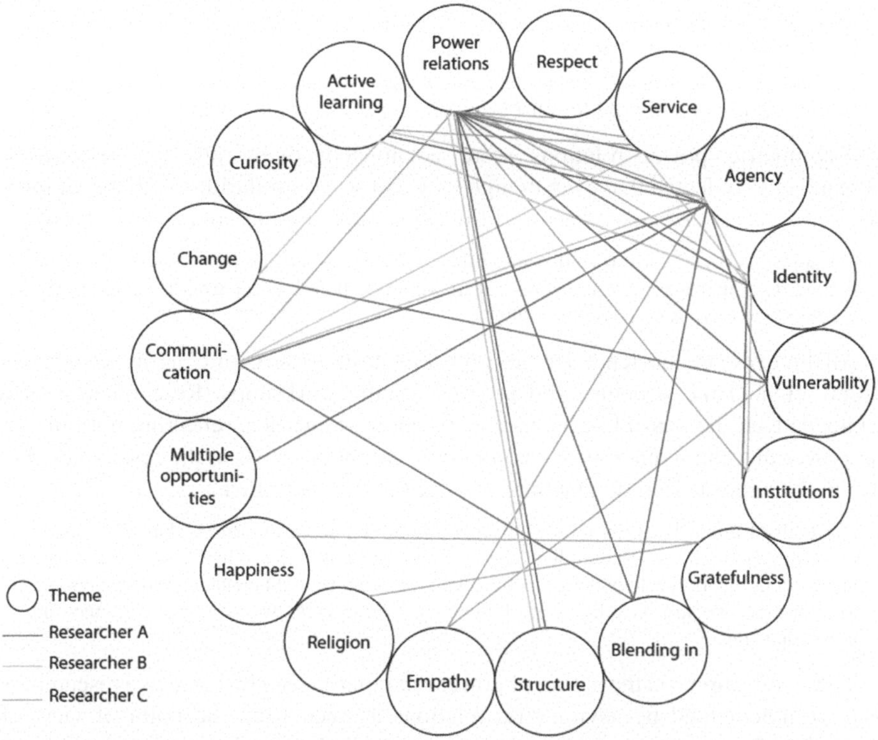

Fig. 1 Themes and connections between themes as mentioned by each researcher

and asymmetric power relations as being present (Fig. 1). Interaction with other themes allowed for tentative connections to be drawn where themes appeared to link to each other.

Symmetrical power relations were perceived to exist in relation to the experience the researchers were having of both challenges and togetherness:

> …we can talk at the same level, and at your level. I am at your level because we are in these conditions together. There are things I can give you from my experience, not my knowledge but my experience. There are things that I am willing to learn from you as well.

There was also the idea that some effort to equalise power relations needed to be made by all and that a period of time might be needed to begin to achieve such a goal:

> …I think that yes, they would expect the power relations, because they were saying Dr (researcher C), Dr (researcher C), and then with the time it was (researcher C), right? So it is up to you… mmmh… It's up to both sides, to be open for that communication.

There was a strong belief from the researchers that they had learned things that could be brought back and enacted in their own practice - another example of the way in which power relations at times, felt more equalised:

> one of the things that really struck me was how alive they were… they were so…passionate… and there was this kind of vibrancy in their being…
>
> …I feel I learned from that a lot from them about being really active students and self-driven… managed to absorb that and I feel that I can bring it to the classroom here..

Asymmetric power relations were acknowledged by all the researchers. Researcher C reflected that even though they had some similarities with the women, they felt a power distance in terms of life experiences that might present a barrier:

> …I am aware that I have travelled more than… That was not the point, I didn't want to remind them that I was more this or that. It would be pointless for what I wanted to do, because then there is no communication…

All three researchers tended to link power relations to 'structure' as the researchers largely created the structure and progress of the workshops. Researcher A also acknowledged the possible existence of power asymmetries, reflecting not only on the power that the women in the community might perceive in the researchers but also in the wider notion of an admiration for the 'foreign other':

> …actually I think that even with [colleague] there is a sense of [Researcher C] is really powerful. They all see [Researcher C] as a powerful thing…she brings money she brings professors…she brings knowledge, people who speak a different language… she brings and to be honest, we still have this in Colombia 'ooh foreigners, they are amazing…they do things amazing.

These reflections on the nature of power relations as seen by these two researchers were complemented by asymmetric relations as seen from the point of view of Researcher B who experienced an unexpected difficulty in a dramatic shift of power relations as they interacted with the women:

> [I thought] what am I doing here, why should they listen to me, what do I have to offer, really? And it felt like I had nothing to offer. At that moment I had nothing to offer and…I thought they were looking at me and they were going, you don't have anything to offer us…Because their lives are so different… there was a gulf, there was a big gulf between us and I, I couldn't cross it really.

For researcher A, power relations were linked to agency, identity, service, vulnerability and blending in. For researcher B, power relations were not linked to any other theme apart from structure. For researcher C, power relations were linked to themes such as agency, identity, service, respect, institutions, structure and communication. These differences in the way in which power relations (the most important theme identified by all the researchers) linked to other themes, highlights the different perceptions of each researcher about the interaction with the community. Thus, in order to understand the links, it is important to describe the other main themes.

4.2 Agency

Agency was crucial for the researchers' learning in the Dique Channel project. Here we use a definition of agency as "the socioculturally mediated capacity to act" (Ahearn 2001. p. 112). This suggests that the capacity to act varies in relation to the social and cultural environment in which the action occurs. The researchers emphasised agency as a key factor in the development and implementation of sustainable development projects (Fig. 1). During the reflective dialogue, the researchers thought that their common agency was linked to power relations, active learning, communication, multiple opportunities, empathy, blending in, vulnerability and identity.

However, each researcher linked agency to different themes (Fig. 1). Indeed, agency had the same or fewer interconnections to other themes than active learning and communication for researcher B, and the same or fewer interconnections than service, gratefulness or empathy for researcher C. The researchers discussed that active learning requires agency of the learners to actively engage in learning (see Fig. 1). In addition, in participatory research all participants involved (including researchers) could be in a learning mindset.

Sustainable development projects involving universities based in the Global North and local communities based in the Global South are likely to be subject to historical power relations potentially mirroring neocolonialism (Andreotti 2016). Although it is not always possible to avoid these power relations, the three researchers tried to address this issue by using reflection throughout the project. The three researchers believe that acknowledging power relations and finding ways to balance them is a crucial element for the successful implementation of sustainable development projects involving academia and communities. Agency therefore plays an important role in addressing power relations and throughout the project the researchers reflected on the power relations and discussed ways to balance them. One approach includes a focus on learning rather than teaching and combined with participatory research this can be a tool to help address power relations. However, power relations may often be present in research deemed 'participatory'.

In this particular case, shifting the focus to learning rather than teaching could be done by setting out to learn from and with the Dique Channel Leaders. This might feel more natural for some researchers than for others.

> Researcher A: hopefully we have managed to create a balance. Because I think I am the opposite of paternalistic, just throw yourself to it and then you can do it…I am a bit too much like that: I throw you there, you deal with it. Whereas researcher C is a lot more structured and you accompany people in the process and maybe I am too much of throwing people and maybe you are too much of …
>
> Researcher B:… leading
>
> Researcher A: or showing the way..
>
> Researcher C: Or I am deductive and you are inductive…?
>
> Researcher A: oh no…
>
> Researcher B: oh Maybe…yeah.. that's so interesting
>
> Researcher A: I don't get it.. tell me about it
>
> Researcher C: I was thinking about my lectures and passions and things.. why it works here but does not work there…but and I was thinking I give examples and examples and examples and I am telling them so they can do it in different ways…here we were trying different methods we were saying you need to find your journey…and how you can do it

from here and here.... And now with structure and I remember from my PGCE and they were pushing me pushing it...and thinking how to be more inductive and they were telling me 'in an activity...talk less and do less and let them find out more and more'... and eventually I was feeling at the end of the lecture 'oh this is rubbish' and the feedback from the people was 'Wow that was so awesome' and What? really? honestly??!'

Researcher b: That's amazing... because you were letting go...

Researchers can also use their agency to take action to balance power relations. For instance, the researchers mentioned approaches they used to try to blend in within a group for example, through showing their own vulnerabilities and communicating empathy so that unequal power relations could be mitigated. It was challenging for one of the researchers not being able to fluently engage in Spanish with the rest of the participants which affected their own sense of agency, whilst also, potentially, equalising power relations.

4.3 Identity

The researchers used a poststructuralist approach to consider their identities as socio-cultural phenomena where "social actors claim, contest, and negotiate power and authority" (Bucholtz and Hall 2008; p. 154). In this way, identity is seen as a social action that agency can achieve, with interaction being an important component. The researchers' identities were thus engaged with in a multidimensional manner. Identity can be considered as rooted in interaction. Agency, identity and power relations are profoundly interconnected but are reflected on here in the context of each researcher and their professional experience.

In the case of researcher B, identity was linked to institutions, agency, service and active learning. For researcher A, identity was linked to power relations and institutions. Researcher C did not link her own Identity to other themes (see Fig. 1). The following extract indicates how the researchers linked their identities with contexts in the global south and global north:

Researcher A: Maybe academia (in the developed world) does something to the beings ... which is about separating those things.

Researcher B: It does exactly that, it puts it in a box doesn't it? That's why those women [the colombian communit] weren't in the box...that's why you got that kind of passion in the classroom because they weren't separating out the academic self and their real human self. They were together.

Researcher A: Because separating [identities] helps you create more power relations maybe...

Researcher B: Or it helps people recognise which bit of you they can gain power over... so they can gain power over the academic because they've got the hierarchy, they've got the institutional hierarchy, they've got the teaching hierarchy ... I am the teacher you're the student, so those hierarchies reveal if you've managed to identify if you've got the academic side or if you've got the whole person...or you don't know how powerful they might be, or they might have all sorts of powers that you know nothing about.

Researcher A: But it's very scary to look at them because suddenly you've got to...

> Researcher A: You do whatever… I don't know if it's tough love… I feel that I really left the group because I was ill [This refers to a WhatsApp group where the researchers and the women leaders had discussions and shared information. The group was created after the residency. Researcher A left the group due to a period of illness] I didn't come back to the women's group because I feel I need to let them go and I feel they need to find their own thing and maybe at some point we can come back but it is time for them to go and do their thing, and I feel that I don't want to intervene in that process…they are powerful enough to go and do their own thing …I feel hopefully they are at that stage… and if they need something they can come to us rather than us asking them if they need something.
>
> Researcher B: Hmm definitely.

Thus, whilst agency was discussed extensively in the reflection, identity seemed to be a theme that was not as developed for the three researchers. Due to the funders' requirements the project had tight time constraints. Therefore, it is possible that the researchers were focusing more directly on agency because they had to make decisions and take action in short periods of time. This focus on practicalities and logistics may not have left enough space for reflection in other areas requiring deeper thinking, such as identity. As outputs are measured, at least in the UK, not only for their quality but also for their quantity, developing reflection on researchers' identities and their links to agency and power relations, might be difficult.

5 Discussion and Conclusions

The results showed that the researchers were able to discern specific learning about their own motivations towards designing and implementing international sustainable development projects in the context of the public good/global citizenship mission of higher education.

5.1 Agency and Identity and Their Links to Asymmetrical and Symmetrical Power Relations

The researchers agreed that their agency was a fundamental concept linked to power relations and through agency, some neo-colonial approaches could be avoided. However, trying to avoid neo-colonial approaches is not enough. Asymmetrical relationships were still present and historically, paternalistic characteristics of human interactions are difficult to break (Stein et al. 2019). All three researchers were able to recognise at least some of their own conscious power differentials manifesting as asymmetric power relations, as seen in Sect. 3. Awareness of this is however, only the first step and efforts to mitigate possible asymmetries in power relations should be considered. Stein et al. (2019) highlight the reproduction and retrenchment of economic and epistemic power relations that clearly exist between the global north and the global south, with epistemic dominance of a eurocentric curriculum, and the still remote notion that the global north might actually have things to learn from the global south.

The researchers agreed that the awareness of one's identity played a crucial role in the design and implementation of the sustainable development project in the Dique Channel. The researchers also felt that this could potentially be the case in other projects that try not to engage with historically ingrained power relations between the Global South/Global North power but to create a more balanced dialogue. The discomfort felt, for example, particularly by Researcher B was, as Andreotti (2016) discusses, a discomfort and challenge to a frame of reference and worldview that is inevitable if we continue to ask searching questions regarding our own motivations—in fact, if we do not experience some discomfort, one could argue, we are not challenging ourselves sufficiently.

5.2 *The Role of HE as Global Citizen, Considerations and Potential Impact*

Accountability in higher education institutions in the UK is increasingly linked to measurable results (Schmidt and Günther 2016). For colleagues in research active positions, this is linked to the quantity and quality of research, with a tendency towards focusing on the quantity rather than the quality (Schmidt and Günther 2016). This means that research projects are ruled by strict time constraints. Therefore, the time to reflect and build awareness around researchers' identities is limited. Also, reflection around researchers' identity is not common practice in all research approaches, or in all disciplines. With sustainable development requiring multidisciplinary approaches, reflection on researchers' identities and motivations may not always be necessarily part of these projects.

One possible way to ensure that this type of reflection is embedded throughout a project, is to include this in the design of the research. Andreotti (2016) devised a pedagogical tool for critiquing paternalistic and ethnocentric soft global citizenship approaches to education for international development. The tool was designed for education for a critical global citizenship, and could be used to interrogate the motivations and reflections of the researchers involved in a 'service learning' project. The tool asks the user to consider seven patterns of engagement and representation: hegemony, ethnocentrism, ahistoricism, depoliticisation, self-serving attitudes, simplistic solutions, and paternalism. It provides questions to interrogate the attitudes of the user to each of these and asks whose knowledge is being privileged.

One of the main problems for universities in developed countries (representing Higher Education and Research) when working with communities in developing countries (Global South) is how to engage with the communities in International Experiential Service Learning -IESL and Global Citizenship Education-GCE without becoming one more example of the typical neo-colonialist (we, the powerful), hegemonic (we, the universal knowledgeable people), ethnocentric (we, the good, moral and desirable whites from the north), a-historicist (we, forgetting local history when addressing problems), de-politicist (we, forgetting the imbalanced power relations), self-focus (our self-affirmation and CV building), and paternalistic (we the helpers for these poor souls) organisations that have passed by the area in the past, getting plenty of information and leaving nothing meaningful for the community (Andreotti 2014, 2016).

This kind of structured global citizenship assessment could be part of the formative assessment of sustainable development projects from the outset. Through this assessment, researchers could reflect and aim at working towards global citizenship 'otherwise' (Andreotti 2015) and critical global citizenship by recognizing and surfacing their neo-colonial and soft global citizenship. This could provide awareness of our own neocolonial and paternalistic assumptions and behaviours and would not only improve the outcomes of the projects in terms of equality, but also may encourage change in and around us.

Researchers in international projects could thus pursue a position of 'constant vigilance and compassion' in order to recognise patterns of complicity in perpetuating past colonialisms and oppressions. Researchers' attention to their agency in the project would be regularly and critically interrogated, using this kind of framework. To attempt to equalise power relations is a complex and difficult path to tread and risks opening up new inequalities. Treading carefully is perhaps the best we can do.

> As we realize our wider complicity and vested interests in social hierarchies and principles of separability the auto-pilot position is to reproduce these same patterns precisely while declaring our innocence or transformation (Andreotti 2015: 224)

Another approach could come from the reflection of researchers who, like some of the authors of this paper, grew up in developing countries, but have lived and worked in a developed country for decades. The process of working with marginalised communities in Colombia has allowed them to understand that in order to engage with the local communities they need to go and live (at least temporarily) with their everyday needs (how to access the area, how to adapt to normal working days under local weather conditions, local food, culture, wages etc.), as well as the benefits of living in the area (access to more ecosystem services and quality of life) to identify opportunities for local community development. More than "feeling responsible for changing or saving the world there" (Andreotti 2014: 22), or assuming that the developed world is "better", the question is how to improve the local conditions and make the most of the benefits that those conditions offer to local communities, from their own perspective, without promoting our "developed country" beliefs as universal. The collaboration then works when both sides (universities in developed countries and communities in developing countries) understand that working together allows them to identify more opportunities and more solutions to problems.

Rather than sympathy or pity from the "developed" world (Andreotti 2014), developing countries could be acknowledged as offering a service urgently needed: more natural resources (forests, rivers, mountains, seas) with greater and richer services (agriculture, recreation, tourism, culture, fish stocks, minerals, etc.). These ecosystem services are degrading due to the mistakes already made. Communities in developing countries are inheriting these treasures, and it is everyone's responsibility to respect them. Under this principle, everybody may benefit in the long term and work as equals. Or as the Kogis in Sierra Nevada-Colombia suggest (Edwards 2010: 18–19): the native communities in the south are the Big Brothers who are waking up to support the Younger Brothers (western societies and developed countries) to stop destroying the natural world. This type of approach could promote a more realistic focus and promote a more democratic global governance and sustainability (Huckle and Wals 2015).

References

Andreotti VDO (2014). Chapter 2. Soft versus Critical Global Citizenship Education. In: McCloskey S (ed) Development education in policy and practice. Palgrave Macmillan, London, UK. pp 21–31. https://doi.org/10.1057/9781137324665_2

Andreotti VDO (2015) Chapter 18. Global citizenship education otherwise: pedagogical and theoretical insights. In: Abdi AA, Shultz L, Pillay T (eds). Decolonizing global citizenship education. Brill Sense, pp 221–229

Andreotti VDO (2016) The educational challenges of imagining the world differently. Can J Dev Stud Rev Canadienne d'études du développement 37(1):101–112. https://doi.org/10.1080/02255189.2016.1134456

Ahearn LM (2001) Language and agency. Annu Rev Anthropol 30(1):109–137. https://doi.org/10.1146/annurev.anthro.30.1.109

Baker S, Kousis M, Young S, Richardson D (eds) (2005) The politics of sustainable development: theory, policy and practice within the European Union. Routledge, London, UK, p 219p

Balasz CL, Morello-Frosch R (2013) The Three Rs: how community-based participatory research strengthens the rigor, relevance, and reach of science. Environ Justice 6(1):9–21. https://doi.org/10.1089/env.2012.0017

Barrios LM, Prowse A, Vargas RV (2020) Sustainable development and women's leadership: a participatory exploration of women's capabilities in Colombian Caribbean fisher communities. J Clean Prod 264:121277. https://doi.org/10.1016/j.jclepro.2020.121277

Brookfield SD (2002) Using the lenses of critically reflective teaching in the community college classroom. New Dir Commun Coll 118:31–38

Bucholtz M, Hall K (2008) Finding identity: theory and data. Multilingua 27:151–163

Chiu LF (2006) Critical reflection: more than nuts and bolts. Action Res 4(2):183–203

Centre for Global Development (2019) The UK needs a new formula for ODA-funded research, 9 Jul 2019. Accessed on: 10-01-2020. Available at: www.cgdev.org/blog/uk-needs-new-formula-oda-funded-research

Dawson MC (2020) Rehumanising the university for an alternative future: decolonisation, alternative epistemologies and cognitive justice. Identities 27(1):71–90

Edwards AR (2010) Chapter 1: Lesson from our ancestors. In: Edwards A (ed) Thriving beyond sustainability—pathways to a resilient society. The Kogi: Guardians of the planet, pp 7–25. New Society Publishers, Gabriola Island, Canada, 227p. ISBN: 978-0-86571-641-4

Engel LC, Siczek MM (2018) A cross-national comparison of international strategies: global citizenship and the advancement of national competitiveness. Comp A J Comp Int Educ 48(5):749–767. https://doi.org/10.1080/03057925.2017.1353408

Erickson M, Hanna P, Walker C (2020) The UK higher education senior management survey: a statactivist response to managerialist governance. Stud High Educ 15:1–8

Estes RJ (2010) Part one: chapter 5. Toward sustainable development: From theory to praxis. In: Negi NJ, Furman R (eds) Transnational social work practice. Columbia University Press, NewYork, USA, p 76

Friedman JZ (2018) The global citizenship agenda and the generation of cosmopolitan capital in British higher education. Br J Sociol Educ 39(4):436–450

Huckle J, Wals AEJ (2015) The UN decade of education for sustainable development: business as usual in the end. Environ Educ Res 21(3):491–505. https://doi.org/10.1080/13504622.2015.1011084

Horowitz CR, Robinson M, Seifer S (2009) Community-based participatory research from the margin to the mainstream. Are researchers prepared? Circulation 119(19):2633–2642. https://doi.org/10.1161/CIRCULATIONAHA.107.729863

Hudzik JK (2014) Comprehensive internationalization: Institutional pathways to success. Routledge, New York, pp 1–279

Kesby M (2005) Participatory diagramming: deploying qualitative methods through an action research epistemology. Area 32(4):423–435. https://doi.org/10.1111/j.1475-4762.2000.tb00158.x

Mansouri F, Johns A, Marotta V (2017) Critical global citizenship: contextualising citizenship and globalisation. J Citizsh Globalisation 1:1–9. https://doi.org/10.1515/jcgs-2017-0001

Nicholls R (2009) Research and Indigenous participation: critical reflexive methods. Int J Soc Res Methodol 12(2):117–126

Oldekop JA, Truelove NK, Villamarın S, Preziosi RF (2012) Information flows in community-based monitoring exercises in the ecuadorian Amazon. Int J Zool 3:1–4. https://doi.org/10.1155/2012/980520

Pashby K, da Costa M, Stein S, Andreotti V (2020) A meta-review of typologies of global citizenship education. Comp Educ 1–21. https://doi.org/10.1080/03050068.2020.1723352

Rajan H, Thornhill K (2019) Dilemmas of feminist practice in transnational spaces: solidarity, personal growth, and potential solutions. Gender Place Cult 26(10):1345–1352

Roepstorff K (2019) A call for critical reflection on the localisation agenda in humanitarian action. Third World Q 41(2):284–301. https://doi.org/10.1080/01436597.2019.1644160

Saavedra-Díaz LM, Pomeroy R, Rosenberg AA (2016) Managing small-scale fisheries in Colombia. Marit Stud 15(6):1–21. https://doi.org/10.1186/s40152-016-0047-z

Spooner M (2017) Qualitative research and global audit culture. In: Denzin NK, Lincoln YS (eds) The SAGE handbook of qualitative research, 5th edn. Thousand Oaks: Sage Publications, pp 894–914

Schmidt U, Günther T (2016) Public sector accounting research in the higher education sector: a systematic literature review. Manag Rev Q 66(4):235–265

Stein S, Andreotti VDO, Suša R (2019) 'Beyond 2015', within the modern/colonial global imaginary? Global development and higher education. Crit Stud Educ 60(3):281–301

United Kingdom Research and Innovation—UKRI (2020) Global challenges research fund (GCRF). Accessed on: 23-03-2020. Available at: https://www.ukri.org/research/global-challenges-research-fund/

United Nations—UK (2015) Transforming our world: the 2030 Agenda for sustainable development. Accessed on 23-02-2020. Available at: https://sustainabledevelopment.un.org/content/documents/21252030%20Agenda%20for%20Sustainable%20Development%20web.pdf

Walker M, McLean M (2013) Professional education, capabilities and the public good: the role of universities in promoting human development. Routledge, New York, USA, p 208p

Wilson S (2007) Guest editorial: what is an Indigenist research paradigm? Can J Native Educ 30(2):193

Geography of Food Summer School: International Exchange for Sustainable Food Value Chains

Isabel Jaisli, Roman Grüter, Bernadette Oehen, Marina Pintar, and Umesh Kontrakere Basegowda

1 Introduction

Providing regular access to sufficient, safe and nutritious food for a growing population without depleting natural resources is one of the key challenges of the twenty-first century (Folke et al. 2016; Steffen et al. 2015). The current global food system has caused serious negative environmental impacts through changes in land use and intensive production of food and feed. It has contributed to the crossing of several planetary boundaries that attempt to define a safe operating space for humanity, including climate change, soil degradation and contamination, the destruction of natural ecosystems, biodiversity loss, depletion of freshwater resources and water pollution through excessive fertilizer and pesticide inputs (Campbell et al. 2017; Springmann et al. 2018; Steffen et al. 2015). At the same time, the different forms of malnutrition still either persist or are increasing. More than 820 million people in the world are undernourished and over 2 billion people suffer from micronutrient deficiencies, while about 2 billion adults are overweight (FAO 2019a).

The strong interlinkages and interdependencies with society, the environment and the economy (Ericksen 2008) make the food system a key driver of sustainability. The United Nations' Sustainable Development Goals (SDGs; Rosa 2017) acknowledge the major importance of our global food system for a sustainable society. Explicitly

I. Jaisli (✉) · R. Grüter
Zurich University of Applied Sciences ZHAW, Campus Gruental, 8820 Waedenswil, Switzerland
e-mail: Isabel.jaisli@zhaw.ch

B. Oehen
Research Institute of Organic Agriculture FiBL, Frick, Switzerland

M. Pintar
Biotechnical Faculty, University of Ljubljana, Ljubljana, Slovenia

U. Kontrakere Basegowda
University of Agricultural Sciences Bangalore, Bangalore, India

W. Leal Filho et al. (eds.), *Universities, Sustainability and Society: Supporting the Implementation of the Sustainable Development Goals*, World Sustainability Series,
https://doi.org/10.1007/978-3-030-63399-8_8

highlighted in SDG 2 ‘Zero Hunger’ but also implicit in all other goals, is the fact that the food system is key to a sustainable future of human kind (FAO 2019b).

The current situation calls for urgent changes in the way food is produced, processed, traded and consumed. The design of sustainable food value chains (FVCs) can offer pathways to fight poverty and malnutrition and to reduce environmental degradation (Neven 2014). Citizens and experts in public, private and non-governmental organizations are called upon to develop innovative solutions to meet these challenges and thus to contribute to the implementation of the SDGs.

Universities play an important role in educating future decisions-makers to design sustainable FVCs and take part in shaping a sustainable global food system. They are therefore challenged to adapt their existing educational formats in order to qualify students for sustainability transformation. For transforming food systems, an understanding of the full FVC and its complexity is needed, as well as the ability to collaborate across disciplinary, national and cultural boundaries. Moreover, as pointed out by Caniglia et al. (2017, 2018), connecting higher education institutions from different parts of the world is key in order to address global sustainability problems.

The question arises as to how educational programmes must be designed to effectively contribute towards intercultural education for sustainability. A list of quality criteria for international sustainability education is provided by Wiek et al. (2013), which includes the following five aspects:

- cross-cultural education
- context-specificity of sustainability problems and potential solutions
- capacity-building in complex problem-solving
- interdisciplinary approach to sustainability problems and potential solutions
- teamwork in cross-cultural settings

This paper presents the ‘Geography of Food’ educational programme as an example for sustainability education through intercultural exchange. It was initiated in order to educate young researchers and future decision-makers from around the world in designing sustainable FVCs. The programme not only focusses on knowledge transfer and participatory problem-solving approaches, but also acknowledges the importance of international exchange to facilitate joint learning and experience capitalization. For this purpose, it has been realized as a blended learning course including e-learning, individual case studies and an international summer school.

After presenting our experiences and learnings from the three-year programme, we will conclude by discussing its quality based on the criteria provided by Wiek et al. (2013). Building upon experiences and theories concerning sustainability education, the results provide a concrete example of how universities can educate those who will ultimately instigate change and are able to contribute towards the transformation of the global food system by designing sustainable food value chains.

2 The Geography of Food Education Programme

2.1 Background and Goals

The programme was initiated by researchers and lecturers of the Zurich University of Applied Sciences (ZHAW) and the Research Institute of Organic Agriculture (FiBL) with the idea of developing a new international summer school in the area of sustainable food systems for students of the ZHAW's Environmental Engineering Bachelor degree. With the goal for international exchange during the summer school, further partners were sought and found with the University of Ljubljana in Slovenia and the University of Agricultural Sciences Bangalore (UASB) in India. Both universities were interested in participating in the programme and offering participation in the summer school to their students.

The programme was designed with the fundamental objective to provide young researchers and future decision-makers in politics, business and society with specialist skills in 'designing food value chains to foster the Agenda 2030', but also to enable them to change perspectives and to work together interculturally. To achieve these objectives, the following key elements of the new programme were identified:

1. Provide students with fundamental knowledge about sustainability in the food system and the relevance of FVCs in the context of the SDGs.
2. Provide students with international exchanges that enhance their ability to understand global challenges in different local contexts.
3. Offer hands-on possibilities to both learn about and solve sustainability issues in real FVCs.
4. Initiate and reflect on intercultural and interdisciplinary teamwork.
5. Involve practitioners and other stakeholders from the food system.
6. Offer free educational resources that can reach and educate a wider public.

With the jointly developed concept, the Mercator Foundation Switzerland was successfully approached for financial support. This did not only allow for the new course to be developed and offered, but also for the provision of full scholarships for students to participate in the programme. The concept was designed and funded for a three-year period from 2017 to 2019.

2.2 Educational Concept

The thematic focus and title of the Geography of Food programme from 2017 to 2019 was 'Designing Sustainable Food Value Chains to foster the Agenda 2030'. This focus was approached from different perspectives, including economic, ecological and social aspects, using the SDGs as guiding principles.

The course was designed with three closely interlinked and mutually reinforcing phases:

Phase 1: E-learning course, which is freely accessible online to impart expert knowledge in the context of the SDGs and designing sustainable FVCs. The e-learning can be used independently of the summer school.

Phase 2: Independent development of a **mini case study** from the country of origin to share challenges and solutions from different parts of the world.

Phase 3: 10-day **study period** at one of the three partner universities for 30 selected students at bachelor and master level, including

a. excursions and inputs to illustrate local aspects in the field;
b. interdisciplinary and intercultural group work to develop new solutions for selected FVCs;
c. presentation and discussion of the mini case study; and
d. testing and reflecting on intercultural cooperation.

Each year, the summer school took place at one of the three participating universities and was dedicated to one specific SDG including SDG 2 (Zero Hunger), 12 (Responsible Consumption and Production) and 15 (Life on Land) with special relevance to the food system (Fig. 1).

For the overall programme, the following learning outcomes were defined. Students were to acquire

- a deeper understanding of sustainability in the global food system and the challenges it faces;
- applicable knowledge for the design of sustainable FVCs;
- a deeper understanding of the SDGs in the context of sustainable FVCs;
- the basics to develop future solutions for sustainable FVCs and to put them into local contexts;
- differentiated knowledge of the respective focus topic (SDG 2, 12 or 15);
- the ability to understand global challenges in a local context (environment, politics, society, economy) and to identify possible solutions; and
- intercultural competence for working in an international context.

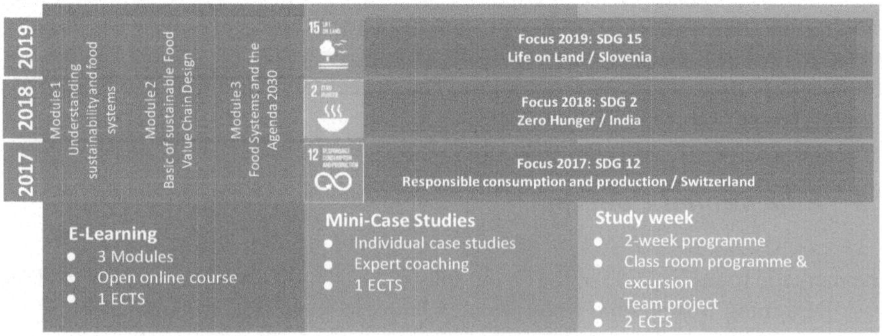

Fig. 1 Overall structure of the Geography of Food education programme

2.3 *Course Phases and Content*

Phase 1: e-learning

The first phase was designed so that students could acquire the basics for designing sustainable FVCs with regard to the UN SDGs through self-study with e-learning modules. The aim here was to level out the previous knowledge of students with different backgrounds and to convey the important basics before students met and worked together during the study period. Three e-learning modules were created for the project and published online (https://sustainable-food.lsfm.zhaw.ch/). The course included introductions to sustainability, FVC design and the Agenda 2030, and was divided into three modules with two to four lectures (Fig. 2). Each module concluded with a multiple-choice test to assess the successful completion of Phase 1. Details on the development of the e-learning course are given in Mariani and Jaisli (2018).

Phase 2: mini case studies

Also prior to the study period, students worked independently on individual mini case studies where they analysed a specific challenge with relevance to (a) sustainable food systems, (b) the specific SDG of the year, and (c) the local context of the students' background. The topic of the project was proposed by the students. The students conducted their projects over a period of four months, supported by virtual coaching by the project staff. Different methodologies were used, such as interviews, surveys or literature reviews.

The goal of this phase was to encourage students to engage with challenges that are specifically relevant in their own local environment. The results of the mini case studies were summarized by the students in a poster and presented during the study period. As a result, students were introduced to various challenges from different regions of the world. Similarities and disparities were discussed during

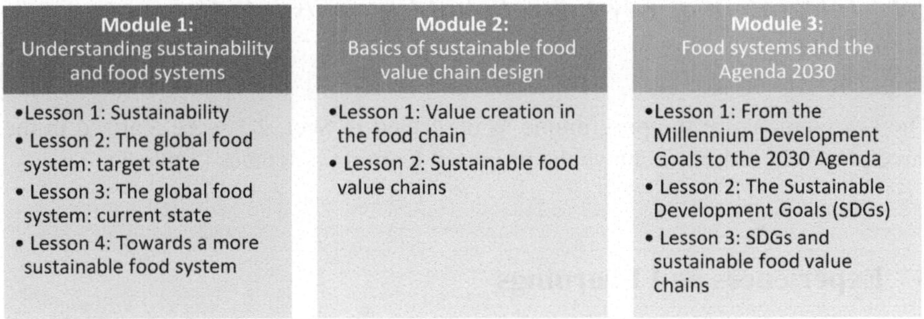

Fig. 2 Structure of the e-learning course Designing sustainable food systems

these sessions. After the individual presentation, students received structured peer and expert feedback.

Phase 3: study period
In the summer following Phase 1 and Phase 2, the ten-day study period took place at one of the three universities involved. During this time, students dealt with the following challenge: 'How should FVCs be designed to positively influence the achievement of the UN SDGs?' Each year was dedicated to a specific SDG with special relevance for the food system, namely SDG 2 (Zero Hunger), SDG 12 (Responsible Consumption & Production) and SDG 15 (Life on Land) (Fig. 1). The detailed content of each year's programme was developed in cooperation with the partner universities. This included different class activities, such as lectures, plenary discussions, workshops and excursions. In addition, students worked in small teams on concrete projects throughout the study period. Interdisciplinary and cultural diversity were taken into account in the team composition. Cooperation in intercultural teams was part of the learning objectives of the programme. In order to support the team process, special exercises and input lectures for intercultural communication were carried out.

The task of the team project was to analyse a specific FVC and to develop a concept or strategy for improving the value chain in terms of its sustainability impact. For this purpose, students collected and analysed data from literature, interviewed experts and consumers, and discussed their ideas with coaches. For each team, several excursions were organized to facilitate an exchange with value chain stakeholders. The result of the project was presented at the final event of the study period. Each year, different FVCs with special relevance to the location and the SDG were selected (Tables 2, 3 and 4).

2.4 *Integration of Key Educational Elements into Course Phases*

The key elements of the programme as described in Sect. 2.1 were realized in the three phases in order to achieve the proposed learning outcomes (Table 1).

3 Experiences and Learnings

3.1 *Implementation*

From 2017 to 2019, a total of 87 students from 20 different countries were able to take part in the exchange. While the overall structure (Sect. 2.3) was followed in all three years, a specific course programme was developed for each year, focussing on

Table 1 Linking key educational elements and implementation phases

Key elements	Phase 1 (e-learning)	Phase 2 (mini case study)	Phase 3 (study period)
Provide students with fundamental knowledge about sustainability in the food system and the relevance of food value chains in the context of the sustainable development goals	Providing basic knowledge on sustainability, SDGs and FVCs	Application and consolidation of knowledge in individual case study	Input sessions during study week
Provide students with international exchanges that enhance their ability to understand global challenges in different local contexts		Exchange of case studies from different locations and contexts (poster presentation and discussion of mini case studies)	Intensive intercultural collaboration on concrete value chain during team project
Offer hands-on possibilities to learn about and solve sustainability issues in real food value chains		Individual analysis of concrete mini case studies with local relevance	Analysis of specific value chain and design of possible improvements in team projects
Initiate and reflect on intercultural and interdisciplinary teamwork			Intercultural and interdisciplinary teamwork in projects
Involve practitioners and other stakeholders from the food system		Opportunity to connect with value chain stakeholders in mini case study	Connect with various stakeholders from food value chains during excursions

the chosen SDG and the local context of the executing university. In the following, selected experiences from the three years are described.

2017: Summer School in Switzerland, SDG 12 'Responsible Consumption and Production'

The first implementation of the course took place on the Gruental Campus of the ZHAW in Waedenswil, Switzerland. SDG 12 was defined as the focus topic. This was selected as especially wealthy countries, such as Switzerland, have the highest per capita emissions from food consumption and food waste. Switzerland also plays an important role as a trader and processor of special food products (e.g. coffee and chocolate) which are known for their high environmental and social impact.

29 students from 11 countries participated in the first year. In the mini case studies, students described different cases from around the world which demonstrated challenges to, but also solutions for, sustainable consumption and production. Three

Table 2 Group work topics and excursions during the 2017 summer school in Switzerland

Value chains	Excursions	Focus of project/proposal
Coffee	ZHAW coffee lab (coffee excellence centre), stoll coffee roasters (small-scale roaster), various coffee shops in Zurich	Increasing income of farmers through shares of roastery (farmer as shareholder)
Cocoa	Garcoa, aeschbach chocolatier, Laederach (B2C chocolatier), Max Felchlin (B2B chocolatier)	Label for chocolate made by cocoa from agroforestry systems
Beer	Waedibraeu (small-scale brewery), Hopfentropfen (hops producer) Feldschoesschen (large-scale brewery), beers of the world (specialty store)	Local versus global beer value chains
Vegetables	Gastrostar (processor of vegetables), Erlenhof (conventional vegetable farm), Bruederhof (organic farm with delivery service), Ortoloco (vegetable cooperative)	Short value chains to reduce food loss
Milk and Meat	A local dairy and an organic farm producing milk and meat	Meat tax
Soybean	Farmer producing Swiss organic soy for food use, a storehouse trading with local soya and a craft tofu producer	Locally produced soya as foodstuff instead of feed

examples show the broad range of topics. Firstly, the aspect of sustainable production was addressed in the mini case study 'Towards more sustainable pineapple production in Costa Rica: Ongoing initiatives and challenges'. The study focussed on the potentials of organic labelling for small farmers and demonstrated the challenges related to prices and access to markets. Secondly, in his mini case study 'Educating fish processors about the Modern Preservation Method to improve nutrition value and extend shelf-life to reduce food wastage', a participant from Nigeria discussed the need for education to reduce food loss in local fish value chains. Local, context-specific challenges such as regular power cuts, low education levels, and lack of access to technology were herby addressed. Thirdly, other mini case studies covered aspects from sustainable consumption such as the 'Sustainable value chain of vegetables in restaurant businesses' from the Czech Republic which provided ideas for making our home consumption more sustainable through, for example, higher vertical integration, local value chains and a root-to-leaf approach.

For the group work which took place during the study period, the following FVCs were selected: coffee, cocoa, beer, vegetables, soybean and milk and meat. The team project included desktop studies, value chain mapping, coaching and expert interviews. Additional excursions were individually organised for each team covering the whole value chain from agricultural production to retail. Based on these insights, the students developed ideas and strategies for more sustainable FVCs which were then presented on the final day of the study period. The experiences and results of the six teams are summarized in Table 2.

2018: Summer School in India, SDG 2 'Zero Hunger'

The second implementation of the course took place at the campus of the University of Agricultural Sciences in Bangalore, India, focussing on SDG 2. On the one hand, this SDG is especially relevant for India because it is among the countries with the highest number of undernourished people and has a high prevalence of child malnutrition. On the other hand, India has an agricultural sector of high importance for the country's economy and employment status.

In 2018, 31 students from 12 countries participated in the summer school programme. To become acquainted with the Indian culture, agriculture and demographic factors, a preparatory workshop was organized for the Swiss participants (the largest group of foreign students). This was conducted by an expert for intercultural communication and business between Switzerland and India. In Phase 2, all students carried out an individual mini case study related to food security, improved nutrition or sustainable agriculture in their home country. One example was concerned with the impact of microfinances through self-help groups on food security and rural empowerment in India. According to the study, especially participation of women in self-help groups led to improved family income and nutritional status of family members. A second mini case study focussed on the potential of pulses value chain integration to improve the nutrition of farmers and the rural population of India. It included socio-cultural, organizational as well as institutional elements necessary to improve the value chains of pulses. In a third study, the status of seed systems for specialty crops in the Central US was investigated. It was found that a greater commitment to a regional seed system would improve genetic and specialty crop diversity, reduce input needs and diversify farmers' income and seed access. These examples highlight the diversity of possible strategies in the agri-food sector necessary to achieve the SDG targets.

Table 3 Group work topics and excursions during the 2018 summer school in India

Value chain	Excursions	Focus of project/proposal
Coconut	Visit to coconut production market, several coconut and copra industries, including a coconut oil refinery, and a tender coconut packing unit	Market development for neera (coconut palm nectar) products
		Introduction of collaborative storage and drying units, copra processing factories
Millets	Visit to millet farms, millet farmer associations, primary processing units, retail outlets, restaurants and canteens	Development and implementation of a millet processing unit
		Incentivizing millet procurement, preparation and delivery of millet-based meals in canteens
Dairy	Visit to dairy farms, milk producing cooperative, and milk collection and cooling centres	Dairy product labelling ('Happy cow, happy farmer')
		Buffalo curd marketing ('Buffalicious')

During the study period, three agricultural commodities of high relevance in Southern India were investigated in the group work activity: coconut, millets and dairy. The respective value chains were introduced by local experts, mapped by the students and the relevant stakeholders visited, and interviewed during the excursions. To reduce organizational and logistic problems, the number of commodities and parallel excursions were reduced to three compared to 2017. Based on this intensive exchange, 6 groups of students developed strategies to improve the existing value chains to foster the Agenda 2030. The experiences and results of the team cases are shown in Table 3.

2019: Summer School in Slovenia, SDG 15 'Life on Land'

In the third year, the summer school took place at the University of Ljubljana in Slovenia with a focus on SDG 15. Soil degradation, biodiversity loss and conflicts between wildlife and agriculture were defined as the three central aspects of SDG 15 to be studied in the Slovenian context. Slovenian agriculture is mainly based on grasslands for milk and beef production. However, orchards, olive groves and vineyards that are located in the hilly areas and along the coast also represent an important Slovenian agricultural sector. With forests covering more than 60% of Slovenia's land surface, it is home to extensive wildlife (bears, wolves and others) and hunting is a well-accepted practice for the regulation of animal numbers.

In the third implementation of the course, 27 students from 9 countries took part in the programme. Topics related to the sustainable use of terrestrial ecosystems, the prevention of land degradation and biodiversity conservation were investigated in the preparatory mini case studies. For example, key factors affecting sustainability and food security of coffee growers in Colombia was the topic of one mini case study. It showed that the environmental costs of coffee production are currently covered by the coffee growers who are already struggling to be food secure and to have a decent livelihood. Another participant investigated how soil biodiversity is affected by management intensity and climate change. She found that the effects on soil biodiversity are very site-specific depending on the type of management and soil bio-chemical properties. Several students focussed on alternative agricultural systems such as agroforestry, community-supported agriculture, organic farming or permaculture and their impacts on environmental integrity, social well-being and economic resilience. They found that such systems generally have the potential for more sustainable agricultural practice, but that the question of scaling-up remains. The mini case studies triggered discussions about the many trade-offs between agricultural production and ecosystem and biodiversity integrity and soil health, all of which are important aspects of SDG 15.

In this year, the value chains of olives and olive oil, grapes and wine and meat were investigated during the group work task. For the meat production, both beef production and the respective conflicts with wildlife (bears) as well as local hunting was looked at. During the first week, the value chains and the SDG 15 subtopics of soil, biodiversity and wildlife were introduced for the Slovenian context by local experts at the University of Ljubljana. The second week was dedicated to field and

Table 4 Group work topics and excursions during the 2019 summer school in Slovenia

Value chain	Excursions	Focus of project/proposal
Olive	Olive and olive oil production and tasting (Istria)	Improving olive oil sales and encouraging ecological cultivation through collaboration among producers
		Professionalization of olive farming by establishing an olive farming school
		Proposal for enhancing sustainable management practices in olive orchards
Grape	Vineyards (Goriška Brda), enology and wine culture (Vinakoper) and tasting	Community-supported agriculture in vineyards
		Marketing campaign for biodiversity-friendly wine
Meat	Dairy farm, beef farm and hunting lodge (Kočevska region)	Strategies for increasing the local consumption of domestic beef

stakeholder visits and group work, and took place at a different location. This translocation to the Slovenian coast offered the opportunity to visit more diverse agricultural systems. In contrast to the first two years, joint excursions were organized for the entire group of participants to give them an insight into the diversity of land use systems. The results of the group case studies are summarized in Table 4. Apart from the excursions related to olives, grapes and meat, the Škocjanski Zatok Nature Reserve (Slovenia's largest brackish wetland) and shellfish production at the coast were visited and discussed within the context of SDG 15.

3.2 Evaluation

Every year, the study programme was evaluated via an online survey at the end of the study period. Students' participation in the evaluation was 66% (2017), 74% (2018) and 78% (2019).

Overall, the content of the e-learning course and the programme of the study period was highly rated. Additionally, the international and intercultural exchange during the study week period was most appreciated by the participants. This was not only related to the team projects where students elaborated on specific cases in international teams, but also to experiences outside the scheduled programme. Many participants highlighted the network of friends and colleagues they gained through the summer school. Secondly, the excursions and field trips related to the group work, including interactions with stakeholders from different FVCs, were very positively rated in all three years. Students also positively emphasized the coaching for their mini case studies and group case studies, and the feedback they received from the experts during the programme.

However, thematic and methodological inputs were generally perceived as being too long, too many, and not interactive or participatory enough. According to overall student opinions, more time could have been dedicated to discussions and group work, including more interactions with FVC stakeholders, in order to gain a deeper understanding of their key challenges and come up with better solutions. More guidance in the creative process to find solutions would also have been appreciated. Additionally, some participants suggested a better integration of students' backgrounds and experiences into the programme to improve joint learning. This was intended by the mini case studies but could have been more intense and integrative.

Apart from these overall evaluations, in 2017, students highlighted the inputs of more practical relevance and the introduction of practical tools. They also appreciated the flexibility that was given in the excursion programme. On the other hand, although they valued the mini case study poster presentations ("The presentations were really nice, the variety of cases from different countries and backgrounds pretty impressive. I think, we have all learnt a lot from them."), they criticised them as being at the end of the day, and also underrated the unsustainable food provided ("Why should we have meat every day? Why are rather unsustainable products like cheese, chocolate and coffee heavily promoted by the staff? Why are we not preparing food together?"). A student also commented on the different cultural contexts ("Please take into consideration that the context of India for both education and living is so different from Switzerland. The expectations of Indian students/lecturers might be very different from that of European and global students.").

In 2018, the programme was evaluated as being too densely packed, which prevented more time from being used for interaction and exchange among the participants. The input lectures were also assessed as being not interactive enough. Participants highlighted the group cases ("It was an amazing experience and my favourite part of the summer school. It was great to work in a multicultural team and we could really incorporate all the knowledge in our group"), the field excursions ("They were by far the most valuable activities of the summer school. Seeing things in the field enabled you to learn practical things that nobody ever teaches you in a lecture theatre") and the exchange among students ("The most memorable and helpful inputs for me actually came from the other participants").

In 2019, a few students raised their concerns about the internationality and sustainability of the programme due to flight emissions. This topic was discussed very controversially with other participants emphasizing that "this experience was one of the most important ones for me. Having people from all around the world was incredibly important and a most valuable experience.", or "for me, particularly the exchange with foreign students was very valuable."

3.3 Quality Assessment

In the following, we will discuss the success factors and challenges of the programme based on the criteria defined by Wiek et al. (2013):

Cross-cultural education
The summer school included lectures and role-plays on intercultural communication. However, a more detailed discussion or reflection on cultural differences or challenges was missing. While few such issues arose during lectures or group work, they became more obvious during social activities. For example, little attention was paid in the organization of the summer school to the cultural dietary habits (e.g. food preferences, abstinence from alcohol), which led to some dissatisfaction among certain students. Opportunities to involve the students from around the world to share and reflect on their cultural insights were not seized.

Context-specificity of sustainability problems and potential solutions
The programme was developed with a strong focus on proving insights and experiences on different sustainability problems and solutions in diverse local contexts. During their mini case studies, students were asked to analyse a specific problem or solution from their local context and share their results during the summer school. Due to the international composition of the group, students had the chance to learn about various cases studies from around the world. Experiencing differences but also similarities of challenges in the different environments was extremely valuable for all participants and was highly appreciated. In addition, the group work provided students with the opportunity to solve sustainability issues in real FVCs in a local context. As part of the excursions, the students themselves initiated contact with local value chain stakeholders. This exchange focussed mainly on knowledge transfer from the stakeholders to the students through explanations and answering questions. However, there was hardly any feedback of the outcomes of the student group work to the stakeholders that could enable joint learning.

Capacity-building in complex problem-solving
During the group work, students were working on specific value chains. Their task was to propose an upgrade of the value chain - a proposal to improve the sustainability outcome of the chain. While students were provided with practical skills to analyse the value chain (e.g. value chain mapping), few problem-solving skills were taught. Rather, students developed solutions based on a 'learning by doing' approach and on creativity.

Interdisciplinary approach to sustainability problems and potential solutions
Students from various disciplines participated in the programme. This included, for example, environmental sciences, agriculture, food sciences, economics, geography and others. Especially during the group work, students had the chance to profit from this diversity, which allowed for interdisciplinary experiences and exchange.

However, this was again more based on a 'learning by doing' approach, and no specific setup was given to seize the full opportunities provided by this diversity.

Teamwork in cross-cultural settings
A large portion of the summer school was dedicated to teamwork. During team composition, a special emphasis was placed on creating multinational teams. This allowed for hands-on experience when working in cross-cultural teams. This opportunity was highly appreciated by the students.

Overall, the evaluation and assessment of the Geography of Food education programme showed that all criteria for international sustainability education as defined by Wiek et al. (2013) were addressed. The set-up of the programme with a combination of e-learning, individual case studies and a joint study period therefore proved to be very suitable. However, room for improvement could still be identified. Opportunities may arise from (i) a better integration of students' backgrounds into the programme to benefit more from their interdisciplinarity, (ii) more directly addressing cultural differences to improve cross-cultural education and (iii) a more structured problem-solving approach to find more creative and elaborated solutions.

4 Conclusion and Outlook

The Geography of Food education programme was developed and realized to provide students with the skills and the ability to design sustainable FVCs as an approach to fostering sustainable development. Three years of experience have provided the participating institutions with deeper insights into the value and success of such programmes.

The programme was successful in providing students with learning objectives for heads (knowing), hands (acting) and hearts (being) (Caniglia et al. 2018; Sipos et al. 2008): knowledge was provided on sustainability, food systems and value chains focussing on the interdependencies of the topics. To provide the necessary skills, students were trained for working interculturally and in an interdisciplinary manner to develop new solutions. Last but not least, students were provided with new perspectives of, and experiences in, different realities. These experiences support students in developing a mind-set and the attitude needed to become 'changemakers' for sustainability and to face sustainability challenges (Svanström et al. 2008). In addition, the overall programme also successfully supported the internationalisation of the universities and established networks between the universities and participants, including both staff and students.

Based on the knowledge acquired, the participating institutions have decided to continue with the programme but in an adapted format. In the future, the summer school will take place at different locations in Europe. As most students are participating from European universities, this decision has been taken in order to reduce emissions from long-haul flights. Students from overseas, especially from developing countries, are nonetheless still invited to participate in the programme to ensure intercultural perspective and cooperation. The future programme will focus more closely on the exchange between participants and local stakeholders, including feedback methods such as peer coaching to improve capacity-building and context-specificity of the challenges addressed. The experience and understanding acquired by this 'knowledge building community' will be documented on an online mediawiki platform.

In addition, a purely 'virtual' summer school is planned between the ZHAW, the UASB and the Universidade Federal de Grande Dourados in Brazil. Following the motto 'Global problems require international cooperation at eye level', students from the three participating universities will work together to develop solutions to challenges related to climate change. This project will extend the learning outcomes with skills for working in virtual teams to overcome distances (John et al. 2017). Again, the context-specificity of sustainability problems will be emphasized. Increased attention will be paid to aspects of cross-cultural education through guided discussion and reflections.

International programmes, such as the programme presented in this paper, are but one way in which universities can play a role in the promotion of sustainability and can contribute towards the achievement of the SDGs (Zamora-Polo and Sánchez-Martín 2019). The increasing complexity and global interconnectedness of sustainability problems call for future leaders who are able to develop solutions in international and interdisciplinary teams. It is therefore crucial that universities equip students with the skills and the mindset needed to successfully achieve this, and with the opportunities to collaborate and learn across borders.

Acknowledgements This project was funded by Mercator Foundation Switzerland. The authors thank all people involved in the project. Special thanks go to Ennio Mariani, Tom Bischof and Lorenz Rieger from ZHAW, Matthias Stolze, Thomas Bernet, Christian Grovermann and Johan Blockeel from FiBL, Ravi SC and Udaykumar MS from UASB as well as Špela Železnikar from University of Ljubljana.

References

Campbell BM, Beare DJ, Bennett EM, Hall-Spencer JM, Ingram JSI, Jaramillo F, Ortiz R, Ramankutty N, Sayer JA, Shindell D (2017) Agriculture production as a major driver of the earth system exceeding planetary boundaries. Ecol Soc 22(4). https://doi.org/10.5751/ES-09595-220408

Caniglia G, John B, Bellina L, Lang DJ, Wiek A, Cohmer S, Laubichler MD (2018) The glocal curriculum: a model for transnational collaboration in higher education for sustainable development. J Clean Prod 171:368–376. https://doi.org/10.1016/j.jclepro.2017.09.207

Caniglia G, Luederitz C, Gross M, Muhr M, John B, Withycombe Keeler L, von Wehrden H, Laubichler M, Wiek A, Lang D (2017) Transnational collaboration for sustainability in higher education: lessons from a systematic review. J Cleaner Prod 168:764–779. https://doi.org/10.1016/j.jclepro.2017.07.256

Ericksen PJ (2008) Conceptualizing food systems for global environmental change research. Glob Environ Change 18(1):234–245. https://doi.org/10.1016/j.gloenvcha.2007.09.002

FAO (2019a) The state of food security and nutrition in the world: safeguarding against economic slowdowns and downturns. Food and agriculture organization of the United Nations
FAO (2019b) Transforming the world through food and agriculture. Food and agriculture organization of the United Nations
Folke C, Biggs R, Norström AV, Reyers B, Rockström J (2016) Social-ecological resilience and biosphere-based sustainability science. Ecol Soc 21(3). https://doi.org/10.5751/ES-08748-210341
John B, Caniglia G, Bellina L, Lang DJ, Laubichler M (2017) The glocal curriculum: a practical guide to teaching and learning in an interconnected world. [sic!] Critical Aesthetics Publishing
Mariani E, Jaisli I (2018) Didaktische Herausforderungen bei der Erstellung eines frei verfügbaren Onlinekurses. In: Lucke U, Strickroth S (eds) E-Learning symposium 2018: innovation und Nachhaltigkeit—(k)ein Gegensatz?! Universitätsverlag Potsdam
Neven D (2014) Developing sustainable food value chains: guiding principles. Food and Agriculture Organization of the United Nations
Rosa W (ed) (2017) Transforming our world: the 2030 agenda for sustainable development. In: A new era in global health. Springer Publishing Company. https://doi.org/10.1891/9780826190123.ap02
Sipos Y, Battisti B, Grimm K (2008) Achieving transformative sustainability learning: engaging head, hands and heart. Int J Sustain High Educ 9(1):68–86. https://doi.org/10.1108/14676370810842193
Springmann M, Clark M, Mason-D'Croz D, Wiebe K, Bodirsky BL, Lassaletta L, de Vries W, Vermeulen SJ, Herrero M, Carlson KM, Jonell M, Troell M, DeClerck F, Gordon LJ, Zurayk R, Scarborough P, Rayner M, Loken B, Fanzo J, Godfray HC, Tilman D, Rockström J, Willett W (2018) Options for keeping the food system within environmental limits. Nature 562(7728):519–525. https://doi.org/10.1038/s41586-018-0594-0
Steffen W, Richardson K, Rockstrom J, Cornell SE, Fetzer I, Bennett EM, Biggs R, Carpenter SR, de Vries W, de Wit CA, Folke C, Gerten D, Heinke J, Mace GM, Persson LM, Ramanathan V, Reyers B, Sorlin S (2015) Planetary boundaries: guiding human development on a changing planet. Science 347(6223):1259855. https://doi.org/10.1126/science.1259855
Svanström M, Lozano-García FJ, Rowe D (2008) Learning outcomes for sustainable development in higher education. Int J Sustain High Educ 9(3):339–351. https://doi.org/10.1108/14676370810885925
Wiek A, Bernstein MJ, Laubichler M, Caniglia G, Minteer B, Lang DJ (2013) A global classroom for international sustainability education. Creative Educ 04(04):19–28. https://doi.org/10.4236/ce.2013.44A004
Zamora-Polo F, Sánchez-Martín J (2019) Teaching for a better world. Sustainability and sustainable development goals in the construction of a change-maker university. Sustainability 11(15):4224. https://doi.org/10.3390/su11154224

The Role of Higher Education Stakeholder Networks for Sustainable Development: A Systems Perspective

Valeria Ruiz Vargas, Alberto Paucar-Caceres, and David Haley

1 Introduction

The United Nations supported the development of the Millenium Goals that were adopted in Millennium Declaration in 2000 (UN 2000). Later, these Goals were developed through a participatory process that became the Sustainable Development Goals (SDGs; UN 2015). Sustainable development has been defined as "development that meets the needs of the present without compromising the ability of future generations to meet their own needs" (WCED 1987: 41). Apart from the Earth Charter, (Earth Charter Commission 2000), sustainable development is the only framework to date that speaks to a wide range of people, including policy makers, internationally (Estes 2010; Baker et al. 2005). Furthermore, like the Earth Charter, the SDGs are considered by many to be a democratic framework to address such transformative global challenges. However, there are large portions of the population that are still not aware of or are confused by the SDGs and there are critics who consider them to be too orientated to the values of the North (Kahn 2010: 20).

The United Nations have suggested that the transition to sustainable development ought to be achieved by the engagement of society at all levels (UN 2015). The

V. R. Vargas (✉)
Department of Natural Sciences, Manchester Metropolitan University, Chester Street, Manchester M1 5GD, UK
e-mail: v.vargas@mmu.ac.uk

A. Paucar-Caceres
Manchester Metropolitan University Business School, All Saints Campus, Oxford Road, Manchester M15 6BH, UK
e-mail: a.paucar@mmu.ac.uk

D. Haley
Zhongyuan University of Technology, Zhengzhou, China
e-mail: davidhaley@yahoo.com

W. Leal Filho et al. (eds.), *Universities, Sustainability and Society: Supporting the Implementation of the Sustainable Development Goals*, World Sustainability Series,
https://doi.org/10.1007/978-3-030-63399-8_9

United Nations has signalled the importance of education, including higher education, in contributing to the achievement of sustainable development goals, as well as achieving the 2030 Development Agenda (UN 2015).

In recent years, higher education institutions have been pressured to consider their ethical investments, procurements and actions. In the United Kingdom, this trend has been largely driven by students holding their universities to account and the universities then being concerned about their marketing profile for student recruitment. One example in 2019 is the University of Manchester, where students staged a week-long occupation in protest against the university's £12 m investments in fossil fuels, forcing it to bring forward its financial review (Wootton-Cane 2019). Indeed, half of the UK public higher education institutions have divested from pension funds and investment portfolios that include identifiable unethical operations (Taylor 2020). Consequently, universities are having to think carefully about the stakeholders they choose and ethically concerned stakeholders are careful about which higher education institutions they wish to associate with.

Also, higher education institutions have developed and signed international commitments to integrate sustainable development into their activities (Lozano et al. 2013). However, there is not yet coherence between international, national and organisational policy frameworks (for instance for the UK where policy frameworks exist; Vargas et al. 2019b).

In addition to these changes, universities are currently seeking funding avenues different from governmental ones (Godemann et al. 2014). Some universities are organisations where the public and private sectors converge, and accountability is linked to result measures (Schmidt and Günther 2016). In addition, Australasian, European and US higher education is progressively seen as a significant sector in the economy (Godemann et al. 2014). Therefore, some universities have developed quasi corporate characteristics (Godemann et al. 2014).

These issues relate directly to the ways in which universities address the SDGs, affecting their policies and implementation of procurement, estate management, curricula, research, professional development training, community relations and stakeholder networks. In other words, the SDGs affect every aspect of university governance and operation, each representing a system of management.

Although leaders, academic and administrative staff, and students have started to work towards the implementation of sustainable development in higher education institutions, embedding it in the everyday practice of higher education is still facing difficulties, especially in terms of systemic change (Dlouha et al. 2018). There are different reasons for this that have been explored in the literature. One of the reasons suggests that higher education institutions could be described as complex systems and this makes driving change particularly difficult (Godemann et al. 2014). Studies have shown that, arguably 'developed', education systems such as the UK, include in their sustainable development policy frameworks a number of organisations with different functions and interactions between them (Vargas et al. 2019a).

According to the UN and echoed in SDG 17, partnership work is crucial for the implementation of sustainable development (UN 2015). There is evidence that suggests that organisational change can be influenced by the pressure of stakeholders

(Radinger-Peer and Pflitsch 2017). Moreover, the collaborative work between stakeholders can be defined through the development of networks (Dlouha et al. 2018; Vargas et al. 2019a). However, only a few papers have studied this phenomenon in the context of sustainable development in higher education, suggesting that in this context, such systems of higher education sector are complex (e.g. Godemann et al. 2014; Dlouha et al. 2018; Vargas et al. 2019a).

Systems thinking (ST) as a way of tackling complexity in management practice has been around for more than four decades but despite its obvious advantage, systems thinking has been overlooked as a means of identifying, analysing stakeholders' involvement in organisations and assessing their role in driving change. One of the key approaches for systems thinking is Soft Systems Methodology (SSM). SSM has widely been used in structural thinking and the intervention into complex organizational problems by addressing management systems that are complex in nature, and it seeks to assess as many diverse possibilities as possible. This approach has been used in many fields that include human resource management, planning of information systems, the planning of health and medical systems and the development of expert systems among many more.

Therefore, this paper aims to explore the potential for stakeholder networks to drive change towards (and beyond) sustainable development. For this the paper will first explore three research questions:

(a) What elements of the role of stakeholder networks support organisational change according to the current academic literature?
(b) How can Systemic Approaches, [i.e., Soft Systems Methodology (SSM)] assess (and measure) the efficacy, efficiency and effectiveness of stakeholders' networks in its role of implementing sustainable development Goals in HEIs?
In order to answer these two research questions, the paper reviews the extant literature on: (1) stakeholder networks as drivers of organisational change; and (2) systemic thinking concepts suitable to tackle complexity and assist organisational change. Based on this review the paper will then discuss (3) future scenarios for the development of higher education stakeholder networks for and beyond the implementation of sustainable development. For this, the following research question will be explored.
(c) How may higher education institutions and their stakeholders support sustainable living as a co-learning system?

2 Stakeholder Networks as Drivers of Sustainable Development Implementation in Higher Education Institutions

Empirical studies have shown that internal and external stakeholders have the potential to influence change in organisations. Internal networks could potentially drive bottom up activity. External networks to the university could potentially drive bottom up activity at national level (Vargas et al. 2019a) and possibly internationally. Although individual stakeholders can facilitate organisational change, this becomes more powerful when stakeholders organise themselves and work together. Collaboration between stakeholders can create formal or informal networks. Networks may be understood as fluid organisations and the interactions between their members tend to change over time. Even in the cases where there are formal structures, networks work differently in practice compared to what might be expected of them. However, there have been some advances in the academic literature which show how network characteristics can facilitate the effectiveness of the network and its aim.

Higher education institutions have been labelled in the literature as loosely coupled systems (Godemann et al. 2014). In simple terms, this refers to the dynamics within organisations that result in elements affecting each other suddenly, occasionally and indirectly (Godemann et al. 2014). Due to these characteristics, Godemann et al., conclude that stakeholder engagement in higher education institutions is particularly challenging due to its fluidity (2014). In addition, academic freedom allows academics to make decisions about their engagement with their preferred topics and specific stakeholders. This means that sustainable development is likely to be difficult to implement through top down approaches only (Vargas et al. 2019a). Top down approaches refer to practices or policies actively promoted by senior leadership who are at the top of the hierarchy in an institution and represent common forms of organisational structures. Therefore, it is crucial to look at how networks could be developed to help drive bottom up change.

Traditionally, networks were not thought of as entities requiring formal leadership and governance models to function (Provan and Kenis 2008). More recently, the literature that is focused on networks shows that networks do require a form of leadership and benefit from clear governance structures suitable for the characteristics of the network (Provan and Kenis 2008). The appropriate model may depend on network characteristics such as the number of members and density (Provan and Kenis 2008). Density refers to the number of actual connections or interactions between members of the network compared to the total number of connections between network members that are possible (Scott and Carrington 2014). The networks as presented in the sustainable development policy frameworks for higher education in the UK have low density (Vargas et al. 2019a). As a result, it is unlikely that this type of network could benefit from shared governance and leadership (Vargas et al. 2019a). Exploring network characteristics could, therefore help improve networks.

One of the few studies, so far, that has looked at this phenomenon in the context of sustainable development implementation in higher education is Vargas et al. (2019a). This paper focused on the stakeholder networks mentioned in the UK's national higher education policy frameworks. The study found that sustainable development networks in higher education as presented in the policy frameworks are vulnerable and fragile (Vargas et al. 2019a). The reasons for this are a lack of vocabulary pertaining to funding routes, low network density and high centrality of few members (Vargas et al. 2019a).

Often in the literature, network leadership models have been highlighted through the use of centrality measures of networks (Hristov et al. 2018). Centrality refers to the number of links between one stakeholder or node in the network and others. The higher number of links equals higher centrality of the stakeholder. The findings from Vargas et al. (2019a) about the vulnerability that a network might have due to the high centrality of one main organisation, may suggest that involving more members of the network in its leadership might be more appropriate for emerging agendas such as sustainable development in higher education.

Traditionally, theories of leadership have had a focus on individuals (Hristov et al. 2018). However, the focus on 'orthodox and heroic' leadership has changed into a focus on 'collective forms of leadership' (Hristov et al. 2018: 127). This change of paradigm has led to the recognition of leadership enacted by groups of individuals between and within institutions. This is sometimes referred to as distributed leadership and it highlights the presence of collaboration in processes requiring leadership (Hristov et al. 2018). There is an opportunity to pursue more studies focused on the practice of distributed models of leadership and the reasons for these models to be used (Hristov et al. 2018).

In addition, distributed leadership can be beneficial for the implementation of the Sustainable Development Goals (Wilson 2018; Kuenkel 2019). Therefore, distributed leadership is crucial to explore stakeholder networks for the implementation of sustainable development in higher education.

One potential avenue to develop the literature on sustainable development in higher education would be to identify and study practices of distributed leadership in inter and intra-organisational stakeholder networks. In order to develop this, perhaps studies could focus on finding out who are the stakeholders with high centrality in higher education networks and what their roles are in terms of supporting organisational change towards the implementation of sustainable development.

To conclude, although the literature on higher education stakeholder networks for sustainable development is underdeveloped (Godemann et al. 2014), the brief literature review presented above suggests that formal or informal, and internal or external stakeholder networks have the potential to play a crucial role in the implementation of sustainable development in higher education. Also, the characteristics of these networks including their leadership model can help estimate some of their potential impact. However, studying universities' internal and external stakeholder networks requires the understanding of different systems e.g. organisational, national and international. The next section, therefore, explores systems thinking and related methodologies for the analysis and assessment of higher education stakeholder networks for sustainable development.

3 Systems Thinking, Analysis and Assessment of Stakeholders' Networks in HEIs

When assessing the performance of a system, the relationships of the components within the system are of crucial importance. Through this premise, systems thinking gains a bigger picture. Systems thinking advocates the importance of making a conscious effort to appreciate other people's perspectives. C. West Churchman, described the systems approach both as a process in which one needs to 'sweep in' as many elements and actors affected as possible; and as a process of looking at the situations from different perspectives, indicating that—*"A systems approach begins when first you see the world through the eyes of another"* (Churchman 1968: 231).

In theory, these are the principles underpinning systemic thinking but when we try to apply them we need to use a methodology to guide the use of these principles. We outline soft systems methodology (SSM) a systemic methodology developed by Checkland (1981, 1999). SSM has been widely used in management to tackle complexity and to bring improvement in a particular complex, messy situation (Ackoff 1993, 1995) and possibly to introduce some changes.

3.1 Soft Systems Methodology (SSM) a Way of Enquiring About Complexity

Peter Checkland's Soft Systems Methodology (SSM) is one of the most developed Systems Methodologies in terms of its theoretical premises and philosophical underpinnings. It is also one of the most widely used in the UK and in other parts of the world (Mingers and Taylor 1992; Ledington and Donaldson 1997; Macadam and Packham 1989; Macadam et al. 1995; Paucar-Caceres et al. 2015). During the 1970s, Checkland and his colleagues at Lancaster University questioned the use of hard (reductionistic) systems thinking to real-world situations and started to test a new methodology that shifted the *systemicity* from the real world to the process of enquiry itself. In other words, we can use systems to enquire about the real world.

In essence, SSM articulates a learning process which takes the form of an enquiry process in a situation that people are concerned about. This process leads to action in a never-ending learning cycle: once the action is taken, a new situation with new characteristics arises and the learning process starts again. The methodology is summarised in Fig. 1. This is the SSM best known methodology and although Checkland has expressed a most flexible way of applying his ideas in his latest book (Checkland and Scholes 1990), the 7-stage methodology is still the most convincing and helpful account of the SSM enquiry.

The basic structure of SSM rests on the idea that in order to tackle real-world situations, we need to make sure that the 'real-world' is separated from the 'systems thinking world'. This distinction is crucial for SSM because that assures us that we

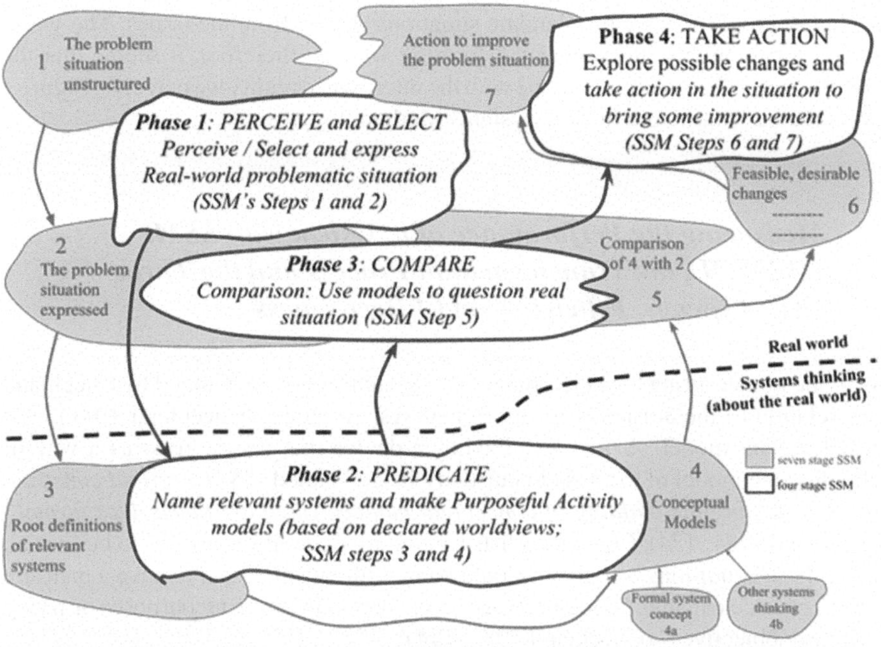

Fig. 1 The basic structure of soft systems methodology: four phases (perceive/select; predicate; compare; and take action) and SSM 7 steps/stages (*source* Jerardino-Wiesenborn, et al (2019), based on Checkland 1981)

won't see systems 'out there'; that is in the real world. SSM urges us to consider 'systems' as abstract concepts (preferably, the word 'holons' should be used) which, when applied to the real-world, can eventually help to bring some improvements to the situation concerned.

3.2 *Soft Systems Methodology (SSM): Managing and Controlling the Transformation Process*

Broadly speaking, the hard, reductionistic approach follows a positivistic epistemology, whereas SSM follows an interpretive perspective (Checkland 1981, 1986; Checkland and Scholes 1990; Jackson 1992, 2003). This can be summarised as follows: According to Checkland, life world is an ever-changing flux of events and ideas and 'managing' means reacting to that flux. We perceive and evaluate, take action(s) which itself becomes part of this flux which lead to next perceptions and evaluations and to more actions and so on. It follows that SSM assumes that different actors of the situation will evaluate and perceive this flux differently creating issues that the manager must cope with. Here, SSM offers to managers the systems ideas as a helpful tool to tackle problematic situations arising from the issues. The world outside seems highly interconnected forming wholes; therefore, it seems that the concept 'system' can help us to cope with the intertwined reality we perceive. Figure 1 shows the basic structure of Soft Systems Methodology.

3.3 Measuring the Performance of a System with SSM: CATWOE, the Transformation Process and the Three 'Es': Efficacy, Efficiency and Effectiveness

The issue of measuring the performance of a system is first mentioned by Checkland with relation to the attributes of the formal system model (Checkland 1981). The formal system model, 'S', is described as a device that can be used as a way of validating the model of the system currently being studied: *"S is a formal construct aimed at helping the building of conceptual models which are themselves formal"* (Checkland 1981: 174). Checkland lists a number of conditions for 'S' to be formal, one of these conditions is that 'S' should be measured; 'S' is said to have a measure of performance capable "to signal progress or regress in pursuing purposes or trying to achieve objectives" (Checkland 1981: 174).

CATWOE is the mnemonic of the six crucial characteristics which should be included in a well-formulated root definition.

C 'Customers'—the beneficiaries or victims of the transformation process.
A 'Actors'—those who would undertake the transformation process.
T 'Transformation'—the conversion of input to output.
W 'Weltanschauung' 'worldview'—the worldview that makes this transformation meaningful.
O 'Owners'—those who could stop the transformation.
E 'Environmental constraints'—elements outside the system which are taken as given.

From these elements, T is at the core of any root definition encapsulating the concepts that inputs are transformed into outputs as the root definition suggests.

T, the transformation process, is the CATWOE element on which Checkland elaborates the notion of measuring the performance of the system. According to Checkland, at the most fundamental level, any purposeful activity may be expressed through a Transformation Process which "changes or transforms some input into some output". In other words, for the *Transformation* to be relevant, Inputs are present in Outputs but in a changed state. Then if the Input is abstract (e.g. 'need for nursing services') then the Output must also be abstract (e.g. 'need met'). If the Input is concrete (e.g. 'a patient') then the Output must be concrete (e.g. 'a treated patient'). This distinction is important because it helps to differentiate between the resources and the inputs of the system. SSM also stresses the fact that there are many ways of expressing a purposeful activity; more ways of expressing the activity in terms of Input-Transformation-Output will enrich the thinking.

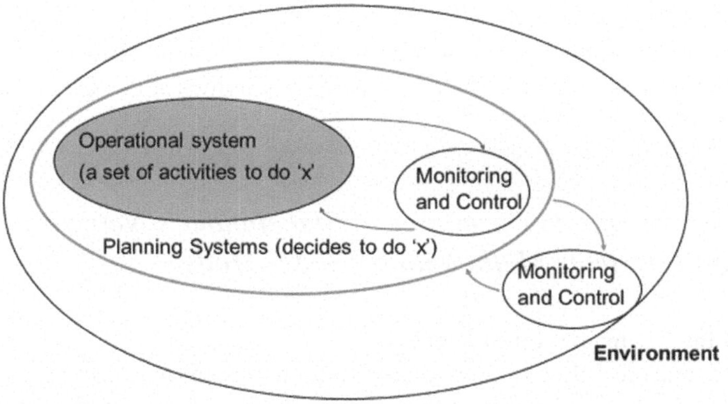

Fig. 2 Monitoring and controlling a system (from Checkland 1986)

According to SSM, when we try to 'manage' purposeful 'systems', it is useful to think of this situation in terms of:

(a) a purposeful system arranged as a set of activities which we may call the "operational system" (a set of linked activities to do 'x').
(b) a set of activities which will inspect the performance of the operational system and eventually will take action to bring it into line with aims and expectations; this is the "monitoring and control" system which monitors and controls the doing of 'x' and shown by the **inner** subsystem in Fig. 2.
(c) The system can be thought of as part of a wider system which decides to do 'x' (the 'what') or decide the way (the 'how') in which 'x' is carried out; these decisions are carried out by its own "monitoring and control" system which monitors and controls the long term objective of the system located on an upper level. This is shown by the **outer** system of Fig. 2. The criteria by which the Transformation can be judged gives the elements by which we can measure the performance of the system.

So, if we think of the two levels expressed in Fig. 2, we should ask the question: **How can the Transformation fail?** For controlling purposes and ultimately for 'managing' this activity, the following reflections and possible answers are useful:

(a) The way chosen to do **T** might not work; therefore, we manage **T** by asking: Does **T** mean selected work? The answer measures the **Efficacy** of **T**, measured by the monitoring and controlling activities at the 'operational system' level.
(b) Is **T** being done with minimum resources (including time)? We manage **T** by asking: Is **T** being done with minimum resources? The answer measures the **Efficiency** of **T**, measured by the monitoring and controlling activities at the 'operational system' level.

(c) **T** could the wrong activity to be doing. We manage **T** by asking: Is **T** the right thing to be doing? The answer measures the **Effectiveness** of the System, measured by the monitoring and controlling activities at the 'planning system' level.

3.4 Assessing the Performance of Stakeholder's Network: A Soft Systems Methodology (SSM) Model

SSM and the Transformation Process
The concept and the different views around issues pertaining to stakeholders' network involvement in HEIs were described using Checkland's Transformation Process. The following is a partial list of how the role of a stakeholders' network is viewed by different members of the network:

(1) A way of exercising HEI staff concerns about SD implementation in HEIs
(2) A way of preparing students to become more aware of the value of SDGs.
(3) A way of exercising pressure on HEI Senior management to be more active in pursuing SD in HEIs
(4) A way of creating and consolidating a SDGs culture in HEIs.

If we use Checkland's model on the third of the above views, we can apply the following control and monitoring concepts to this situation:

In Fig. 3, we have as input: "pressure on HEI Senior management not exercised" and as an output the same kind of input but in a transformed state: "pressure on HEI Senior management exercised", this according to SSM is an important distinction and the right way of stating i/o in systems language (as opposed to say input as 'resources', i.e.: material, money, etc. and as an output: 'result product) SSM insists that the output is just the input but in a transformed state. So, in the diagram above, to successfully manage this process we should ask the following questions:

(1) Do the means selected to do **T** work? In other words does the strategic plan allow the creation of a strong research culture—The answer will be a measure of **Efficacy** of **T**.
(2) Is **T** being done with minimum resources (including time)?—This is relatively easy to measure (cost associated to the consolidation of a research culture). This answer measures the **Efficiency** of **T**.

Fig. 3 Input and output in a transformation process

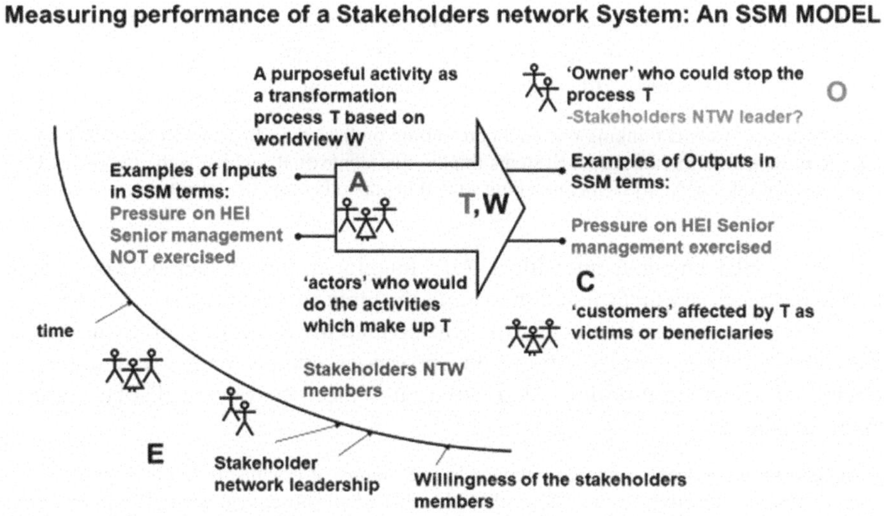

Fig. 4 Measuring stakeholder network involvement and performance: an SSM model (adapted from Checkland and Poulter 2006)

(3) Is **T** the right thing to be doing? Here we need to question the need for a stakeholder network involvement in HEI in the first place. This question can be asked only if we move to the upper level, that is to the planning system. At this level, the planning system, may decide to have a 'strategic plan' and indeed to question involvement in the first place; the planning system of this notional systems could decide to do 'y' instead of 'x' (to move the pressure to government agencies rather than HEI senior management). The answer will be a measure of the **Effectiveness** of the **system**.

Stakeholder networks' involvement in HEIs is a complex situation and we argue that systems thinking helps to clarify the situation and make sense of the elements and its connections in this situation. Figure 4 is an attempt to clarify the role of stakeholder networks and how the role and performance can be assessed using CATWOE analysis tools from soft system methodology.

4 Conclusions

Applied to the transformational challenges we currently face, the SDGs attempt to use both hard and soft systems, in that the planetary crises are atomised into seventeen parts, while the seventeenth SDG recognises the need for holistic systems thinking to make sense of all the other SDGs as a whole. '… (K)nowing, is at the same time separating and connecting, it is to make analysis and synthesis. Both are inseparable…' (Morin 2007: 25). However, most HEIs represent a reductionist structure in their disciplinary separation and hierarchical organisation by faculties, departments, courses and management. Hence the popular view that academia is far removed from the 'real world', but it doesn't have to be this way:

> Accordingly, systems thinking does not concentrate on basic building-blocks but rather on basic principles of organisation. Systems thinking is 'contextual', which is the opposite of analytical thinking. Analysis means taking something apart in order to understand it; systems thinking means putting it into the context of a larger whole.' (Capra 1996: 29)

Perhaps, HEI engagement with a truly integrated stakeholder network could provide the necessary meta-function to unify the units within universities, to become whole systems—similar in structure to multi-cellular life forms (ourselves included), whereby a number of organs, flows and processes operate within a semi-closed system, far from equilibrium. At a different scale, ecosystems display similar structural patterns.

> The ecosystem concept – defined today as 'a community of organisms and their physical environment interacting as an ecological unit' – shaped all subsequent ecological thinking and, by its very name, fostered a systems approach to ecology'. (Lincoln et al. Capra 1996: 33)

In addition to asking how may a stakeholder networks support universities to deliver the SDGs, we may turn the question round to ask how may universities support other stakeholders in this task? Better still, perhaps we should consider how they may support each other, as a sustainable living network. Here the word 'development' has been deliberately dropped and replaced with 'living', as the word development assumes the perpetuation of neoliberal capitalism as a means of addressing the transformative challenges we face. Many would argue that this assumption renders SD to be an oxymoron (Haley 2017) and furthermore, to be linked to the social and environmental crises that have become the climate emergency, species extinction and ecological collapse (Klein 2018; Kahn 2010; Haley 2020: 4).

Given the dominant neoliberal economic context, therefore, how may HEIs find stakeholders to support sustainable living beyond development? Climate emergency declarations (Climate Emergency Declaration 2020) by universities, local government authorities and corporations have in many cases become a double-bind for these institutions (Haley 2017), as their good intentions to support the SDGs are compromised by a political-economic culture that compromises and confounds the situation (Meadows 1999). Indeed, some corporations, national and international bodies that are unable to break the western paradigm, maintain that 'sustainable growth' (Inc.) is the only solution, while others like BP (BP 2020a; Storrow 2020) make unbelievable and unsubstantiated claims to relinquish their carbon industrial interests. The cynicism of such powerful potential stakeholders, the financiers of the majority of UK research funding (BP 2020b), makes things very difficult for universities and other educational institutions (including schools, academies, FE colleges and research centres) to shift from unsustainable networks to sustainable networks.

From a university perspective, much Environmental Education has already been appropriated, as Richard Kahn puts it;

> … this laudatory praise of its environmental literacy program by environmental educators is little more than the present-day technocratic standards movement in education masquerading as a noteworthy "green" improvement. Put bluntly: this environmental literacy as a greenwash. (Kahn 2010: 9)

'(C)hanging is difficult, but it is always possible' (Freire 2016), so maybe we should consider alternative systems for sustainable futures? Such foresight scenarios for the development of higher education stakeholder networks, for and beyond the implementation of sustainable development, may include:

(a) New Radical stakeholder networks
Despite the pressures of the powers that be, since the emergence of Extinction Rebellion (XR) (BBC 2019), UK universities are facing a far deeper and more united call from their students and staff to change to be socially and environmentally accountable. While students have in the past demanded such ethical and moral alignment from their HEIs, over issues like fracking (People and Planet n.d.) and carbon industry divestment (Taylor 2020; Wootton-Cane 2019), the persistence of XR reveals a potential learner stakeholder with real interest in their futures, the students.
Considering the structure, pattern and processes of an HEI(s), how may stakeholders contribute to its development of sustainable living? (This could be plural, as this then raises the question of how autonomous is each HEI is and what impact might they have individually and collectively—i.e. stacked systems).
(b) Co-learning as the basis action underpinning stakeholders networks
Co-learning or collaborative learning could form the basis of stakeholder networks. The idea is akin to the concept of 'post-disciplinarity', developed by ecological artists, Helen Mayer Harrison and Newton Harrison in 2008 (Haley 2012). This form of cooperative governance starts with the image of all the disciplines seated at a circular table; no one discipline has dominance over any other—they must all listen to each other. Climate change, species extinction, ecological collapse and poverty are then placed at the centre of the table for all to address, equally. Of course, this profoundly simple system owes much to Dewey (2015: 19) and Deluze and Guttari (2013: 3), and it has been extended and expanded to include non-colonial, feminist, pro-LGBT+ and other inclusive forms of teaching and learning that go beyond educational institutions to society at large and back to the original Ancient Athenian concept of 'paidiea' (Kahn 2010: 40). For Paulo Freire and Richard Kahn, this is the essential starting point of critical pedagogy for genuine societal transformation that could actually bring about sustainable living through 'ecopedagogy' (Kahn 2010:1; Haley 2009: 306). Here, Ecoliteracy (Capra 1996) takes precedence over fiscal, commercial and industrial development and the cognitive dissonance of the international community's disingenuous response to the collective global crises (Anderson 2015).
"The question is not only ethical in the conventional sense, it is also an ecological question. The means by which one man influences another are a part of the

ecology of ideas in their relationship, and part of the larger ecological system within which that relationship exists." (Bateson 2000: 512).

(c) Eco-pedagogy as the base for resilient network stakeholders

Let us then consider what an ecopedagogical network for sustainable living might look like, how such a network or system might operate and who the stakeholders might be. Well, who could possibly be left out of such a network and why? For the purposes of practicality, let us consider including just the representatives of organisations, that way we may create a network of networks, enabling direct and meaningful communication to many more people; a social ecosystem, as opposed to PR and advertising that lack personal interconnectivity and the potential for criticality. Favouring no stakeholders, the form could be rhizomatic (Deluze and Guttari 2013: 3). Like bamboo or ginger, the pattern of growth is on a horizontal plane, with subgroups (nodes) that send out rootstalks. It is the mass of roots that are the network, the ecology, or 'the pattern that connects' (Bateson 2000: 7). There is no hierarchy and just one aim; to promote sustainable living through ecological literacy. This is the modus operandi—structure, pattern, process and intention as one entity. HEIs would simply be an integral, interconnected, interdependent stakeholder of stakeholders.

Some may recognise the potential for the emergent system of Transdisciplinary Knowledge, a process or method that goes '... at once between the disciplines, across the different disciplines, and beyond all disciplines' (Nicolescu 2008: 2). Here the potential for creative regeneration (Wahl 2016) may be able to design sustainable living futures that incorporate social, cultural, environmental and political conservation, regeneration and reinvention in a non-deterministic fashion. This path to self-determination and self-reliance through ecopedagogy has the potential to cast aside fatalistic attitudes of 'that's how things are', or 'that's human nature' (Freire 2016: 79). Indeed, seen as a self-supporting system of sustainable living, this network of stakeholders would become ecologically resilient (Gunderson and Holling 2002: 28), by virtue of its flexible capacity to withstand impacts and adaptive capabilities to emerge from a state of collapse to another state of becoming.

Finally, as we become increasingly cognisant of the global ecological crisis, we may reflect on the etymological root of the word, crisis. From the Greek verb krinein, it means "to decide"; however in Chinese, the equivalent ideogram for crisis—wei-ji—is formed of the two characters for 'danger' and 'opportunity' (Kahn 2010: 4), While sustainability paradoxes persist and corporate interests dominate our society, opportunities emerge for transformation, or organisational change at a greater scale of magnitude. 'It means the necessity of the introduction of a new order of conceptions leading to new modes of practice' (Dewey 2015: 5). The efficacy, efficiency and effectiveness of stakeholder networks then shifts to a different set of values, as the cultivation of an ecopedagogical stakeholder network could become the primary role for universities intent on sustainable living. Again, as Paulo Freire wrote; 'changing is difficult, but it is always possible'.

References

Ackoff R (1993) The art and science of mess management. In: Mabey C, Mayon-White B (eds) Managing change. Paul Chapman Publishing, London
Ackoff R (1995) 'Whole-Ing' the parts and righting the wrongs. Syst Res 12(1):43–46
Anderson K (2015) Talks in the city of light generate more heat. Nature 528
Baker S, Kousis M, Young S, Richardson D (eds) (2005) The politics of sustainable development: theory, policy and practice within the European Union. Psychology Press
Bateson G (2000) Steps to an ecology of mind. University of Chicago Press, Chicago, USA
BBC (2019) What is extinction rebellion and what does it want? https://www.bbc.co.uk/news/uk-48607989. Accessed 23 Feb 2020
BP (2020a) BP sets ambition for net zero by 2050, fundamentally changing organisation to deliver. https://www.bp.com/en/global/corporate/news-and-insights/press-releases/bernard-looney-announces-new-ambition-for-bp.html. Accessed 15 Feb 2020
BP (2020b) Research and partnerships. https://www.bp.com/en/global/corporate/what-we-do/technology-at-bp/research-and-partnerships.html. Accessed 15 Feb 2020
Capra F (1996) The web of life: a new synthesis of mind and matter. HarperCollins, London
Checkland PB (1986) Some basic Ideas of monitoring and control for managers, University of Lancaster, lecture notes for MA "Systems in Management"
Checkland PB (1981, 1999) Systems thinking, systems practice. Wiley
Checkland PB, Scholes P (1990) Soft systems in action. Wiley
Checkland PB, Poulter J (2006) Learning for action: a short definitive account of soft systems methodology, and its use for practitioners, teachers and students. Wiley
Churchman CW (1968) The systems approach. Dell, New York
Climate Emergency Declaration (2020) Climate emergency declarations in 1,385 jurisdictions and local governments cover 815 million citizens. https://climateemergencydeclaration.org/climate-emergency-declarations-cover-15-million-citizens/. Accessed 22 Feb 2020
Deluze G, Guttari F (2013) Ten thousand plateaus. Bloomsbury, London
Dewey J (2015) Experience and education. Free Press, New York
Dlouha J, Henderson L, Kapitulčinová D, Mader C (2018) Sustainability-oriented higher education networks: Characteristics and achievements in the context of the UN DESD. J Clean Prod 172:4263–4276
Earth Charter Commission (2000) Earth charter. https://earthcharter.org/invent/images/uploads/echarter_english.pdf. Accessed 25 Feb 2020
Estes RJ (2010) Toward sustainable development: from theory to praxis. Trans Soc Work Pract 76
Freire P (2016) Pedagogy of indignation. Routledge, Oxford
Godemann J, Bebbington J, Herzig C, Moon J (2014) Higher education and sustainable development: exploring possibilities for organisational change. Account Audit Accountability J 27(2):218–233
Gunderson LH, Holling CS (eds) (2002) Panarchy: understanding transformations in human and natural systems. Island Press
Haley D (2009) Steps to an art of ecology: an emergent practice. Ph.D. Thesis. British Library
Haley D (2012) Seeing the whole: art, ecology and transdisciplinarity. Revista Arte y Politicas de Identidad, Arte como terapia social 4(3). https://revistas.um.es/api/article/view/146081
Haley D (2017) 'Undisciplinarity' and the paradox of education for sustainable development. In: Leal Filho W (ed) Handbook of sustainable science and research. Series, climate change management. Springer
Haley D (2020) Unconventional educational approaches: an eco-pedagogy to address our transformative challenges. In: Leal Filho W, Azul A, Brandli L, Özuyar P, Wall T (eds) Quality education. Encyclopedia of the UN sustainable development goals. Springer
Hristov D, Scott N, Minocha S (2018) Distributed leadership typologies in destination management organisations. Tourism Manage Perspect 28:126–143
Jackson M (1992) Systems methodology for the management sciences. Plenum Press

Jackson MC (2003) Systems thinking: holism for managers. Wiley, Chichester
Jerardino-Wiesenborn B, Paucar-Caceres A, Ochoa-Arias A (2019) A conceptual framework based on Maturana's ontology of the observer to explore the checkland's soft systems methodology. Syst Pract Action Res. https://doi.org/10.1007/s11213-019-09502-y
Kahn R (2010) Critical pedagogy, ecoliteracy, and planetary crisis: the ecopedagogy movement. Peter Lang, New York
Klein N (2018) Capitalism killed our climate momentum, not "human nature". https://theintercept.com/2018/08/03/climate-change-new-york-times-magazine/?comments=1. Accessed 23 Feb 2020
Kuenkel P (2019) Global transformation as a collective leadership challenge. In: Stewarding sustainability transformations. Springer, Cham, pp 43–58
Ledington P, Donaldson J (1997) Soft OR and management practice: a study of the adoption of SSM. J Oper Res Soc 48:229–240
Lozano R, Lukman R, Lozano FJ, Huisingh D, Lambrechts W (2013) Declarations for sustainability in higher education: becoming better leaders, through addressing the university system. J Clean Prod. 48:10–19. https://doi.org/10.1016/j.jclepro.2011.10.006
Macadam RD, Packham RG (1989) A case study in the use of SSM restructuring and academic organisation to facilitate education of Systems agriculturists. Agric Syst 30(4):352–367
Macadam RD, Van Asch R, Hedley B, Pitt E, Carrol P (1995) A case study in development planning using a systems learning approach—generating a master plan for the livestock sector in Nepal. Agric Syst 49(3):299–323
Meadows DH (1999) Leverage points: places to intervene in the system. Academy for systems change. https://donellameadows.org/archives/leverage-points-places-to-intervene-in-a-system/. Accessed 23 Feb 2020
Mingers J, Taylor S (1992) The use of soft systems methodology in practice. J Oper Res 43(4):321–332
Morin E (2007) Restricted complexity, general complexity. In: Gershenson C, Aerts D, Edmonds B (eds) World Scientific, pp 1–25. https://philpapers.org/rec/MORRCG. Accessed 23 Feb 2020 (in Press)
Nicolescu B (2008) Transdisciplinarity: theory and practice. Hampton Press Inc., New Jersey
Paucar-Caceres A, Hart D, Roma i Vergés J, Sierra-Lozano D (2015) Applying soft systems methodology to the practice of managing family businesses in Catalonia. Syst Res Behav Sci33:312–323
People and Planet. Fracking with education: public money, research and fracking. https://peopleandplanet.org/system/files/resources/Fracking_Education.pdf. Accessed 23 Feb 2020
Provan KG, Kenis P (2008) Modes of network governance: Structure, management, and effectiveness. J Publ Admin Res Theor 18(2):229–252
Radinger-Peer V, Pflitsch G (2017) The role of higher education institutions in regional transition paths towards sustainability. Rev Reg Res 37(2):161–187
Schmidt U, Günther T (2016) Public sector accounting research in the higher education sector: a systematic literature review. Manage Rev Q 66(4):235–265
Scott J, Carrington PJ (2014) The SAGE handbook of social network analysis. SAGE Publications Ltd. London. https://doi.org/10.4135/9781446294413
Storrow B (2020) BP pledges to go carbon-neutral—how remains an open question. https://www.scientificamerican.com/article/bp-pledges-to-go-carbon-neutral-how-remains-an-open-question/. Accessed 15 Feb 2020
Taylor M (2020) Half of UK universities have committed to divest from fossil fuel. The Guardian 13 Jan 2020. https://www.theguardian.com/environment/2020/jan/13/half-of-uk-universities-have-committed-to-divest-from-fossil-fuel. Accessed 17 Feb 2020
United Nations Millennium Declaration (2000) General assembly resolution 55/2 of 8 Sept 2000. https://www.ohchr.org/EN/ProfessionalInterest/Pages/Millennium.aspx. Accessed 23 Feb 2020

United Nations (2015) Transforming our world: the 2030 agenda for sustainable development. https://sustainabledevelopment.un.org/content/documents/21252030%20Agenda%20for%20Sustainable%20Development%20web.pdf. Accessed 23 Feb 2020
Vargas VR, Lawthom R, Prowse A, Randles S, Tzoulas K (2019a) Sustainable development stakeholder networks for organisational change in higher education institutions: a case study from the UK. J Clean Prod 208:470–478
Vargas VR, Lawthom R, Prowse A, Randles S, Tzoulas K (2019b) Implications of vertical policy integration for sustainable development implementation in higher education institutions. J Clean Prod 235:733–740. https://www.sciencedirect.com/science/article/pii/S0959652619323613
Wahl DC (2016) Designing regenerative cultures. Triarchy Press, Axminster, England
WCED (1987) Brundtland report 1987 our common future: report of the world commission on environment and development, United Nations. https://sustainabledevelopment.un.org/content/documents/5987our-common-future.pdf. Accessed 15 Feb 2020
Wilson CH (2018) Building the structures for change. In: Designing the purposeful world. ROUTLEDGE in association with GSE Research, vol 87, no 95, pp 87–95
Wootton-Cane N (2019) Students stage week-long occupation in protest against university's £12m investment. The Guardian 27 Nov 2019. https://www.theguardian.com/education/2019/nov/26/university-of-manchester-to-review-fossil-fuel-shares-after-student-protest. Accessed 15 Feb 2020

Corporate Sustainability Benchmarking in Academia: Green Campus, Living Labs, Socioeconomic and Socioenvironmental Initiatives in Brazil

Mario Corrêa de Sá e Benevides, José Baltazar Salgueirinho Osório de Andrade Guerra, Robert Samuel Birch, and André Borchardt Deggau

1 Introduction

This study explores a benchmarking scheme for the evaluation of the Brazilian higher education community in its performance towards an environmentally sustainable future. The scheme is tested using a sample group of higher education institutes

M. C. de S. Benevides
PowerSB Sustainability and Business, Florianópolis, Brazil
e-mail: mario.benevides@unisul.br
URL: https://www.unisul.br; https://www.powersb.com

Centro Internacional, University of Southern Santa Catarina (UNISUL), Florianópolis, Brazil

J. B. S. O. de Andrade Guerra
Management and Environmental Sciences, University of Southern Santa Catarina (UNISUL), Florianópolis, Brazil
e-mail: baltazar.guerra@unisul.br
URL: https://www.unisul.br; https://www.ceenrg.landecon.cam.ac.uk/

J. B. S. O. de Andrade Guerra · A. B. Deggau
Research Centre on Energy Efficiency and Sustainability (GREENS), University of Southern Santa Catarina (UNISUL), Florianópolis, Brazil
e-mail: andre.deggau@gmail.com
URL: https://www.unisul.br

J. B. S. O. de Andrade Guerra
Department of Land Economy, Cambridge Centre for Environment, Energy and Natural Resource Governance (C-EENRG), University of Cambridge, Cambridge, United Kingdom

R. S. Birch (✉)
School of Engineering, University of Liverpool, Liverpool, United Kingdom
e-mail: r.s.birch@liverpool.ac.uk
URL: https://www.liverpool.ac.uk

W. Leal Filho et al. (eds.), *Universities, Sustainability and Society: Supporting the Implementation of the Sustainable Development Goals*, World Sustainability Series,
https://doi.org/10.1007/978-3-030-63399-8_10

Fig. 1 Location of Santa Catarina State in Brazil

(HEIs) located in the Brazilian state of Santa Catarina and compares these results with those gathered from similar HEIs in Europe and North America. The principal aim of the scheme is to capture data relating to a HEI's approach to the concepts and practices of green campuses, living laboratories and other economic and socioenvironmental sustainability initiatives.

Benchmarking itself has a useful place in society and there are good examples where this is applied to sustainability performance in HEIs,[1] but the potential disadvantage of this exercise degenerating into a process of simply collecting indicators concerned only with establishing a position on a league table must be recognized (Mayle et al. 2002). Instead, it is intended that this process should offer a useful means of collecting and collating ideas with a view to revealing and importing best practices in sustainability. The aim is to enhance the contribution of academia to a society that is more focused towards sustainable development rather than a means of gaining a competitive advantage. In this respect, the terms indicator-benchmarking provides a more explicit distinction. Yet, it is acknowledged that a sense of competition can be an important incentive to improvement, especially where complacency predominates. However, it can be argued that the contacts established in face-to-face benchmarking exercise rather than collecting and collating performance indicators offers a more powerful driver for discovery and exploitation of good practice (Mayle et al. 2002). In applying a benchmarking exercise to a group of HEIs in Santa Catarina, this study attempts to highlight some of these issues arising from the experience.

Brazil's state of Santa Catarina (SC—see Fig. 1) has one of the best set of socioeconomic indicators among the 27 Brazilian federative units. SC has 26 HEIs, which accounts for just over 1% of the total number of the HEIs in Brazil. On the other hand, SC is host to two of the 13 Brazilian universities to reach the maximum grade (grade five) General Course Index, as assessed by the Brazilian Ministry of Education in 2019. With a territorial area corresponding to 1.12% of Brazil's 8,516 million-km2

[1] An example is: https://peopleandplanet.org/university-league—accessed on April 5, 2020.

territory, SC has a Human Development Index (HDI) of 0.840—lower only than that of the capital Brasília (0.874), higher than the one of Brazil's largest cities, São Paulo (0.833), and higher than the country's HDI as a whole (0.699). Brazil itself ranks 73rd in the HDI world ranking.

A brief survey among Brazilian universities and of the literature in general shows a lack of such basic comparative data and, therefore, a further aim of this paper is to prompt a deeper understanding of sustainability issues, especially in Brazil. This impression is reinforced when consulting the websites of a selection of universities in Santa Catarina, which showed evidence of the embryonic growth around the concepts of green campuses and living laboratories. In this way, collecting and compiling data from HEIs and comparing this with like institutions that may be recognized as having a more established approach to sustainable development, a greater awareness will permeate the sector in Brazil. The following sections contextualize the main aspects of this study and represent some of the most relevant issues under discussion today.

1.1 Climate Change

Climate change poses the single biggest threat to humanity in 2020 and over the next ten years, according the Global Risks Report (World Economic Forum 2020). While society is increasingly concerned with this problem, especially among the younger generation, there is still little political will to act. However, this crisis is striking faster and harder than previously expected: the last five years are on track to be the five warmest on record set and global temperatures are to increase by at least 3 °C at the end of this century. The potential impacts of climate change include loss of life, increased migration, stress of ecosystems, food and water crisis, economic impacts and the exacerbation of geopolitical tensions.

In this context, there is growing pressure to advance sustainability in all environments including HEI campuses. University campuses work as small cities, consuming water, energy and food and producing waste. Potentially, better management of these resources can prove less costly to a HEI with less impact on the environment. Implementing environmentally sustainable processes in the same location where environmental research happens makes sense, as universities are highly appropriate places for researching and developing objectively those new advanced approaches and methods to reduce environmental impact.

1.2 Green Campuses

Due to growing concerns over climate change, HEI campuses are being used increasingly as an experimental platform for greening experiences (Leal Filho et al. 2015) and, in this respect, the concept of a Green Campus has emerged. Green campuses minimize their environmental impact, create public awareness to the issue of sustainability while performing their daily activities—i.e., teaching and research—in a sustainable way (Tezel et al. 2018). One possible way of measuring environmental impact is by calculating carbon footprints. Today, carbon footprints are one of the most used measures of environmental and climatic impact of human action and consist of calculating the amount of CO_2 emitted, directly or indirectly, by human activities. In HEIs, electricity,[2] transportation and waste management are the main drivers of greenhouse gas emissions. Assessing carbon footprints creates an initial step towards greening a campus (Ridhosari and Rahman 2020).

Universities are in the forefront of spreading awareness and directly addressing environmental concerns (Ridhosari and Rahman 2020). The first joint commitment for sustainability activities in universities dates back to 1972, in the form of the Stockholm Declaration. In 1990s, a global green campus movement began but not meaning, however, that all universities are in this same stage yet. Green Campus efforts are still being spread worldwide with the United States and the United Kingdom leading the world in the number of green campus projects followed by Australia, Canada and China (Tezel et al. 2018).

1.3 Living Labs

Green Campuses can be living laboratories (Living Labs) for sustainability initiatives and influence the everyday lives of staff and students (Leal Filho et al. 2015). The concept of a Living Lab started initially through business initiatives as a space designed to validate projects of innovation, including technological, social and cultural innovation. The concept is based around setting up these innovations in real situations before introducing them into the market or society in general. To this extent, HEIs provide this near-real environment in which Living Labs can generate data and new ideas and concepts (Colobrans 2019).

Living labs also turn out to be a practice-driven set of experiences in real-life environments and arenas making it possible to study and experiment with new ideas and processes. Living labs may have different approaches aimed at several purposes, but they share a common representation of a real-life setting using a multi-method approach with multi-stakeholder engagement and the purpose of co-creation. The methodology of Living Labs can serve for bridging gaps between the academia and the society as well as enhancing knowledge and idea sharing (Malmberg et al. 2017). Typically, ideas and importance of Living Labs is their use of this in scientific experiments applied to medicine, as exemplified through published case studies (Malmberg et al. 2017).

[2]It is important to note that in Brazil the electric energy grid is predominantly fueled by renewable primary sources.

1.4 Economic and Socioenvironmental Sustainability Initiatives

Academia collaborates frequently with companies, governmental and non-governmental institutions in initiatives for the purpose of research and innovation aimed at the improvement in economic and socioenvironmental sustainability in society. This paper presents something of these in the following sections.

2 Methodology

The methodology applied to this study is based on themes and indicators (or indices) chosen from theoretical and/or practical approaches available in the literature. The criteria for the selection of HEIs to participate in this data survey are explained in detail in the following sections.

2.1 Themes

The principal reference for this study is the Green-Campus Guidebook 2018–2019—Green-Campus Programme (2018) issued by The National Trust for Ireland, a well-known reputable environmental organization. This reference establishes a set of themes, which the authors consider important for guiding assessments and action plans relating to the Green Campus initiative. Based on these, the authors selected the following analyses and directions for evaluation of the Green Campus theme considering the following:

- Litter and waste
- Energy
- Water Conservation and Protection
- Travel and Transport
- Biodiversity

In addition, the following themes mentioned previously complement the goals of this study:

- Living labs
- Socioeconomic and socioenvironmental sustainability Initiatives

2.2 Indicators

To facilitate the collection of data from each HEI the authors created a set of survey questions grouped into three thematic tables titled Green Campus, Living Labs and socioeconomic/socioenvironmental sustainability initiatives listing their individual themes and indicators as shown in Tables 1, 2 and 3.

Table 1 Green campus

Theme	Indicators
Litter and waste	kg waste per annum per full time equivalent (FTE) student
	kg waste per course
	Is the waste separately disposed? (E.g. metal, paper, plastic, organic, non-recyclable)
	% of waste materials recycled
	kg of waste prevented per FTE student
	Is there a correct disposing of used electrical equipment?
Energy	Is there an energy management system on site?
	What sources of energy are used? (E.g. electricity, oil, gas, solid fuels, bio/renewable fuels…)
	Is there an adequate storage of fuel, oils, diesel and others, so posing no environmental threat?
Water conservation and protection	Is there a water management on site?
	Is water metered and how is it metered?
	Is there a leak detection programme in place?
	Where is water used? (Ex.: swimming pools, gyms, accommodation, cleaning and cooking, labs…)
Travel and transport	How is travel to and from the college managed?
	What are the main travel and location issues?
	How are goods and services delivered to the campus?
	Are there ways to reduce deliveries to/from the campus?
	Are there linkages with local and/or transport authorities to promoting sustainable transport?
Biodiversity	Are there grounds management methods regarding herbicides, pesticides, fertilizers…?
	Is there any risk management of invasive species?

Indicators based on reference Greens Campus Program (2018)

Table 2 Living-labs

Theme	Indicators
Living labs	Is there a living lab in the campus, or an action plan to have one?
	Main focus and objectives of the living lab
	Is there a multi-method approach to share user-centered, co-creation methodology with other living labs?
	Are the living lab's users engaged? How?
	Is there a multi-stakeholder participation, with public and private sector, academia and community representatives?
	What is the real-life setting of the living lab?
	How the co-creation is practiced?

Indicators based on Ref. Malmberg et al. (2017)

Table 3 Socioeconomic and socioenvironmental sustainability initiatives

Theme	Indicators
Socioeconomic and socioenvironmental sustainability initiatives	Does the University develop this kind of initiative?
	What are the focuses of this kind of initiative?
	(a) Innovation, research and development
	(b) Economic growth
	(c) Environmental management improvement
	(d) Social inclusion
	(e) Education
	(f) Others, as…

Based on the experience of the authors

2.3 *Obtaining Information*

The information regarding the universities approached in this paper was obtained via the HEI's websites, complemented through follow-up consultations with professionals in the relevant institutions.

2.4 *Selecting the Reference Universities*

Although, at present, some governments in Europe and the United States of America are reticent and resistant to the anthropic influence on climate change, research initiatives aiming for a more sustainable development are significant in both continents.

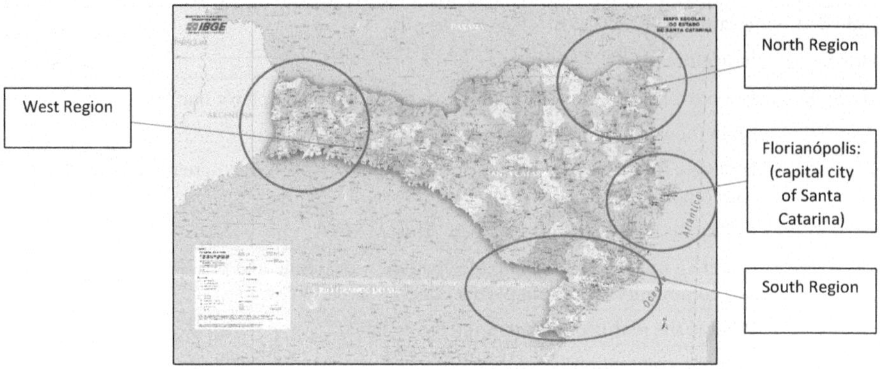

Fig. 2 Political map of Santa Catarina. *Source* IBGE—The Brazilian institute of Geography and Statistics

Besides this, Europe and the United States of America have traditional and secular universities with high reputation all over the world. Thus, in order to set some kind of datum for the analysis of Brazilian universities in Santa Catarina in tackling the sustainability issues as outlined in Tables 1, 2 and 3, data from a typical European university and a typical North American university were included in the survey.

2.5 Selecting the Universities in Santa Catarina

A geographic subdivision of the State of Santa Catarina determined the selection HEIs in the assessment. Four HEIs with campuses located in four regions of Santa Catarina where selected as shown in Fig. 2.

3 Profile of the Selected Universities

3.1 European and North American Campuses (EACs)

For the purposes of comparison, two typical university campuses were selected; one European university and one North American. The European campus is based in a large city, and was founded in the late 1800s. It comprises three faculties organized into 35 departments and schools. This HEI is recognized as having a strong research focus. Even based in the heart of a large city, its main campus occupies 100 acres that comprises nearly 200 non-residential buildings that house lecture theatres, teaching areas, accommodation and research facilities. The university has 33,000 students pursuing over 450 programs, spanning 54 subject areas. The university has received various awards for its implementation of environmental policies on transport, waste management, sustainable procurement, emissions, water recycling and energy usage.

Similarly, the North American university was founded in the late 1800s and is based in the suburbs of a large city. It has high-profile policies on sustainability with green buildings incorporating energy efficient technologies and a focus on human health. Environmentally conscious grounds management pursues initiatives to plant local and adaptive species. The university as a whole prides itself on the use of solar energy and efforts on carbon neutrality. It strives towards zero waste and works with community partners to offer sustainable transportation. This culture of sustainability is evident through faculty research efforts, student involvement, and staff practices.

3.2 Santa Catarina Campuses (SCCs)

Considering the four regions indicated in Fig. 2, five SCC sites were selected for the survey comprising one from each of these regions plus a second from Florianópolis, the capital city of Santa Catarina State. Typically, these were first established between 40 and 60 years ago and have gone through processes of modernization and expansion during this time ranging from new faculty buildings to additional campuses builds on green-field sites. Each campus provides teaching, research, catering and recreation resources for between 10,000 and 30,000 students and staff. Unlike the EACs, the SCCs do not offer student accommodation, which is typical of most Brazilian HEIs.

Santa Catarina's universities offer graduate, MSc and PhD courses over a broad range of subject areas that tend to be closely identified with the needs of their local communities. Typical focus areas include sustainable production alternatives for family farming, process optimization for cleaner production, environmental studies and integrated design of the built environment and energy. Their interests lie not only in the training of professionals but also in the production and publication of new knowledge, the development of community actions aimed at promoting economic, social, cultural, sports and health development. Their outreach activities include social assistance initiatives, seminars, scientific meetings and active engagement with the community. More recently, these institutions have invested heavily in expanding postgraduate studies and research, the creation of technological centers and internationalization.

4 Corporate Sustainbility Bench Marking

Using the methodology set out in Sect. 2, the benchmarking survey yielded results that clearly demonstrate that all the HEIs have interests in all the themes outlined in this paper. However, the information that is available via their open websites varies greatly, ranging from high-profile publicity statements on corporate sustainability policy and performance to almost no information that can be accessed easily. Similarly, the perception of professional staff who were approached in the survey varied from a detailed well informed knowledge of their HEI sustainability policies to very little knowledge or awareness of them at all. Surprisingly, in some cases there was resistance to sharing such information as it is seen as being strategic to their organization and, therefore, confidential. Some of the more important observations are detailed in the following sections.

4.1 Green Campus

(a) **Litter and waste**

The survey reveals that the EACs appear to have well established management processes for monitoring and control of their waste which caters for the separation of recyclable (paper, plastic, organic) and non-recyclable materials. Nevertheless, there remain opportunities to improve waste handling further with one institution reporting less than 30% being recycled from a total annual disposal of seven tons of waste representing approximately 80 kg per full-time student per year. Typically, the drivers to the waste management are:

- High-profile communication and engagement with staff and students about waste handling services has resulted in a greater response towards recycling programs.
- This continues to be emphasized through targeted campus-wide waste reduction campaigns spearheaded by a student lead initiatives such as returning to reusable cutlery instead of disposable plastic utensils.
- Paperless classrooms where students access lecture notes and submit work electronically through a central teaching hub.
- A continuing education, awareness and behavior modification of the entire campus community by measuring and setting targets to be accomplished.
- The disposing of used electrical equipment according to recognized regulations such as WEEE.[3]

Survey data from the SCCs indicates that all are making efforts to separate, recycle and avoid waste but there appears to be less emphasis on engagement with the stakeholders. There is variability in the way waste is managed with few institutions being able to report the total amount of waste they produce per annum. However, the indications are that they produce waste in the region of approximately 50 kg per full-time student per year, which is not too dissimilar to the EAC experience. However, one SCC reports only 2% of this being recycled. Used electrical equipment is stated as being correctly disposed of but it is not clear what criteria or regulation is being adopted in this respect.

---

[3]WEEE: European Waste Electric and Electronic Equipment Regulations 2013.

(b) **Energy**

The EACs have energy management systems on site and strive or have obtained carbon-neutrality through the use of energy saving measures such as low power consumption devices and on-campus combined heating/power generation plants based on natural gas. Efforts to reduce costs through energy saving, reduced emissions and potential opportunities to energy export electricity to the national systems are the main influencers in their initiatives.

SCC report their sole source of energy to be electricity from the national grid, with no campus' energy management systems being used. All indicate interests in renewable energy research such as solar photovoltaic energy, but little of this appears to be contributing to the corporate energy requirements of the individual campuses. It should be noted that these campuses (as the most of facilities and residences in the country) benefit from a largely predominance of renewable energy that makes up the Brazilian power matrix. Currently, 83% is generated from renewable primary sources, with 63% alone being supplied by hydroelectric power plants.[4]

(c) Water conservation and protection

In line with most developed universities, EACs provide not only the normal toileting and hygiene stations across the campus but also accommodation, sport facilities including swimming pools & showers, restaurants, research laboratories that are all significant consumers of water. As with energy consumption, cost saving is a significant driver to conservation of water with well-known practices being employed such as:

- Early detection of leaks through metering and regular maintenance checks and repairs.
- Water saving devices such as low-flow faucets and showerheads, dual flush toilets, and waterless urinals across campus.
- Replacement of older equipment such as washing machines with newer, energy star models, which use nearly 50% less water.
- Planting of native and adaptive plant species used across campus to minimize irrigation water usage, and a computerized irrigation system capable of adjusting water usage based on weather and rainfall patterns.
- Wastewater is deposed of according to the regulations of the local water management authority.

The established SCCs indicate that there are initiatives focused on avoiding water wastage especially where swimming pools, hygiene, cleaning, cooking and other high-usage facilities are present. Nevertheless, it is not clear if any rigorous management system for metering and leak detection is in place.

[4]Source: the National Electrical Energy Agency (Aneel).

(d) Travel and Transport

The EACs are located in or near to a large city where traffic congestion is a concern for the city hall officers. These universities adopt city transport policies specific to their locations and actively encourage staff and students to use public transport and other sustainable transport schemes, such as:

- Promoting metro-passes, bicycle purchase/rental assistance, e-scooters, bicycle share, secure bicycle parking and commuter ride sharing.
- Offering no student car parking and only limited chargeable car parking for staff and visitors.
- Home working.

Goods and services are delivered to the campus via dedicated vehicles, but with no obvious means to optimize these.

The SCCs rely heavily on public transport as the main way of moving staff and students between home and the campuses. Car parking is provided but there are no obvious efforts to encourage environmental friendly transport habits. These institutions have sustainability groups pursuing lines of research in modifying transport habits in the community but it is not clear how this contributes to their transport management policies.

(e) Biodiversity

EACs operate management methods in their estates regarding the control of herbicides, pesticides, fertilizers by ensuring these are compliant with national regulations. They also implement policies to promote native and adaptive species and reduce the use of harmful chemicals.

The SCCs appear to be more active in this area through schemes, for example:

- Maintaining a forest nursery with native species.
- Management methods regarding herbicides, pesticides and fertilizers.
- Risks management regarding invasive species are kept under control.
- Initiatives to protect natural systems through direct interventions.

4.2 Living Labs

All institutions have Living Lab-style learning employing problem solving and applied real-world learning through a multi-method approach and co-creation methodology with other Living Labs through online campaigns. The same applies to a multi-stakeholder participation, with public and private sector, academia and community representatives. Typically focus objectives of the living labs are:

- Solar thermal panel for water purification.
- Software development.
- Behavioral research.

- Peace and political violence.
- Understanding, sexual orientation, gender, and racial/ethnic disparities in health, transgender and racial/ethnic minorities.
- Technological incubation with industries and popular cooperatives.
- Transfer of technology and knowledge in favor of social, economic, cultural and environmental development.

4.3 Economic and Socioenvironmental Sustainability for Society Initiatives

All institutions have research groups dedicated specifically to sustainability and community engagement services that broadly reflect the Sustainable Development Goals of the United Nations and the demand for sustainable engineering solutions to mitigate global problems such as climate change and declining natural resources. All institutions aim to strengthen their institutional identity with such activities. However, whereas EACs research efforts tend to bias towards innovation, environmental management and sustainable development, SCCs appear to place more emphasis on social inclusion, health, education, art, culture, sport, science and the promotion of citizenship. This is facilitated through programs for the voluntary participation of staff and students at local community schools, especially those teaching ethical and legal principles, citizenship rights and duties, as well as support for homeless people.

5 Conclusions

In this study, the authors examined performance benchmarking as a means of evaluating and improving corporate sustainability in the Brazilian higher education community. Data from a group of HEIs in the Brazilian state of Santa Catarina was gathered and compared with similar institutions in Europe and North America. The aim is to establish a process by which performance can be measured and ideas shared rather than simply creating another league table or tick-box mentality.

The proposed benchmarking questionnaire set out in Tables 1, 2 and 3 yielded information that would guide HEIs towards a more environmental sustainable operation. Some important highlights from the themes are:

- **Litter and waste**: The survey highlights the need to improve and enlarge the measurements, establish goals and targets to minimize avoid litter and waste in the SCCs. The EAC model of high-profile communication and engagement with staff and students on corporate waste handling services generates a greater awareness and response towards recycling programs.

- **Energy**: There is no doubt that the EAC models for energy usage on campus have evolved out of a need to reduce costs and emissions from fossil fuel generated power. Conversely, the Brazilian national power generation has, already, a large contribution from renewable sources making this a near sustainable source of energy. Nevertheless, cost remains a principal driver to applying energy saving measures for the SCCs.
- **Water conservation and protection**: Concern about water usage is common in all the institutions but it is not clear that the SCC campuses uses any systematic approach to measurement of this and establishing goals and targets in order to minimize water consumption or avoiding waste.
- **Travel and transport**: Unlike the EACs, the SCCs do not appear to address this theme by, for example, developing policies in partnership with local authorities and private transport services to minimize journeys and encourage ecofriendly modes of transport with less pollution in the movement of people and goods to and from their campuses.
- **Biodiversity**: The SCCs efforts appear to exceed those of the EACs in this theme through initiatives to protect natural systems and direct interventions to control invasive species.
- **Living Labs**: SCCs are highly active in promoting a culture of innovation, entrepreneurship, intellectual property, technology and knowledge transfer in their internal and external communities. An approach to share user-centered, co-creation methodology happens through innovation, research and programs developed with companies, research centers in Brazil and abroad.
- **Economic and socioenvironmental sustainability for society initiatives**: All institutions have research groups dedicated specifically to sustainability and community engagement services. EACs research efforts tend to bias towards innovation, environmental management and sustainable development, SCCs appear to place more emphasis on social inclusion, health, education, art, culture, sport, science and the promotion of citizenship.

This study provides only the initial framework for much benchmarking activity. Nevertheless, this simple attempt at benchmarking HEIs in Santa Catarina with respect to their sustainability has revealed some significant differences in the performance between the institutions and with the activities of like institutions in Europe and North America. It is not clear why some of these differences exist, but more significant information is yet to be gained from understanding and refining the process of benchmarking. More importantly, it lays the groundwork for the exchange and evaluation of ideas between HEIs for the development of greener campuses in general.

Acknowledgements This study integrates the project BRIDGE - Building Resilience in a Dynamic Global Economy: Complexity across scales in the Brazilian Food-Water-Energy Nexus. BRIDGE is under the responsibility of the Centre for Sustainable Development (Greens) of the University of Southern Santa Catarina (Unisul). The following institutions are responsible for BRIDGE's funding:

• Newton Fund.

• Fundação de Amparo à Pesquisa e Inovação do Estado de Santa Catarina (FAPESC)—the Santa Catarina State Research and Innovation Support Foundation.

• *Coordenação de Aperfeiçoamento de Pessoal de Nível Superior (CAPES)*—the Brazilian Institution for the Coordination for the Improvement of Higher Education Personnel.
• *Conselho Nacional de Desenvolvimento Científico e Tecnológico (CNPq)*—the Brazilian National Council for Scientific and Technological Development.
• The Research Councils United Kingdom (RCUK).

References

Colobrans J (2019) MINDb4ACT: living lab guide: guidelines. Available at https://mindb4act.eu/wp-content/uploads/2019/03/Living-Lab-Guide_web.pdf

Green-Campus Programme (2018) The green campus guidebook, an Taisce, green campus office. Smarter sustainable campus communities: a guide for campuses embarking on the green-campus programme. Available at https://www.greencampusireland.org/wp-content/uploads/2018/11/Green-Campus-Guidebook-2018-2019.pdf

Leal Filho W, Shiel C, do Paço A, Brandli L (2015) Putting sustainable development in practice. In: Sustainability in higher education, pp 1–19). https://doi.org/https://doi.org/10.1016/B978-0-08-100367-1.00001-9

Malmberg K, Vaittinen I, Evans P, Schuurman D, Ståhlbröst A, Vervoort K (2017) Living lab methodology handbook. Available at https://u4iot.eu/pdf/U4IoT_LivingLabMethodology_Handbook.pdf

Mayle D, Hinton M, Francis G, Holloway J (2002) What really goes on in the name of benchmarking? In: Neely A (ed) Business performance measurement. Cambridge University Press, pp 211–224

Ridhosari B, Rahman A (2020) Carbon footprint assessment at Universitas Pertamina from the scope of electricity, transportation, and waste generation: toward a green campus and promotion of environmental sustainability. J Clean Prod 246:119172. https://doi.org/10.1016/j.jclepro.2019.119172

Tezel E, Ugural M, Giritli H (2018) Towards green campuses: student's perceptions and expectations. In: 5th international project and construction management conference, pp 859–866

World Economic Forum (2020) The global risks report 2020, insight report, 15th edn. Available at https://www3.weforum.org/docs/WEF_Global_Risk_Report_2020.pdf

What About the Human Side of Climate-Induced Migration? The Role of Universities in Improving Refugees' Wellbeing by Implementing the Sustainable Development Goals

Issa Ibrahim Berchin, João Marcelo Pereira Ribeiro, Brenda Caroline Geraldo Castro, Jonas Moraes-Filho, and José Baltazar Salgueirinho Osório de Andrade Guerra

1 Introduction

The world is facing large scale involuntary migration as never before; however, these refugees are treated as numbers, not as human beings. According to the International Organization for Migration (IOM 2018), in the past few years, international migration corresponded to more than 250 million people annually, from which 25.4 million were conventional refugees (contemplated by international law) and 18.8 million displaced by sudden-onset environmental disasters. Refugees frequently migrate to cities, being in less developed zones, or refugee camps. In both situations, migrants are exposed to countless threats like zoonotic diseases, diseases caused by poor sanitation, violence, hunger and malnutrition.

I. I. Berchin (✉) · J. Moraes-Filho
Faculdade ANCLIVEPA, São Paulo, Brazil
e-mail: issa.berchin@faculdadeanclivepa.edu.br

J. Moraes-Filho
e-mail: jonasfilho@faculdadeanclivepa.edu.br

I. I. Berchin · J. M. P. Ribeiro · B. C. G. Castro · J. B. S. O. de Andrade Guerra
Centre for Sustainable Development (Greens), from the Universidade do Sul de Santa Catarina (Unisul), Florianopolis, Brazil
e-mail: brenda_carolyne@hotmail.com

J. B. S. O. de Andrade Guerra
e-mail: baltazar.guerra@unisul.br

J. B. S. O. de Andrade Guerra
Cambridge Centre for Environment, Energy and Natural Resource Governance (C-EENRG), from the University of Cambridge, Cambridge, UK

W. Leal Filho et al. (eds.), *Universities, Sustainability and Society: Supporting the Implementation of the Sustainable Development Goals*, World Sustainability Series,
https://doi.org/10.1007/978-3-030-63399-8_11

"Already, one-third of the world's urban population, about 1 billion people, live in slums: in poor quality housing with limited clean water, sanitation and education services. By 2030 it is estimated that this number will rise to 1.7 billion people. High population densities and high contact rates help to spread disease, while health and education services are often inadequate. In India, for example, unplanned urbanization has been associated with the spread of dengue fever" (IOM 2008, p. 32).

The United Nations recognizes migration as a means for many people in vulnerable situations to find better living conditions, and to seek job opportunities, study or escape conflicts that put their lives at risk, such as terrorism and the effects relating to climate change and natural disasters. Migration is a major challenge in the twenty-first century, threatening sustainable development by exposing millions of people.

In 2015, the United Nations implemented the Agenda 2030 for the 17 Sustainable Development Goals (SDGs) and its 169 targets. The SDGs represent the continuation and improvement of the Millennium Development Goals (MDGs), aiming to promote a holistic and more comprehensive approach for sustainable development, which establishes indicators and targets that can be measured and implemented with greater cooperation between organizations and countries around the world (UN 2020b).

To achieve its goals the world community must overcome the obstacles to the education of all sorts of migrants, including refugees (UNESCO 2016). Within the reach of SDG 4, regarding quality education, immigrants have access to better opportunities and to achieve other goals such as SDG 1 (No poverty), SDG2 (Zero Hunger), SDG 3 (Good Health and Well Being). It also contributes to the reduction of gender inequalities established in SDG 5 and to promote climate change strategies proposed in SDG 13 (Nicolai et al. 2016). The SDGs address targets and indicators for migration (e.g. SDG 10.7) regarding the goal to reduce inequalities to "facilitate orderly, safe, regular and responsible migration and mobility of people, including through the implementation of planned and well-managed migration policies" (UN 2020a).

"Sustainable development is a development that meets the needs of the present without compromising the ability of future generations to meet their own needs" (United Nations General Assembly 1987). This concept is strongly linked to the quality of life of societies in the face of major changes that impact people's daily lives. The literature on sustainability still has a major focus on environmental studies, while the social and economic dimensions are rarely discussed (Lozano 2008).

By ignoring the human side of migration, governments and civil society are condemning entire generations of people to a life of misery, segregation, emotional imbalances and lack of possibilities for the future (Crea 2016). This perverse side of migration is particularly concerning when about half of any refugee population consists of children. Which requires additional efforts from the refuge country and the international community to promote an educational experience of quality for all, including the challenges imposed by language barriers, social wellbeing, housing and post-trauma (Hirano 2014).

Universities are important to raise awareness about the difficulties and needs of refugees, being knowledge-disseminating institutions and key actors in the relationship between society, organizations and governments. Universities also promote technical aid and resources that help refugees' reception and contribute to their integration in the host countries. This assistance ranges from the availability of housing, food, access to health and education (Leal Filho 2011).

Refugees' educational challenges are particularly difficult in higher education, which is the educational level with least adherence among refugees due to many challenges, i.e. academic literacy, language barriers and the need to work on jobs that require low qualification (Crea 2016). The number of studies related to the role of universities to promote refugees' wellbeing is still insufficient. Therefore, it is important to discuss their realities and challenges, aiming to find and expand the actions that contribute to improving their quality of life (Nicolai et al. 2016). Accordingly, this study aims to contribute to the literature by analyzing how universities can contribute to improving refugees' wellbeing by implementing the Sustainable Development Goals. To reach this goal, a qualitative approach was used to collect and analyze the necessary data, followed by interviews with seven university managers in Brazil.

2 Literature Review

2.1 Climate Change-Induced Migration, the Climate Refugees

Since the 1951 Convention Relating to the Status of Refugees, Refugees are defined by the United Nations as "any person who owing to a well-founded fear of being persecuted for reasons of race, religion, nationality, membership of a particular social group or political opinion, is outside the country of his nationality and is unable or, owing to such fear, is unwilling to avail himself of the protection of that country; or who, not having a nationality and being outside the country of his former habitual residence as a result of such events, is unable or, owing to such fear, is unwilling to return to it" (UNHCR 1951).

Despite its fundamental importance to ensure the protection of refugees, the 1951 convention does not cover a different category of forced migrants, failing to recognize and protect the climate refugees (Eckersley 2015; McNamara and Gibson 2009). There is no common agreement on the concept of climate refugees; however, a comprehensive definition was proposed in 2017, defining climate change refugees as "any person who has been forced to leave their home, or their country, due to the effects of severe climate events, being forced to rebuild their lives in other places, despite the conditions to which they are subjected" (Berchin et al. 2017).

As a common ground, refugees are international migrants forced to flee their home countries seeking protection or better living conditions. Refugees migrate due to violence, persecution of many kinds, social and political unrest, famine, poverty, war, natural disasters (Loescher 1996). Despite its increasing numbers, climate change refugees are not protected by international policies due to the lack of policies covering their category (Hodgkinson et al. 2009). However, its importance is represented by the rising number of people internally and internationally displaced due to the irreversible effects of climate change (e.g. extreme weather events, desertification, sea-level rise, salinization of the soil and groundwater) (Berchin et al. 2017).

Forced migration is an emergency action, which is not planned by those fleeing their homeland. Therefore, climate change refugees are more susceptible to suffer financial restraints and social exclusion in their destinations, this process condemns entire generations to social vulnerability and inhibits the promotion of the Sustainable Development Goals (SDGs).

Climate change is increasing the frequency and intensity of environmental phenomena that induce human migration (Warner et al. 2009), which increase local disputes over natural resources (Mazo 2010; Matthew et al. 2009; Welzer 2010) and improve the conditions to the proliferation of infectious diseases, particularly those related to warm weather (Moran 2011).

2.2 Universities for the Sustainable Development Goals: Assisting the Climate Refugees

Universities are secular institutions with a key role as social organizations (de Andrade et al. 2018). They influence economic, political, social, technological and environmental elements of our societies, preparing future leaders to develop new methods to meet societal demands (Aktas 2015).

Universities have the intrinsic mission to endorse democratic processes throughout society and inside its campuses (de Andrade et al. 2018). According to the United Nations Economic Commission for Europe (UNECE 2009), besides being a human right, education "is a prerequisite for achieving sustainable development and an essential tool for good governance, informed decision-making and the promotion of democracy".

For decades universities advocate for a more active role of education on sustainability and climate change; e.g. the Stockholm Declaration in 1972, the Belgrade Charter in 1975, the Tbilisi Declaration in 1977, the Talloires Declaration in 1990, the Agenda 21 in 1992, the Millennium Development Goals (MDGs) in 2000, the Rio +20 in 2012, and the SDGs in 2015 (Lozano et al. 2013; Berchin et al. 2018).

Large-scale climate change migration compromises the MDGs and SDGs by affecting all their goals but, particularly, it interrupts the provision of education and health services (IOM 2008). "Forced migration hinders development in at least four ways; by increasing pressure on urban infrastructure and services, by undermining economic growth, by increasing the risk of conflict and by leading to worse health, educational and social indicators among migrants themselves" (IOM 2008, p. 32).

The International Organization for Migration (IOM) stated that climate change-related events, particularly climate change-forced migration, threatened the achievement of the MDGs and, consequently, the SDGs as well (IOM 2008). The world is facing unprecedented large-scale migration due to socioeconomic, political and environmental challenges, forcing people to flee their homeland; and failure to adapt to climate change is inducing even more displacement and migration due to severe droughts, floods, sea-level rise, salinization and natural catastrophes (Berchin et al. 2017).

From the 8 MDGs to the 17 SDGs, education, health, food security and the environment are at the core of discussions. Both MDGs and SDGs support stronger multi-stakeholder partnerships to promote sustainable development, and universities should be at the core of these arrangements, leading by example. Particularly, SDG 10 "Reduced Inequality", aiming to "facilitate orderly, safe, regular and responsible migration and mobility of people, including through the implementation of planned and well-managed migration policies"; and SDG 4 "Quality Education", aiming to create an "immigration status parity index for achieving at least a fixed level of proficiency in functional skills, by numeracy/literacy skills".

Universities have a unique capacity to innovate and to promote social inclusion at the same time, developing creative environments to induce cognitive learning by exposing students to a daily routine of good practices. Innovative universities, through its campuses, are frequently referred by the literature as a living laboratory (Evans et al. 2015; Berchin et al. 2020; Crea 2016).

There are many cases of universities supporting the inclusion of refugees in their activities, promoting education, cultural activities and social inclusion—e.g. the University of South Santa Catarina in Brazil, which has a program to "welcome migrants", providing capacity building and education, and promoting their inclusion with the students and local community (Unisul 2016); among many other universities that have committed to support refugee students (https://refugeeswelcomemap.eua.be/Editor/Visualizer/Index/34).

Students' engagement in universities' initiatives is key for integrating refugees and making them feel welcomed. Guilford College (USA) relies on its students to host many refugee families and in enrolling in activities on campus, such as developing artwork and podcasts that discuss the challenges experienced by immigrants and refugees. Guilford College started the "Every Campus a Refuge" (ECAR) initiative in 2015, for universities and colleges to invest their resources with municipal agencies to guarantee housing, food and other assistance needed by refugees until they can become independent (Guilford College 2017).

Language is an important barrier for refugees to adapt to their country of destination. Therefore, understanding the language spoken in this country is crucial to refugee's integration in the new community and their access to job opportunities that can contribute to their quality of life. In this regard, the Federal University of Santa Catarina (Brazil) created the PLAM project, which teaches the Portuguese language to immigrants and refugees with the help of volunteer professors and academics from the Language Department (UFSC 2020). As discussed in SDG 4, quality education offers greater chances of finding jobs and it also helps to reduce poverty (SDG 1) and hunger (SDG 2) whilst improving the families' well-being (SDG 3) (UN 2020a). However, prejudice and xenophobia are still a big challenge.

As another example, the Pontifical Catholic University of Paraná (Brazil) integrates the Scholars at Risk network that unites academic institutions aiming to support and protect the human rights of academics in a situation of vulnerability due to threats against academic freedom. The network operates in 40 countries with the participation of nearly 400 higher education institutions, which grant academic refugees temporary positions in the signatory institutions. In addition to encouraging researches, that promotes freedom of expression worldwide, the Scholars at Risk network also helps to hold seminars and workshops that integrate refugees into the local communities (Scholars at Risk Network 2019).

Local and international networks are great allies in assisting Universities in the design of projects and fundraising that help refugees in their transition within hosting countries. Europe has several partnerships due to the large flow of immigrants and refugees it receives annually, one of them is the inHERE project, in partnership with United Nations High Commissioner for Refugees, the University of Barcelona (Spain) and the Sapienza University (Italy). The project aims to share knowledge and offer training to academic and administrative staff within Universities, ranging from legal knowledge to administration courses. It does so for them to be able to understand the context experienced by refugees and assist in their economic and social integration. It also strengthens the integration capabilities of their institutions and allows them to serve as a model for other universities (inHere Project 2020).

In Austria, Universities joined forces in 2015 to develop the MORE Project, which is currently applied in more than 21 universities across the country. They provide housing, food and medical assistance to refugees, as many are injured or ill after arriving at the chosen location of refuge. According to SDG 3, this concern for refugees' access to health and well-being is essential for them to establish themselves in the hosting community (UN, 2020a). Refugees are also allowed to attend classes to expand their knowledge of the country's language and other areas of interest, without the need to be registered as university students or pay tuition fees (Uniko, 2020). The actions and efforts of universities' initiatives strengthen the importance of SDG 17, which discusses partnerships to meet the goals, which accelerate their achievement.

3 Methods

This study uses a qualitative approach to collect and analyze the necessary data to achieve its goal, "to analyze how universities can contribute to improving refugees' wellbeing by implementing the Sustainable Development Goals".

This study used the semi-structured interview method to collect the data. In order to be representative, seven university managing directors were interviewed, from Private, Public and Community universities in Sao Paulo (4), Florianopolis (2) and Sao Bernardo do Campo (1), Brazil.

The open-ended questions of the questionnaire (Table 1) were based on the literature, aiming to understand how universities can contribute to improving refugees'

Table 1 Interview questionnaire

1—What does sustainable development mean to you?
2—Are you aware of the Sustainable Development Goals of the United Nations?
3—In your opinion as a university manager, do universities have a responsibility for sustainable development?
4—How is your university promoting sustainable development? What are the key-programs and actions towards sustainability developed by your university?
5—Are you aware of the current refugee crisis? If yes, what is your opinion on how this issue is being addressed by the international community?
6—Have you heard about climate change refugees? Can you define them?
7—Do you think universities can support those refugees? If yes, how?
8—Does your university contribute to support refugees? If yes, how? If not, what could be done?

wellbeing by implementing the Sustainable Development Goals (Berchin et al. 2018; de Andrade et al. 2018).

To analyze the data from the semi-structured interviews, we followed the indications of Bardin (2011), which are pre-analysis, when the interviews are transcribed and reviewed; material exploration, when an in-depth analysis of the collected material takes place; and subjective interpretation of the research outcomes, through the analysis and comparison among the propositions stated on the current scientific literature.

4 Results and Discussions: The Contributions of Universities to Implement the Sustainable Development Goals for Climate Refugees

As living laboratories, universities have the knowledge and the means to promote effective initiatives for sustainable development. These initiatives include social and economic wellbeing, environmental preservation, technology and infrastructure innovations.

Questionnaire 1st Question: What Does Sustainable Development Mean to You?

interviewees understand sustainable development as "an ethical duty of all", which promotes "economic development, whilst being inclusive and environmentally responsible". According to another interviewee, sustainable development integrates "environmental preservation and economic growth, ensuring social wellbeing and the longevity of human beings, whilst respecting the planetary boundaries".

Questionnaire 2nd Question: Are You Aware of the Sustainable Development Goals of the United Nations?

Three out of the seven interviewees considered that they were deeply aware of the SDGs and four of them had some knowledge of the SDGs. Universities have a key role to create and disseminate knowledge, using sustainable practices (Leal Filho 2011). However, the number of university managers holding a deep knowledge of the SDGs is insufficient and warns of the need for further training of academic staff. The world community is increasing the pressure for leaders to make responsible decision-making towards sustainable development. Therefore, university managers have great knowledge at their disposal (i.e. specialists that integrate the academia) and the tools (i.e. students, alumni, the campus) to promote sustainable development paths in their surrounding communities.

Questionnaire 3rd Question: In Your Opinion as a University Manager, Do Universities Have a Responsibility for Sustainable Development?

Although the seven interviewees confirmed that universities have a responsibility to promote sustainable development, two of them understand that their institutions are not doing enough to achieve the SDGs. One of the interviewees indicated that among the greatest challenges for the promotion of sustainable development in universities is the administration; management systems used by these institutions are usually outdated and difficult to change.

Transforming organizational systems creates barriers due to the employees' resistance to change (Leal Filho et al. 2017). This is often endorsed by the lack of commitment or the lack of awareness of the employees to matters related to sustainability and the environment. The fact is that these barriers have a negative impact on the effectiveness of universities' sustainable practices, such as the integration of immigrants in the universities' academic and social routines. The interviewee also stated that there is more economic pressure than scientific awareness available to support decision-making to implement sustainability in organizations.

Another interviewee argued that universities are key social agents that hold the responsibility to promote sustainability and awareness throughout society; and to improve their contribution to sustainable development these institutions should implement education for sustainable development in their courses, using the knowledge and technologies at their disposal to promote local development and environmental quality.

Questionnaire 4th Question: How is Your University Promoting Sustainable Development? What Are the Key-Programs and Actions Towards Sustainability Developed by Your University?

One of the interviewees argued that universities should promote the SDGs "through subjects that discuss environmental education and propose alternative technologies of development that respect the environment". Also "promoting projects to engage students with the local community through outreach and initiatives on campus, providing the students with critical knowledge and awareness of real-life challenges".

The sustainability challenges are not only technical but ethical and involve the development of individual and collective awareness of society at the global level. There is an urgent need for a change in the economic organization of institutions, but also in the personal habits of individuals (Lozano 2013; Wright and Wilton 2014).

According to one of the interviewees, "students are strategic to the promotion of sustainable initiatives, as they will be the mentors and leaders of the future". However, the lack of students' engagement in the universities' processes can be linked to several challenges, such as distance between the university and the student, lack of interest in several campus projects, low investments in communication platforms and little control over the performance of the initiatives (Ribeiro et al. 2018). The literature on universities and sustainability indicate that students often neglect the social aspect of sustainability, for they are taught to consider the environmental aspect as the major concern. Therefore, "university campuses must engage students to promote social activities", including the integration and capacity building of refugees.

Among the initiatives developed by the universities managed by the interviewees, two institutions have structured specific programs for the SDGs. One of them created a holistic program developed by an innovation center, which encompasses Zero Hunger (SDG 2), Clean Water and Sanitation (SDG 6), Affordable and Clean Energy (7), Sustainable Cities and Communities (SDG 11), Responsible Consumption and Production (SDG 12); and the other maintains a holistic environmental education program based on the SDGs, which includes education, research, outreach, campus operations and management.

Other institutions promote general initiatives for energy efficiency, waste management, social responsibility and environmental awareness. One of the interviewees also encouraged the "participation of universities and their research centers on international networks", which contributes to "share experiences and gather information and resources to promote the SDGs locally". Finally, knowing that more than 60% of human diseases are zoonotic, one of the institutions also maintain public veterinary hospitals that offer "nearly a thousand free veterinary services and procedures for the local communities, particularly to those in a vulnerable situation", which helps to promote health, wellbeing and education; the interviewee also informed that the institution applied to the SDGs Knowledge Platform.

Questionnaire 5th Question: Are You Aware of the Current Refugee Crisis? If Yes, What is Your Opinion on How This Issue is Being Addressed by the International Community?

Universities, as the main promoters of sustainable development, need to strengthen their engagement with social causes, including refugees (Crea 2016). All the interviewees were aware of the current refugee crisis and agreed that the international community is not doing enough to deal with the causes of the problem, nor with the challenges faced by the migrants.

Two of the interviewees stated that the international community is neglecting the refugee crisis and are failing to encourage countries to implement efficient pubic policies and aid to refugees worldwide; also failing to resolve the causes of forced migration. Two interviewees did not feel prepared to talk about the actions of the international community on the subject.

Questionnaire 6th Question: Have You Heard About Climate Change Refugees? Can You Define Them?
When it comes to climate change refugees, an analysis of the interviewees' speech indicates that anthropogenic climate change is a pressing challenge for humanity and forces people to migrate seeking better living conditions in other territories. However, one of the interviewees argued that, although of critical importance, climate change "is not the main concern for the current global refugee crisis".

One of the interviewees argued that climate change is a pressing issue for every country and territory, which has the most negative effects on vulnerable agricultural communities "forcing people to migrate for better living conditions, in which they can produce food, access quality water resources and raise their families". Other interviewees understand that human activities (environmental disasters) also force people to flee their homes.

Questionnaire 7th Question: Do You Think Universities Can Support Those Refugees? If Yes, How?
All interviewees agreed that universities can support refugees in their adaptation to the country of destination. Among the suggestions of the interviewees to improve the lives of the climate refugees in the host country, are: (1) the organization of events to promote awareness regarding the challenges faced by these migrants, (2) promoting more research and outreach to spread the importance of the topic, (3) to offer scholarships to support the education of the refugees and integrate them in the market, and (4) to give job opportunities for their inclusion. Thus, one of the interviewees argued that universities must "provide an environment to recognize the refugees as citizens and provide study opportunities for them. Even with basic subjects like the local language. Also, encouraging student groups that can work as social support".

Questionnaire 8th Question: Does Your University Contribute to Support Refugees? If Yes, How? if not, What Could Be Done?
Only one out of the seven universities has an established program to support climate change refugees, which promotes social activities on the campus to engage the students with the refugees. The main goal of this university from southern Brazil is to promote social inclusion and cultural diversity, by assisting the process of social inclusion of international migrants and refugees from the State of Santa Catarina, South Brazil. Also, promoting language classes and technical support to assist refugees. Figure 1 illustrates the proposed initiatives for the university to promote refugees' wellbeing.

Although the other six universities do not have a specific program for refugees' wellbeing, they have other social support programs that can be used by them. Among the social programs suggested by the managers in the questionnaire, were: (1) a program of scholarship for people in need; (2) the development of researches to

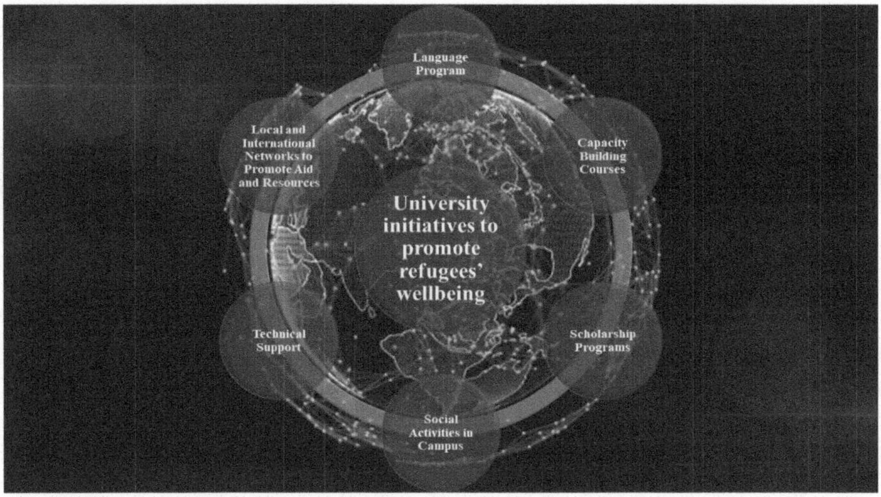

Fig. 1 University initiatives to promote refugees' wellbeing

investigate the main causes of human migration and the challenges faced in the process of migration and in the country of destination, which could generate relevant data to support policymakers through a science-policy dialogue; (3) addressing the theme of climate change refugees and their challenges in classes, to raise awareness of local students on the subject; (4) to create job opportunities for refugees in university campuses; (5) to create training programs to capacitate refugees for the market; (6) to promote language programs to teach refugees the local language.

5 Conclusions and Recommendations

Universities have the knowledge and the resources to promote education and capacity building, to teach the local language, and technically and culturally prepare these refugees to fit in the host country. However, there are many challenges faced by universities when trying to engage refugees in higher education, including fragile educational literacy, difficulties to communicate in the refuge country's language (both oral and written), social and economic exclusion, including in job opportunities.

Although there is a lack of literature exploring the role of universities to support refugees, it is possible to draw some recommendations based on the existing literature on the social role of universities and the interviewees' answers. Universities can contribute to improving refugees' wellbeing by:

- Developing a language program to teach refugees;
- Promoting methods to facilitate the inclusion of refugees in higher education;

- Developing a vocational training program to rehabilitate refugees to the labor market;
- Promoting social activities on the campus to improve the social life of refugees and engage students with extracurricular activities;
- Offering technical support to assist refugees in the bureaucratic processes;
- Encouraging Local and International Networks to Promote Aid and Resources.

Universities must encourage a deeper engagement of students in initiatives to support refugees. These institutions must raise critical thinking and awareness among students, supporting transdisciplinary learning that emancipates them to use their knowledge background to promote sustainable development. Besides the integration of refugees in universities, courses and activities, these institutions must provide opportunities for refugees to integrate the job market.

Future studies could explore more cases of university programs on refugees, investigating the structure of these programs to fully understand the potential of universities to promote the development of refugees' capacities and inclusion in refuge countries.

Acknowledgements This study was conducted by the Centre for Sustainable Development (Greens), from the University of Southern Santa Catarina (Unisul), in the context of the project BRIDGE—Building Resilience in a Dynamic Global Economy: Complexity across scales in the Brazilian Food-Water-Energy Nexus; funded by the Newton Fund, Fundação de Amparo à Pesquisa e Inovação do Estado de Santa Catarina (FAPESC), Coordenação de Aperfeiçoamento de Pessoal de Nível superior (CAPES), National Council for Scientific and Technological Development (CNPq) and the Research Councils United Kingdom (RCUK).

References

Aktas CB (2015) Reflections on interdisciplinary sustainability research with undergraduate students. Int J Sustain Higher Educ 16:354–366. http://dx.doi.org/https://doi.org/10.1108/IJSHE-11-2013-0153

Bardin L (2011) Content analysis. São Paulo: Edições 70:279. ISBN: 978-8562938047

Berchin II, Valduga IB, Garcia J, de Andrade JBSO (2017) Climate change and forced migrations: an effort towards recognizing climate refugees. Geoforum 84:147–150. https://doi.org/10.1016/j.geoforum.2017.06.022

Berchin II, Sima M, de Lima MA, Biesel S, dos Santos LP, Ferreira RV, de Andrade JBSO, Ceci F (2018) The importance of international conferences on sustainable development as higher education institutions' strategies to promote sustainability: a case study in Brazil. J Clean Prod 171:756–772. https://doi.org/10.1016/j.jclepro.2017.10.042

Berchin II, de Amorim WS, Valduga IB, Heerdt ML, de Andrade JBSO (2020) Sustainable campuses as living labs for sustainable development: an overview of a Brazilian Community University. In: Universities as living labs for sustainable development. Springer, Cham, pp 87–102. https://doi.org/https://doi.org/10.1007/978-3-030-15604-6_6

Crea TM (2016) Refugee higher education: contextual challenges and implications for program design, delivery, and accompaniment. Int J Educ Develop 46:12–22. https://doi.org/10.1016/j.ijedudev.2015.11.005

de Andrade JBSO, Garcia J, de Andrade Lima M, Barbosa SB, Heerdt ML, Berchin II (2018) A proposal of a balanced scorecard for an environmental education program at universities. J Clean Prod 172:1674–1690. https://doi.org/10.1016/j.jclepro.2016.11.179
Eckersley R (2015) The common but differentiated responsibilities of states to assist and receive 'climate refugees'. Euro J Polit Theor 14(4):481–500. http://dx.doi.org/https://doi.org/10.1177/1474885115584830
Evans J, Jones R, Karvonen A, Millard L, Wendler J (2015) Living labs and co-production: university campuses as platforms for sustainability science. Curr Opin Environ Sustain 16:1–6. https://doi.org/10.1016/j.cosust.2015.06.005
Guilford College (2017) Every campus a refuge: a PPSE (principled problem solving experience) minor. https://www.guilford.edu/sites/default/files/2017-06/EveryCampusARefugeAPPSEPrincipledProblemSolvingExperienceMinor.pdf. Last accessed 27 Apr 2020
Hirano E (2014) Refugees in first-year college: academic writing challenges and resources. J Second Lang Writ 23:37–52. https://doi.org/10.1016/j.jslw.2014.01.001
Hodgkinson D, Burton T, Coram A, Dawkins S, Young L (2009) Towards a convention for persons displaced by climate change: key issues and preliminary responses. IOP Conf Ser Earth Environ Sci 6(56):562014. https://iopscience.iop.org/article/https://doi.org/10.1088/1755-1307/6/56/562014/pdf. Last accessed 12 Mar 2020
InHere (2020) About project https://www.inhereproject.eu/project/about. Last accessed 27 Apr 2020
IOM. International Organization for Migration (2008) Migration and climate change. IOM Geneva, pp 56. https://www.iom.cz/files/Migration_and_Climate_Change_-_IOM_Migration_Research_Series_No_31.pdf. Last accessed 12 Mar 2020
IOM. International Organization for Migration (2018) Global migration indicators 2018. IOM Berlin, pp 64. https://publications.iom.int/system/files/pdf/global_migration_indicators_2018.pdf. Last accessed 12 Mar 2020
Leal Filho W (2011) About the role of universities and their contribution to sustainable development. Higher Educ Policy 24(4):427–438. https://doi.org/10.1057/hep.2011.16
Leal Filho W, Wu YCJ, Brandli LL, Avila LV, Azeiteiro UM, Caeiro S, Madruga LRDRG (2017) Identifying and overcoming obstacles to the implementation of sustainable development at universities. J Integr Environ Sci 14(1):93–108. https://doi.org/10.1080/1943815X.2017.1362007
Loescher G (1996) Beyond charity: international cooperation and the global refugee crisis: a twentieth century fund book. Oxford University Press. ISBN: 978-0195356076
Lozano R (2008) Envisioning sustainability three-dimensionally. J Clean Prod 16(17):1838–1846. https://doi.org/10.1016/j.jclepro.2008.02.008
Lozano R, Lukman R, Lozano FJ, Huisingh D, Lambrechts W (2013) Declarations for sustainability in higher education: becoming better leaders, through addressing the university system. J Clean Prod 48:10–19. https://doi.org/10.1016/j.jclepro.2011.10.006
Matthew R, McDonald BL, Barnett J, O'Brien K (2009) Global environmental change and human security. MIT Press, Cambridge, pp 35–76. ISBN: 978-0262513081
Mazo J (2010) Climate conflict: how global warming threatens security and what to do about it. Routledge. ISBN: 978-0415591188
McNamara KE, Gibson C (2009) We do not want to leave our land': Pacific ambassadors at the United Nations resist the category of 'climate refugees. Geoforum 40(3):475–483. https://doi.org/10.1016/j.geoforum.2009.03.006
Moran EF (2011) Meio Ambiente e Ciências Sociais: interações homem-ambiente e sustentabilidade. São Paulo: SENAC, 2011. ISBN: 978-8539600908
Nicolai S, Wales J, Aiazzi E (2016) Education, migration and the 2030 agenda for sustainable development. https://doc.rero.ch/record/308985/files/28-6._ODI_Education.pdf. Last accessed 12 Mar 2020
Scholars at Risk Network (2019) Membership directory. https://www.scholarsatrisk.org/membership-directory/. Last accessed 15 Mar 2020
UFSC (2020) Projeto PLAM. https://projetoplam.paginas.ufsc.br/. Last accessed 15 Mar 2020

UNECE. United Nations Economic Commission for Europe (2009) Learning from each other: the UNECE strategy for education for sustainable development. United Nations New York and Geneva, pp 161. https://www.unece.org/fileadmin/DAM/env/esd/01_Typo3site/LearningFromEachOther.pdf. Last accessed 12 Mar 2020

United Nations General Assembly (1987) Report of the world commission on environment and development: our common future. United Nations General Assembly, development and international co-operation: environment. Oslo, Norway. https://sustainabledevelopment.un.org/content/documents/5987our-common-future.pdf. Last accessed 26 Apr 2020

UN. United Nations (2020a) Sustainable development goals. https://www.un.org/sustainabledevelopment/inequality/#tab-cb85489469cb0d48293. Last accessed 12 Mar 2020

UN. United Nations (2020b) Transforming our world: the 2030 agenda for sustainable development. https://sustainabledevelopment.un.org/post2015/transformingourworld. Last accessed 26 Apr 2020

UNHCR. United Nations High Commissioner for Refugees (1951) Convention and protocol relating to the status of refugees. https://www.unhcr.org/3b66c2aa10.html. Last accessed 14 Mar 2020

UNESCO (2016) Global education monitoring report. Education for people and planet: creating sustainable futures for all. https://unesdoc.unesco.org/images/0024/002457/245752e.pdf. Last accessed 14 Mar 2020

Unisul (2016) Project welcoming migrants: social inclusion and cultural diversity. https://www.unisul.br/wps/wcm/connect/700aab92-109f-4bee-a8c7-38a4868cab4c/projeto-acolhida-ao-migrante-extensao-2016-tb.pdf?mod=ajperes. Last accessed 12 Mar 2020

Uniko (2020) MORE initiative https://uniko.ac.at/themen/more/index.php?lang=EN. Last accessed 27 Apr 2020

Warner K, Ehrhart C, Sherbinin AD, Adamo S, Chai-Onn T (2009) In search of shelter: mapping the effects of climate change on human migration and displacement. Columbia University. UNHCR, 2009. https://www.ciesin.columbia.edu/documents/clim-migr-report-june09_media.pdf. Last accessed 14 Mar 2020

Welzer H (2010) Guerras climáticas: por que mataremos e seremos mortos no século 21. São Paulo: Geração Editorial, 2010. ISBN: 978-8561501419

Issa Ibrahim Berchin Professor at the Graduate Program of Innovation and Sustainability Project Management at the Faculdade de Tecnologia Termomecanica. Director of Research and Outreach at Faculdade ANCLIVEPA. Researcher at the Centre for Sustainable Development (GREENS).

João Marcelo Pereira Ribeiro Ph.D. Student in Public Policies and Development at Santa Catarina State University (UDESC), Brazil. Researcher at the Centre for Sustainable Development (GREENS); and Research Group on Public Policies and Development (Callipolis/UDESC)

Brenda Caroline Geraldo Castro Student in International Relations at the Universidade do Sul de Santa Catarina (UNISUL), Brazil. Researcher at the Centre for Sustainable Development (GREENS).

Jonas Moraes Filho Professor of the postgraduate course at Santo Amaro University, São Paulo, SP. Coordinator of Research and Outreach at Faculdade ANCLIVEPA.

José Baltazar Salgueirinho Osório de Andrade Guerra Professor at the Graduate Programmes of Management and Environmental Sciences at the Universidade do Sul de Santa Catarina (UNISUL). Director of the Centre for Sustainable Development (GREENS). Fellow at the Cambridge Centre for Environment, Energy and Natural Resource Governance (CEENRG), Department of Land Economy, University of Cambridge, United Kingdom.

Critical Thinking, Moral Courage, and Sustainable Development Goals. Universities as a Driving Force for Societal Change

Daniela Dumitru

1 Introduction

The main danger for the UN's Sustainable Development Goals (SDGs) is that they become mere rhetoric, another "glass bead game" (Hesse 2002) which will drop in the end, disappearing in the sands of forgetfulness. The SDGs are straightforward elements for change and they can be pursued by everyone, either separately or integrated into more complex sustainability plans. However, they are endangered by ideology and by just merrily talking about them, without taking any concrete action to fulfil them.

This paper argues in favor of critical thinking as creating an environment for the development of the courage that is crucial in attaining the SDGs. Furthermore, the university is the starting place for the construction of the human personality. The university should exit its current state of researching in a void by taking its true role in society: to research for the common good. Hence, the university should reassume its Humboldtian creed and become teleological and moral. The paper will attempt to demonstrate that only by placing sustainability as its moral purpose can the university truly serve society. Moreover, in view of this, courage is always needed to serve the moral good when facing uncertainty.

The paper brings together three concepts rarely put in one place: critical thinking (CT), courage, and sustainable development. This adds to its original approach to the sustainability issue, in the context of encouraging the university to change and to reassume its moral ground in society. This is a rare approach in the literature concerning sustainability. The three concepts are often connected two-by-two: for

D. Dumitru (✉)
Teacher Training Department, Bucharest University of Economic Studies, Piața Romană 6, Bucharest, Romania
e-mail: daniela.dumitru@ase.ro

W. Leal Filho et al. (eds.), *Universities, Sustainability and Society: Supporting the Implementation of the Sustainable Development Goals*, World Sustainability Series,
https://doi.org/10.1007/978-3-030-63399-8_12

example, courage with sustainability, especially with regard to action toward sustainability (Sidibe 2016); critical thinking as a spur for courage, particularly intellectual courage (Paul and Elder 2020), and as a condition for education for sustainable development (Straková and Cimermanová 2018; Concina 2019; Thomas 2009); or CT is simply connected to sustainability (Minott et al. 2019) or directed towards society in the form of critical action (Barnett 1997). However, it is hard to find all three, courage, CT and sustainability, in one place in the literature.

Moreover, for society to benefit from courage, CT, and sustainable behavior, *education* must be involved. And not just any education, but specifically higher education. It may be argued that the beginning of transformation lies in higher education. Higher education should not be allowed to escape its formidable responsibility—to act (morally) toward sustainable development.

It is necessary first to establish the definitions of some concepts that will facilitate the discussion of a complicated topic which may seem somewhat counterintuitive, viz. the relation between critical thinking and courage. After defining the main concepts, the connection between engaging in critical thinking and the development of courage and fortitude will be demonstrated.

What is the link between these concepts and sustainability? World-wide, the notion of sustainability is currently still in its early stages, confronted with unjustified skepticism and abusive refutations. Unverified rejection of scientific evidence creates a dangerous situation should it also be endorsed by political leaders or political–decisional bodies. When governments deny obvious facts and withdraw from organizations fighting against climate change, the result is delay in the implementation of urgent measures to stop climatic deterioration and social turmoil.

Reasoning and critical thinking are conditions for action for sustainability. Higher education should have as its primary goal the development of these skills. Moreover, sustainability itself should be included in all educational programs as a result of the current situation in which everybody's contribution is needed. No higher education institution should evade its responsibility as an initiator of change. For this reason, higher education institutions (HEIs) should also be responsible for the fulfillment of the Sustainability Goals. Goal 4 addresses education directly, through target 4.7: "*by 2030, ensure that all learners acquire the knowledge and skills needed to promote sustainable development.*" The SDGs are connected to the 2030 Agenda (Shulla et al. 2020), which shows continuity and hope, at least at the level of world-wide statements and declarations.

The question is: when does change start? Where does transformation begin? It will be argued here that the process should begin at the top, with higher education, the last link of the education chain, chronologically speaking, but the first in line for change. The most flexible link is HE because it is the most research-oriented sector and it should embed new knowledge in its curricula. Change should start from higher education. Consequently, it is the responsibility of universities to implement the SDGs, ensuring the acquisition of sustainable development knowledge and skills.

Being something very new, which means changing lifestyles and habits, this will generate a lot of resistance. Courage, grit, and determination are essential to start the change and to fulfill the SDGs and the aims stated in the 2030 Agenda for Sustainable Development.

2 Critical Thinking and Courage as Hallmarks of Higher Education[1]

Two concepts must now be addressed, with an explanation of why they are crucial for HE, in the twenty-first century. Critical thinking (CT) is a new concept, which was introduced in the early twentieth century. However, it draws its essence from Ancient Greek philosophy, being traceable back to Socrates' discussions as captured in Plato's *Dialogues* (e.g. *The Republic*, 2007). Aristotle (2014) came to theorize rationality as a critical tool.

In the Middle Ages, in his famous work, *Summa Theologica,* Thomas Aquinas (1981) wanted his reasoning to meet all the requirements of critical thinking, always to be systematically exposed and always to answer criticisms of the ideas expounded.

Some centuries later, in England, we find a number of representatives of empiricism who had the active idea of criticism and reflection in their work. Bacon (2008) was concerned with the way in which our mind gathers information about the world. He said it was important for us to study the world empirically, meaning that epistemology and its outcomes (knowledge) are constrained by the capacity of the cognitive system. This can be considered as the first work about critical thinking, although it is not named like this.

Fifty years later, in France, René Descartes wrote what could be called the second book of critical thinking, a lesser known, incomplete, and posthumously published work, *Regulae ad directione ingenii* (*Rules for the Direction of the Natural Intelligence* 1619–1628, 1998). In this work, Descartes intended to describe thirty-six rules, but he only succeeded in writing twenty-one before his death. By stating the rules of the mind, Descartes demonstrated that it must be systematically disciplined when it reasons about anything.

In the seventeenth century, Hobbes (2019) and Locke (1996) talked about the critical attitude. They did not accept what was considered "normal" in their own time or what was traditionally accepted as being true. Hobbes had a naturalistic vision of the world, in which everything needed to be explained in terms of proofs and reason. Locke supported the application of critical thinking at the level of common

[1]The history of critical thinking presented in this chapter is partially published in my article of 2019, "Creating meaning. The importance of Arts, Humanities and Culture for critical thinking development", *Studies in Higher Education*, 44(5), 870–879, https://doi.org/10.1080/03075079.2019.1586345.

sense and everyday thinking. He laid the foundations of critical thinking applied to fundamental human rights. Thus it was John Locke who brought into discussion and highlighted the necessity of critical thinking first in a democracy and second in social life.

Kant (1999) wrote the *Critique of pure reason*, one of the most significant works in philosophy, in which critical thinking turns to itself, to its own powers and limits.

The twentieth century brought the term "critical thinking," and it was the most prolific in establishing this concept through the works it inspired. Dewey (1909) showed us the instrumental side of critical thinking in relation to the aims, objectives, and targets we plan to reach. Wittgenstein (1998) showed us the need to take into consideration the importance of concepts to human thinking, as well as the need to evaluate the power of thinking and its limits. Piaget (1969) specified the importance of the egocentric and socio-centric tendencies of our thinking, and drew attention to the need to develop reflective thinking, which is capable of judging from the perspective of several points of view, until it reaches a level called "the conscious acknowledgment" of thinking.

Nowadays, we have many works called either *Critical Thinking* or *Argumentation* or *Informal logic*. Our times have added a clarification of the concept and of its utility to the ancient, modern, and enlightenment views presented above. Specifically, critical thinking has become critical toward itself. The *modern concept* of critical thinking was outlined by the philosopher, psychologist, and educator John Dewey. In his book, *How we think*, he suggested the term "reflective thinking" (Dewey 1909). Under this name he groups a number of features that differentiate the superior use of human beings' rational faculty from its minimal and routine functioning (Stoianovici 2005, 123).

Dewey used the term "reflective thinking" to denote "*Active, persistent, and careful consideration of any belief or supposed form of knowledge in the light of the grounds that support it, and the further conclusions to which it tends,*" (Dewey 1909, quoted in Fisher 2001, 2). Subsequent generations of authors, including contemporary ones, whether philosophers, educators, or other researchers interested in CT development, are the heirs to Dewey's work.

The Delphi panelists (Facione 1990) describe CT as a set of skills and dispositions. Analysis, Interpretation, Evaluation, Inference, Explanation, and Self-regulation are the dimensions of CT as a skill set. The Delphi Project experts claim that thinking critically is not equivalent to thinking correctly. The person who builds an argument is imbued with this content because something motivates them to argue one way or another. They have a certain purpose and relate affectively to the content used and also to the position adopted.

This may be considered proof that critical thinking is both a complex of high-level cognitive skills and a complex of affective and motivational dispositions, intrinsically bound to a cognitive and rational cluster. The Delphi panelists call this reality the disposition to think critically, and it has several dimensions: truth-seeking, open-mindedness, analyticity, systematicity, self-confidence, inquisitiveness, and cognitive maturity (Facione 1990).

Paul and Elder (2006), Bailin and Battersby (2016), and Hamby (2014 apud Hitchcock 2018) propose the term "virtues" for these components of critical thinking. They are virtues of character, concerning a person's ways of thinking rather than their ways of behaving. They are thus not moral virtues but intellectual virtues or cognitive values (Kuhn 2008).

Thinking dispositions or intellectual virtues are real properties of thinkers. They are general tendencies, propensities, or inclinations to think in different ways in particular circumstances. Dispositions are personality traits of a complex psychological reality named critical thinking (Dumitru 2012).

Facione (1990) divides the dispositions into two categories: initiating dispositions (those that contribute to starting to think critically about an issue) and internal dispositions (those that contribute causally to doing a good job of thinking critically once one has started) (Facione 1990). The disposition is a person's consistent internal motivation to act toward, or to respond to persons, events, or circumstances in habitual, and yet potentially malleable, ways (Facione 2000).

Although there is some overlap between the two categories, it is among the initiating dispositions that we find courage. Paul and Elder (2006) explain that fear of thinking for oneself can stop one from doing it. Thus, willingness to think critically requires intellectual courage (Paul and Elder 2006).

In the ancient world, courage, moderation, and justice were prime species of moral virtue. A virtue is a settled disposition to act in a certain way (Parry 2014). Aristotle says that courage is the just measure between fear and confidence. So, courage cannot exist in the absence of fear. Courage means that a person should act despite their fear. An excessive disposition to confidence is rashness and an excessive disposition to fear and a deficiency in confidence is cowardice (Aristotle, apud Parry 2014). In the *Nicomachean Ethics*, Aristotle defines fear as the expectation of evil things (III6, 10), so people should avoid encountering these (disgrace, poverty, disease, friendlessness, death). It is actually noble to fear some of them, like disgrace and, especially, death, for which neither good nor bad exists.

Courage involves pain, and is justly praised, since it is more difficult to endure what is painful than to abstain from what is pleasant (Aristotle, III7b). Courage also means knowledge, as the Stoics maintained. It is simply knowledge of what is to be endured: the impulse to endure or not (Parry 2014). It might appear as strange that courage, a virtue separated from reasoning, depends upon knowledge, which is an aspect of reasoning. However, courage is not in fact separated from reasoning at all. Aristotle points out that ignorance seems courageous, but when the person under a delusion finds out that the situation is different from what they imagined, they run (Aristotle III7a, 20). So, a courageous *decision, rationally chosen* (we are getting close to CT) can be made based on foreseen actions.

This is the link between courage and critical thinking (deciding on what to believe or to do). Critical thinking needs all the dispositions, including courage, for its development and purposeful use, and courage needs CT for right action based on truth and morality.

Observe the pictures below.

Which is courage, and which is critical thinking?

Image 1 (Right). The thinkers from Hamangia, Eneolithic culture, Romania. Fifth millennium B.C. *Source* Cristian Chirita—Operă proprie, CC BY-SA 3.0, https://commons.wikimedia.org/w/index.php?curid=27999795. Image 2 (Left). The Tiananmen Square protests, China, June 5, 1989. *Source* Associated Press, originally photographed by Jeff Widener, Fair use, https://en.wikipedia.org/w/index.php?curid=239713.

An answer will be attempted here, but for the reader is encouraged to find their personal response to the question in the light of their own expertise and experience. The first picture shows two statues from the Stone Age (c. 7,000 years old), a man sitting on a little chair, with his chin in his hands, a typical position of somebody thinking, and a woman sitting beside him, in a posture that invites us to imagine them talking, wondering, discussing, etc. They look like people trying to figure out some puzzling dilemma or perplexed by some new events.

The second picture is one of the most iconic images of the twentieth century: the man who briefly stops a column of tanks during the crackdown on the Tiananmen protests on June 5, 1989. He is clearly not there to talk. It is not the place for dialog, a much needed and necessary action of critical thinkers. Is he a critical thinker? Maybe. His motives for standing in front of the tanks could have been determined by reasoning and certitude, but the precise moment captured by the picture is not about critical thinking. Is he courageous by Aristotle' definition? It depends on one's interpretation. Is this man taking the middle way between confidence and fear?

3 The Purpose of the University. Sustainability as the Teleological, Greater Good

What is the connection between critical thinking, courage, and sustainability? The university.

Universities have in their mission statements the values of Wilhelm von Humboldt when he described the university as the "pinnacle" of the nation's "moral culture" (apud Reitter and Wellmon 2020). The mission of the Humboldtian university is to lead people to a higher level of moral consciousness. But Max Weber, in his 1917–1919 lectures on vocation, points out that in the modern world, specialization is a reality of the university, of scholarly work. So, the idealism and the moral ground of the scholar is no longer to be found. In this "disenchanted" world, the purpose of universities is to educate students by pursuing knowledge in an open-ended way. Scholars are to exercise self-restraint and not transmit beliefs and ideals to their students. If they act differently, they will undermine the legitimacy of academic freedom. The purpose of the university is the acquisition and creation of new knowledge without questioning the use and social impact of the knowledge thus generated.

However, in the light of the overwhelming scientific proof that the environment is suffering irreparable damage, it is argued here that the university should put sustainability at the top of its moral culture, and ask itself how, if at all, the knowledge taught by its scholars and their new research findings contribute to sustainability.

Nowadays academia is "disenchanted" from its greater purpose; it has to "let the facts speak for themselves", independent of morals or values. Max Weber (2020) identified the essential values of scholarship, among them being: inclusiveness, intellectual integrity, courage, and a principled commitment to intellectual and value pluralism. For Weber, training in research should include the imparting of these values, which are at the same time both scholarly and moral (Reitter and Wellmon 2020).

The university should have a moral ground for its activity and purpose. This moral infrastructure of academia should be based on sustainability. In the interests of intergenerational justice, it is necessary to nurture and promote programs and research projects committed to sustainability, and which contribute to the fulfillment of the seventeen sustainability Goals.

If universities were to seek to develop critical thinking skills without developing dispositions, the result "would be deficient and perhaps dangerous" (Ennis 1996).

As Reitter and Wellmon (2020) argue, "reason [has] failed to guide us to universal peace and prosperity. To read Weber's vocation lectures today is to be reminded of the moral urgency of sober, unglamorous, disciplined thinking in times of crisis. It is to be reminded, as Weber put it in 'The Politician's Work,' that ethics can be and often is 'used in morally disastrous ways' in both the academic and political spheres, ways that preclude an honest and responsible reckoning with the world in which we find ourselves" (Reitter and Wellmon 2020, p. 288/2312).

The SDGs are far easier to pursue than any other declarations of commitment, signed or assumed publicly, but ignored in actions and policies. Based on the 2030 Agenda, the Goals are concrete pieces of action and they can be embedded in universities' educational programs as learning outcomes and in university presidents' policies in their twofold roles: as leaders of educational institutions and as managers of educational businesses.

We must not forget about this bizarre status of HEIs, as educational agents (with a humanistic orientation, sometimes incompatible with obtaining profit) and as business and market-oriented agents (oriented toward profit). It is time for universities to assume the role of change-makers even though this task means renouncing profit attainment. The first institution that should have courage is the university. To inculcate courage in its students and faculty, the university's leaders must themselves have courage first. And not any type of courage, but the courage driven by knowledge and critical thinking skills and dispositions. The teleological, purposeful university must have sustainability.

4 Conclusions

This essay proposes a view in which the university is the starting point of change and the point zero of the implementation of the SDGs. These are far easier to pursue than any other declarations of commitment. The paper brings together three concepts rarely put in one place: critical thinking (CT), courage, and sustainable development, thus proposing an original approach to the sustainability issue. The university should assume its moral ground, a teleological humanistic role, in which sustainability is the prime principle.

It has been argued that critical thinking nurtures courage, which is necessary for the implementation of the SDGs, institutionally and at personal level, in everyday life. Without a good education for critical thinking, education for sustainability suffers. Without an education for critical thinking, courage will come short. Without a strong university, as a pillar of society, we cannot understand and implement the SDGs. According to Max Weber, academia is "disenchanted" from the greater purpose and has to "let the facts speak for themselves"; it has no morals or values. The university should exit its current state of researching in a void by taking its true place in society: researching for the common good. Thus, the university should reassume its Humboldtian creed and become teleological and moral. What universal good is today more important and urgent than the Sustainable Development Goals?

Bibliography

Aquinas T (1981) Summa theologica, revised. Christian Classics, Allen, TX

Aristotle (2014) Nicomachean ethics (revised edition), vii–xxxv. In: Crisp R (ed) Cambridge University Press, Cambridge

Bailin C, Battersby M (2016) Fostering the virtues of inquiry. Topoi 35(2):367–374. https://doi.org/10.1007/s11245-015-9307-6

Bacon F (2008) The major works. Oxford University Press, Oxford

Barnett R (1997) Higher education: a critical business. Open University Press, Buckingham

Concina E (2019) Critical thinking methods for sustainable development. In: Leal Filho W (eds) Encyclopedia of sustainability in higher education. Springer, Cham. https://doi.org/https://doi.org/10.1007/978-3-319-63951-2_205-1

Descartes R (1619–1628, 1998) Rules for the direction of the natural intelligence/Regulae ad directione ingenii. Rodopi, Amsterdam-Atlanta

Dewey J (1909) 2012 How We think. Mansfield Centre, Martino Publishin, CT

Dumitru D (2012) Critical thinking and integrated programs. The problem of transferability. Proc Soc Behav Sci 33:143–147

Facione PA (2000) The disposition toward critical thinking: its character measurement, relationship to critical thinking skill. Inform Log 20(1):61–84

Facione PA (1990) Critical thinking: a statement of expert consensus for purposes of educational assessment and instruction. Research findings and recommendations. The California Academic Press, Millbrae

Fisher A (2001) Critical thinking. An introduction. Cambridge University Press, Cambridge

Hesse H (2002) The glass bead game. Picador, NY

Hitchcock D (2018) Critical thinking. Stanford encyclopedia of philosophy. https://plato.stanford.edu/entries/critical-thinking/. Accessed on 27 Jan 2020

Hobbes T (2019) Complete works. Kindle Book

Kant I (1999) Critique of pure reason. Cambridge University Press, Cambridge

Kuhn D (2008) Education for thinking. Harvard University Press

Locke J (1996) Essay concerning human understanding. Abridged edition.Kindle Book.Hachett Classics, IN

Minott D, Ferguson T, Minott G (2019) Critical thinking and sustainable development. In: Leal Filho W (eds) Encyclopedia of sustainability in higher education. Springer, Cham. https://doi.org/https://doi.org/10.1007/978-3-319-63951-2

Parry R (2014) Ancient ethical theory. Stanford encyclopedia of philosophy. https://plato.stanford.edu/entries/ethics-ancient/. Accessed on 27 Jan 2020

Paul RW, Elder L (2006) The miniature guide to critical thinking: concepts and tools, 4th edn. Foundation for Critical Thinking, Dillon Beach, CA

Paul R, Elder L (2020) The miniature guide to critical thinking concepts and tools, 8th edn. Rowman & Littlefield, London

Piaget J, Inhelder B (1969) The psychology of the child. Basic Books, NY

Plato (2007) The Republic. Revised Edition. Penguin Classics, London

Reitter P, Wellmon C (2020) Introduction. In: Weber M (ed) Charisma and disenchantment: the vocation lectures. Kindle Book, New York Review Books Classics

Shulla K, Leal Filho W, Lardjane S, Sommer JH, Borgemeister C (2020) Sustainable development education in the context of the 2030 agenda for sustainable development. Int J Sustain Develop World Ecol. https://doi.org/10.1080/13504509.2020.1721378

Sidibe M (2016) Universal health coverage: political courage to leave no one behind. Lancet Glob Health 4(6):e355–e356. https://doi.org/10.1016/S2214-109X(16)30072-9

Stoianovici D (2005) *Argumentare şi gândire critică*[Argumentation and Critical Thinking]. Editura Universităţii din Bucureşti, Bucharest, RO

Straková Z, Cimermanová I (2018) Critical thinking development—a necessary step in higher education transformation towards sustainability. Sustainability 10:3366

Thomas I (2009) Critical thinking, transformative learning, sustainable education, and problem-based learning in universities. J Transform Educ 7(3):245–264. https://doi.org/10.1177/1541344610385753

Wittgenstein L (1998) Tractatus Logico-Philosophicus, 471st edn. Dover Publications, New York, NY

Sustainable Development: An Analysis Based on Conference Papers

Carlos Rogério Montenegro de Lima, André Borchardt Deggau, Gisele Mazon, Thiago Coelho Soares, and José Baltazar Salgueirinho Osório de Andrade Guerra

1 Introduction

The issue of sustainability in the context of higher education is a relatively recent theme. More than 600 universities around the world have committed to sustainability by signing international agreements and conventions, such as the Bologna Charter, the Halifax Declaration, the Talloires Declaration and the Copernicus Charter for Sustainable Development. More recently, the 2030 Agenda established by the UN General Assembly in 2015 and the associated Sustainable Development Objectives (SDO) mean that higher education institutions should, now more than ever, engage in the implementation of sustainable development in all thematic areas.

Considering that sustainable development is a rising global concern, the growing range of concepts, methods and tools in the field of sustainability requires a unifying and operational definition of sustainability to increase the robustness and scientificity

C. R. M. de Lima
Center for Sustainable Development (Greens), Agência de Fomento do Estado de Santa Catarina (Badesc), Florianópolis, Brazil
e-mail: calmontenegro@gmail.com

A. B. Deggau (✉) · G. Mazon · T. C. Soares · J. B. S. O. de Andrade Guerra
Center for Sustainable Development (Greens), University of Southern Santa Catarina (UNISUL), Florianópolis, Brazil
e-mail: andre.deggau@gmail.com

G. Mazon
e-mail: gisele.mazon@unisul.br

T. C. Soares
e-mail: Thiago.soares@unisul.br

J. B. S. O. de Andrade Guerra
e-mail: baltazar.guerra@unisul.br

W. Leal Filho et al. (eds.), *Universities, Sustainability and Society: Supporting the Implementation of the Sustainable Development Goals*, World Sustainability Series,
https://doi.org/10.1007/978-3-030-63399-8_13

of the term (Missimer et al. 2017a, b). However, the polysemy of the term 'sustainability' has caused the concept to be disregarded and makes it difficult to translate debates into action (Bolis et al. 2014; Holden et al. 2014). According to Ramsey (2015), Berchin et al. (2018), the concept of sustainability must be coherent and applied in both theory and practice. Lukman and Glavic (2007) indicate that sustainability varies according to the specific needs and visions of a specific socio-political-economic context, integrating environmental protection, economic performance and social welfare, driven by political will, and ethical and ecological imperatives.

Promoting societal awareness regarding sustainability issues, as well as facilitating dialogue among academia, community, authorities and students, are important goals of HEIs and the international conferences promoted by them. Global sustainability and sustainable development conferences have become worldwide phenomena, and often enlist ever-broadening inter/multidisciplinary research to address environmental challenges. These conferences can unite disciplines and global universities under a common goal, and thus create space for new kinds of partnerships within HEIs and broader networks (Berchin et al. 2018).

Universities must assist in the sustainability process in society, supporting the implementation of the Sustainable Development Goals. Educational institutions receive public and private investments and impact the cities where their units are present. Therefore, it is essential that universities help the community around them, even as a way of giving back to society what is invested in education and promoting actions to reduce disparities in access.

This research aimed to analyze the significance of international conferences for higher education institutions' (HEIs) promotion of sustainability. The bibliometric method was used, considering articles found in the Web of Science database related to the subject of sustainable development in higher education from 2015 to 2019, for detecting tendencies, possible paths and the research front about this field. The main authors, concepts and theories identified in 648 articles related to this topic were analyzed.

This article is structured into five parts. The introduction is followed by the second part which contains a review of the literature on sustainable development in higher education and bibliometric studies on the subject. In the third part, the methods used are discussed, including data collection, sample research and analysis procedures. In the fourth part, the results of the research are examined using techniques including the analysis of Web of Science Core Collection and World Sustainability Symposium, Bridge International Conference 2019. The conclusions are presented in the fifth and final part.

2 Literature Review

The literature review begins by addressing the theme of sustainable development of higher education and then bibliometric and revision studies on the subject.

Universities have a special responsibility concerning the sustainable development of society (Kräusche and Pilz 2018). Increasingly, universities and the higher education sector are seeking to incorporate education for sustainable development. To assess the progress measurement and monitoring systems are needed (Tierney et al. 2015). There are several challenges associated with the development of the tools necessary to achieve this goal, which is why sustainability capacity assessment methods merit specific attention from researchers (Sandri et al. 2018).

Sustainability has become a global concern in the struggle to deal more successfully with complex and unprecedented survival, social, political and peace issues. Higher education institutions (HEIs) play a key role in this transformation (Veiga Ávila et al. 2019).

In general, the HEIs focus on helping people who live around the campus with social projects, health issues and preparation for work. Further areas of practice are new forms of energy, selective waste collection and new standards of plastics consumption. HEIs that receive a higher evaluation from the Education Ministry focuses on educational programs and projects aimed at improving sustainability inside and outside the campus (Moura et al. 2019).

While many business schools around the world are discussing the importance of integrating sustainability into their curricula, and although employers and students are demanding the same, few business schools have actually progressed in meeting these demands (Barber et al. 2014).

Universities have an important role to play in articulating the transformation of society, in the formation of individuals capable of acting in the processes that will allow the sustainable development goals (SDGs) to be reached, in theoretical areas and in practical and concrete actions (Frandoloso and Gasparetto Rebelatto 2019).

In general, topics related to sustainability (e.g., environmental protection, waste management and sustainable development) are largely addressed and, to some extent, applied in faculties dealing with earth sciences (geography, ecology, etc.) and technical sciences, such as environmental engineering (Sima et al. 2019).

In the past few decades, addressing issues related to sustainability has become ever more prevalent in higher education, although the pace with which the actions required are adopted varies and there is a wide range of interpretations and practices (Vargas et al. 2019).

There is a widely held belief that sustainable development policies are essential for universities to successfully engage in matters related to sustainability, and are an indicator of the extent to which they are active in this field (Leal Filho et al. 2018). Concerns about sustainability, prompted by the Brazilian federal legislature, have led federal HEIs to change their internal processes, infrastructure and organizational culture toward sustainability (Berchin et al. 2017).

The potential for energy efficiency and the deployment of renewable energy at universities is substantial, and these represent promising ways to meet an institution's energy needs, on the one hand, and reduce its contribution to climate change, on the other. However, in order to optimize investment and achieve successful results, it is imperative to understand the level of current commitment to energy actions (Leal Filho et al. 2019).

Humans are at the center of global climate change and The United Nations Sustainable Development Goals (SDGs) are promoting sustainability with proactive, global, social goals, moving us away from the Brundtland paradigm of 'do nothing today to compromise tomorrow's generation'. This promotes a regenerative shift in the sustainability concept, no longer only considering resources and energy, but also significant human-centric attributes (Sonetti et al. 2019).

The role of universities in climate change education (CCE) is of great importance to the scientific, social, environmental and political challenges facing the world. Future leaders must make decisions from an informed standpoint, including climate change mitigation tools in their work and private life. Therefore, it is essential to understand the scope of CCE strategies adopted globally by higher education institutions (HEIs) and to explore and analyze ways in which HEIs can better address this challenge (Molthan-Hill et al. 2019).

Cooperation between HEIs through collaborating in international conferences is an important measure to promote and disseminate research regarding global sustainability and sustainable development. Moreover, in addition to cooperation, participation and collaboration between institutions, international conferences sponsored or organized by HEIs have the potential to promulgate interdisciplinary and multidisciplinary approaches to addressing global sustainability challenges. The implementation of interdisciplinary and multidisciplinary approaches within HEIs is important for the discussion and promotion of sustainability and sustainable development. International conferences may be used for communicating the scientific results to stakeholders and to send strong messages on sustainability-related issues (Berchin et al. 2018).

3 Method

Firstly, the authors performed a literature review. The data considered were collected from the Web of Science Core Collection (http://webofknowledge.com), an electronic database made available by Clarivate Analytics, which is one of the most widely used in applied social sciences. The Web of Science (WoS) database was used because of its wide coverage of social science publications, as it indexes the main international periodicals in the field. Furthermore, WoS is compatible with the Bibexcel software used in the cocitation analysis. Persson et al. (2009) state that only Web of Science and Scopus database include the bibliographic references needed to quantify the author or paper citations, through the computerized processing and analysis of these references. The selection of the sample occurred through research of keywords, used from 2015 to 2019, for detecting tendencies, possible paths and the research front about this field, resulting in 3590 articles. The keywords used by Boolean expression were: ("higher education" OR "campus" OR "university *" OR "college *") AND ("sustainab *" OR "sustainable development"). Using the asterisk at the end of the keyword captures all possible variations of these in the titles, abstracts and keywords ("topic" option) of the selected articles.

To refine the results of the research, only the studies classified (document type) as articles of congresses (proceedings paper) were considered as search criteria. In addition, as it is a comprehensive research, only articles in the category (Web of Science Categories) Education Educational Research were included. With the application of these search criteria, 648 articles were found. These articles cited 8073 references. Thus, for this study, the most cited references in the sample were analyzed.

Finally, after the analysis of all academic conferences from the Web of Science Core Collection, one conference hosted by the University of Southern Santa Catarina (hereafter UNISUL), "World Sustainability Symposium, Bridge International Conference 2019", was selected for a more in-depth analysis, exploring its results and contributions to promote sustainability. The website of the conference has been analyzed for information like conference goals, profiles of participants, and partners. Likewise, several details of the papers approved for presentation were analyzed: authors and their higher education institution's geographic location, paper objectives and results, and main terms.

4 Results

4.1 Citations Analysis from Conference Papers on Sustainable Development

A list of the most frequently cited articles identified using the global citation score (GCS), which is the total number of citations of a paper in the Web of Science, is shown in Table 1. The cut-off point was based on the top 20 articles on the list. This table also shows the percentage of citations in relation to the total citations of the 648 articles used as a sample.

The most frequently cited study was carried out by Chen in 2015, with twenty-two citations. The next four most cited articles in the sample were: Pashby and Andreotti (2016), Dzienkowski et al. (2016).

The conferences highlighted by the number of papers, considering the 648 in the sample, are shown in Table 2. The ones that appeared most frequently were: 9th International Conference on Education and New Learning Technologies with 39 12th International Technology, Education and Development Conference with 38 articles; and 9th International Conference of Education, Research and Innovation with 34 articles.

Table 1 Most frequently cited articles on conference papers on sustainable development

	Articles	Citations (n)	Citations (%)
1	Chen et al. (2015)	22	10.1
2	Pashby and Andreotti (2016)	17	7.8
3	Sierpinski et al. (2015)	9	4.1
4	Dzienkowski et al. (2016)	8	3.7
5	Artal-Sevil et al. (2015)	8	3.7
6	Plechawska-Wojcik et al. (2015)	7	3.2
7	Tian (2017)	6	2.8
8	Katiliute and Daunoriene (2014)	6	2.8
9	Tasci (2014)	6	2.8
10	Martins et al. (2017)	4	1.8
11	Alabugin et al. (2016)	4	1.8
12	Dulama et al. (2016)	4	1.8
13	Pena et al. (2014)	4	1.8
14	Platonova et al. (2015)	4	1.8
15	Mikhailova and Chorosova (2017)	3	1.4
16	Palaigeorgiou et al. (2017)	3	1.4
17	Matei et al. (2016)	3	1.4
18	Ferrer-Marti et al. (2015)	3	1.4
19	Hagen and Hasleberg (2015)	3	1.4
20	Momete, DC (2015)	3	1.4

Source Authors

Table 3 shows the educational institutions with the highest number of articles in the sample, the most notable being Universidad Politecnica de Madrid and the Universidad Politecnica de Valencia, both in Spain, with 13 and 9 articles, respectively.

The most prominent authors, their countries and institutional affiliations, with more papers on sustainability in higher education in the sample of 648 papers are listed in Table 4: Arraiza MP with 12 papers, Ioras F, Garcia JL and Gimenez MC with 10, 8 and 8 respectively.

Figure 1 presents the most cited terms in the papers of the sample, along with their frequency on the abstracts. The terms represent themes that reflect the needs, resources or concerns in the papers on sustainable development in higher education of international conferences of our sample.

Table 5 presents the ten most-cited terms in the international conferences, along with their frequency on the abstracts. The terms on this Table represent themes that reflect the main global needs, resources or concerns in these conferences.

The interest in some of the most cited terms in the conferences was investigated in the searches made at the Google search engine, using Google Trends—search trends

Table 2 Conferences that appeared most frequently as publishers of the papers on sustainability in higher education included in the sample

	Conference	No. of papers
1	9th International Conference on Education and New Learning Technologies (Edulearn17)	39
2	12th International Technology, Education and Development Conference (Inted)	38
3	Iceri2016: 9th International Conference of Education, Research and Innovation	34
4	Iceri2015: 8th International Conference of Education, Research and Innovation	31
5	10th International Conference of Education, Research and Innovation (Iceri2017)	30
6	Inted2017: 11th International Technology, Education and Development Conference	30
7	Inted2015: 9th International Technology, Education and Development Conference	24
8	Edulearn16: 8th International Conference on Education and New Learning Technologies	21
9	Inted2016: 10th International Technology, Education and Development Conference	20
10	Edulearn15: 7th International Conference on Education and New Learning Technologies	19

Source Authors

Table 3 Frequency data on universities with published articles on sustainability in higher education included in the sample

	Institution	Country	No. of articles
1	Univ Politecn Madrid	Spain	13
2	Univ Politecn Valencia	Spain	9
3	Univ Novi Sad	Serbia	8
4	Buckinghamshire New Univ	England	7
5	Univ Basque Country UPV EHU	Spain	7
6	Silesian Tech Univ	Poland	6
7	Tianjin Univ Finance & Econ	Peoples R China	6
8	Univ Jaume	Spain	6

Source Authors

feature that shows how frequently a given search term is entered into Google's search engine relative to the site's total search volume over a given period. According to the data presented in Fig. 2, the most searched terms on Google platforms from 2015 to 2019 were "mobile" and "university". One interesting point for analysis in the terms

Table 4 Frequency data on authors of articles on sustainability in higher education included in the sample

	Author	HEI	Country	No. of articles
1	Roots MP	Univ Politecn Madrid, Madrid	Spain	12
2	Ioras F	Buckinghamshire New Univ, Wycombe	England	10
3	Garcia JL	Univ Politecn Madrid, Madrid	Spain	8
4	Gimenez MC	Univ Politecn Madrid, Madrid	Spain	8
5	Lopez JV	Univ Politecn Madrid, Madrid	Spain	6
6	Alcala M	Univ Girona, Girona	Spain	5
7	Barber E	Univ North Carolina Wilmington, Wilmington, NC	USA	5
8	Calderon-Guerrero C	Univ Politecn Madrid, Madrid	Spain	5
9	Cordon C	Univ Politecn Madrid, Madrid	Spain	5
10	Delgado-Aguilar M	Univ Girona, Girona	Spain	5
11	Potts A	Univ North Carolina Wilmington, Wilmington, NC	USA	5
12	Sedlak O	Univ Novi Sad, Fac Econ Subot, Novi Sad	Serbia	5
13	Snyder J	Alamance Community Coll, Graham, NC	USA	5
14	Tarres Q	Univ Girona, Girona	Spain	5

Source Authors

identified in the conferences is that these two terms have a decreased interest along this period, and the other eight terms have a constant average. Some of the most cited terms at conferences (education and development) are not among the ones of greatest interest in searches in the analyzed period.

From the Similitude Analysis (Fig. 3), based on the graph theory, performed with the aid of the Iramuteq software, it is possible to identify the connections between words, which helps to identify the structure of the content of the textual corpus. Five words that stand out in the articles were identified: development, university, education, and learn. From them, terms like environment, system, knowledge, model, cooperation, sustainable, student, mobile, and technology, branch out.

In Fig. 3, it is possible to see some of the themes of the articles most cited in the selected sample: mobile application development for environment monitoring, international development and sustainability, dissemination of sustainable development on universities' websites, education for social innovation, cooperation and human development, cooperation between businesses and universities, computer science and environmental engineering, and environmentally friendly travel behavior.

Fig. 1 Word cloud of the predominant words in the most cited articles in the sample. *Source* Authors

Table 5 The ten most-cited terms in international conference papers on sustainable development

Terms	Frequency
Education	53
University	43
Development	42
Student	41
Learn	38
System	31
Mobile	25
Social	23
Sustainable	21
High	21

Source Authors

4.2 Case Study of One International Conference in Brazil

The analysis of the papers in proceedings (4.1) uses statistics based, for example, on frequencies, the number of citations and authors, whereas the case study of one

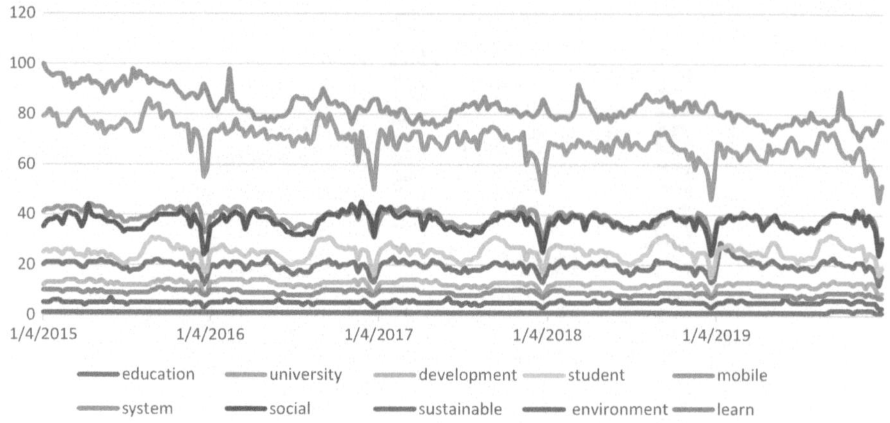

Fig. 2 Terms addressed in international conferences. *Source* Authors

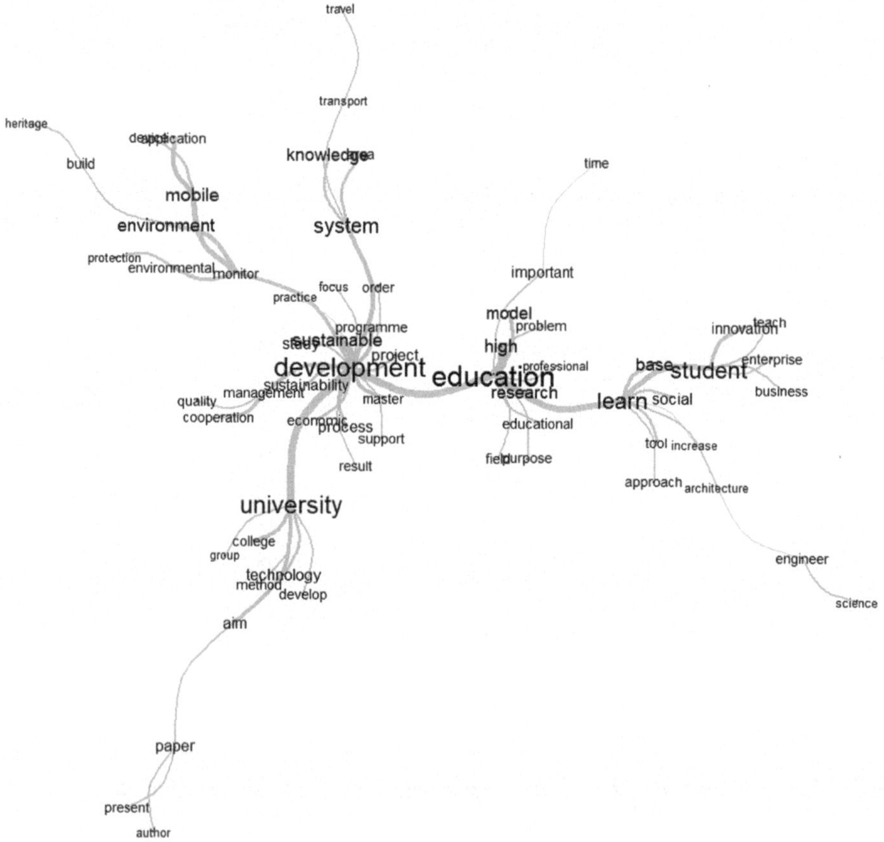

Fig. 3 Similarity analysis of the most cited articles in the conferences. *Source* Authors

conference (4.2) considers the themes covered and the objectives of the conference since qualitative studies complement quantitative studies (Doorenbos 2014; Fawcett 2015).

This chapter analyzing data from one international conference organized and hosted by UNISUL, World Sustainability Symposium, Bridge International Conference 2019, organized by University of Southern Santa Catarina (Unisul); Cambridge University; Hamburg University of Applied Sciences, Germany Manchester; Metropolitan University, United Kingdom; Center for Sustainable Development (Greens), Unisul; Cambridge Center for Energy, Environment and Natural Resources Governance (Ceenrg); Inter-University Sustainable Development Research Program (IUSDRP); University of Exeter; and Open University.

Objectives of the event:

- Promote the exchange of information, ideas and experiences acquired in the execution of projects, based on successful initiatives and good practices related to technological changes in the Brazilian energy sector and their impacts on altering the use of land, food, water and emissions of greenhouse gases;
- Discuss methodological approaches and projects that aim to integrate the study of the links between water, energy and food in supporting projects and implementing action plans in communities and rural areas;
- Highlight the importance of innovation in the generation of technologies and in the promotion of sustainable development;
- Develop permanent knowledge exchanges between conference participants on the links between water, energy and food sustainable development;
- Establish research cooperation between the universities participating in Bridge, in order to contribute to a permanent scientific collaboration on sustainable development.

The World Sustainability Symposium, BRIDGE 2019 International Conference, featured a set of presentations divided into three themes.

- Theme 1: The connections between water, energy and food in Brazil.
- Theme 2: Technological change in the Brazilian energy sector and its impacts on land use, food production, energy, water and the generation of greenhouse gases.
- Theme 3: Political and legal responses to conflicting/competing demands for energy, food and water.

Participation in international conferences proved to be a positive way for HEIs to establish participants' networks for sustainability, strengthen global ties, share best practices, experiences, methods, and sustainability researches. However, most papers presented in The Bridge International Conference were from Brazil. Figure 4 presents the geographic origin of papers submitted by Brazilian authors to this Conference.

Most papers from Brazil came from researchers in the southern region, representing 64% of the Brazilian articles—this could be explained by their geographic proximity to UNISUL.

The Conference accepted 245 papers, with the most cited words shown in the word cloud (Fig. 5). Of these, 215 articles were presented at the conference, 222 had

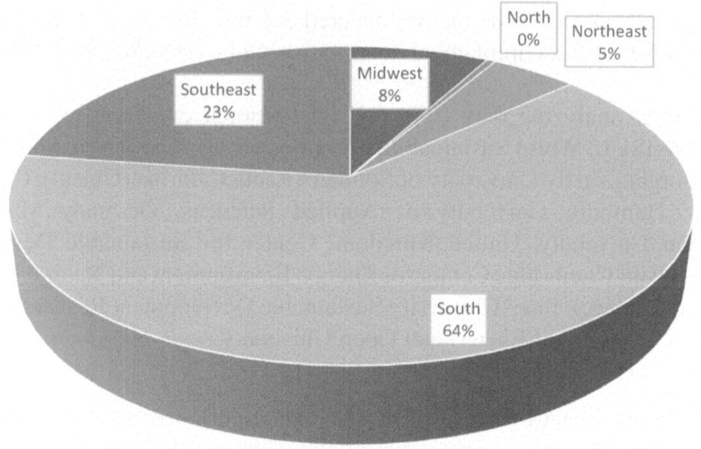

Fig. 4 Geographic origin of papers submitted by Brazilian authors. *Source* Authors

Fig. 5 Word cloud of the most cited words in the abstracts of the papers in the World Sustainability Symposium, Bridge International Conference 2019. *Source* Authors

Table 6 The ten most-cited terms in the Bridge International Conference

Terms	Frequency
Energy	248
Environmental	207
Study	194
Water	183
Waste	140
Result	138
Production	138
Sustainable	128
Present	124
System	118

Source Authors

their abstracts published in the Book of Abstracts, and 182 were published in special editions of Revista Gestão e Sustentabilidade Ambiental.

Table 6 presents the ten most-cited terms in the conference, along with their frequency on the abstracts. The terms in Table 6 represent themes that reflect the local needs, resources or concerns in the Bridge International Conference in Brazil.

The interest in some of the most cited terms in the conferences was investigated in the searches made at the Google search engine, using Google Trends. "Energy" and "environmental" are the most cited terms in the Bridge International Conference, but according to the data presented on Fig. 6, the most searched terms on Google platforms from 2015 to 2019 were "result" and "water", and both had an increasing interest in the period.

With the Similitude Analysis (Fig. 7), performed with the aid of the Iramuteq

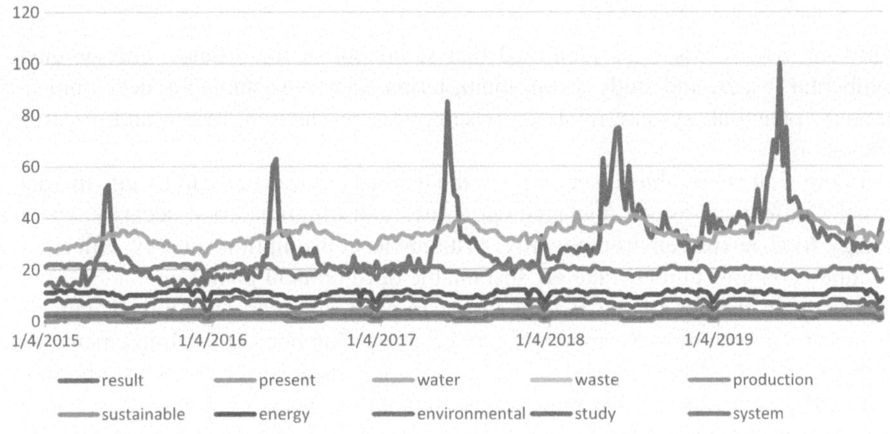

Fig. 6 Terms addressed in the Bridge International Conference. *Source* Authors

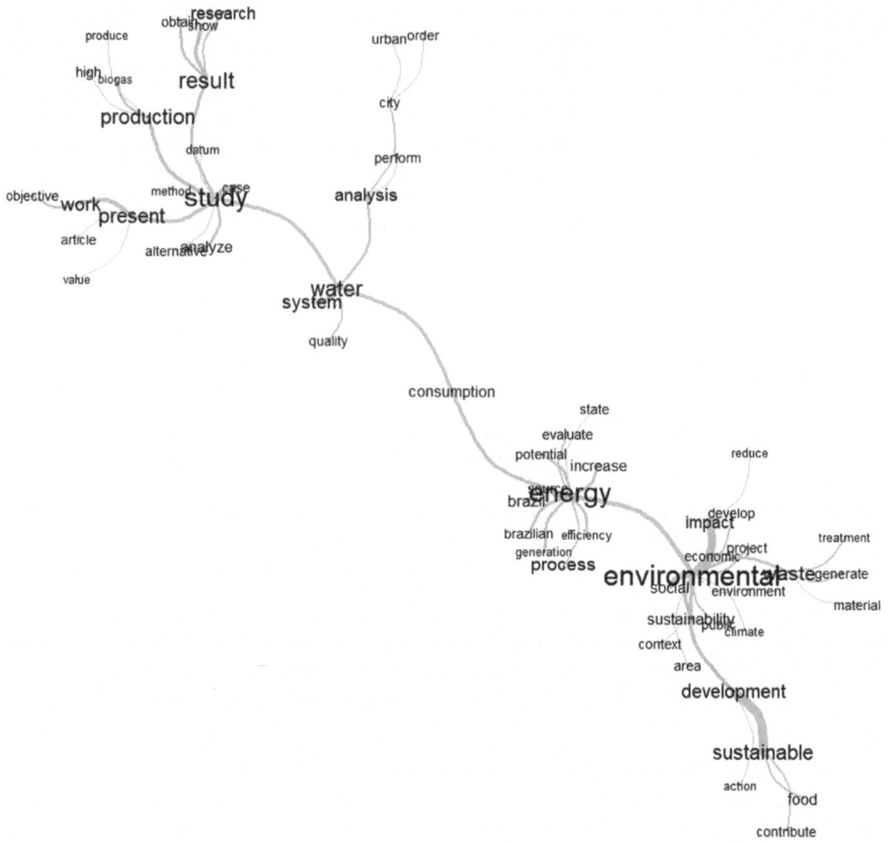

Fig. 7 Similitude analysis of the most cited articles at the Bridge International Conference. *Source* Authors

software, five words were identified that stand out in the articles: energy, environmental, water, and study. From them, terms such as sustainable, development, process, potential, system, quality, consumption, production, result, and research, branch out.

In Fig. 7, it is possible to see some of the themes of the articles in Bridge International Conference: waste management, water consumption, food security, water-energy-food nexus, environmentally sustainable consumption, energy efficiency, circular economy, climate change, sustainable development goals.

The main themes of the Bridge International Conference are linked with the theme of "Universities, Sustainability and Society: Supporting the Implementation of the Sustainable Development Goals", as shown in the next section. This is because conferences are ways of disseminating sustainability practices. In this way, educational actions, practices in higher education on other aspects of sustainability, also climate change education and outreach are ways of promoting sustainability in the university environment and in society.

4.3 Relevance for Society and Implementation of the SDGs

The term "sustainable development" is the basis for the SDGs and appears in a large portion of the articles. It is up to events on the theme to develop research and promote the discussion of ways of implementing sustainable development as a contribution to society. Several themes are also discussed in the articles and are adherent to the SDGs, as discussed below.

The terms "university" and "education" appear in the articles most cited in the sample of articles from international conferences and also in the case study articles and are related to SDG 4—Ensuring inclusive and equitable and quality education, and promoting opportunities for lifelong learning for all, and Goal 4.3—By 2030, ensure equal access for all men and women to quality technical, vocational and higher education at affordable prices, including university.

The word "energy" appears among the most cited as a research item discussed in the articles of the World Sustainability Symposium, Bridge International Conference 2019 and is related to SDG 7—Ensuring access to cheap, reliable, sustainable and renewable energy for all.

The word "waste" appears as a research item discussed in the articles of the World Sustainability Symposium, Bridge International Conference 2019 and is related to Goal 11.6—By 2030, reduce the negative environmental impact per capita of cities, including paying special attention to the quality of air, municipal waste management and others and Goal 12.5—By 2030, substantially reduce the generation of waste through prevention, reduction, recycling and reuse.

The term "water" appears as a research item discussed in the articles of the World Sustainability Symposium, Bridge International Conference 2019 and is related to SDG 6—Ensure availability and sustainable management of water and sanitation for all.

The terms "consumption" and "production" appear as a research item discussed in the articles of the World Sustainability Symposium, Bridge International Conference 2019 and are related to SDG 12—Ensure sustainable production and consumption patterns, and Goal 8.4—Progressively improve, by 2030, the efficiency of global resources in consumption and production, and strive to decouple economic growth from environmental degradation, according to the Ten-Year Plan of Programs on Sustainable Production and Consumption, with developed countries taking the lead.

Terms that appeared to a lesser extent in the articles of the World Sustainability Symposium, Bridge International Conference 2019, but that are directly related to the SDGs are "food", which is linked to SDG 2—End hunger, achieve food security and improve nutrition and promoting sustainable agriculture, and "health", related to SDG 3—Ensuring a healthy life and promoting well-being for all, at all ages.

A term that appears to a lesser extent in the sample articles in the database, but which is directly related to the SDGs is "build", which is linked to SDG 9—Building resilient infrastructure, promoting inclusive and sustainable industrialization, and fostering innovation.

The construction of these articles demonstrates that society has developed its own studies and established priorities in its research agendas regarding the SDGs. The publication of these in scientific events will contribute to debates on the national implementation challenges of the SDGs. Society articulates to take the SDGs as a guiding axis for the action of subnational governments.

To develop an agenda like that of the SDGs in a sustainable way, the commitment of governments, companies, society and academia is necessary, each with its role and responsibilities and, above all, in collaboration with each other. By adopting actions that assist in achieving the goals of the SDGs, companies, society, academia and governments help to disseminate these objectives in communities and society. Thus, when 2030 arrives, many of these objectives will have been achieved.

The universities incorporated sustainability as a social issue and as a resource for teaching, research and outreach, and university conferences disseminate sustainable posture, skills, development, knowledge and technologies.

5 Conclusions

Cooperation between HEIs through collaborating in international conferences is an important measure to promote and disseminate research regarding global sustainability and sustainable development. Such efforts may also foster opportunities for future research and joint projects. Sharing knowledge, experience, projects, initiatives and methods can hasten and deepen the research process, potentially contributing to greater global impacts.

For example, the World Sustainability Symposium, Bridge International Conference 2019 analyzed in Brazil resulted from collaborative projects between multiple HEIs (from Europe and Brazil). In establishing international partnerships and promoting international conferences on sustainability themes, UNISUL is implementing strategies toward sustainability and sustainable development. The goals of such endeavors are to disseminate knowledge, create and enhance an international network of HEIs committed to environmental education, and to engage multiple stakeholders in projects under the broad thematic umbrella of sustainability. Thus, UNISUL and the community benefited from these collaborative projects and conferences by increasing sustainability awareness locally and internationally.

Moreover, the findings suggest that in addition to cooperation, participation and collaboration between institutions, international conferences sponsored or organized by HEIs have the potential to promulgate interdisciplinary and multidisciplinary approaches to addressing global sustainability challenges. In the international conferences analyzed, the diverse array of disciplines and research teams illustrate this point.

The implementation of interdisciplinary and multidisciplinary approaches within HEIs is important for the discussion and promotion of sustainability and sustainable development. Thus, it is important to note that the analysis of the papers presented at the selected international conferences created a picture of the main terms addressed by them, enabling the identification of the most important themes discussed.

There is still enough room to increase the inter and multidisciplinary character of the sustainability conferences organized by HEIs by more actively involving relevant stakeholders at local and regional scales and to debate the most pressing societal problems these actors raise. In this respect, international conferences may be used for communicating the scientific results to stakeholders and send strong messages on sustainability-related issues.

The use of a single database is a limitation of the paper. Another limitation is not to consider studies related to sustainability outputs by HEIs and presented at conferences not only about Education but also about sustainability research by practitioners in HE on other aspects of sustainability, also CE, outreach, etc. Future research will be necessary to develop a better understanding of how universities' conferences can promote the implementation of the Sustainable Development Goals.

References

Alabugin A, Aliukov S, Alabugina R (2016) Managing cooperation between business and university: quality assurance using sustainable development indicators from a knowledge management. In: 9th annual international conference of education, Research and Innovation (iCERi), 14–16 Nov 2016. Seville, Spain

Artal-Sevil JS, Bernal-Agustin JL, Dominguez JÁ (2015) M-learning (mobile learning) in higher education. The impact of smartphone as interactive learning too. In: Edulearn15: 7th international conference on education and new learning technologies, 6–8 July 2015. Barcelona, Spain

Barber N, Wilson F, Venkatachalam B, Cleaves S, Garnham J (2014) Integrating sustainability into business curricula: University of New Hampshire case study. Int J Sustain High Educ 15(4):473–493

Berchin I, Grando V, Marcon G, Corseuil L, Guerra J (2017) Strategies to promote sustainability in higher education institutions. Int J Sustain High Educ 18(7):1018–1038

Berchin II, Sima M, Lima MA, Biesel S, Santos LP, Ferreira RV, Guerra JBSA, Ceci F (2018) The importance of international conferences on sustainable development as higher education institutions' strategies to promote sustainability: a case study in Brazil. J Clean Prod 171:756–772

Bolis I, Morioka SN, Sznelwar LI (2014) When sustainable development risks losing its meaning. Delimiting the concept with a comprehensive literature review and a conceptual model. J Clean Prod 83:7–20

Chen L, Lee G, Chen L (2015) Research on sense of belonging of young teachers in private colleges and universities. In: 3rd international conference on education reform and management innovation, 5–6 Sept 2015. Bangkok, Thailand

Doorenbos AZ (2014) Mixed methods in nursing research: an overview and practical examples. Kango Kenkyu 47(3):207–217

Dulama, ME, Vana VM, Ilovan OR (2016) Assessing Territorial Planning M.Sc. Students using Facebook. In: 11th international conference on virtual learning (ICVL), 29 Oct 2016. Craiova, Romania

Dzienkowski M, Plechawska-Wojcik M, Milosz M, Stryczewska H (2016) Mobile application development for environment monitoring—a new programme of master studies in English. In: Inted2016: 10th international technology, education and development conference, 7–9 Mar 2016. Valencia, Spain

Fawcett J (2015) Invisible nursing research: thoughts about mixed methods research and nursing practice. Nurs Sci Quart 28(2):167–168

Ferrer-Marti L, Garfi M, Ferrer I (2014) Cooperation and human development projects as Bachelor, Master and PhD Thesis: evaluating an internship program. In: International conference on university teaching and innovation (CIDUI), 2–4 July 2014. Tarragona, Spain

Frandoloso M, Gasparetto Rebelatto B (2019) The participatory process of planning social and environmental responsibility at a Brazilian University. Int J Sustain High Educ 20(5):917–931

Hagen ST, Hasleberg H (2015) Entrepreneurship in higher education—new teaching method. In: 7th international conference on education and new learning technologies (EDULEARN), 6–8 July 2015. Barcelona, Spain

Holden E, Linnerud K, Banister D (2014) Sustainable development: our common future revisited. Glob Environ Change 26:130–139

Katiliute E, Daunoriene A (2014) Dissemination of sustainable development on universities website. In: Proceedings of 6th world conference on educational sciences, 6–9 Feb 2014. Valletta, Malta

Kräusche K, Pilz S (2018) Integrated sustainability reporting at HNE Eberswalde—a practice report. Int J Sustain High Educ 19(2):291–312

Leal Filho W, Brandli L, Becker D, Skanavis C, Kounani A, Sardi C, Papaioannidou D, Paço A, Azeiteiro U, de Sousa L, Raath S, Pretorius R, Shiel C, Vargas V, Trencher G, Marans R (2018) Sustainable development policies as indicators and pre-conditions for sustainability efforts at universities. Int J Sustain High Educ 19(1):85–113

Leal Filho W, Salvia AL, do Paço A, Anholon R, Quelhas OLG, Rampasso IS, Ng A, Balogun AL, Kondev B, Brandli LL (2019) A comparative study of approaches towards energy efficiency and renewable energy use at higher education institutions. J Cleaner Prod 237(10):1–9

Lukman R, Glavič P (2007) What are the key elements of a sustainable university? Clean Technol Environ Policy 9(2):103–114. https://doi.org/10.1007/s10098-006-0070-7

Martins AMT, Carlos JS, Nepomuceno MCS (2017) Built heritage research and education. In: 9th international conference on education and new learning technologies (EDULEARN), 3–5 July 2017. Barcelona, Spain

Matei A, Savulescu C, Antonovici CG (2016) Education for social innovation. Fal model and transfer of best practices. In: 8th international conference on education and new learning technologies (edulearn), 4–6 July 2016. Barcelona, Spain

Mikhailova, EI; Chorosova, OM (2017) Federal university: synchronising trends of sustainable development of a university and region. In: 3rd international forum on teacher education (ifte), 23–25 May 2017. Kazan, Russia

Missimer M, Robèrt K-H, Broman G (2017a) A strategic approach to social sustainability—Part 1: exploring the social system. J Clean Prod 140:32–41

Missimer M, Robèrt K-H, Broman G (2017b) A strategic approach to social sustainability—Part 2: a principle-based definition. J Clean Prod 140:42–52

Molthan-Hill P, Worsfold N, Nagy GJ, Leal Filho W, Mifsud M (2019) Climate change education for universities: a conceptual framework from an international study. J Clean Prod 226(20):1092–1101

Moura M, Frankenberger F, Tortato U (2019) Sustainability in Brazilian HEI: practices overview. Int J Sustain High Educ 20(5):832–841

Palaigeorgiou G, Malandrakis G, Tsolopani C (2017) Learning with drones: flying windows for classroom virtual field trips. In: 17th IEEE international conference on advanced learning technologies (ICALT), 3–7 July 2017. Timisoara, Romania

Pashby K, Andreotti VD (2016) Ethical internationalisation in higher education: interfaces with international development and sustainability. In: 1st symposium of the environmental-and-sustainability-research-network (ESER) at the European Conference on Educational Research (ECER). Set, 2014. Porto, Portugal

Pena OFC, Llanos RA, Coria MD, Perez-Acosta AM (2014) Multidimensional model of assessment of economic thinking in college students. In: Proceedings of 6th world conference on educational sciences, 6–9 Feb 2014. Valletta, Malta

Persson O, Danell R, Schneider JW (2009) How to use Bibexcel for various types of bibliometric analysis. In: Åström F, Danell R, Larsen B, Schneider J (eds) Celebrating scholarly communication studies: a Festschrift for Olle Persson at his 60th Birthday. International Society for Scientometrics and Informetrics, Leuven, Belgium, pp 9–24

Plechawska-Wojcik M, Milosz M, Stryczewska HD (2015) Building bridge between computer science and environmental engineering—the new master programme. In: Edulearn15: 7th international conference on education and new learning technologies, 6–8 July 2015. Barcelona, Spain

Platonova E, Bogomolova J, Musarskiy M, Igumnov O (2015) Various approaches to financing russian higher education institutions integrating into the global educational environment. Worldwide trends dev educ acad res. 15–18 June 2015. Sofia, Bulgaria

Ramsey JL (2015) On not defining sustainability. J Agric Environ Ethics 28(6):1075–1087

Sandri O, Holdsworth S, Thomas I (2018) Assessing graduate sustainability capability post-degree completion: why is it important and what are the challenges? Int J Sustain Higher Educ 19(1):2–14

Sierpinski G, Staniek M, Celinski I (2015) New methods for pro-ecological travel behavior learning. In: Iceri2015: 8th international conference of education, research and innovation, 16–20 Nov 2015. Seville, Spain

Sima M, Grigorescu I, Bălteanu D (2019) An overview of campus greening initiatives at universities in Romania. Int J Sustain High Educ 20(3):410–422

Sonetti G, Brown M, Naboni E (2019) About the triggering of UN sustainable development goals and regenerative sustainability in higher education. Sustainability 11:254

Tasci BG (2014) Project based learning from elementary school to college, tool: architecture. In: 5th world conference on learning, teaching and educational leadership, WCLTA 2014, 29–31 Oct 2014. Prague, Czech Republic

Tian YY (2017) Research and practice on the model and method of funds management in colleges. In: Proceedings of the 7th international conference on management, education, information and control (meici 2017), 15–17 Sept 2017. Shenyang, Peoples Republic of China

Tierney A, Tweddell H, Willmore C (2015) Measuring education for sustainable development: Experiences from the University of Bristol. Int J Sustain High Educ 16(4):507–522

Vargas L, Mac-Lean C, Huge J (2019) The maturation process of incorporating sustainability in universities. Int J Sustain High Educ 20(3):441–451

Veiga Ávila L, Beuron T, Brandli L, Damke L, Pereira R, Klein L (2019) Barriers to innovation and sustainability in universities: an international comparison. Int J Sustain High Educ 20(5):805–821

Strategic Planning for a Sustainable Development Centre Using the Balanced Scorecard

Mauricio de Andrade Lima, Gisele Mazon, Brenda Caroline Geraldo Castro, Stephane Louise BocaSanta, and José Baltazar Salgueirinho Osório de Andrade Guerra

1 Introduction

The main objective of strategic management is to provide companies with guidance in pursuit of goals and a better future. Strategy, in organizations, can be broken down into strategy formulation, planning, implementation, evaluation, and when necessary, returns to the formulation phase. These steps can be understood as the flow of strategic management (Stoner et al. 1995).

Nowadays, organizations operate in more competitive and dynamic environments, compared to years ago, regardless of their industry/sector. In the past customer satisfaction was sought, today retention is sought; if before there was a standard management model, today the search is for a model proper to the organization; if before the focus was strategic planning, today the focus is strategic management. Due to

M. de Andrade Lima
Center for Sustainable Development (Greens), Universidade Alto Vale do Rio do Peixe (Uniarp), Florianópolis, Brazil
e-mail: mauricioadelima@icloud.com

G. Mazon · S. L. BocaSanta · J. B. S. O. de Andrade Guerra
Center for Sustainable Development (Greens), University of Southern Santa Catarina (Unisul), Florianópolis, Brazil
e-mail: gisamazon@gmail.com

S. L. BocaSanta
e-mail: stephanelou.bs@gmail.com

J. B. S. O. de Andrade Guerra
e-mail: baltazar.guerra@unisul.br

B. C. G. Castro (✉)
Researcher at the Research Centre for Energy Efficiency and Sustainability (Greens), University of Southern Santa Catarina (Unisul), Florianópolis, Brazil
e-mail: brendacarolyne19@gmail.com

W. Leal Filho et al. (eds.), *Universities, Sustainability and Society: Supporting the Implementation of the Sustainable Development Goals*, World Sustainability Series,
https://doi.org/10.1007/978-3-030-63399-8_14

these changes in the organizational environment, the definition and development of management strategies (company diagnosis, analysis, and definition of strategic objectives) by companies are no longer a "fad" becoming a factor of value creation for the organization (Mintzberg et al. 2005).

It is noteworthy that the development of management strategies, as well as strategic planning, fits any type of environment or organization. This research shows the reality of the Energy Efficiency and Sustainability Research Group—Greens. Greens aim to promote sustainability through excellent research that contributes to science and the lives of citizens. This research group, being linked to a university, understands that universities have enormous potential to contribute to the formulation of public policies and the implementation of the SDGs (explain what an SDG is before using it as an abbreviation) In addition, the training of teachers and new researchers plays a central role in the process of social change and the direction of a sustainable future (Nousheen et al. 2020).

It should be noted that teachers and students who undergo a training process within their universities, where they develop education for sustainability, have a more positive attitude towards sustainable development compared to colleagues who do not undergo this training process. To reach this stage, it is necessary to implement forms of performance measurement derived from their strategies and capabilities. This brings the Balanced ScoreCard—BSC, whose measures should be used to articulate organizational strategy then communicate it to the community and help align interdepartmental and individual initiatives, to achieve a higher goal synergistically (Kaplan and Norton 1996).

In this study, we aimed at the application of BSC as a research group management tool. In this application, called BSC Greens Group, the traditional BSC perspectives will be renamed "Dimensions" as well as changes being made to their traditional positions and their guiding questions. The uniqueness of this study highlights the contribution, as well as the adaptation and/or adequation of organizational management tools for research group management, where the planning, implementation, alignment and deployment for strategy are relevant to a qualified scientific production. Universities contribute a large part of the production of new knowledge, as well as the improvement of existing literature. However, the implementation of sustainable development, in a practical way, is not direct and requires the contribution of different actors, university, and society. The integration of the university with society, through public policies, is crucial to support the effective implementation of sustainable development.

2 Theoretical Reference

Making a rescue of the definition of strategy Stoner et al. (1995) brings that, when it comes to the meaning of the word strategy, it served to designate the administrative function of *generalate* (general, as in Greek), the art of the general. The strategist (the strategist) was the military commander responsible for formulating and designing the necessary maneuvers to achieve a goal. The concept of strategy, from its inception, was associated with the military view, as the application of forces against an enemy. After the Renaissance, strategy began to be used as the art of war, no longer just a function of the *generalate*. Although, the Greeks were aware that strategy was more than just fighting in battles.

The duty generals were responsible for properly planning their actions. Their responsibilities also included deciding when to fight, when not to fight, and managing the army's relations with politicians, diplomats and citizens. Effective generals could not only plan, but above all had to act. The concept of strategy since the time of the Greeks had elements of decision making or action. These two elements together were the basis of the great strategic planning (Stoner et al. 1995).

Strategic competition is nothing new. Elements of strategy have been recognized and used since humans combined intelligence and imagination, with resources and coordinated behavior to direct their wars. Strategic business competition is a fact, and it can certainly have as massive an impact on the productivity of organizations as the Industrial Revolution brought on individual productivity (Montgomery and Porter 1998).

A strategy is the set of initiatives and indicators that companies need to formulate to achieve their goals. Planning can be understood as a basic process that managers use to determine the company's strategic objectives and what it will do to achieve them (Stoner et al. 1995; Montgomery and Porter 1998).

Mintzberg et al. (2005) add that strategic planning can be understood as a formalized process to generate and articulate results, as an integrated decision system. The technique called "strategic planning" arose basically from the union of two concepts: strategy and planning.

Strategic planning management requires evaluation processes that include measurement, but do not end with it, that is, evaluation is the next step to the measurement process. To understand the importance of measuring performance and evaluating an organization, Kaplan and Norton (1996) express an example, which is briefly presented below.

Corrêa and Hourneaux Jr. (2008) think further by describing that the belief that what is measured is managed must be exchanged for what is measured and used in assessments is managed. Evaluating organizational performance consists in valuing what an organization considers important, considering its strategic objectives, informing what level of performance it is in, with the purpose of improvement actions.

In other words, the organizational performance evaluation deals with the process of identifying aspects considered important. It does this in an organizational context by evaluation of these aspects, visualization of organizational performance, and as a result is an indication of improvement actions.

2.1 Balanced ScoreCard-*BSC, Perspectives and Premises*

As a basic premise of the BSC, rather than a control system, its measures should be used to articulate organizational strategy, communicate it, and help align interdepartmental and individual initiatives. They should do this in order to achieve a larger goal in a more effective and synergistic manner. The BSC measures operational performance using four perspectives: financial, customer, internal business processes, and learning and growth (Kaplan and Norton 1996). The connection of the four perspectives of the BSC can form a cause and effect relationship. The four collaborate to define project priorities to be developed and implemented to achieve the desired financial results (Dias-Sardinha et al. 2007).

The term "balanced" refers to a "balance between external measures for shareholders and customers, and internal measures of critical business processes, innovation and learning and growth" (Möller and Schaltegger 2005). Financial and non-financial measures, external indicators and internal measures of critical business processes, innovation, learning and growth should be part of BSC. For Epstein and Wisner (2001), all relevant aspects must be included in the BSC in order to achieve a permanent competitive advantage by explaining the value proposition established in the organizational strategy. Creating and preserving competitive advantages serves to ensure the economic success of a company.

Therefore, it is a strategic management system that uses performance indicators organized around four dimensions. The BSC's properties as a measurement system establish a focus that provides a sense of direction shared throughout the organization (Kodrin 2016). Finally, it should be noted that BSC is not a tool for strategy formulation. It should serve to describe an existing strategy consistently to improve its execution and make it successful (León-Soriano et al. 2010). Thus, the BSC enables the identification of potential value-added areas and prepares the strategic process by defining the strategic map. This, in turn, will make it possible to understand the impacts of other aspects of these dimensions, not only being linked to operational aspects, but also to strategic strategies, development, and value creation.

The development of the BSC requires that the organization's mission and vision are well defined and consolidated. The BSC methodology reveals a continuous interaction between these concepts. The entire established structure: strategy, strategic map, objectives and indicators derive from the organization's mission statements and vision. In the original BSC, for each perspective there is a guiding question for selecting the most appropriate strategic objectives. Therefore, the BSC has the premise that there is "a way" for the organization to go from formulating its strategy, which is stated in its mission, vision and values, to the results that will come from them.

The main questions asked in each perspective for the construction of the management system according to Kaplan and Norton (1996, pp. 26–29) are (Table 1).

For Kaplan and Norton (1996) the success of implementing a BSC lies in articulating the objectives with coherent mutually reinforcing measures. For the authors

Table 1 BSC Dimensions

Financial perspective	For financial success, how should we be seen by our shareholders?
Customer perspective	To achieve our vision, how should we be seen by our customers?
Internal processes perspective	To satisfy our shareholders and customers, which business processes should we excel at?
Learning and growth perspective	To achieve our vision, how will we preserve our ability to change and improve?

Source Kaplan and Norton (1996)

"strategy is a set of hypotheses about causes and effects […] and must permeate all four perspectives of a BSC."

2.2 BSC and Its Relationship to the Strategic Map

The first step for the use of BSC in organizations is the design of the Strategic Map—SM. The purpose of the SM is to decode the complex processes of choice against which managers at all levels are exposed daily. It is through the SM that senior management monitors can comply with the strategy. According to Kaplan and Norton (2004), the SM spells out the strategy hypothesis, and each objective becomes an integral part of a logical cause-and-effect chain that connects the desired outcomes of the strategy. The SM describes the process of transforming intangible assets into tangible results for customers and therefore into financial results as shown in the Fig. 1.

Other authors (Rohm and Montgomery 2011), in their studies confirm the purpose presented by Kaplan and Norton (2004), stating that such a tool "translates" the complex processes of choice against which managers at all levels are exposed daily. It enables the identification of potential areas of added value and prepares the strategic process by defining the strategic. This, in turn, will make it possible to understand the impacts of other aspects of these dimensions, not only linked to operational aspects, but also to strategies for the development and creation of strategies with sustainable value. In addition, it will serve to help organizations view their strategies in a cohesive, integrated, and systematic manner, as well as build the foundation of the management system for effectively and efficiently implementing the strategy. This demonstrates the link between strategy formulation and execution, and are tools for shaping, narrating, visualizing, communicating, and aligning organizational participants around the adapted strategy.

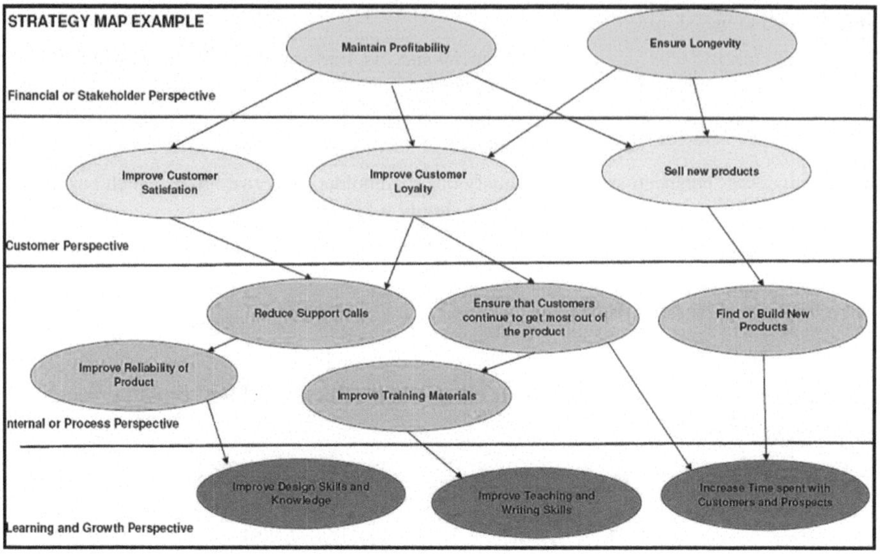

Fig. 1 Strategic map example (Kaplan and Norton 2004)

3 Methodology

The research objectives themselves are descriptive, though it is qualitative in nature, and based off of a singular case study. As for the data collection technique, there were interviews, unsystematic observations, documentaries, and bibliographic research used. As for the data analysis technique, content analysis and triangulation were used.

BSC is a tool that enables the practical deployment of organizational strategy, translating the company's mission and values into practical, tangible goals that can be measured and optimized throughout the organization's existence and how such concepts are prerequisites.

While not specifically addressing a BSC geared towards an organization as a whole, Kaplan and Norton (2000) do make some reflections on adapting to other specific applications. Some comments are of interest and may serve as a guide for the adaptations of the model focused on the purpose of this research, for example, commenting on BSC's adherence to the public sector, the authors state that:

> Most nonprofit and government entities face difficulty with the original BSC architecture, where the financial perspective is placed at the top of the hierarchy. Considering that financial success is not the main objective of most of these organizations, it is possible to reformulate the architecture. (Kaplan and Norton 2000, p. 147)

In this sense, Table 2 was made with the new structure, but with the same philosophy as the original BSC. In this approach it was called the BSC Greens Group, renaming Traditional BSC's perspectives by "Dimensions" as well as changing their positions and their guiding questions.

Table 2 BSC Greens Group

Dimensions	Guiding questions
Scientific research perspective	How do we want to be seen by our maintainers and the academic community?
Academic community perspective	What scientific means should be present for the dissemination of our research?
Enabling internal processes perspective	To satisfy our stakeholders and the academic community, which internal processes should we perform optimally?
Staff learning and development perspective	For the best results in our processes, how will we align and develop our human, information and organizational capital?

Source Authors (2020)

In the traditional BSC model, the economic-financial perspective aims to maximize shareholder value, that is, to evaluate the contribution of the implementation and execution of the company's strategy, being measured by the financial results with the aid of other perspectives (Huang 2009). According to Kaplan and Norton (2000) the financial perspective reflects the shareholder's point of view. For companies to be financially successful, they must satisfy their shareholders, so it is placed at the top of the BSC hierarchy. Therefore, this is an aspect that cannot be disregarded from the context, quite the opposite. However, the purpose of this paper is to make it a "perspective" that is supporting the process and not the protagonist, since the purpose here is to offer a mechanism where the main result to be sought is to promote sustainability through excellence on research, contributing to science and citizens' lives.

For this purpose it was decided to create the **Scientific Research Perspective** being the first, and having as its guiding question "How do we want to be seen by our maintainers and the academic community?", where their objectives will succeed when the objectives of the other perspectives were achieved, in a cause and effect relationship. For the fulfillment of this process, the **Academic Community Perspective** was put in support of the first perspective, having as a guiding question, "What scientific means should be present with a view to disseminating our research?", **Customer Perspective**, in the original BSC, is identified as companies meeting their customers' needs through a value proposition (Kaplan and Norton 1996). The second perspective will aim to clarify the key and relevant issues that will highlight the new value proposition and high impact publications.

Next, in place of the **Internal Process Perspective** (original BSC) where, according to Kaplan and Norton (2000), allows managers to identify processes that are critical to meeting customer and stakeholder goals, we have the **Enabling Internal Processes Perspective** with the following research question: "To satisfy our stakeholders and the academic community, which internal processes should we perform optimally?" where similarly to the original BSC, the group should identify which internal processes to be able to perform. to provide high impact research.

BSC's fourth and final perspective develops objectives and measures to guide learning and organizational growth. For Kaplan and Norton (1996), this is considered the scorecard result vector, as it offers support, change support, innovation and organizational growth for the first three perspectives to achieve their goals. Thus, the main objective of the learning and growth perspective is to be the driver of good results from other BSC perspectives. For this purpose, the **Staff Learning and Development Perspective** was created, with the guiding question: "To obtain the best results in our processes, how will we align and develop our human, information and organizational capital?" In this perspective, investment is directed at people, information systems infrastructure and Research Group alignment. Thus, human capital is of great relevance to strategic objectives, as it recognizes that the individual in work teams or business areas is the most important factor for the survival and renewal of companies at all levels of activity (Kaplan and Norton 1996).

Kaplan and Norton (2004) also point to a "second layer" of BSC details (Descriptors) illustrating the strategy's time dynamics, and also add a level of detail that enhances clarity and focus while BSC translates objectives into indicators and goals. Thus, organizations should launch a set of programs that will create value and conditions for achieving the goals and objectives of all indicators.

For this work, based on the "Dimension" and its respective "guiding question", actions or practices that corresponded to the function/objective of each Dimension were identified in the literature. The adequacy presented in this paper is justified by "facilitating" the guidelines/keywords that should contemplate each dimension from the guiding question. Thus, based on the literature, this process contributes to better structuring and/or flexibility for those who wish to make use of this proposal. Thus, the guidelines and strategic objectives can be better adapted according to the strategy and maturity of the organization (Fig. 2).

4 Results and Analysis

4.1 BSC Greens Group: The Strategic Map

For deployment and implementation, the Mission, Vision and Values were defined based on the Group's Strategic Planning. Mission: to promote sustainability through research of excellence contributing to science and the lives of citizens. Vision: To be a research center of world reference in sustainable development through teaching, research and extension and obtaining successful publications in Science or Nature until 2030. Values: high performance in scientific publication; value proposition to transform lives; joint and integrated work; Freedom of thought and expression; and ethics (Guerra 2019).

According to Guerra (2019), Greens is currently international and therefore intends to maintain and expand its objectives. Still, the idea of reaching dated goals is a good one, since after a while, there is a need to review them. It also considers 2030

Scientific Research Perspective *How do we want to be seen by our maintainers and the academic community?*
[Intellectual Production] [Social Insertion] [Scientific Collaboration]
Academic Community Perspective *What scientific means should be present for the dissemination of our research?*
[Qualified Journals] [International & National Events] [Media Reference] [Exchanges]
Enabling Internal Processes Perspective *To satisfy our stakeholders and the academic community, which internal processes should we perform optimally?*
[Profile Selection] [Planning & Scheduling] [Portfolio Themes / Research] [Periodic Portfolio] [Editing, Layout & Translation] [Submission]
Staff Learning and Development Perspective *For the best results in our processes, how will we align and develop our human, information and organizational capital?*
[Human Capital] [Information Capital] [Organizational Capital]

Fig. 2 Guidelines and strategic objectives (Authors 2020)

the ideal year, as it will be aligned with the objectives of sustainable development (SDGs). Still, according to Guerra (2019), one goal that Greens has is to transform the lives of students and through that transform Brazil, and then keep training competent scientists, undergraduate students, masters, doctors, and post-doctors. By doing this, Greens hopes that their researchers are free to propose studies and improvements, free to speak about any subject.

Once the mission, perspectives, guiding questions and strategic themes of the BSC Greens Group are defined, it is understood that the objectives of the "Scientific Research Perspective" are linked to the strategy, which is the main objective, that is, from this perspective. It is understood that the results to be obtained are those identified by the scientific community as a group that presents intellectual production, social insertion and excellent and high impact scientific collaboration. The reason it has the function of being the focus for the other objectives and following prospects by means of causal relationships. Thus, it is relevant to understand the correlation between each of them.

In this sense the second perspective (**Perspective of the Academic Community**) it's directly linked and aligned with the first perspective, because in order to be identified as a research group of scientific excellence by the academic community, there is need to publish in qualified journals, participate in events internationally and nationally, as well to be recognized by society as an opinion maker on the subject and conduct exchanges to improve research.

Once the desired objectives are set for the group's scientific research (Perspective 1) and how Greens should introduce itself to the academic community (Perspective 2), the third perspective (**Perspective of Enabling Internal Processes**), will point to which group internal processes should perform better, as these processes from this perspective will underpin previous perspectives.

Finally, there is **Staff Learning and Development Perspective**, this perspective concerns the people who make up the organization analyzed. That is, if we need to have excellent processes, and since these processes are made up of people, they must be aligned and developed to act on them. For this we need to have excellent human capital (skills, training and knowledge), information capital (systems, databases and networks) and organizational capital (culture, leadership, strategic alignment and teamwork).

In order to better understand these connections, it is considered essential to know the general structure of the SM where are presented the dimensions, guiding questions, and descriptors identified from the literature and the strategic objectives, as shown in Fig. 3.

From the **Scientific Research Perspective**, Guerra (2019) points out that Greens aims to be seen as one of the most productive research centers, in relation to high impact publications, with the participation of undergraduate, master and doctoral students (which is the current structure of the center).

The insertion of the social area is one of the main concerns, and the center seeks to make projects that unite them to the community. Currently, two projects are active with the creation of a Hydroponic Greenhouse that works at the Unisul Technological Center and another at the Soares School in Tubarão, where everyone benefits from the coexistence of clean technologies, locally produced food and also for students' environmental education.

Regarding scientific collaborations, Greens has several projects with international universities such as Cambridge University, Hamburg University, Manchester University, University Institute of Lisbon with the University of Chile. Guerra (2019) wishes that the center will maintain the approval of projects in international calls for proposals from international agencies from the United Kingdom and the European Union, and thus be able to provide the center's researchers with knowledge exchange networks in international research networks.

Currently, the center has national and international funding to carry out its projects, such as Fapesc (Santa Catarina State Research Support Foundation) and abroad, Global Challenges Round Tables of the United Kingdom and Research Councils United Kingdom.

As for the **Academic Community Perspective**, Guerra (2019) stresses the center's concern with publishing its work in high-impact journals. The center currently wants to publish at least four articles per year in journals with an impact factor of over 2,5. Currently, Greens publications are present in at least ten different journals such as the Journal of Cleaner Production and Natural Resources Forum. The center hopes to increase its level of publication and someday publish in Nature or Science.

As for event participation, Greens organizes every two years a major international event related to sustainability, such as conferences or symposiums. Externally, it seeks to participate in at least two international events annually, and hopes to continue with these numbers. The center participation in the media is mainly on local television

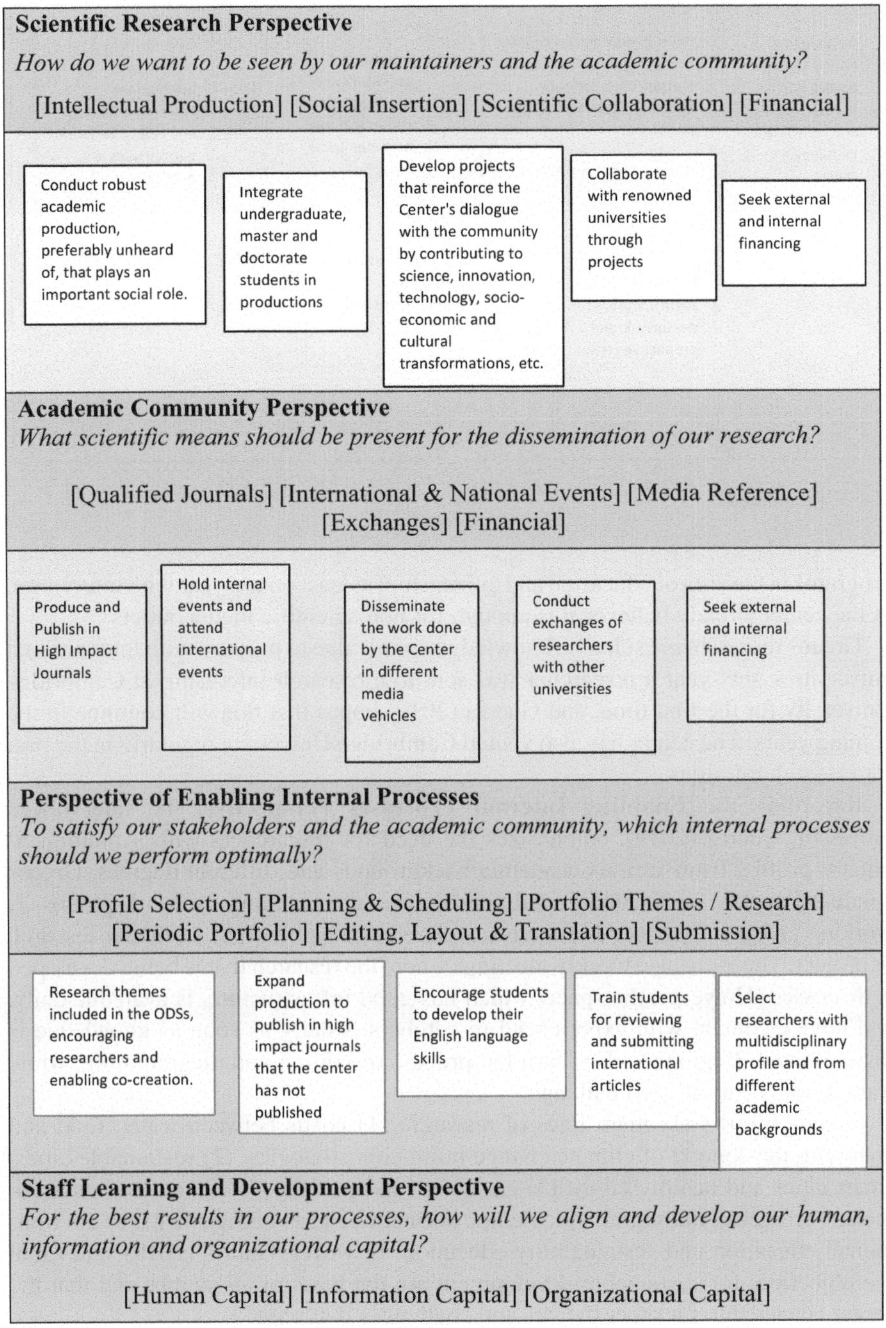

Fig. 3 Dimensions perspectives (Authors, 2020)

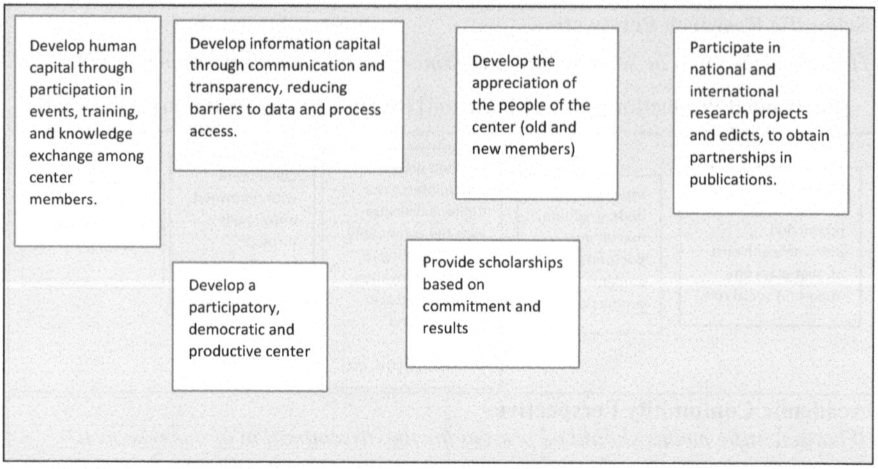

Fig. 3 (continued)

programs in the area of education and citizenship, at least once a year with interviews, as the center already believes it is enough for non-scientific media outlets.

Greens researchers exchange knowledge mainly due to projects with international universities, this year a researcher was sent to a research internship at Cambridge University for the first time, and Guerra (2019) hopes that this will continue in the coming years. The center has also visited Cambridge University regularly in the past for research missions.

Regarding the **Enabling Internal Processes Perspective**, the interviewed professor, Guerra (2019), emphasizes the need for researchers with a multidisciplinary profile, from various academic backgrounds and different degrees. Greens conducts weekly, semi-annual, and annual schedule planning for work groups. A working group is composed of the researchers responsible for a particular research or project. There are also weekly meetings where the research that is being developed is discussed. However, the space, which has good infrastructure, is available daily, and novice contact with experienced researchers is daily as soon as knowledge is disseminated. Beginners also learn the process of editing and diagramming, while translation is currently a challenge.

Greens follows six main lines of research: (1) nexus between water, food and energy in the context of climate change mitigation strategies; (2) sustainable cities, smart cities and healthy cities; (3) environmental justice; (4) climate and environmental refugees and migrants; (5) empowerment of girls and women; (6) environmental education and sustainability education. However, Guerra (2019) states that the objectives for sustainable development are the basis of all studies and that the center is available for other themes and challenges that arise.

The research center has published in several high-impact journals such as Land Use Policy, Journal of Cleaner Production, Natural Resources Forum, International Journal of Sustainability in Higher Education, and Energy Policy, but the intention is to move to other journals. Greens aims to publish in renowned journals such as Science and Nature. As for the **Staff Learning and Development Perspective**, researchers are constantly improving. Guerra (2019) believes that researchers with varied academic backgrounds bring success to the center. All researchers at the center have access to information from the working groups, because the purpose is to create a co-creation space. That is why researchers are motivated to participate democratically and productively. Guerra (2019) states that the center has a deliberate anarchic structure that allows freedom to create. Thus, people feel more comfortable and therefore produce more.

The center already participates in research projects and edicts, and aims to continue participating and strengthening partnerships, as well as constantly seeking new partnerships. Finally, the biggest financial benefit that Greens generates is scholarships, made available according to students' commitment and results. All of these factors allow the visualization of a strategic Greens map as shown (Fig. 4).

5 Conclusions

In order to help Greens achieve its goals and manage the research group, the study aimed to apply the BSC, called BSC Greens Group. Following the steps of the BSC, the group's mission, vision and values were defined and validated with the coordinator and responsible for Greens. Subsequently, strategic maps were defined based on the following perspectives: (1) Scientific Research; (2) Community; (3) Enabling Internal Processes; (4) Staff Learning and Development.

Greens has an adequate physical structure. In addition, undergraduate, master's, doctoral students and professors develop and promote research and extension at the university and in the community in which they operate. The study contributes for practical purposes, to structure the goals and assist in the management of the group. The study also contributes regarding scientific purposes, bringing an adaptation and/or adequacy of the BSC to a research group.

This research is limited to the theory and methods used. The research is also limited to the research group studied, however, just as it was possible to adapt the BSC to the Greens, it may be possible to adapt the BSC to other research groups. For future studies, the adaptation of the BSC in other research organizations would be valid, as well a study a few years later to analyze how much the BSC implementation contributed to the organizations.

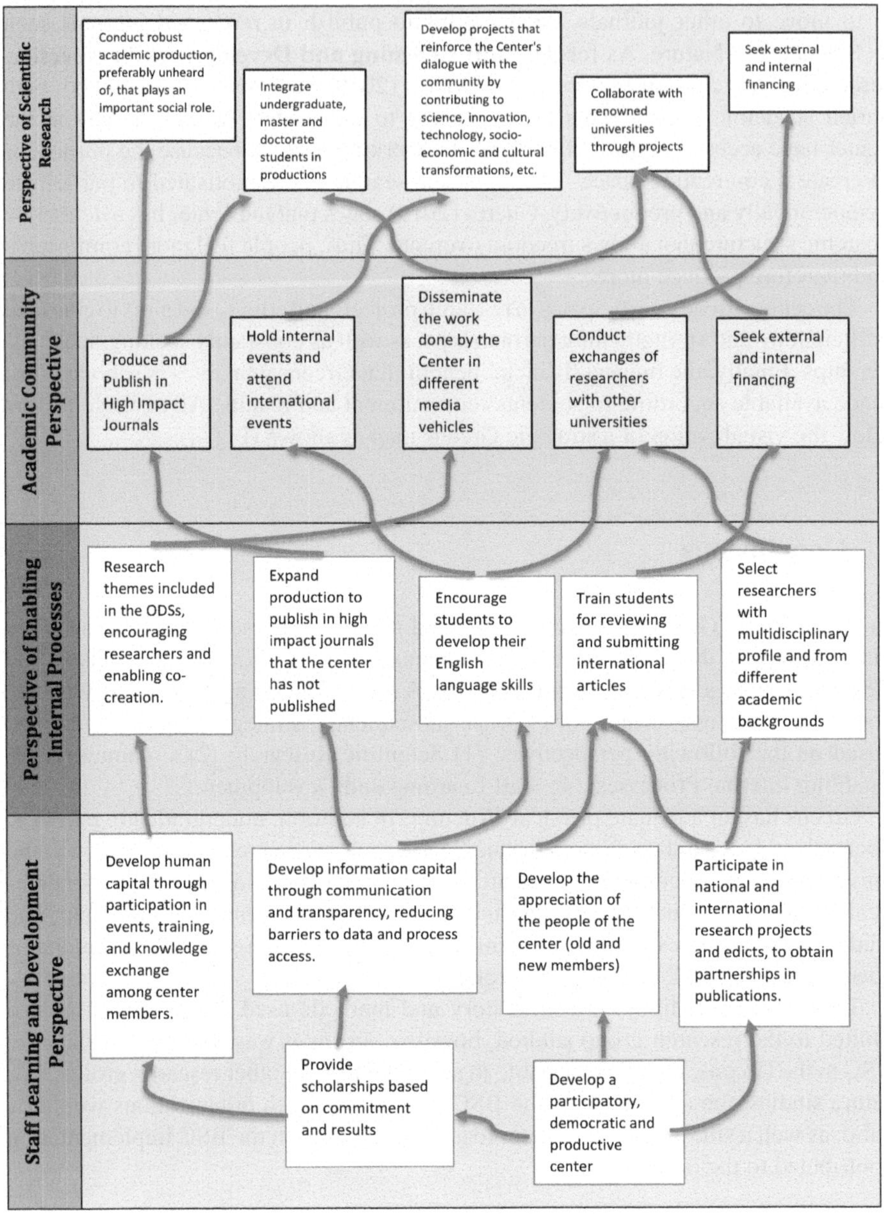

Fig. 4 Greens strategic map (Authors 2020)

Acknowledgements We would like to thank Michael Scalia from Palm Beach Atlantic University for the contributions with translation and review of this article.

References

Corrêa HL, Hourneaux F Jr (2008) Measurement systems and organizational performance evaluation: case studies in the chemical sector in Brazil. R Cont Fin Sao Paulo 19(48):50–64

Dias-Sardinha I, Reijnders L, Antunes P (2007) Developing sustainability balanced scorecards for environmental services: A study of three large Portuguese companies. Environ Qual Manage 16:13–34

Epstein MJ, Wisner PS (2001) Using a balanced scorecard to implement sustainability. Environ Qual Manage 11:1–10

Guerra JBSOA (2019) Interview. In: Boca Santa SL, Castro BCG (eds) Interviewers. File .m4a (41 min). Greens, Florianópolis

Huang HC (2009) Designing a knowledge-based system for strategic planning: a balanced scorecard perspective. Expert Syst Appl 209–218 (Elsevier)

Kaplan RS, Norton DP (1996) Using the balanced scorecard as a strategic management system. Harvard Bus Rev 74(1):75–85

Kaplan RS, Norton DP (2000) The strategy-focused organization: how balanced scorecard companies thrive in the new business environment. Harvard Business School Publishing Corporation, Boston

Kaplan R, Norton D (2004) Strategy maps: converting intangible assets into tangible outcomes. Harvard Business School, pp 43–55

Kodrin A (2016) The balanced scorecard approach for sustainable performance in an uncertain future: the case of climate change: Doctoral Dissertation. https://pdfs.semanticscholar.org/b113/59da9c5c17506c29632d2f9895fcdc34faa0.pdf. Accessed 23 Mar 2020

León-Soriano R, Jesús Muñoz-Torres M, Chalmeta-Rosalen R (2010) Methodology for sustainability strategic planning and management. Ind Manage Data Syst 110(2):249–268

Mintzberg H, Ahlstrand B, Lampel J (2005) Strategy Safari: a guided tour through the wilds of strategic management. Simon and Schuster

Möller A, Schaltegger S (2005) The sustainability balanced scorecard as a framework for eco-efficiency analysis. J Ind Ecol 9(4):73–83

Montgomery CA, Porter ME (1998) Estratégia: a Busca da Vantagem Competitiva. Campus, Rio de Janeiro

Nousheen A, Zai SAY, Waseem M, AliKhan S (2020) Education for sustainable development (ESD): effects of sustainability education on pre-service teachers' attitude towards sustainable development (SD). J Cleaner Prod 250: 119537 (2020, March 20)

Rohm H, Montgomery D (2011) Link sustainability to corporate strategy using the balanced scorecard. The Balanced Scorecard Institute, Cary, NC, USA. http://test.balancedscorecard.org.tempdomain.com/wp-content/uploads/2019/08/BSI-linking-sustainability-to-corporate-strategy-using-balanced-scorecard.pdf. Accessed 24 Mar 2020

Stoner JA, Freeman RE, Gilbert DR Jr (1995) Management, vol 17, issue no 1. Prentice-Hall, Anxiety, Stress and Coping, USA, pp 103–122

Drawing Sustainability: Helping Students to Know What They Know

Kay Emblen-Perry

1 Introduction

The traditional adage that "a picture is worth a thousand words" has been long held as true by the scientific community; consequently, 'drawing' is a well-established teaching tool for science within Higher Education (HE). However, the recognition that students can understand and communicate complex issues through drawing (Boehm and Lees-Spalding 1993) and recognise learning in the process of drawing (Gobert and Clement 1999; Roam 2008; Quillin and Thomas 2015) suggests that it could also be a valuable and innovative learning and teaching approach for students studying non-scientific subjects such as sustainability and society in business management. The simplicity of the approach and its flexibility for learners could promote learning recognition within Education for Sustainability (EfS) within the business curriculum so that students come to know what they know, i.e. they develop a conscious recognition of their own knowledge and, through this acknowledgement of self-knowledge, appreciate their personal and professional knowledge, skills and values. Consequently, an innovative in-class activity has been adopted for Business Management students; drawing sustainability.

The activity of drawing sustainability, incorporated within sustainability modules for Business Management students which introduce them to social justice, environmental protection and economic legitimacy and their roles within the Sustainable Development Goals, is designed to achieve two purposes. Firstly, to provide a simple way for students to recognise their learning and secondly, to provide a benchmark for tutors to recognise students' knowledge at the start of the module and extent of learning through the module.

K. Emblen-Perry (✉)
University of Worcester Business School, Castle Street, Worcester WR1 3AS, UK
e-mail: k.emblenperry@worc.ac.uk

W. Leal Filho et al. (eds.), *Universities, Sustainability and Society: Supporting the Implementation of the Sustainable Development Goals*, World Sustainability Series,
https://doi.org/10.1007/978-3-030-63399-8_15

Traditionally, students' recognition of learning within EfS has relied on tutors' formative feedback and assessment grades. However, relying on students to act on their formative and post assessment feedback means that by the time tutors realise students have fallen behind or have misunderstood the complex interrelated topics of sustainability it is frequently too late for them to catch up (Webb 2020). Drawing sustainability as an in-class reflexive and self-analytical learning activity may therefore offer an alternative approach for both tutors to know what students know (i.e. benchmarking learners' knowledge) and for students to know what they know (i.e. learning recognition through a judgement of self-knowledge).

In this study, drawing sustainability is defined as by the author as 'an activity that engages students in a process of visually representing an understanding of the topic of sustainability to produce a pictorial output'. The drawings, which are the pictorial output of the activity, are not subject to content analysis in this study, rather they are used as indicators of knowledge and learning. Drawings can demonstrate the extent of students' knowledge through the number of different elements of sustainability included.

The idea for incorporating drawing sustainability to promote students' recognition of learning emerged from the combination of three influences: firstly, the author's preference for creative and visual thinking and belief that having fun will support deeper learning, secondly literature that stresses the importance of students recognising their own learning and thirdly, a conference presentation highlighting the use of ecological hand printing to engage students in alternative ways of thinking. Together these suggested that drawing could be a potential way to encourage students to participate in reflection, peer-to-peer learning, and self-analytical learning, which may positively influence learning motivation, (Schunk and DiBenedetto 2016), enhance learning recognition and develop self-efficacy and confidence by making thinking visible (Vygotsky 1978).

This paper explores the value of encouraging students to recognise their learning through drawing sustainability. The activity of drawing sustainability is described in detail to enable it to be implemented by those seeking innovative approaches to learning recognition within EfS. The effectiveness of the activity is then presented and its' potential impacts on students and tutors reflected upon. This analysis and reflection incorporate students' drawings and the students' voice, through examples of post-activity feedback, to illustrate the outcomes of the activity. The study finds that whilst supporting students' recognition of learning to enable them to they know what they know, drawing sustainability may also promote higher order cognitive skills and encourage students to relate personal sustainability practices to society and the environment, all of which are key skills and behaviours required to promote sustainable futures and achievement of the Sustainable Development Goals.

This study will be of interest to EfS practitioners seeking innovative means and methods of active learning that enable them to benchmark students' learning and help students to develop a conscious recognition of their own sustainability knowledge skills and values. The study also builds on existing pedagogic discourse on the importance of learning recognition and contributes to research into innovative learning and teaching approaches to sustainability and society and the Sustainable Development Goals within the business curriculum.

2 The Role of Business Schools in Education for Sustainability

As demonstrated within Sustainable Development Goals 4 (Quality Education) and 12 (Responsible Consumption and Production), there has been a growing expectation that educational systems should contribute to developing a sustainable society (Sterling et al. 2013; Higher Education Funding Council for England 2013; Higher Education Academy 2015; United Nations 2017) such that EfS is becoming mainstream in HE (Figuero and Raufflet 2015; Franco et al. 2018). Lonzano et al. go further and suggest that it is a university's social responsibility to help students acquire the required competencies to enable them to develop sustainable futures.

The momentum for EfS has never been stronger (UNESCO 2017) but to align it to the Sustainable Development Goals and to achieve the skills, values and attitudes required to develop sustainable futures HE practitioners need to significantly change curricula and their learning and teaching practices (Navarro 2008; Molthan-Hill 2017; UNESCO 2017; Franco et al. 2018). Education systems must introduce pedagogies that empower learners to transform the way they think and act (Howlett et al. 2016) to support practical achievement of a sustainable future for society, industry and organisations (Boron et al. 2017).

Business schools are well placed to address this. They can adopt teaching practices that develop knowledge and mirror the real-world decision making that students will face in their future role(s) as business leaders (Storey et al. 2017). This will provide businesses with graduates who possess the appropriate sustainability knowledge and employment skills (Drayson 2015), and reflexive ability (Hind et al. 2009) to become architects of socially and environmentally sustainable futures. To achieve this, Laurie et al. (2016) suggest tutors should adopt practices that go beyond the business perspective and promote learning of the skills, knowledge and values that are necessary to foster sustainable societies.

Although Business Schools are recognised as potential enablers of change (Weybrecht 2017), a challenge remains; how to prepare students to both recognise and act upon the knowledge gained through participation in Higher Education (Storey et al. 2017). This study seeks to contribute to overcoming this challenge by presenting an innovative way to encourage students to reflect upon and recognise their learning so that they know what they know to become agents of change.

3 The Importance on Learning Recognition

Over the last 30 years the need for learners to recognise their learning has gained importance as it has become clear that being able to judge self-knowledge enhances the quality of a learner's education (Kurnaz and Cimer 2010). Those able to reflect on and become aware of their abilities are likely to perform better than those who do not (Flavell 1976) and become lifelong learners (McDonald and Boud 2003). This may be due to the fact that becoming aware of one's own knowledge promotes the self-analysis and recognition of strengths and weaknesses which can be addressed thus facilitating further learning (Kurnaz and Cimer 2010). Hannon et al. (2004) and Kurnaz and Cimer (2010), therefore recommend that tutors should introduce self-assessment strategies into their learning and teaching practice to promote students' development of reflective skills and recognition of learning. This may be achieved by engaging students in knowledge construction practices (Reiser 2004), challenging learners to focus around sense making, process management, articulation and reflection (Quintana et al. 1999), and developing students' feedback literacy (Carless and Boud 2018). In this study drawing sustainability is adopted as an activity to facilitate reflection, knowledge construction, feedback literacy and self-analytical learning.

In spite of the long-standing recognition of the benefits of encouraging students to know what they know, graduates' knowledge and attributes remain difficult to measure, frequently by graduates themselves (Storey et al. 2017). This may take the form of under or over confidence (de Bruin et al. 2017). Davis (2003) recognises that learners are not always effective in analysing whether they have understood and may therefore be overconfident in their self-assessment of learning. Caleon and Subramaniam (2010) suggest this confidence may be related to the familiarity of topics; where topics are less familiar confidence levels are lower. Therefore, approaches to promoting reflection on knowledge in new subjects such as sustainability require this to be taken into account.

To overcome this lack of understanding related to familiarity and overconfidence, Reiser (2004) suggests students should be encouraged to be more reflective and focus on understanding rather than performance to promote skills of reflection and articulation. This may allow tutors to effectively monitor and evaluate students' progress and encourage them to learn from their experience and recognise their learning. These are key processes designed into the activity of drawing sustainability.

The importance of students' learning recognition provides additional benefits to tutors. Asking students to explain a concept in their own pictures may benchmark knowledge, help to identify misunderstandings and offer opportunities to provide high quality, individualised feedback which can guide students to greater success (Webb 2020).

4 The Value of Drawing

Webb (2020) suggests an effective way to check students' understanding is to ask them to write down everything they know about a topic. This study develops this approach further by adapting it into a visual representation of knowledge thus incorporating the value of drawing that science recognises can provide an opportunity for students to recognise their own learning i.e., to know what they know. To do this, students are asked to draw sustainability.

Drawing is a means by which individuals can create, develop, record and communicate their thoughts and ideas (Van Meter and Garner 2005); it therefore presents an opportunity to support thinking and learning. (Hope 2008; Adams 2017) by allowing students to apply what they have learned and recreate it in another format (Griffith and Burns 2014).

In primary school, drawing as a teaching tool has been found to promote pupils' analytical skills through use of perception and critical thinking by introducing pupils to new and increasingly complex subjects (Quillin and Thomas 2015). It can also build self-esteem and self-confidence (Jones 1997). In Higher Education's scientific subjects, drawing has been widely adopted as a tool to test ideas, elaborate knowledge, and imagine new relationships (Ryan and Stieff 2019) through which learners are encouraged to create their own visual representations which can deepen understanding. (Ainsworth et al. 2011). As concepts within EfS such as social justice, environmental protection and economic legitimacy are frequently new to Business Management students and expose them to a level of complexity that they have not previously faced within their university career, it may be beneficial to utilise the same teaching tool of drawing. This may offer students the opportunity to see their previously unseen knowledge on paper (Vygotsky 1978; Quillin and Thomas 2015), thus helping them know what they know. Drawing gives students an opportunity to ask themselves *What do I know?* and *What else do I need to know?*

Drawing can be a collaborative tool that promotes thinking (Hay et al. 2008), the visualisation and interpretation of data (Ainsworth et al. 2011), and communication of results (Ryan and Stieff 2019). The reflection required to enable this visualisation and interpretation has a key role in learning (Schon 1987). Ainsworth et al. (2011) also suggest that drawing encourages students to make their thinking explicit to exchange and clarify meaning between their peers, which Vygotsky (1978) considers is due to the ability of drawing to make ideas tangible and thinking visible. In turn this may motivate students and make them more self-aware of their own learning (Quillin and Thomas 2015). Drawing may therefore help students to know what they know.

Drawing sustainability as a teaching tool may have a particular value for Business Management Students; drawing can encourage the learner to see the world differently (Unicheck 2015). This ability to see the world differently is a vital requirement for effective EfS (UNESCO 2017) as sustainable futures require students to transform the way they think and act (Howlett et al. 2016).

In addition to using drawing as a tool for learning specific knowledge (Gobert and Clement 1999; Hay et al. 2008) and communication (Roam 2008) it may also have a number of other positive benefits for learning and teaching. Drawing is considered to expose misconceptions (Dikmenli 2010) as it can demonstrate understanding in an inspectable form (Van Meter and Garner 2005), to enhance motivation (Glynn and Muth 2008), and to connect concepts and ideas (Long et al. 2014). Brooks (2005) also suggests drawing can be a valuable mode of exchange as the drawing can be presented as a record of learners' thinking that can be reviewed and revisited by the tutor and student, thus helping students to recognise their learning and tutors to know what students know, i.e. a learning benchmark. All of these pedagogic benefits will help staff to support students to understand the interconnected problems, complexity and opportunities that are inherent in EfS, which the author recognises is generally a problem for students.

As drawing caters for individual learners' differences, with the task being shaped by the students' ideas (Ainsworth et al. 2011), it can be effective as a formative exercise as it can scaffold learning and facilitate targeted feedback (Quillin and Thomas 2015) which in turn can prepare students for future learning (Schwartz and Martin 2004). This is valuable as EfS seeks to promote students' success and encourage lifelong learning.

5 Drawing Sustainability as a Learning Activity

Drawing sustainability attempts to benchmark student's knowledge and understanding of sustainability to support future teaching inputs and encourage students' reflection on learning. The activity is therefore incorporated into the first and last lectures of second- and-third year undergraduate sustainability modules for Business Management students. In the first iteration of the activity no information on the module structure, content or assignments is given until after the students have completed their drawing to ensure that no unintended prompts are offered. Students are informed that the purpose of the task is to benchmark their knowledge and that their drawings would be retained for use later in the module.

In the last lecture of the module, students are asked to repeat the activity to provide a new record of their sustainability knowledge. This second iteration of the activity provides students with an opportunity to analyse the changes to their drawings and recognise the extent of their knowledge. Although Webb (2020) suggests that learning cannot be seen, drawing sustainability challenges this view as it encourages students to see both learning and progression of learning.

In both occurrences of the activity, the small self-selected groups are given a piece of flip chart paper and a set of marker pens in four colours. The use of pen and paper encourages students to think in a different way, apply their knowledge and reflect on knowledge to select, organise and integrate information (Quillin and Thomas 2015), all of which are required to promote employment skills for sustainable futures. Students are given the flexibility to draw sustainability in any way they choose but are asked to put away phones and laptops so that their own knowledge is presented and discussed rather than an internet opinion.

The activity of drawing sustainability is kept simple, with clear instructions given by the tutor. These are provided on a slide which is left on display throughout the activity. These instructions are, "In small groups and using the paper and pens provided (1) draw your understanding of 'sustainability' and (2) share your thoughts with the class". The first element of the task is given so that students demonstrate their knowledge and provide a knowledge benchmark. The second element has a number of purposes: it engages students in groupwork, collaboration, peer-to-peer learning and peer feedback and indicates to the students that they are required to participate as an outcome is expected. This feedback requirement is incorporated to encourage students to take greater interest in, and responsibility for, their drawing, its content and process of creation in an attempt to overcome the current trend for "presenteeism".

As learning is primarily a social activity (Dewey 1916) and current business management students increasingly prefer active, collaborative learning (Oblinger and Oblinger 2005; Abdel Meguid and Collins 2017), students are encouraged to participate in the activity of drawing sustainability in small groups of between 2 and 5 colleagues to support the development of collaboration and negotiation skills. These groups are self-forming and tend to relate to the friendship groups that are sought by students upon entering the classroom at the start of the first lecture. However, to ensure inclusivity is protected students preferring to work on their own may do so.

Many current students are risk adverse, which Quillin and Thomas (2015) suggest can emerge as insecurity about their ability to draw. Therefore, to overcome the frequent response of "I can't draw", the purpose of the activity and intended use of drawings is highlighted (benchmarking knowledge), and the unimportance of drawing skills is emphasised. Additionally, the tutor praises each group's drawing in some way (e.g. a particular piece of the content or the recognition of a sustainability challenge) during the early stages of the activity to promote activity and drawing confidence. After a few minutes, students appear to forget their initial reluctance and fully participate in the task, even if it is only drawing with simple shapes or by contributing ideas that another member of the group draws.

After the students complete their drawings one or more self-selected members of each group present their group's drawing to the class. They are asked to describe the content of their drawing and share why they chose these items. The tutor identifies and praises their recognition of one significant element of sustainability included in the drawing and expands upon it to emphasise its' relevance for sustainable futures and its' importance to businesses. Examples of the students' initial drawings are shown in the results section below.

After students present their drawings to the class, peer reflection and feedback is encouraged through a brief class discussion. This can promote learning from critiquing what their peers have drawn (Linn et al. 2000) and provide criteria with which to recognise the effectiveness of a task and reflect on learning (Brooks 2005).

The same activity is conducted for the second iteration of drawing sustainability in the last lecture of the modules. However, an additional task is included. After the groups' drawings are presented and discussed, the students are given their original drawings and asked to reflect upon them and reflect on the differences. From this the students may recognise how their drawing content has developed thus realising what they have come to know. Group findings are then fed back to the class to further encourage students to know what they don't know through peer-to-peer learning. To further support students to learn from their activity experience, as advocated by Reiser (2004), a group discussion on the developed content and its implications for businesses is held. Following this, students are asked to reflect on their experience of learning through drawing sustainability, the feedback from which is utilised within this study.

6 Research Process

This study into the learning recognition promoted by drawing sustainability was undertaken in one second year and one third year business sustainability module over three academic years from 2016–17 to 2018–19. The modules used within this study are both one semester long and scheduled in the second semester of the academic year.

A total of 163 Business Management students participated in this study over the three years, with an average of 36 third- and 18 second-year undergraduates participating in each year. Only two third-year students in 2017–18 and one in 2018–19 had previously studied the second-year module so should have had more knowledge than classmates. To more accurately test the development of students' knowledge, the drawings from the groups containing these students have been excluded from this study.

As students work in the same groups throughout the module, the drawings completed in the last lecture can be directly compared to those completed in the first lecture. In this study the drawings were not subject to formal content analysis, rather the elaboration of content reflecting the wide-ranging topics contained within 'sustainability' and the complexity, context and interrelationships between these topics were used as indicators of learning. Examples of students' drawings from their first and last lectures of both second- and third-year modules and quotations taken from students' activity reflection are used to illustrate the findings of this study.

To further evaluate the value of drawing sustainability for promoting reflection, self-analysis and recognition of learning, the inputs (i.e. activity design elements) and the students' reflections on learning through drawing (collected in post activity feedback in the last lecture of the module) were mapped against the Indicators of Learning Recognition Framework. Additional reflections from the tutor are also added to illustrate the activity design outcomes. This framework adopts the eleven Indicators of Learning proposed by Hughes (2006) as the criteria to evaluate the effectiveness of drawing sustainability as a teaching tool with which to develop students' sustainability knowledge and recognition of learning.

7 Results

7.1 Drawing Sustainability as a Tool to Benchmark Students' Knowledge

Examples of drawings completed in the first lecture of the modules shown in Figs. 1, 2 and 3 highlight the limited knowledge of the contents and complexity of sustainability.

As shown in Figs. 1, 2 and 3 over the course of this study students have consistently shown limited knowledge of elements of sustainability at the beginning of the module. Their drawings suggest their knowledge of sustainability is mostly limited to environmental sustainability such as recycling, renewable energy and plastic pollution rather than social sustainability. It might be expected that these environmental sustainability challenges are presented as they are widely publicised in the news and on social media.

Fig. 1 First lecture drawing Group 1 (second-year students 2016–17)

Fig. 2 First lecture drawing Group 2 (third-year students 2018–19)

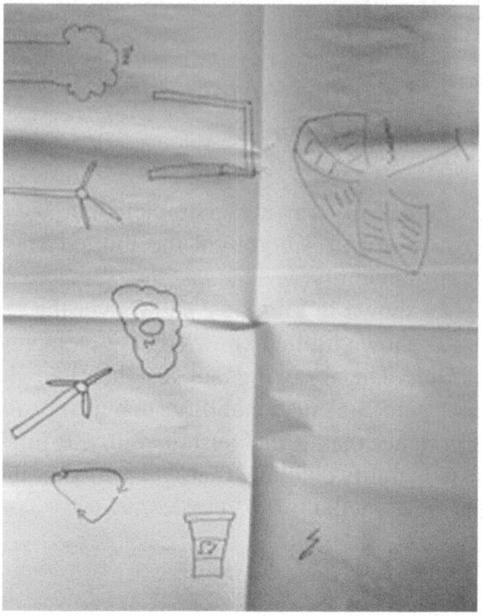

Fig. 3 First lecture drawing Group 5 (third-year students 2017–18)

7.2 *Recognition of Learning Through Drawing Sustainability*

Over the course of the three-year study, both second- and third-year undergraduates' drawings created in the first and last lectures of their modules consistently demonstrated the growth of sustainability knowledge and its application to business environments (Figs. 4a, b, 5a, b and 6a, b). The visual representations of this knowledge shown in Figs. 4a, b, 5a, b and 6a, b suggest students have conducted a self-analysis of learning (perhaps subconsciously) so that they can demonstrate a wider range of social, environmental and economic sustainability knowledge, Engaging students in making learning visible in the first and last lectures of their modules appears to facilitate the learning recognition that tutors hope for. Including the post activity refection on drawing sustainability in the last lecture of their module has given students a further opportunity to self-analyse learning. Their feedback confirms that they can recognise their learning; students reported:

> "*Using pictures contributed to my sustainability knowledge*" (third year student, 2018–19)
>
> "*I know I have gained new skills and knowledge*" (third year student, 2016–17)
>
> "*Now I know I learned a lot*" (second year student, 2018–19)
>
> "*Drawing helped me test my current knowledge*" (second year student, 2017–18)
>
> "*I started with no knowledge and now know I am quite good*" (second year student, 2016–17)

Post activity feedback also suggests drawing sustainability may have encouraged alternative ways of thinking which can develop the perspectives and values that Laurie et al. (2016) consider necessary to foster sustainable societies, which can contribute to the achievement of the Sustainable Development Goals. Students reported:

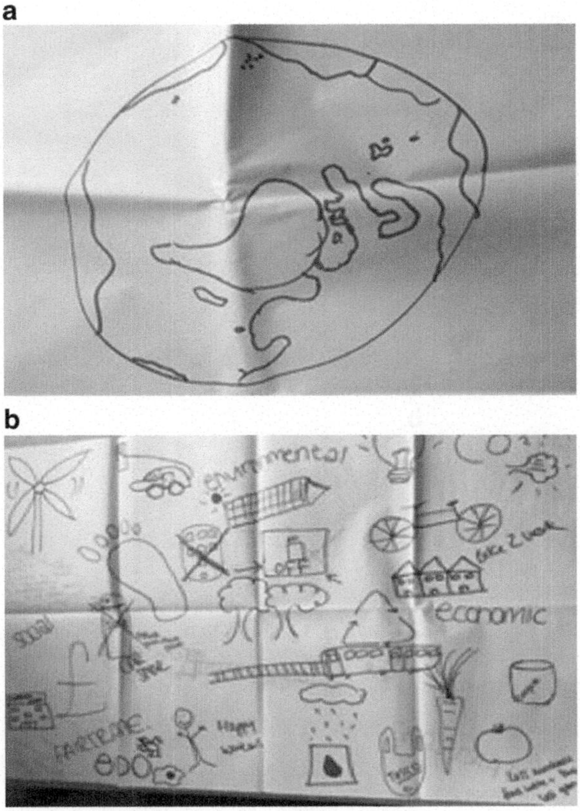

Fig. 4 **a** First lecture drawing Group 4 (second-year students 2016–17). **b** Last lecture drawing Group 4 (second-year students 2016–17)

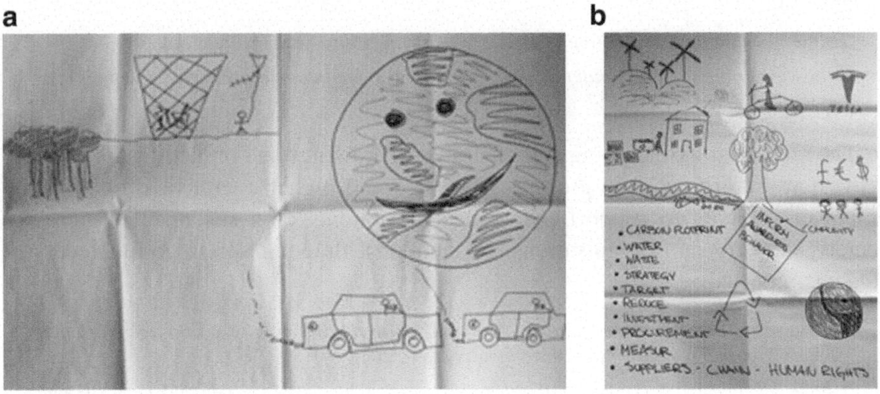

Fig. 5 **a** First lecture drawing Group 4 (Second-year students 2017–18). **b** Last lecture drawing Group 4 (second-year students 2017–18)

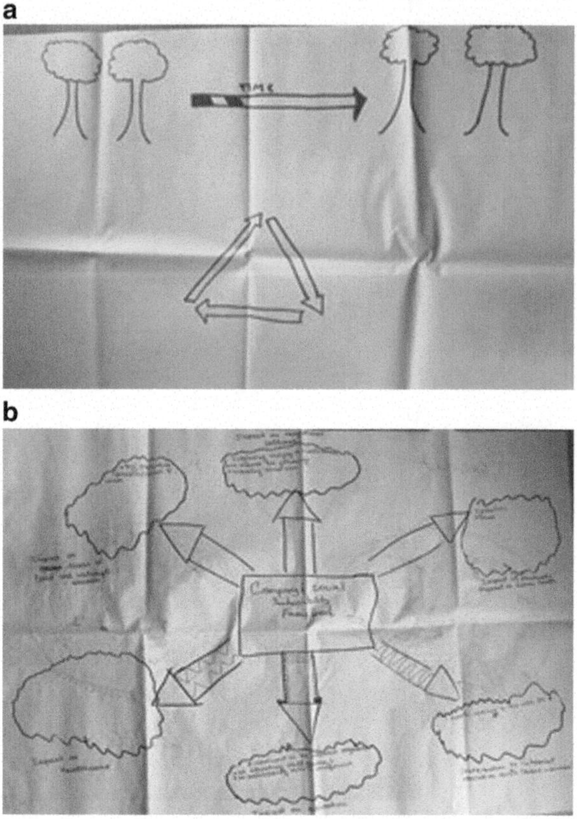

Fig. 6 **a** First lecture drawing Group 1 (third-year students 2017–18). **b** Last lecture drawing Group 1 (third-year students 2017–18)

> "*Drawing can help you think outside the box and challenge yourself*" (third year student, 2018–19)
>
> "*Drawing helped me to feel free to express my ideas*" (third year student, 2017–1)
>
> "*It has given me a better understanding of how to communicate my ideas*" (second year student, 2017–18)

The recognition of learning demonstrated in drawings and reflections on drawing skills may in turn promote greater confidence and higher cognitive skills that can empower students to support practical achievements of a sustainable future for society, industry and organisations. In their post drawing sustainability feedback students reported:

> "*It helped me improve my critical thinking and analysis as I now take into account other people's arguments and views*" (third year student, 2018–19)
>
> "*I have become more confident about demonstrating my opinion*" (second year student, 2017–18)
>
> "*I have developed more innovative ways of thinking*" (second year student, 2016–17)

7.3 Other Learning Outcomes Recognised

In addition to the effective achievement of the two learning outcomes designed into drawing sustainability, the research findings suggest it can also encourage students to relate personal sustainability practices to society and the environment. The recognition of learning gained appears to inspire students to reflect upon and recognise personal behaviour changes. Students reported:

"*I now take a personal view*" (second year student 2017–18)

"*I now care about better solutions*" (second year student 2017–18)

"*Don't be shy to speak about global topics*" (second year student 2018–19)

As well as gaining some of the knowledge, skills and values required to promote sustainable futures and support the achievement of the Sustainable Development Goals, students also appear to have recognised the importance of different learning approaches which may enhance academic achievement. For example, students reported:

"*I've been able to collaborate with individuals who provide interesting views*" (third year student 2080–19)

"*I know I learned a lot from my friends*" (second year students 206–17)

"*The power of observation is an important skill*" (second year student 2017–18)

Overall the research findings suggest that drawing sustainability can effectively deliver the two initial purposes the activity: benchmarking learning and encouraging students to know what they know. However, the findings also suggest drawing can encourage students to form an image of sustainability to reflect upon and, through this, relate personal sustainability practices to societal and environment challenges.

8 Drawing Sustainability and Indicators of Learning

The inputs and the students' and tutor's reflections on drawing sustainability are mapped onto the Indicators of Learning Recognition Framework to further explore the value of drawing sustainability. The framework (Table 1) adapts the Indicators of Learning proposed by Hughes (2006) as the criteria against which to evaluate the effectiveness of drawing sustainability in promoting students' self-analysis of learning to know what they know.

The results of this analysis (Table 1) suggest that all Indicators of Learning Recognition may be achieved by the activity of drawing sustainability through the inputs to the activity and the activity itself. In addition, students' reflections on drawing sustainability suggest that learning can be recognised in relation to all Indicators of Learning Recognition.

Table 1 Indicators of learning recognition framework

	Indicators of learning recognition	Achievement of Indicator recognised through drawing sustainability design elements and *tutor reflections (in italics)*	Achievement of Indicator recognised through students' post activity reflections on learning through drawing
1	Explaining something in their own words	Post activity reflection and feedback *Feeding back the drawing contents and its rationale promotes wider self-analysis of what other students know*	"Drawing helped me to feel free to express my ideas" "I have become more confident about demonstrating my opinion"
2	Asking questions	The process of agreeing drawing content *Group members discuss and agree what to put in the drawing which drives peer-to-peer learning and higher order cognitive skills*	"I've been able to collaborate with individuals who provide interesting views"
3	Recreating (rather than reproducing) information	The process of drawing sustainability *Students overcome their belief they can't draw to recreate their knowledge visually and defend it*	"The power of observation is an important skill"
4	Justifying their decisions	Providing a rationale for the inclusion of items during in-group planning, discussions with the tutor and presentation to the class *Students gain confidence in defending their drawings through praise and support from the tutor*	"It helped me improve my critical thinking and analysis as I now take into account other people's arguments and views"
5	Explaining their thinking	Feeding back their drawing to class at end of task *Students gain confidence in presenting their own ideas to the class*	"I have developed more innovative ways of thinking"
6	Talking to each other	Group members discussion and agreement of how to approach the task and what to include in the drawing *Drawing encourages most students to grab a pen and create their own input*	"I know I learned a lot from my friends" "It has given me a better understanding of how to communicate my ideas"

(continued)

Table 1 (continued)

	Indicators of learning recognition	Achievement of Indicator recognised through drawing sustainability design elements and *tutor reflections (in italics)*	Achievement of Indicator recognised through students' post activity reflections on learning through drawing
7	Active—doing something with the information	Creating a drawing with prior knowledge (pre-module survey) and in-module learning (post-module survey) *Using pens and paper encourages students to present their own ideas rather than looking them u*	"Drawing helped me test my current knowledge"
8	Reflecting at a conscious level	Group discussion in preparation of their drawing and post activity class discussion *The process of drawing requires students to reflect openly and collaboratively on the information*	"Drawing can help you think outside the box and challenge yourself" "I have become more confident about demonstrating my opinion"
9	Offering analogies and metaphors of their own	The activity requires students to present their own interpretation of sustainability *Frequently students incorporate personal experiences in group discussions*	"I now take a personal view"
10	Re-drafting, revising, rethinking and so on	Students to undertake an initial planning discussion and draw their ideas *Visiting a group during the activity can encourage further reflection and additions to their drawin*	"Using pictures contributed to my sustainability knowledge"
11	Frowning (the penny is stuck) … and then smiling (as the penny drops)	Each group is required to interpret the task for themselves and agree the drawing process *Once the process is agreed it appears that the group members understand the task and find contributing topics of sustainability relatively easy*	"I know I have gained new skills and knowledge" "I now care about better solutions"

9 Impacts and Implications of Drawing Sustainability

The activity of drawing sustainability is designed for two purposes. Firstly, to promote learning recognition through students' reflection on their own knowledge in order to know what they know and to recognise what they don't know through peer-review and reflecting on what colleagues say when they present their drawing to the class. Secondly drawing sustainability is designed to enable the tutor to benchmark students' knowledge to pitch lecture content correctly and recognise students' progress. Drawing sustainability may therefore contribute to the innovative learning and teaching practice that Navarro (2008), Boron et al. (2017), Molthan-Hill (2017), UNESCO (2017) and Franco et al. (2018) suggest is required to equip students with the skills, values and attitudes needed to promote sustainable futures for society, industry and organisations and contribute to the achievement of the Sustainable Development Goals.

The research findings presented above suggest that drawing sustainability may offer an innovative learning and teaching practice that can provide a knowledge benchmark for tutors and promote students' recognition of learning. Recognition of learning has become increasingly important over the last three decades and it is now recognised that judging self-knowledge can enhance the quality of an individuals' education (Kurnaz and Cimer 2010) and boost academic achievement (Webb 2020). This may be achieved in three ways through drawing sustainability. Firstly, by students' developing an understanding of what they know through reflecting on their knowledge to create a visual representation of sustainability. Secondly, by understanding how their knowledge is different to what others know through reflection on other classmates' drawings, and thirdly by appreciating what they still need to know through tutor-to-peer and peer-to-peer feedback and in-class discussions. This activity may therefore be of interest those seeking teaching tools that provide a knowledge benchmark and a platform from which to introduce additional sustainability challenges, interconnections, and contexts to help students know what they know.

Although the drawings in this study are not subject to formal content analysis, the development of drawings shown in Figs. 4a, b and students' reflections on the activity of drawing suggest that learning has taken place and that students recognise it. The difference in students' drawings between the first and last lectures of both second and third year modules over the three years of the study and the comments in students' post activity reflections suggest that the processes within the activity of drawing sustainability may encourage students to reflect upon their development of knowledge to enable them to know what they know.

In addition, the different styles of the drawings and substance within them suggests drawing sustainability is sufficiently flexible to enable students to create their own visual representations of social, environmental and economic sustainability, which may challenge the students to elaborate their knowledge, imagine new relationships in their drawings and test their ideas as advocated by Vygotsky (1978) to make thinking visible. As students are not always able to evaluate whether they have understood in-class and independent study activities (Davis 2003), drawing sustainability may provide a practical tool to encourage reflection on, and recognition of, learning. Mapping the inputs to drawing sustainability and students' and tutor's post activity reflections against the Indicators of Learning Recognition Framework suggests that learning can take place through drawing and that the learning is recognised by the students and the tutor. Hopefully this recognition of what they know and what they need to know will enhance the quality of the students' education (Kurnaz and Cimer 2010), promote their higher attainment (Flavell 1976), and facilitate students' further learning as advocated by Schwartz and Martin (2004). Together these may contribute to the achievement of the Sustainable Development Goals.

As well as providing an opportunity to deepen understanding, as suggested by Ainsworth et al. (2011), the research findings suggest that drawing sustainability may also increase students' confidence and higher cognitive skills. Students' post activity reflections suggest that drawing sustainability can build the self-esteem and self-confidence advocated by Jones (1997) and promote the critical thinking proposed by Quillin and Thomas (2015). Both of these higher order cognitive skills may contribute to development of work-ready graduates with the sustainability knowledge and employment skills (Drayson 2015), and reflexive ability (Hind et al. 2009) to become architects of socially and environmentally sustainable futures.

As EfS requires students to gain and understanding of the complex interconnectivity of, and challenges inherent in, developing social and environmental sustainability practices, the task of drawing sustainability may offer tutors an effective teaching tool to support them through this. The research findings suggest drawing sustainability has effectively promoted the learning recognition of a new subject proposed by Quillin and Thomas (2015) and enabled students to apply and recreate their knowledge in another format as recommended by Griffith and Burns (2014). The research findings suggest that through presenting their ideas visually students can see their previously unseen knowledge as proposed by Quillin and Thomas (2015), thus challenging them to focus on the reflection and articulation of knowledge which can guide them to know what they know (Quintana et al. 1999). The research findings presented here therefore suggest drawing sustainability may present the opportunity to enhance thinking and learning advocated by Hope (2008), Adams (2017).

In addition to achieving the hoped-for benchmark of knowledge and recognition of learning outcomes, a number of additional outcomes have emerged from this study. For example, students' reflections on the drawing activity suggest the development of critical thinking is recognised. This may be promoted through the drawing activity's process and presentation of the drawing to the class involving the communication of students' own ideas. In addition, students' reflections upon the activity of drawing sustainability suggests it can encourage them to form an image of sustainability to reflect upon and relate to personal societal and environmental sustainability practices, thus contributing to the development of their personal values which Howlett et al. (2016) suggest can empower learners to transform the way they think and act to become change agents for sustainable futures who can contribute to the achievement of the Sustainable Development Goals.

The research findings also suggest drawing sustainability can offer benefits for tutors as drawings are a record of students' thinking (Brooks 2005). By asking students to connect concepts and their ideas as advocated by Long et al. (2014) through drawing and explain them in their own words, tutors are helped to identify misunderstandings and highlight valuable insights. In turn, this can provide opportunities to offer high quality, individualised feedback which can help overcome misconceptions and promote module attainment (Dikmenli 2010; Webb 2020). Fostering participation in drawing sustainability through clear instructions and encouragement may support students to learn from the experience as well as the drawing process as advocated by Reiser (2004).

10 Limitations Recognised in This Research

The study has some limitations which need to be recognised. As this is a small-scale study using 163 students over 3 years, the research findings cannot be stated conclusively. It is recognised that post activity reflections and feedback used to provide the students' voice in this study are taken at the end of the second iteration of drawing sustainability which may not have given time for extensive reflection. No interviews were conducted with students to explore their choice of drawing content or their perceptions of drawing sustainability. This is to be undertaken in a future study. Despite these imitations, the drawings and reflections used to illustrate drawing sustainability suggest the outcomes of the activity proved valuable to students' recognition of learning.

11 Conclusions

This paper presents the findings of a small, three-year study into the ability of drawings to promote learning recognition within EfS for Business Management students. It assesses the effectiveness of drawing sustainability, an activity designed to encourage students to know what they know and provide tutors with a benchmark of students' knowledge. The findings suggest that drawing sustainability can offer an innovative learning and teaching approach and can promote students' recognition of learning which is increasingly seen as a way to enhance the quality of an individuals' education and their academic achievement.

The teaching tool of drawing sustainability and its' associated scaffolding activities such as presenting drawings to the class and reflecting on the activity also offer educators a tool to benchmark students' sustainability knowledge, recognise and intercept misconceptions and provide individual feedback which may all support students to become change agents for sustainability and lifelong learners, both of which can contribute to the achievement of the Sustainable Development Goals. Drawing sustainability may therefore be of interest to other educators seeking teaching tools that enable them to benchmark students' knowledge and help students know what they know.

References

Abdel Meguid E, Collins M (2017) Students' perceptions of lecturing approaches: traditional versus interactive teaching. Adv Med Educ Pract 8:229–241

Adams E (2017) Thinking drawing. Int J Art Des Educ 36(3):244–252

Ainsworth S, Prain V, Tytler R (2011) Drawing to learn in science. Science 333(6046):196–1097

Boehm K, Lees-Spalding J (1993) The student book: the indispensable applicant's guide to UK colleges and universities. Papermac, London

Boron S, Murray K, Thomson G (2017) Sustainability education: Towards total sustainability management teaching. In: Filho W, Brandi L, Castro P, Newman J (eds) Handbook of theory and practice of sustainable development in higher education, vol 1. Springer, Switzerland, pp 37–52

Brooks M (2005) Drawing as a unique mental development tool for young children: interpersonal and intrapersonal dialogues. Contemp Issues Early Child 6(1):80–91

Caleon I, Subramaniam R (2010) do students know what they know and what they don't know? using a four-tier diagnostic test to assess the nature of students' alternative conceptions. J Res Sci Educ 40:313–337

Carless D, Boud D (2018) The development of student feedback literacy: enabling uptake of feedback. Assess Eval High Educ 43(8):1315–1325

Davis E (2003) Prompting middle school students for productive reflection: Generic and directed prompts. J Learn Sci 12:91–142

de Bruin A, Kok E, Lobbestael J, de Grip A (2017) The impact of an online tool for monitoring and regulating learning at university: overconfidence, learning strategy, and personality. Metacognition Learn 12:21–43

Dewey J (1916) Democracy and education. An introduction to the philosophy of education. Macmillan, New York, NY

Dikmenli M (2010) Misconceptions of cell division held by student teachers in biology: a drawing analysis. Sci Res Essays 5:235–247

Drayson R (2015) Student attitudes towards and skills for sustainable development, executive summary: employers. https://www.iau-hesd.net/sites/default/files/documents/executive-summary-students-2015.pdf

Figuero P, Raufflet E (2015) Sustainability in higher education: a systematic review with focus on management education. J Clean Prod 106:22–33

Flavell J (1976) Metacognition and cognitive monitoring: a new area of cognitive development inquiry. Am Psychol 34:907

Franco I, Saito O, Vaughter P, Wheraet J, Kanie N, Takemoto R (2018) Higher education for sustainable development: actioning the global goals in policy curriculum and practice. Sustain Sci 14:1621–1642

Glynn S, Muth K (2008) Using drawing strategically: drawing activities make life science meaningful to third- and fourth-grade students. Sci Child 45:48–51

Gobert J, Clement J (1999) Effects of student-generated diagrams versus student-generated summaries on conceptual understanding of causal and dynamic knowledge in plate tectonics. J Res Sci Teach 36:39–53

Griffith A, Burns M (2014) Teaching backwards. Crown House Publishing, Bancyfelin

Hannon S, McBride H, Burns B (2004) Developing creative and critical thinking abilities in business graduates: the value of experiential learning techniques. Ind High Educ 18(2):95–100

Hay D, Kinchin I, Lygo-Baker S (2008) Making learning visible: the role of concept mapping in higher education. Stud High Educ 33:295–311

Higher Education Funding Council for England (2013) Sustainable development in higher education: Consultation on a framework for HEFCE. http://www.hefce.ac.uk/workprovide/Framework/

Higher Education Academy (2015) Education for sustainable development (ESD). https://www.heacademy.ac.uk/workstreams-research/themes/education-sustainable-development

Hind P, Wilson A, Lenssen G (2009) Developing leaders for sustainable business. Corp Gov 9(1):7–20

Hope G (2008) Thinking and learning through drawing in primary classrooms. Sage, Los Angeles

Howlett C, Ferreira J, Blomfield J (2016) Teaching sustainable development in higher education: building critical, reflective thinkers through an interdisciplinary approach. Int J Sustain High Educ 17:305–321

Hughes M (2006) And the main thing is… Learning: keeping the focus on learning—for Pupils and Teachers, (Jigsaw Pieces). Education Training and Support, London

Jones J (1997) A lesson in teaching art self-confidence from drawing on the right side of the brain. Art Educ 50(2):33–38

Kurnaz F, Cimer S (2010) How do students know that they have learned? An investigation of students' strategies. Procedia Soc Behav Sci 2:3666–3672

Laurie R, Nonoyama-Tarumi Y, Mckeown R, Hopkins C (2016) Contributions of education for sustainable development (ESD) to quality education: a synthesis of research. J Educ Sustain Dev 10(2):226–242

Linn M, Lewis C, Tsuchida I, Songer N (2000) Beyond fourth-grade science: why do U.S. and Japanese students diverge? Educ Researcher 29(3):4–14

Long T, Dauer J, Kostelnik K, Momsen J, Wyse S, Speth E, Ebert-May D (2014) Fostering ecoliteracy through model-based instruction. Front Ecol Environ 12:138–139

McDonald B, Boud D (2003) The impact of self-assessment on achievement: the effect of self-assessment training on performance in external examinations. Assess Educ 10:209–220

Molthan-Hill P (2017) The business student's guide to sustainable management: principles and practice, 2nd edn. Greenleaf Publishing, Sheffield

Navarro P (2008) The MBA core curricula of top-ranked US business schools: a study in failure? Acad Manag Learn Educ 1:108–123

Oblinger D, Oblinger J (2005) Educating the net generation. https://www.educause.edu/educatingthenetgen

Quillin K, Thomas S (2015) Drawing-to-learn: a framework for using drawings to promote model-based reasoning in biology. Life Sci Educ 14(1):1–16

Quintana C, Eng J, Carra A, Wu H-K, Soloway E (1999) Symphony: a case study in extending learner-centered design through process space analysis. In: Proceedings of CHI 99 conference on human factors in computing systems, pp 473–480

Reiser B (2004) Scaffolding complex learning: the mechanisms of structuring and problematizing student work. J Learn Sci 13(3):273–304

Roam D (2008) Back of the Napkin: solving problems and selling ideas with pictures. Penguin, New York

Ryan S, Stieff M (2019) Drawing for assessing learning outcomes in chemistry. J Chem Educ 96:1813–1820

Schon D (1987) Educating the reflective practitioner. Jossey-Bass, San Francisco, CA

Schunk D, DiBenedetto M (2016) Self-efficacy theory in education. In: Wentzel K, Miele D (eds) Handbook of motivation at school. Routledge, London, pp 34–54

Schwartz D, Martin T (2004) Inventing to prepare for learning: the hidden efficiency of original student production in statistics instruction. Cogn Instr 22:129–184

Sterling S, Maxey L, Luna H (2013) The sustainable university: progress and prospects. Earthscan, London

Storey M, Killian S, O'Regan P (2017) Responsible management education: mapping the field in the context of the SDGs. Int J Manag Educ 13:93–103

UNESCO (2017) Education for sustainable development goals learning objectives. http://unesdoc.unesco.org/images/0024/002474/247444e.pdf

Unicheck (2015) Teaching students to think using drawing and sketching. https://unicheck.com/blog/drawing-and-sketching/

United Nations (2017) Sustainable development knowledge platform: sustainable development goals. https://sustainabledevelopment.un.org/sdgs

Van Meter P, Garner J (2005) The promise and practice of learner-generated drawing: literature review and synthesis. Educ Psychol Rev 17:285–325

Vygotsky L (1978) Mind in society: the development of higher psychological processes. Harvard University Press, Cambridge, MA

Webb H (2020) How do you know your students are learning? http://www.sec-ed.co.uk/best-practice/how-do-you-know-your-students-are-learning/

Weybrecht G (2017) From challenge to opportunity—management education's crucial role in sustainability and the sustainable development goals—an overview and framework. Int J Manag Educ 15:84–92

We Can Only Do It Together: Addressing Global Sustainability Challenges Through a Collaborative Paradigm

Helen Avery and Birgitta Nordén

1 Introduction

A consensus is emerging in both scientific circles and international policy discourse that business-as-usual approaches to sustainability have led to a critical state, and that urgent measures are required to reverse these dynamics. At the same time, national and international commitments to address the crisis remain reluctant to reconsider the assumption that growth can continue unchecked (Alexander 2012; Gough 2017; Adloff and Neckel 2019). Indeed, in the Sustainable Development Goals (SDGs), the argument is even made that growth should be encouraged to provide the necessary resources for implementing the goals, and efforts for sustainability in higher education seldom question that paradigm (Ruiz-Mallén and Heras 2020).

In recent decades, we have seen that growth, as such, does not address inequity within or across states, and instead tends to reinforce an extractivist model that concentrates control over resources and key technologies in a small number of hands (Acosta 2013). Technological advances have not offset environmental impacts and emissions continue to rise. Climate change and other environmental impacts continue to disproportionately affect the poorest, destroying the basis of their livelihoods. Both armed conflicts and economic warfare, aimed at control over water, minerals and hydrocarbons, drive internal displacement and forced migration, increasing the number of people worldwide that no state is willing to take responsibility for. It

H. Avery (✉)
Centre for Advanced Middle Eastern Studies and Centre for Environmental and Climate Science, Lund University, Lund, Sweden
e-mail: helen.avery@cme.lu.se; helen.avery@lnu.se

Linnaeus University, Växjö, Sweden

B. Nordén
Faculty of Education and Society, Malmö University, Malmö, Sweden
e-mail: birgitta.norden@mau.se

W. Leal Filho et al. (eds.), *Universities, Sustainability and Society: Supporting the Implementation of the Sustainable Development Goals*, World Sustainability Series,
https://doi.org/10.1007/978-3-030-63399-8_16

has been constantly repeated, since Rio and before, that the only way to break this vicious cycle is with concerted international action and a paradigm shift (Waddell 2011; Waddock et al. 2015), but we seem no closer to this today than thirty years ago. Economic arguments continue to prevail, and few higher education institutions (HEIs) have chosen to orient their efforts to benefit societies (see Beynaghi et al. 2016).

This study addresses the important issue of how HEIs can achieve sustainability transitions within their own organisations (Miller et al. 2011; Baker-Shelley et al. 2017). Although HEIs contain expertise that is crucial for sustainability, few universities today see sustainability as their main purpose. Each department typically drives its own core mission. As long as the majority of departments and those that are best funded have other main objectives, limited resources will be available for concerted efforts towards sustainability. Indeed, numerous specialisations drive technologies and education that consolidate vested interests and aggravate existing problems, thus contributing to lock-in effects (Payo et al. 2015) and business-as-usual. For HEIs as a whole, this typically creates a conflict between marginalised fields of research or education that see sustainability transitions as their core mission, and all the other sections of the HEI, which strive to maintain the status quo and their current positions within the institutional or disciplinary hierarchies.

Both policy and the literature stress the role of collaboration for sustainability, but relatively little attention has been paid to how we can address the structural drivers in HE that reduce opportunities for collaboration. The present study contributes to this discussion, where we argue that not only competition for funding, but also aspects such as formalised communication or performance assessment have adverse effects. We further argue that ethics and social justice cannot be peripheral considerations. Engagement with actors outside academia has been called for, but often translates into collaboration with large companies rather than citizens. Importantly, most resources in HEIs are still devoted to activities that consolidate societal unsustainability. Rapid changes in higher education are thus urgent and must be coupled with strategies to deal with the disruption that such changes cause, both within academia and in society at large.

What strategies are open to HEIs to create the capacity for real change globally? How can we collectively avoid the points of no-return that our societies appear to be racing towards? Who will be given a say in deciding our future, and which voices are allowed to represent "society"?

HEIs today suffer from much the same paradigm of fragmentation, competition (Shore and Wright 2000; Marginson 2006) and short-sightedness reflected in the formulation of the SDGs, leading to lock-in effects, incrementalism, formalism and the absence of systemic strategic vision. Radical transformation is needed to move beyond the status quo at the speed necessary to address planetary challenges (Vogt and Weber 2020). To achieve this, HEIs need to overcome many obstacles, including funding structures, power imbalances, the disciplinary organisation of knowledge and the absence of future-oriented methodologies. Nevertheless, HEIs still have options available that can open pathways towards a collaborative paradigm (Waddell 2011), enabling them to play a positive role in equitable, inclusive and sustainable societal transformation.

2 The Role of Higher Education in Addressing Global Challenges

Informed decision-making is an essential component in strategic capacity-building for resilience and sustainable economies, and capacity is needed to generate and provide access to accurate and relevant research. Equally essential is the capacity to creatively use available information; discern causalities, risks and points of leverage; understand complexity and dynamic principles; exercise caution, integrity, benevolence and moral judgement; integrate across disciplines and sectors; take initiative and continuously develop and plan ahead. Foresightful policies will not produce results unless they are combined with the capacity to develop socially and environmentally sustainable technologies, attitudes and organisational forms. New professional competences are needed for the new structures, which means also reforming higher education structures (Avery and Nordén 2017). New practices have to be tried out and refined, resources need to be mobilised, and collaborations and alliances have to be established or consolidated.

These are the classical elements of governance, and by no means new. What is new is the extent to which technology and scientific understanding play a role. Action and access to information is mediated through multiple layers of technology, but also multiple layers of global interaction and institutional configuration. Causes and consequences are no longer transparent, in the sense that they cannot be evaluated based on personal experience alone. Indeed, even collective experience and scientific inquiry only provide partial indications of either causes or consequences, making intelligent and strategic reflection based on dialogue increasingly difficult. Facing such challenges, we need approaches that allow us to make wise decisions (Vogt and Weber 2020) and adequate plans under conditions of great uncertainty, such as working with hypothetical or future scenarios (Schwartz 1991; Gallopín 2002; Goldstein 2009). We are also witnessing the development of other approaches to jointly finding solutions to highly complex problems, for instance in the field of systemic design (Sevaldson 2015).

To function well in times of major changes, the capacity to collaborate and communicate across contexts and collectively envision solutions become central issues. Leadership and initiative, as well as access to resources, are needed to restructure HEIs. But due to the opaqueness of existing knowledge systems (alongside the uncertainties concerning global developments), no single individual, and no single institution, is in a position to take adequate decisions on its own. Open "in-between" spaces are thus singularly important (see Power 1997; Jary 1999; Baker-Shelley et al. 2017; Vogt and Weber 2020), in which joint reflection, dialogue and re-negotiation of roles can take place, independent of particular bias or power relationships. Such in-between spaces need to be sufficiently funded, while continuity in interpersonal relationships (Shore and Wright 2000) plays a key role in ensuring effective communication and enabling investment in long-term, deeper reflections.

Crucially, the development of such spaces is obstructed by the paradigm of competition in HEIs (Marginson 2006) that fosters mistrust and a lack of collaboration. Competitive pressures lead to a focus on streamlining and prioritising the here-and-now to generate immediate revenue, rather than investing in the necessary open spaces between institutions, sectors and knowledge fields. An overarching paradigm of collaboration rather than competition is therefore needed to sufficiently prioritise work for the "commons" (Waddell 2011) and the public good.

3 Blueprint for a New Paradigm in HEIs

Despite global efforts towards sustainability, the principal drivers of our economies and national policies remain substantially the same. There is, therefore, reason to believe that in the coming years we will see an increase in the adverse effects of business-as-usual, including risks of financial collapse, currency shocks and bankruptcies; the deterioration of working conditions and loss of livelihoods and the incapacity of institutions to provide services or social insurance. We can also expect to see an increase in loss of functionality of automated systems and "hi-tech" as well as an increase in accidents and disasters. Direct climate impacts include the loss of food security and loss of health and adequate living conditions. Among further impacts are increasing inequity, causing extreme stress on family and social support networks, particularly if there is a loss of collective solidarity. Finally, we are already witnessing increasing militarisation, control and surveillance, resulting in a loss of democracy.

In the face of these developments, what kind of knowledge, capacity and competences do we need, and where should HEIs place their priorities?

- To enable action, both trust (Power 1997; Shore and Wright 2000) and hope, as well as the ability to coordinate and mobilise resources, are fundamental conditions (socially, technically and economically in developing new knowledge and capacity).
- For resilience within our institutions and in society (Berkes 2007; Sriskandarajah et al. 2010), we need to develop solidarity, empathy, caring and understanding. Importantly, we must strive to build alliances and teams, prioritise the reduction of risk and vulnerability, and build real capital to maintain margins of action. Essential elements to any strategy are therefore to maintain redundancy and free capacity, rather than to "slim" our organisations. We must plan for multiple scenarios and have the capacity to do so in ways that are in line with facts rather than wishful thinking and rhetoric.
- For agility and creativity, there is an urgent need to develop future-oriented methodologies and stimulate collective action-oriented imagination. Critical

thinking is key in both teaching and research, which requires us to encourage diversity and complementarity, as well as a sense of initiative and responsibility, rather than conformism and obedience to authority.

- Science has long lived on its reputation for integrity. At a time when science is driven by funding rather than integrity, ethics and trust must again become our top priority if we are not to lose all credibility (Vogt and Weber 2020). New ethical codes of conduct for our professions must be developed to address extreme challenges and new technologies. All activity must be informed by a sense of responsibility, foresight and understanding the consequences of action or inaction while prioritising respect for life, social justice and humanity.

Finally, employees and students at all levels need to develop the capacity to collectively resist inadequate decisions and defend basic conditions for life. We must urgently learn to organise local economies, services and insurance; develop international networks of solidarity and exchange of experience; as well as develop the capacity to invent new technologies and actions needed in regenerative economies.

4 Knowledge and Policy—Which Knowledge Is Needed in a Time of Systemic Shifts?

At present, the evaluation of whether a particular policy measure might be suitable for achieving a desired effect tends to be based on evidence with limited relevance for future situations. When we are looking at the most important questions for the survival of our societies, these involve transition and deep-reaching changes in societal and economic structures. It is therefore problematic that a large portion of the evidence on which we base decisions is geared to manage business-as-usual within a given system, rather than considering the issues that arise in transitions between systems.

An important point to remember is that actions that are difficult and costly within one system may be easy and profitable within another, and vice versa. In other words, the relationship between action and impact can change depending on the system. Consequently, any analysis of costs and benefits only applies within the conditions of a particular system. Evidence-based suppositions about the effects of any set of policy measures are also constructed using data from past experiences. The knowledge that we use to drive a certain development is, in other words, valid only within a particular system. This makes it difficult to assess future developments that span across a longer period, particularly if we are interested in transitions across periods with different systemic characteristics, as we are looking at both external and internal uncertainties.

Developing competences, capacity, experience and know-how for any new technology or societal model takes time. By the time resources are mobilised, needs may have already changed and proposed solutions may be outdated. If the required competence and capacity cannot be mobilised sufficiently quickly, countries run the risk of a permanent gap and mismatch between needs and how they are addressed.

Higher education thus faces a fundamental double constraint: to provide new generations of professionals and researchers with adequate competences and to develop the specific types of scientific knowledge that are needed (see Miller et al. 2011), we must know in advance what these needs will be. At the same time, our forecasting tools cannot provide such answers in an era of rapid and unprecedented change. As in ecosystems, sufficient diversity is therefore essential for HEIs to increase resilience and to create hotbeds in which solutions can reach maturity and be ready for use at the precise moment they are needed.

Such creativity presupposes experimenting at a small scale, and the willingness to accept that the large majority of experiments will not lead to viable solutions. Driving HEIs through "performance" criteria that mainly reward "success" thus blocks one of the most important services HEIs can render. In terms of methodologies, emphasising predictability gears our forecasting tools towards regularity and systemic conditions relevant to the past, rendering them increasingly inadequate to describe a future characterised by systemic instability, variability and conditions that differ from any witnessed before (see Miller et al. 2011; Spangenberg 2019).

5 Limits to and Risks with the "Technological" Approach

It has been argued that increasing efficiency, above all through technological means (see, for instance, von Weizsäcker et al. 2014), offers a way out of the situation so that growth can continue through decoupling. While the intelligent use of technology is certainly needed to reduce environmental impacts, over-reliance on technology to solve the major challenges we face today may instead generate new sets of problems that we are not able to solve (see also Adloff and Neckel 2019). Seen globally, improvements through the use of less resource-consuming technologies have been offset by economic growth, demographic growth and the new challenges that appear as various tipping points are reached. Even at the local level, many improvements correspond to outsourcing negative environmental and social impacts to other parts of the planet. Although we may not yet have entirely exhausted all our potentials for growth and expansion, there are limits to indefinite growth (Alexander 2012; Gough 2017; Raworth 2017), and denial amounts to placing the burden of undoing our mistakes on the next generation. Transitions are therefore not simply a matter of developing sustainable alternatives and supporting their implementation, but also involve working proactively with phasing out unsustainable structures and technologies.

All transitions involve deep-reaching rapid change. Although the precise future impacts of different decisions are difficult to assess, certain conclusions can nevertheless be drawn regarding the disruptive effects of change (Miller et al. 2011). One of the fundamental aspects to consider is how to develop the necessary expertise, and how to develop institutions that are capable of continuous development, to respond adequately to the demands that emerge. In times of environmental, technological and societal shifts, much former expertise loses its relevance, and the organisation of labour—including the various interrelationships between professions—consequently loses its functionality. Theorists have thus stressed the need for transdisciplinarity (Florin et al. 2015; Mochizuki and Yarime 2015; Vogt and Weber 2020) and epistemological pluralism (Miller et al. 2011) in knowledge construction for the future. Interprofessional collaboration is equally necessary, as well as the capacity to reconfigure organisations and structures. Several approaches additionally suggest more interactive processes, as well as involving concerned populations actively in research (see, for instance, Lynch 2006; Lindhult 2016; Marquard et al. 2016).

6 Particular Challenges for Universities

The main challenge that HEIs face today is likely that of funding, regardless of whether they are private- or state-funded. In most contexts, education is oriented towards immediate employability, rather than societal, global or future needs (see Lynch 2006). Similarly, research is dependent on industry funding (Beynaghi et al. 2016) and is largely oriented towards immediate applicability and commercial interests.

The second challenge is the pervasive structure based on competition among students, staff, departments or disciplines, researchers or research environments. Competition systematically prevents available resources from being used effectively and tends to reduce the necessary diversity.

A third major challenge for addressing global challenges is how knowledge is shared and developed. The language of instruction and modes of publication systematically put the Global South at a disadvantage, marginalising Southern perspectives and knowledge production on topics relevant for low-income countries. Hallinger and Chatpinyakoop (2019) have found that in the area of higher education for sustainable development, only a small fraction of publications come from developing countries, and among these, only the larger economies are represented. The issue is aggravated by the fact that scientific methodologies discourage future-oriented explorative approaches and systematically prioritise the aggregation of data rather than context-sensitive descriptions and observations. Not only does this render us collectively blind to emerging trends, but it precludes the development of responses that are relevant and possible to implement in local low-income contexts.

In the short term, HEIs cannot change such underlying structures. But by refusing the competitive paradigm (Chan 2004) and prioritising collaboration—within each HEI as well as with other HEIs nationally and globally—existing resources can be used much more effectively, benefitting from scale, synergies and complementarities. Risks inherent in explorative and innovative research can be shared among partners, while the costs caused by lack of continuity can be reduced, such as those incurred in building and demolishing capacity for disconnected, externally-funded research projects. Importantly, creating a culture of trust and integrity enhances knowledge sharing, engagement and teamwork, and enables HEIs to become learning organisations that are open to creativity (Cooke 1987; Pahl-Wostl 2009; Baker-Shelley et al. 2017).

In the current system, funding is earmarked mainly for educational programmes and research projects that do not have sustainability at their core, or which may contribute to global injustice and environmental threats. As academics and academic institutions, we can and should collectively lobby to change such policies. In the meantime, we still have options for action. Existing courses can be restructured to prioritise the ethics and crosscutting competences that are fundamental to sustainability, reading lists and cases can be selected to address actual challenges, and joint courses can be built into programmes to enable collaborative work and transdisciplinary understanding (see the Roskilde model, Andersen and Kjeldsen 2015). Similarly, research projects can be clustered to enable collaboration in directions that are relevant for sustainability transitions, departmental accounting and performance evaluation can be geared to disincentivise competition, while longer-term investments and recruitment can be primarily oriented towards key issues for sustainability.

7 Why Do We Need Agile Learning Organisations in Higher Education?

In times of rapid change, institutions alone do not provide a sufficient basis for the trust that enables people to participate in emerging processes, since the stabilising effect of routinised behaviour can interfere with the need to develop new practices (Power 1997; Shore and Wright 2000). Current discussions that phrase these issues in terms of resilience instead stress the significance of a strong social fabric and interpersonal relations (Berkes 2007; Sriskandarajah et al. 2010). Interpersonal relations also play a key role in *agility* in orienting the development of knowledge and competences in new directions (cf. Miller et al. 2011). This is because knowledge and know-how in contemporary societies have become progressively more "opaque" due to strong specialisation and divisions of labour. Opaqueness includes not only knowledge as a product, but also the instruments, methodologies and structural drivers that underlie knowledge construction. Institutions, therefore, come to mediate relations and communication between professions and between different segments of society. While this regime of work organisation can function as long as changes are slow and incremental, it is subject to collapse if changes accelerate.

In times of rapid and unpredictable change, individuals who can play the role of knowledge brokers and mediate between contexts come to have an increasingly strategic function. Formalised communication becomes increasingly inadequate, since formalised systems are geared to address past conditions rather than current and emerging challenges. At the same time, the need for crosscutting communication increases within, between and beyond HEIs to coordinate new responses to new situations. Brokering carried out by human beings is thus needed across contexts, but such brokering presupposes continuity, to enable trust, engagement and quality in communication (Holste and Fields 2010). Equally importantly, trust and continuity in interpersonal relationships are necessary within individual workplaces to deal with change in constructive ways. Strong and continuous interpersonal relationships are required to communicate and renegotiate practices when the initial assumptions underlying formalised steering systems no longer apply. These relationships also make it possible to collaborate effectively and to focus on the tasks at hand.

Inversely, changes that concern roles and performance at the workplace, and possibly also job security, expose employees to considerable stress (Shore and Wright 2000; Miller et al. 2011). Relationships also become “opaque”, since the implications of a given action are no longer clear. Responsibilities tend to become confused so that a disproportionate amount of time and attention is needed to find information and to assess developments in the surrounding contexts. With continuous changes and reforms, there is, therefore, a risk that people lose confidence and minimise investment in the future—a process of alienation.

Besides the risk of disengagement, a further serious consequence of continuous change, mobility and weakening of interpersonal relationships is a loss of trust. To compensate, control mechanisms are formalised, but this formalisation tends to lead to rigidity, paralysing constraints and, ultimately, to dysfunctional organisations that block intelligent change (Cooke 1987). In times of structural pressures, when agility, creativity and collaborative intelligence are most needed, a vicious circle of top-down control, formalisation and rigidity can instead occur. The formalised aspects of institutions are frozen solid, while teachers and researchers are treated as dispensable and interchangeable workers on a chain. Higher education fundamentally differs from many other fields of activity to the extent that it is about developing and imparting new knowledge, rather than merely reproducing a pre-defined set of actions according to instructions. Researchers and teachers do their jobs best when they have intrinsic motivation (Marginson 2008) and when their work is in alignment with their values and convictions. Driving academia with principles of Taylorism, top-down steering and external motivation only leads to alienation (Oleksiyenko 2018).

In other words, one of the paradoxes that needs to be addressed in transitions towards sustainability is that to construct relevant decision-making capacity, as well as necessary knowledge and expertise applicable to new situations, existing professional roles and knowledge systems need to be reorganised and re-negotiated, but such changes may also create confusion and anxiety, with risks of top-down steering through formalisation. Thus, although it has been argued that a focus on resilience can delay efforts for mitigation (see Adger 2006; Fiksel 2006; Olsson et al. 2014 for an overview of the debate), maintaining cohesion, autonomy (Vogt and Weber 2020) and continuity are crucial to enable structural changes in HEIs.

8 A Role for the SDGs in Higher Education?

The SDGs are comprehensive, covering most areas of human activity and impacts, and designed for ease of communication across contexts. The SDGs also set concrete targets and provide indicators to measure progress. However, they do not engage in binding commitments for working towards sustainability; rather, they are intended as soft power and as a common vision towards which efforts can be directed. Through mechanisms of monitoring, states worldwide may be incentivised to work more systematically towards the various targets. Within states, the various criteria could be used for evaluations or in decisions on resource allocation. A major merit of the SDGs is the broad emphasis on global equity, but from the perspective of sustainability, the goals do little to reduce the immanent risk of environmental collapse. Additionally, several of the goals are premised on the idea of continued growth (see Ruiz-Mallén and Heras 2020; Vogt and Weber 2020) and the argument that economic growth will liberate the resources needed to realise the SDGs.

Soft power and policy documents at supra-national levels, including the EU, the OECD and UNESCO, have been shown to have a considerable impact on education systems, notably through formalising qualifications that give access to professions in the labour market. It can be expected that although they are not binding, the SDGs will also affect HEIs through the way they are evaluated and possibly through financial incentives. Higher education and research are most directly concerned with SDG 17, which includes knowledge sharing globally (17.6).

Despite the limitations of the SDGs as a basis for supporting transitions towards sustainability and global justice, two of the targets could be used as leverage in higher education. The first is target 4.7 on education for sustainable development. The ethical stance outlined in this target is in line with the values required to build a collaborative paradigm (Waddell 2011) and can, therefore, be used to support transformation in this direction.

The second is target 17.6, on knowledge sharing and access to technology. Unfortunately, in terms of implementation, the indicator for this target is limited to ICTs. Nevertheless, although target 17.6 does not specify that global knowledge sharing should focus on areas that will enable transitions to sustainable societies within planetary boundaries, HEIs can use this target within their respective national evaluation systems to systematically support internationalisation, specifically aiming at global justice and addressing sustainability challenges globally. To make actual contributions, however, internationalisation needs to go beyond its current forms, which are largely geared to serve dominant Northern partners (Chan 2004; Holmarsdottir et al. 2013; Melber 2015). Furthermore, to the extent that mobility is also involved, benefits need to go beyond individual students, staff or research environments, which supposes structures dedicated to spreading knowledge and know-how within the sending and receiving institutions.

9 Conclusions

Regardless of which strategies and pathways are chosen in HEIs, planetary boundaries (Steffen et al. 2015; Raworth 2017) must be respected. The material/ecological limits for sustainable modes of production and consumption are non-negotiable. These limits concern impacts on the planet as a whole, but also the limits of each country and locality. Collapses at any point in our life systems may have irreversible effects and trigger cascades, which means that planetary boundaries need to be considered qualitatively and holistically from their most vulnerable points, rather than only assessed through indicators and geographical or chronological averages. It follows that among the most urgent steps in transforming HEIs is to move away from formalised systems of evaluation, to instead consider capacity, education, collaboration and research based on real impacts on our societies and planetary life systems.

From the arguments outlined in this paper, it would seem that there is a paradox in the need for innovation, mobility and change, on the one hand, and the need for continuity on the other. But if we look closer, we can see that this paradox only exists to the extent that trust, social position, decision-making or innovation are primarily mediated through institutional structures and configurations, and to the extent that communication and reflection are carried by formalised information systems that serve a competitive paradigm. If these various functions are instead carried by humans and mediated through stable teams and interpersonal relationships, collective intelligent action becomes possible.

Despite the urgency of transforming higher education, there is a real danger in reducing academia to serving purely instrumental purposes. Academic freedom with space for critical thinking can prevent blind spots and miscalculations and is essential to preserving democratic values and maintaining credibility in scientific judgement as independent from particular political or commercial interests. Academic freedom is additionally crucial for maintaining foundational and exploratory research, which may provide key insights in the longer term. Diversity of fields of knowledge and academic actors, as well as pluralism in perspectives and methodologies, are vital for the longer-term resilience of our knowledge systems, enabling creativity (Marginson 2008) and preventing group think or perceptual "bubbles". Considering the great uncertainties in how political or environmental issues will unfold in the coming years, critical voices are even more important.

A further key consideration in any strategy to transform HEIs for sustainability, is that although challenge-driven, future-oriented research requires transdisciplinarity (Mochizuki and Yarime 2015), transdisciplinarity in turn depends on functioning and sufficiently stable environments in which disciplinary perspectives can be developed. There is otherwise a risk that research will be reduced to a series of disconnected endeavours locked within specific projects. To create the radical transformation that is needed in the face of global challenges, sufficient attention must be devoted to maintaining diversity, academic freedom, disciplinary expertise and continuity.

Finally, in any redefinition of priorities and restructuring of HEIs, the most important question is who defines what "sustainability" is, and through which processes new agendas are shaped. All fields and all employees must be involved, and have the time and means to do so—surveys and isolated consultations will not lead to engagement and determination. At the same time, those academics and research environments that are already committed to sustainability and global justice are an invaluable resource and can no longer be marginalised in agenda-setting (Avery and Nordén 2017).

HEIs today are constituted by academics whose disciplinary orientation, expertise and identity have been vested into an unsustainable paradigm. These are the people we have to work with, this is the expertise we have available. Paradoxically, therefore, to be radical, the transformation of HEIs must also be "soft", shaping realistic pathways and incentives for a shift in priorities and space to develop new skills for teachers, administrators and researchers. However, transformation will not be possible as long as sustainability and global justice are seen as add-ons and additional demands on time and resources, on top of a long list of other requirements. The focus must, therefore, lie on dismantling those drivers that otherwise keep sustainability on the margins.

References

Adger WN (2006) Vulnerability. Glob Environ Change 16(3):268–281

Adloff F, Neckel S (2019) Futures of sustainability as modernization, transformation, and control: a conceptual framework. Sustain Sci 14(4):1015–1025

Alexander S (2012) Planned economic contraction: the emerging case for degrowth. Environ Politics 21(3):349–368

Andersen AS, Kjeldsen TH (2015) Theoretical foundations of PPL at Roskilde university. In: Andersen AS, Heilesen SB (eds) The Roskilde Model: problem-oriented learning and project work. Springer, Cham, pp 3–16

Avery H, Nordén B (2017) Working with the divides: two critical axes in development for transformative professional practices. Int J Sustain High Educ 18(5):666–680

Acosta A (2013) Extractivism and neoextractivism: two sides of the same curse. In: Lang M, Mokrani M (eds) Beyond development: alternative visions from Latin America. Transnational Inst., pp 61–86

Baker-Shelley A, van Zeijl-Rozema A, Martens P (2017) A conceptual synthesis of organisational transformation: how to diagnose, and navigate, pathways for sustainability at universities? J Clean Prod 145:262–276

Berkes F (2007) Understanding uncertainty and reducing vulnerability: lessons from resilience thinking. Nat Hazards 41(2):283–295

Beynaghi A, Trencher G, Moztarzadeh F, Mozafari M, Maknoon R, Leal Filho W (2016) Future sustainability scenarios for universities: moving beyond the United Nations decade of education for sustainable development. J Clean Prod 112(4):3464–3478

Chan W (2004) International cooperation in higher education: theory and practice. J Stud Int Educ 8(1):32–55

Cooke RA (1987) The organizational culture inventory. Human Synergistics, Plymouth

Fiksel J (2006) Sustainability and resilience: toward a systems approach. Sustain Sci Prac Policy 2(2):14–21

Florin P, Guillermin M, Dedeurwaerdere T (2015) A pragmatist approach to transdisciplinarity in sustainability research: from complex systems theory to reflexive science. Futures 65:45–56
Gallopín GC (2002) Planning for resilience: scenarios, surprises, and branch points. In: Gundersen LH, Holling CS (eds) Panarchy. Understanding transformations in human and natural systems. Island Press, Washington, pp 361–392
Goldstein BE (2009) Resilience to surprises through communicative planning. Ecol Soc 14(2):article 33
Gough I (2017) Heat, greed and human need: climate change, capitalism and sustainable wellbeing. Edward Elgar Publishing, Cheltenham and Northampton
Hallinger P, Chatpinyakoop C (2019) A bibliometric review of research on higher education for sustainable development, 1998–2018. Sustainability 11(8):2401
Holmarsdottir HB, Desai Z, Botha LR, Breidlid A, Bastien S, Mukoma W, Nomlomo V (2013) COMPARE forum: the idea of North-South and South-South collaboration. Compare: J Comparative Int Educ 43(2):265–286
Holste JS, Fields D (2010) Trust and tacit knowledge sharing and use. J Knowl Manag 14(1):128–140
Jary D (1999) The implications of the audit society? The case of higher education. In: Dent M, O'Neill M, Bagley C (eds) Professions, new public management and the European welfare state. Staffordshire University Press, Stoke on Trent, pp 29–52
Lindhult E (2016) Co-Production, Action Research and the Movement towards Mode III Co-Production of Knowledge. Paper developed from contribution to Action Research Seminar, Aalborg University. October 7, 2016
Lynch K (2006) Neo-liberalism and marketisation: the implications for higher education. Eur Educ Res J 5(1):1–17
Marginson S (2006) Dynamics of national and global competition in higher education. High Educ 52(1):1–39
Marginson S (2008) Academic creativity under new public management: foundations for an investigation. Educ Theor 58(3):269–287
Marquard M, Sempler M, Mikkelsen H (2016) Transformative learning circles. http://www.creativeuniversity.aau.dk/digitalAssets/217/217625_maria-marquard–marianne-sempler–hans-mikkelsen.pdf. Last accessed 11 Mar 2020
Melber H (2015) Knowledge is power and power affects knowledge: challenges for research collaboration in and with Africa. Afr Dev 40(4):21–42
Miller TR, Muñoz-Erickson T, Redman CL (2011) Transforming knowledge for sustainability: towards adaptive academic institutions. Int J Sustain High Educ 12(2):177–192
Mochizuki Y, Yarime M (2015) Education for sustainable development and sustainability science: re-purposing higher education and research. In: Barth M, Michelsen G, Rieckmann M, Thomas I (eds) Routledge handbook of higher education for sustainable development. Routledge, pp 35–48
Oleksiyenko A (2018) Zones of alienation in global higher education: corporate abuse and leadership failures. Tert Educ Manag 24(3):193–205
Olsson P, Galaz V, Boonstra WJ (2014) Sustainability transformations: a resilience perspective. Ecol Soc 19(4):Art. 1
Pahl-Wostl C (2009) A conceptual framework for analysing adaptive capacity and multi-level learning processes in resource governance regimes. Glob Environ Change 19(3):354–365
Payo A, Becker P, Otto A, Vervoort J, Kingsborough A (2015) Experiential lock-in: Characterizing avoidable maladaptation in infrastructure systems. J Infrastruct Syst 22(1):02515001
Power M (1997) The audit society: rituals of verification. Oxford University Press, Oxford
Raworth K (2017) Doughnut economics: seven ways to think like a 21st-century economist. Chelsea Green Publishing
Ruiz-Mallén I, Heras M (2020) What sustainability? Higher education institutions' pathways to reach the Agenda 2030 goals. Sustainability 12(4):1290
Schwartz P (1991) The art of the long view. Doubleday, New York
Sevaldson B (2015) Gigamaps: their role as bridging artefacts and a new Sense Sharing Model. In: Relating systems thinking and design 4. Systemic design research network. Banff, pp 1–11

Shore S, Wright S (2000) Coercive accountability: the rise of audit culture in higher education. In: Strathern M (ed) Audit cultures: anthropological studies in accountability, ethics and the academy. Routledge, London, pp 57–89

Spangenberg JH (2019) Scenarios and indicators for sustainable development: towards a critical assessment of achievements and challenges. Special Issue. Sustainability ix–x

Sriskandarajah N, Tidball K, Wals AEJ, Blackmore C, Bawden R (2010) Resilience in learning systems: case studies in university education. Environ Educ Res 16(5–6):559–573

Steffen W, Richardson K, Rockström J, Cornell SE, Fetzer I, Bennett EM, Folke C (2015) Planetary boundaries: guiding human development on a changing planet. Science 347(6223):1259855

Vogt M, Weber C (2020) The role of universities in a sustainable society. Why value-free research is neither possible nor desirable. Sustainability 12(7):2811

von Weizsäcker EU, de Larderel J, Hargroves K, Hudson C, Smith M, Rodrigues M (2014) Decoupling 2: technologies, opportunities and policy options. A Report of the Working Group on Decoupling to the International Resource Panel. UNEP

Waddell S (2011) Global action networks: creating our future together. Palgrave Macmillan

Waddock S, Meszoely GM, Waddell S, Dentoni D (2015) The complexity of wicked problems in large scale change. J Organ Change Manag 28(6):993–1012

Change Project Approach for Reorienting University Teaching Towards the Implementation of Sustainability Principles

Oleksandra Khalaim, Tatjana Tambovceva, Lovísa Eiríksdóttir, and Shepherd Urenje

1 Introduction

The World Commission on Environment and Development proposed in 1987 a long-term strategy for achieving sustainable development by the year 2000 and beyond (World Commission on Environment and Development 1987). The report emphasized that attempts to address environment and sustainability problems require a holistic approach, which embraces the socio-ecological dimension in tandem with economy changes. Nowadays we can pose the question, why 33 years later the world is still talking the talk, and finding it almost impossible to walk the talk. Propelled by technological advances and global integration, the so-called economic growth experienced over the last century has been accompanied by increased pressures on socio-ecological resources. The list of the most pressing environmental, social and cultural issues has continued to grow including the supply of clean energy, availability of freshwater, ecological restoration, combating global climate change, increased human migration, increased urbanization and industrialization. Most recently, the

O. Khalaim (✉) · S. Urenje
Swedish International Centre of Education for Sustainable Development, Uppsala University—Campus Gotland, 621 67 Visby, Sweden
e-mail: oleksandra.khalaim@swedesd.uu.se

S. Urenje
e-mail: shepherd.urenje@swedesd.uu.se

T. Tambovceva
Faculty of Engineering Economics and Management, Riga Technical University, Kalnciema Street 6-213, 1048 Riga, Latvia
e-mail: tatjana.tambovceva@rtu.lv

L. Eiríksdóttir
Department of Business Studies, Uppsala University, Campus Gotland, 621 67 Visby, Sweden
e-mail: lovisa.eiriksdottir@fek.uu.se

W. Leal Filho et al. (eds.), *Universities, Sustainability and Society: Supporting the Implementation of the Sustainable Development Goals*, World Sustainability Series,
https://doi.org/10.1007/978-3-030-63399-8_17

seriousness of confronting COVID-19 has spawned a myriad of military metaphors including, the "greatest challenge since the Second World War", "working on the frontline", the virus is an "invisible enemy", and several other terms (Harris 2020). Never in the history of the planet has the world been united to deal with one common cause. It is therefore clear that global action is needed to create a more sustainable future. One problem that sticks out for higher education is in challenging the values of current growth trends and to decouple economic growth from environmental pressures and social justice.

The primary role of institutions of higher education is that of producing knowledge. It is therefore obvious that these institutions are a powerful means by which society can create a more sustainable future and this can happen when teaching, learning and research are reoriented to incorporate the concept of education for sustainable development to address many of the problems associated with human development. As we enter into the twenty-first Century, we are confronted with a supersonic rate of globalization resulting in snowballing challenges that demand a different educational approach, an approach that matches this complexity in order to meet the local and global challenges (Urenje 2019). According to UNESCO (2015b), a good quality education empowers people to change the way they think and act in relation to a sustainable future. When sustainable development issues are integrated into all aspects of teaching, learning and research, institutions of higher education have the potential of transforming society and developing students who are change agents. Since the current environment and sustainability challenges are linked to the way the current generation has been educated, our priorities should focus on reorienting our education system to help students think and behave in different ways that foster sustainable lifestyles, human rights, gender equality, promotion of a culture of peace and non-violence, global citizenship and appreciation of cultural diversity. Higher education institutions are therefore charged with the task of equipping students with essential knowledge, skills, attitudes and values empowering them to participate in creating a sustainable future and this can be achieved through transformative learning approaches.

Society agrees that scientific and technological innovations are essential in order to help us break free from unsustainable trends. However, without changes in social norms, habits, practices, and structures technology alone is inadequate. In our response to the current and future "wicked problems" which are characterized by high levels of volatility, uncertainty, complexity, and ambiguity, we also need multifaceted cultural changes that are long lasting. Education research acknowledges that not much is known about how social transformations emerge from the learning process, or how it contributes to the development of transformative agency in community contexts (Lotz-Sisitka et al. 2015). Wals (2007, 2010) has argued for transformative social learning in which a communicative process includes learners (a) critically examining their own values, habits, and norms (deconstruction), (b) listening to what others have to say (confrontation), and (c) co-constructing new viewpoints, values, and action repertoires (reconstruction).

What do we mean by transformative learning? Despite increased technological developments and wealth, massive social injustices and inequalities persist and are getting worse. When we engage in the discourses around our assumptions and expectations, i.e. our social world, community, and culture, as oppressive, unfair or unsustainable we can enter into a process of learning for social transformation. Transformative learning is therefore the process of examining, questioning, and revising those perspectives that we have otherwise taken for granted. Transformative learning also draws on the humanistic assumption that people have the ability to change deliberately their frames of reference and in this way have an inherent potential for personal growth and development (Ojala 2016).

The transformative learning perspective generally agrees with the common assumption of the education for sustainable development that openness and pluralism are necessary, i.e. learning processes that are collective, open-ended and reflexive are a vital part of sustainable development in which dominant ways of knowing can be challenged (Boström et al. 2018). Hampson and Rich-Tolsma (2015) summarized symptoms of transformative learning in three ways regarding the changing roles of learner, teacher and learning environment, i.e. the transformative learner, the transformative teacher and the transformative occasion. Firstly, the transformative learner; the student becomes transdisciplinary, a scientific investigator, a scholar and a practitioner all wrapped in one package. Secondly the transformative teacher; the educators serve the roles of mediator and companion; they become the enabler of the students' self-determination and social emancipation: teacher as catalyst for transformation. This includes facilitating emotional intelligence, and critical awareness. Thirdly, the transformative occasion; the learning environment becomes transformed into one in which the planetary, the complex, and the holistic are welcomed, encompassing multiple perspectives and states of awareness. The student encountering transformative learning is no longer a mere educational consumer, but rather a transdisciplinary scientist-scholar-practitioner deeply engaged in self-determination and the apt transformation of their world (Hampson and Rich-Tolsma 2015).

The challenge for higher education is to create an integrated and systemic response that addresses sustainability issues meaningfully, consistently and effectively. Four different responses to the challenge of sustainability have been identified by Sterling (2004):

(a) Denial (it's a hype that will go away)
(b) Bolt on (add a 'green aspect' to a curriculum or program)
(c) Built-in (important enough to integrate in all we do)
(d) Whole system re-design (we need to re-think the very foundations of what we currently do).

In most universities that we have worked with to address the sustainability challenge, initiatives are (b and c) of the 'bolt-on' and/or 'built in' variety. This has been due to a variety of reasons including policy limitations, incentive constraints and individuals lack a clear and consistent sustainability dimension among others (Wals and Benavot 2017). It is the last response that suggests a paradigm shift from "business as usual" (education as usual) and a transition towards doing better things differently (transformation) rather than doing what we do (business as usual) better (optimization). The ultimate goal for the Change Project Approach is (d), the 'whole institution' approach in which the university makes concurrent changes to curriculum, extracurricular activities, staff development, human resources and infrastructure operations and processes (Mcmillin and Dyball 2009).

The aim of the research is to describe and analyze the change project approach (CPA) as an enabler (triggering/starting point) of transformative learning process in higher education. It initiates innovations towards implementation of Agenda 2030 and sustainable development in interdisciplinary university teaching, supported by case study analysis from three universities of Baltic region.

2 Change Project Approach (CPA) as a Response to Socio-Economic Challenges of Sustainable Development

The Change Project Approach introduces Sustainable Development Goals (SDGs) implementation into university teaching and learning, specifically SDG 4, which seeks to "*ensure inclusive and equitable quality education and promote life-long learning opportunities for all*" (UNESCO 2015a). The emphasis is placed on the relationship between Education for Sustainable Development (ESD) and educational quality (target 4.7 of SDG 4) that requires all governments to:

> By 2030, ensure that all learners acquire the knowledge and skills needed to promote sustainable development, including, among others, through education for sustainable development and sustainable lifestyles, human rights, gender equality, promotion of a culture of peace and non-violence, global citizenship and appreciation of cultural diversity and of culture's contribution to sustainable development. (UNESCO 2015a)

The Change Project Approach is also expected to contribute to SDG 12—*ensure sustainable consumption and production patterns* and SDG 13—*take urgent action to combat climate change and its impacts*. The CPA engages participants in responding to the sustainability inadequacies in their current university curriculum. The emphasis is on how the institution can aligning their teaching and learning to SDGs and at the same time to build their ESD story of learning and change. We describe the Change Project as any project that brings about change in an institution's policy, strategy, curriculum, learning programs, learning environments, or institution-community relationships, with implications for ESD 'on the ground' (Lotz-Sisitka et al. 2015). The Change Project must be co-defined and should be collaboratively developed and implemented in a small community of practice with colleagues from the same or another department.

2.1 Change Project Action Research Cycle and Stages

The Change Project Approach (CPA) is an action learning intervention designed to stimulate transformative innovations towards sustainability in Higher Education. The intervention is contextually defined and implemented through a specific Change Project. We present the Change Project as a collectively developed initiative within the institution. The CPA stimulates and supports change processes beyond individuals, such that learning processes and the momentum of change remain in the institution even after the individuals move to other places. Each partner institution will be challenged to respond to their institutional demands in line with national and global sustainability trends. These methods and processes are applied in the iterative Action Research Cycles (Fig. 2). CPA Research Cycle presented at the Fig. 1 is an example of the processes taking place in cycle one. Similar actions will be repeated in cycles two and three as shown in Fig. 2.

The CPA begins with an assessment of current educational practices by in addressing twenty-first century challenges which help to align intended changes with national and institutional policies that relate to ESD and that change is within the resource capability (human and financial) of the institution. The following steps are suggested:

Step 1: Assessment of practice Question
How can current teaching and learning be transformed in order to respond to sustainability challenges of the twenty-first century?
Step 2: How is our Institution/Department meeting these demands?
What is working well?
What is not working so well?
Step 3: Change Project ideas
What is possible (probable/desirable) to change?
Develop a vision for change—what kind of a future do we want?

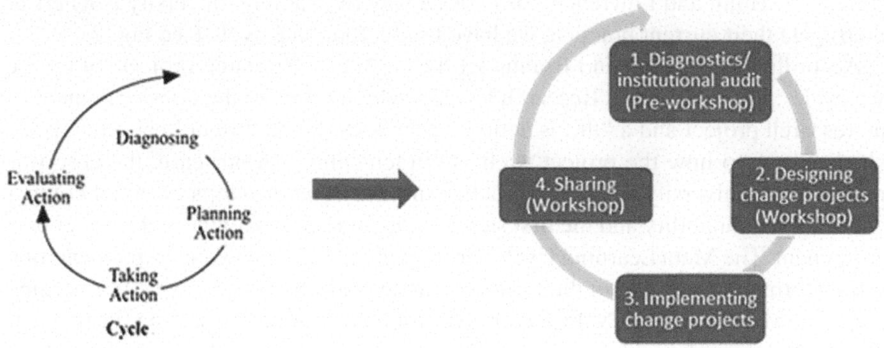

Fig. 1 Change project action research cycle (Rumjaun and Urenje 2017)

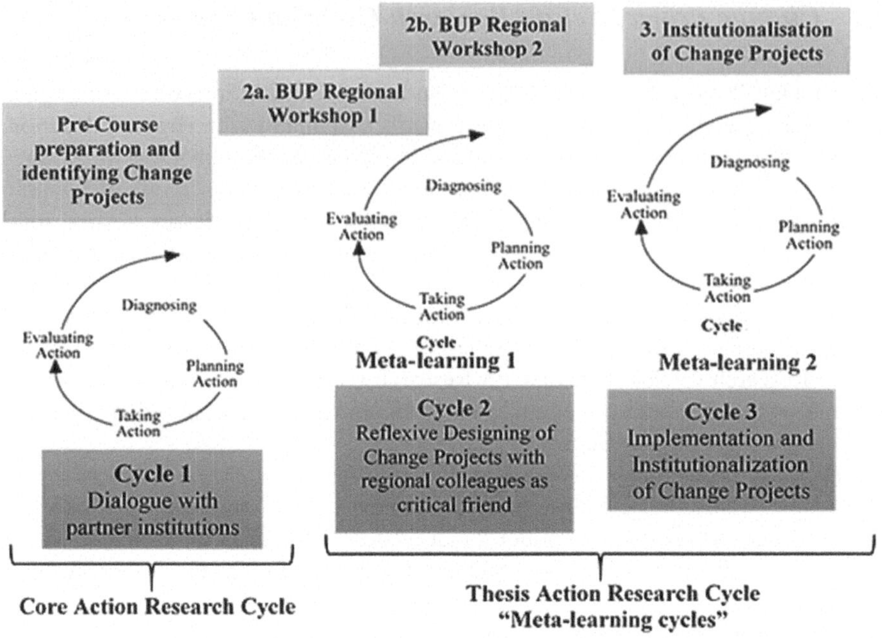

Fig. 2 Core and meta learning cycles adapted from Coghlan and Brannick (2001) (Rumjaun and Urenje 2017)

We chose Action Research because the methodology is the best way for practitioners to integrate their innovative actions while at the same time developing an understanding of the effectiveness of this implementation in an iterative manner. It is possible for university teachers to closely examine their role as change agents and decision-makers through their own problematisation of the teaching and learning processes within their work contexts, and through their own reflexive practice (Alsop et al. 2007; Hong and Lawrence 2011). As a way of enabling university teachers to interrogate their current practice, we have used a four-step cycle (see Fig. 2).

According to Coghlan and Brannick (2001), Action Research projects often use two cycles: a "Core Action Research Cycle", which refers to the aims or content of the research project and a "thesis action research cycle" or "Meta Learning Cycle" which relates to how the project itself is implementing. In our case, the core was transforming university education practice in a way that develops essential competences for sustainability and the first step was institutional dialogue and a pre-course assignment. The Meta Learning Cycle foresees the development and implementation of transformative learning in the form of change projects. A third cycle is emerging within the Meta Learning Cycle, which is the implementation and scaling (institutionalization) of their change project ideas (Rumjaun and Urenje 2017). CPA facilitators should be in direct contact with participating institutions in cycles one and two. After the second and final Regional Workshop in Cycle 2(b), CPA facilitators are no longer in direct control of the innovations, but continue supporting institutional collaboration remotely. In Cycle 3, the change continues to be strengthened with engagement of leadership and colleagues to form an institutional team for change. A second round of dialogue and critical reflection is suggested within the institution involving participants, colleagues and other interested parties to deliberate and present how the Change Project is being situated in the institution to be shared in the form of a poster, video or document. Generally, participants are encouraged to publish their project work and achievements.

3 Materials and Methods

3.1 Materials: Description of Cases

In this paper, we analyze three CPA cases of the Baltic region countries. Case coordinators directly participating in this paper (Lovísa Eiríksdóttir and Tatjana Tambovceva) provided the description of their cases from Uppsala University and Riga Technical University. The CPA case from Sumy State University is reflected and analyzed based on its description derived from Kostyuchenko and Smolennikov (2018). Additional information on further case progress was obtained in a personal interview with CPA case leader Nadiya Kostyuchenko in January 2020 by Oleksandra Khalaim.

CPA case study 1 "Active learning techniques for education for sustainable development"
Nadiya Kostyuchenko, Sumy State University, Oleg Balatskyi Academic and Research Institute of Finance, Economics and Management, Faculty of Teacher Training, Ukraine.

Sumy State University (SSU) has been functioning in Sumy city in the North-East of Ukraine since 1948. It hosts about 14,000 students who are pursuing bachelor and master degrees in 55 majors and 23 fields of knowledge, with about 1750 foreign students representing 50 countries worldwide. According to the worldwide Internet ranking Webometrics, SSU takes the third place among Ukrainian universities, and is placed in TOP-300 universities worldwide by the societal and economic impact According to the University Impact Rankings by Times Higher Education (SSU 2020).

The aim of the Sumy University change project was to develop a teacher-training program, "Active learning techniques for education for sustainable development" that focuses on active learning techniques with sustainable development issues inclusion as a key aspect of the content. The general purpose of the change project was to promote education for sustainable development through teacher training, addressing teachers from Sumy State University and other Ukrainian educational institutions.

The expected outcomes of the change project were focused mainly on sustainable development issues: dissemination of the idea in teachers' pedagogical practice through active learning techniques as well as increasing the awareness of both teachers and students.

The change project foreseen four realization stages (Kostyuchenko and Smolennikov 2018):

1. Organizing a new teacher-training program in active learning techniques.
2. Establishing a creative educational space "IdeaLab" as a venue for the change project.
3. Framing the change project through participation in Baltic University Program teachers' course.
4. Carrying out a workshop on Sustainable Development Goals for university teachers.

CPA case study 2 "From Control to Care"

Lovísa Eiríksdóttir, Dep. of Business Studies, Uppsala University, Campus Gotland, Sweden.

Uppsala University (2020) is the oldest university in the Nordic region, founded in 1477. It is ranked among the top 100 universities in the world with more than 40,000 students and 5000 researchers and teachers conducting world-leading research. The university is quite conventional and has not been known for its sustainability efforts, but still has several initiatives around the issues such as;

(1) *CEMUS*, a student-led centre for environment and development studies,
(2) *SWEDESD*, an international centre of Education for Sustainable Development, and
(3) *Zennström professorship*, a 10-year series of international visiting professors that are in the forefront of climate change leadership research (UU official website).

At Uppsala University, The Department of Business Studies was formally launched in 1958, but carries a tradition of Business Studies at the University dating back to the eighteenth century. It is one of the biggest departments at the university, with about 3500 students every year and 150 employees (UU official website). The university has a small campus on the island of Gotland. At Campus Gotland, the department has a small unit, facilitating one bachelor program and one master program, where the master level is focused on sustainability.

The master program, called *Sustainable Management*, started in 2013 and has constantly been in development for sustainable development since then, with Liberal Arts approach towards teaching and learning. However, the program is a part of a bigger and more traditional department that remains within the scope of disciplinary boundaries where management, organization, finance, marketing and international business are the main subjects. This means that even if the program on Campus Gotland focuses on sustainability it has remained a program of management and organization.

Getting the opportunity to discuss more deeply what ESD really entails, like in the process of the Change Project course, one can go beyond his/hers discipline and use sustainability as a lens instead of his/her own discipline, take a look with a lens from outside the field and see what needs to be *changed* instead of *managed* with the 'status quo' theories and methods. Consequently, the first introductory course of the master program, called *Managing Sustainability*, makes one wonder whether we should actually be *managing* sustainability or whether we should rather imagine how it could be *managed* according to the global challenges humanity faces today.

This change project focuses on implementing a new and altered perspective around sustainability issues to the introductory course in the master program of Sustainable Management at Campus Gotland, where you look at sustainability as a 'frame of mind' instead of as a 'management problem' like any other. The first tasks were to:

1. Change name: The name is important where it sets the standard of expectation. The name went from *Managing Sustainability* to *Business, Society and Nature: Reflective Inquiry*. Instead of assuming that sustainability is something that can be managed we include nature and society as equal to the business mind, where all are interconnected in relation to wicked world problems. Reflectivity and inquiry also give a hint to a different ontological view and perspective towards sustainability issues that are more open for imagination and change.
2. Course description: The description became more emphatic in a way that words like managing and control were taken away and words like care, creativity and critical were put instead. Wicked issues were also introduced to depart from a more open and complex perspective.
3. Sharing: The thought process and changes made were shared with teachers at the program and the management team at the department, such as chair professors and others to explain the importance of change.
4. Future tasks: In this new framing, the course will be held for the first time in 2020. The focus will be on seminars and discussions using different kind of perspectives including classical business theories, historical roots of the concept of sustainability and ethical considerations, where all areas will have room for reflection and critical discussion.

So far, the process of change has gone rather smoothly. Uppsala University is a traditional academic and collegial institution where the bottom up approach is very strong. It was challenging however to get new concepts in the description such as; (a) *wicked problems*, taken from the discipline of environmental studies and the field of education for sustainable development or (b) *care*, taken from the discipline of philosophy, where the management wondered about, both their academic legitimacy and their relevance to the subject. Nonetheless, explaining and emphasizing the importance of including new vocabulary to the field for an actual implementation of change and interdisciplinary perspectives seemed to be accepted as a legitimate argument to move forward.

CPA case study 3 "New course development"

Tatjana Tambovceva, Faculty of Engineering Economics and Management Riga Technical University, Latvia.

Riga Technical University (RTU) is a modern internationally recognized university, the only polytechnic one in Latvia and the largest one in the country; it educates and trains almost 15 thousand students at nine faculties. RTU is focused on becoming a third-generation university that not only provides high quality education, but also conducts advanced research and ensures innovation and technology transfer. Evaluating the performance of RTU in the implementation of the United Nations (UN) Sustainable Development Goal 17, RTU is ranked 101–200 in the highest-ranked group of universities in the ranking «The Times Higher Education Impact Rankings 2020». RTU efforts to achieve the goal of «Climate Action» are highly valued—RTU is recognized as the 14th best in the world in this field. RTU has shared 34th place reaching the goal «Affordable and Clean Energy», shared 45th place in implementing activities of goal «Decent Work and Economic Growth», and 70th place in fulfilling the intentions of goal «Industry Innovations and Infrastructure» (RTU 2020). Many study programs in RTU include different study courses connected with sustainable development issues. Some programs do not provide a full sustainability-related course, but mostly cover specific topics of particular study programme. In overall, RTU has a good basis for development of ESD at the university, because of ESD delivery is being realized both in forms of its integration into existing courses/programs or university common core requirements and development new sustainability courses/programs. The Faculty of Engineering Economics and Management (FEEM) is the biggest faculty at RTU with approximately 4000 students (RTU 2020).

Despite the fact that separate questions and topics connected to SD have been included in many study programs and courses, the purpose of the project was to develop a study course covering different SD questions for bachelor study programs at FEEM, for enhancing a systemic approach and broader vision of future economists. Specific tasks were to:

- Audit of existing courses at Faculty of Engineering Economics and Management.
- To review/improve curricula/to create new course on SD covering different SD questions for bachelor study programs at Faculty of Economics and Management.
- To improve teaching tools and materials.
- To use project-based learning activities for students.
- To use problem-based cases from real life.
- To use cross and multi-disciplinary cases.
- To use new technology and interactive methods in educational activities.

The project is addressing the following SDGs: SDG 4 (ensure inclusive and quality education for all and promote lifelong learning); SDG 8 (promote inclusive and sustainable economic growth, employment and decent work for all); and SDG 12 (ensure sustainable consumption and production patterns).

The course "*Eco-economics*" was developed in cooperation with my colleagues as an academic course of free choice for six ECTS credits under the bachelor study program "Economics". The goal of the course is to prepare a competent specialist who understands the impact of external and internal environment on economic processes in the national economy.

The insight into opportunities and prerequisites for the successful synergy of economic processes and problems of maintenance of natural resources and distribution of financial resources are given to students within the framework of the study course. Students have the opportunity to analyze specific situations faced by EU member states' companies, which aims to promote understanding of the importance of environmental protection. Global environmental problems and their prevention are emphasized as well as application of international quality standards in eco-economics.

The project helped to improve curricula and to include modern and necessary topic in the study program, as a result to have more competitive study programs. To build new society according to all SDGs we have to educate our students and give them the newest information and use the most modern methods for education. The study course is developed and is currently being taught once within the academic year. Since the semester is not over yet, it is impossible to generalize the results, but it is evident that the course can help developing a systematic and integrated vision of environmental problems among students and can potentially help them find solutions.

3.2 Methods: Case Study Analysis and Value Creation Conceptual Framework

In order to reflect on change project approach as a key component of transformative learning process in higher education towards sustainable development in interdisciplinary university teaching, we used a basic case study analysis method (Yin 1994) applied to the three cases from universities of Baltic region described above.

The value created by change projects was identified based on the framework provided by Wenger et al. (2011). The value creation methodology foresees five estimation cycles. First cycle estimates a productivity of certain community or network activity (for instance a community meeting or a project). The second cycle highlights a resource: an idea, a piece of advice, a document, a procedure, a model, or a relationship that came out of the activity described in cycle 1. The third cycle is an explanation of how this resource was applied in the practice of the storyteller and with what effects. A practical effect can be linked to an outcome at the cycle 4 (such as a measure of performance in the organization or for a person). Finally, a reflection on the definition of success and new considerations to frame the expectations of value creation are involved at the last fifth cycle (Wenger et al. 2011).

Based on the method, we developed a table with estimation of value created by three change projects (CPA cases) with further comparative analysis of the results (see Annex).

4 Results and Discussion on CPA Case Study Analysis

At the initial stage, we have formulated a number of starting discussion questions, serving as key points in the process the results interpretation:

- How did the CPA cases influence teachers on individual level, and the working environment on departmental/university level?
- Does the CPA induce changes by its own (as an external idea), or it serves primarily as a catalyst/platform/framework for changes, based on personal intension of a teacher to change the system? In other words, does it shape the way we are changing things?
- Do we need "grassroots-induced" changes in university ESD? What are institutional enablers and hindrances for CPA cases realization in universities of Baltic region?

The value creation stories and comparative analysis of change projects allowed reflecting on each case following the discussion questions.

The change project from Sumy State University is a good case highlighting the importance of small bottom-up initiated actions designed according to CPA that resulted in a significant system change in the quality education on the university level in a long-time perspective. The starting barriers for the initiative described by CPA case leader Nadiya Kostyuchenko in a personal interview included primarily non-willingness of experienced university teachers, professors and senior lecturers, to explore and practice the methods of active learning. "Why should I play these games at an obligatory teacher-training?"—that was a typical question during the first training sessions under a new program [interview with Nadiya Kostyuchenko 2020]. The first resistance was on adopting an interactive learning environment. Since the very idea of an open space equipped with flexible furniture and necessity to be part of constantly changing group models (from working in pairs to mini group discussions etc.) was not regarded as "business as usual" for many teachers, it provoked some opposition at the first stage. That is why the overall group of teachers was split in two independently working groups: a smaller one expressed their interest to be involved in teacher training under the new program, and a bigger one decided to follow the original program (designed in a format of "passive listening"). The new program became popular in three years of running through many university departments. Currently, the active learning techniques promoted by it were approved by the whole university along with a new design of classrooms that enables using these methods in teaching.

The second barrier in new program implementation was related to engaging trainers for the new program. Unexpectedly, those teachers who were practicing active learning techniques in their teaching finally were refusing to transfer their knowledge to other colleagues. The main reason was as well in a general lack of university community justification of these methods, especially from experienced elderly professors and some senior lecturers. Thus, the big value of the CPA case here was in conversion this "out-of-box" practices to a completely new "business as usual" paradigm, based on sustainability principles.

The main aim of ***the change project of Uppsala university*** was to enable a more open, interdisciplinary and attentive dialogue and methods for the students to be able to experience the world from many different perspectives, that is not only instrumental but also thought provoking. Teachers in higher education are not only for 'producing' people for the status-quo society but should also facilitate different thought processes that create agents of change for the common good. This cannot be done by solely implementing a standard or initiative like the SDGs where the students pick one box of seventeen to work with. In the business context the SDGs have begun to dominate the whole narrative and often ignore the fact that the goals are full of contradictions and ambiguity that needs to be discussed and put in context with various perspectives. Thus, a conscious focus on critical thinking and an open dialogue, that often is taken for granted in higher education, is a fundamental factor for ESD and cannot be replaced with a standard or a model. Going from the question of 'how can we control the situation' to 'how can we take care of the situation' is a crucial move towards a deeper implementation of ESD and starts the conversation with open-ended questions about the future—where the educators and the students meet in the process of learning and creating good judgement. This process is not solely about knowing but about deciding what knowledge deserves salvation, because questions of ESD are not only scientific, they are moral. Moral questions and answers cannot be transmitted from one person to another, but they need to be discussed in a space and developed within a human being, where everybody's voice and human conviction is heard. The course focuses on transformation for the new rather than transmitting the old. To enable this, it was important to introduce new concepts into the business vocabulary, that is oftentimes very instrumental and solution oriented. Concepts such as (1) 'wicked problems' that cannot be solved but tamed. 'Increasingly, these are the problems strategists face—and for which they are ill equipped' (Camillus 2008, p. 1) because of the solution and control-oriented ways the business field tends to look at the world. When writing the concept into the course description, the faculty wondered what the word meant and if it was of any relevance to the context. This started a conversation about the meaning and relevance for the concept of wicked issues that has become increasingly important in the vocabulary of ESD.

(2) Care and heart work were also placed there, and got questions of relevance from the management team at the business department in Uppsala. The intention with integrating this vocabulary into the description is to illustrate that ESD in business education is not solely about learning how to 'tackle' the sustainability challenges we face and gain knowledge about models to 'deal' with it. It is more about realizing your own values around the challenges and discover what one personally thinks is worthwhile taking care of in the future, as well as taking part in discovering what knowledge and practices it is that deserves salvation.

The CPA gives teachers a platform to start this thinking process by themselves and with other teachers. It gives courage and support to change things one did not dare to rethink before. The CPA course forces teachers to 'stop and think' for realization on what is meaningful and important to enable their responsibility and role towards their students and society. Moreover, it gives change and creativity the legitimacy it deserves.

The change project of Riga Technical University made it possible for the CPA coordinator and her colleagues to rethink the goals of education and the organization of a training course in a specific program. The project allowed revising seriously the ideas and approaches in teaching, as well as enabled understanding of what knowledge, practical skills, value attitudes and behavioral habits will allow students to become direct participants in the process of sustainable development. Interactive and student-centered practice-transforming teaching methods were used to encourage students' independent work, critical thinking, individual and collective projects, and a problem-oriented inter- and transdisciplinary approach to cognition. It can help to promote the formation of students' skills in making informed decisions and reasonable measures in order to ensure economic viability, the preservation of the environment, and a fair basis for the existence of society for present and future generations.

Thus, the analysis of three cases shows that despite the common methodological background of Change Project Approach applied to shape and design the projects, they appeared to be quite different regarding the ideas, aims, and ways of realization. At the same time, all three cases served the general idea of spreading and enforcing sustainability actions at the university/department level, spreading ESD principles and values. The key common outcomes of all cases were related to (1) scale effect (ability to apply the project at other departments/universities), (2) renewed values (rethinking on basic sustainability and ESD principles), and (3) converting the project results to a new "business-as-usual" practice.

5 Conclusions

The CPA cases presented in this paper illustrate that small individual actions are powerful enough to initiate system changes at the university level. The CPA design allows shifting the way of perceiving sustainability related problems that occur in universities and come "out-of-box" in solving them. With CPA methodology we can re-estimate the basic sustainability principles taken for granted from the times of Brundtland Report (World Commission on Environment and Development 1987) and convert the idea of "control" through SDGs to "care" with the help of sustainability redefining. Here coordinators of change projects act as "agents of change", creating new trends and alternative solutions that empower the sustainability vision for the rest of the university community.

What does CPA provide for ESD in higher education? Firstly, it encourages the participants to introduce changes—namely that are defined by themselves as the most relevant and urgent for a given academic community, not top-bottom imposed from university administration. Secondly, CPA raises effectively the awareness on quality education both for teaching and administrative staff of a university, by redefining it under sustainability vision. Furthermore, it provides a smoothly functioning way of a constant rethinking and reevaluation of CPA participants' own work and self-evaluation of themselves as reflective practitioners.

Importantly, CPA methodology designs an additional time and common platform for reflections for participating teachers. Under the conditions of constant routine working burden, the majority of university teachers as a rule cannot barely afford involving personal energy and additional efforts in the organization of a reflection and re-evaluation process related to the quality education. In these circumstances, CPA framework could serve as a good solution for such process design. It unites the participants for short working sessions and thus provides a possibility of common co-designing of change project cases. This approach brings out the reflective practitioner in university teachers, an opportunity to be reflexive in re-thinking and re-evaluation of our practice to meet the current demands of dealing with complexity.

Acknowledgements Acknowledgements to the Swedish Institute (SI) for the financial support for the postdoctoral research of Oleksandra Khalaim under the supervision of Dr. Shepherd Urenje at Uppsala University (Campus Gotland).

Annex: Change Project Value Creation Stories

Typical cycles	Latvia, RTU	Ukraine, SSU	Sweden, UU
Cycle 1: Immediate value **Activity**: Describe a meaningful activity you participated in and your experience of it (e.g., a conversation, a working session, a project, etc.)	Participation in the international "BUP Teachers Course on ESD in Higher Education" in 2018. All course participants performed tasks and received evaluations remotely on various aspects of sustainable development using Moodle system. During 2 off-line seminars in Kaunas and Riga, there was an opportunity to gain new knowledge and share experiences with colleagues and like-minded people from different BUP universities. The participants used various methods, such as games, group work, case studies, etc., which made the learning process interesting and exciting	Participation in the international "BUP Teachers Course on ESD in Higher Education" in 2018. In parallel under coordination of CPA case leader Nadiya Kostyuchenko, a new teacher-training program in active learning techniques was developed and the creative educational space IdeaLab was established with the financial support of home university	Participation in a CPA course at Uppsala University (UU), Campus Gotland in 2019 gave rise and empowerment to act on a necessary ontological and epistemological shift in an introductory course at the master programme Sustainable Management, given by the Department of Business Studies at UU. First step was a change in the course description that was inspired by a critical discussion and literature provided in the CPA course. New course description gave room for a more open, personal and interdisciplinary dialogue The changes already made have raised awareness and started a conversation at the business department around ESD and the meaning for the context

(continued)

(continued)

Typical cycles	Latvia, RTU	Ukraine, SSU	Sweden, UU
Cycle 2: Potential value **Output**: Describe a specific resource this activity produced for you (e.g., an idea or a document) and why you thought it might be useful	The course was not only interesting and fascinating, but also informative. As a result, SWOT analysis was carried out and a draft of changes was developed. A feedback was received from colleagues and moderators of the course, which allowed rethinking many points that did not occur before and improving the draft of the study program. New colleagues and friends appeared, with whom we continue to work on various project applications and articles	The BUP course provided new view on solving the traditional problems of university quality teaching and inspired to design specifically the change project. Both new teacher-training program and the creative educational space IdeaLab formed a promising basis for systemic change in teaching methods towards more innovative ones for university teachers under periodical series of obligatory teacher-trainings for all university teaching staff	The CPA course gave room for a deep discussion on the meaning of ESD and how it connects to critical thinking. This has resulted in an attempt to consciously implement critical thinking in the introductory course in the business program and intentions to have a more personal dialogue with students on existential questions that will furthermore create a better room for creative methods and vision for the future
Cycle 3: Applied value **Application**: Tell how you used this resource in your practice and what it enabled that would not have happened otherwise	The course was created in collaboration with colleagues in our department in the Faculty of Engineering Economics and Management at Riga Technical University. We worked as a team, discussed topics that need to be included in the syllabus, developed case studies and assignments	Teachers engaged in the new training program in a new learning space shared their experience of new innovative teaching methods. It appeared that there was an essential number of them already using "non-traditional" teaching methods, but they did not have opportunity to share the experience and felt themselves being not legitimated to use such "out-of-box" approach The newly designed learning space IdeaLab provided more physical space for group work, active gaming techniques, flexibility of switching the type of group work arrangement (mini groups vs one big group vs pair work)	The change project realization is still in progress

(continued)

(continued)

Typical cycles	Latvia, RTU	Ukraine, SSU	Sweden, UU
Cycle 4: Realized value **Outcome**: a. **Personal**: Explain how it affected your success (e.g., being a better professional, job satisfaction) b. **Organizational**: Has your participation contributed to the success of your organization	It is doubtful if the change project somehow particularly influenced personal success or career development, but it allowed improving cooperation with university colleagues and creating a new modern course that meets the changing conditions and priorities of sustainable development A series of new open-access publications were released based on the change project results (Khalaim and Tambovceva 2018; Tambovceva et al. 2019; Tambovceva and Titko 2020)	The new teacher-training program allowed university teachers to "legitimate" and enrich their individual teaching experience of active learning techniques, as well as find connections with other colleagues using same innovative approach. The new approach successfully replaced an old one with one time per five years traditional obligatory teacher-trainings with passive listening about innovations in pedagogy. A series of new open-access publications were released based on the change project results (Kostyuchenko 2017a, b), Kostyuchenko and Smolennikov 2016, 2018)	
Cycle 5: Reframing value **New definition of success**: Sometimes, such a story changes your understanding of what success is. If it happened this time, then include this here	The university colleagues appreciated the importance of the subject and the achievement of the goals of sustainable development. It essentially improved the understanding of what matters first of all, which topics should be considered and discussed As a result, it was decided to exclude from the study program two old subjects and replace them with a newly developed one	Scale effect. The active learning techniques were included in other university study programs in addition to the teacher-training program after its successful three-years running Renewed values. The creative learning space approach has been supported at the university level and is being introduced in many other university classrooms New business-as-usual. The innovative teaching methods are not treated by teachers as something suspiciously non-traditional and thus not worth replicating and positively recognized by colleagues for now	The change project realization is still in progress

Methodological framework adapted from Wenger et al. (2011)

References

Alsop S, Dippoa D, Zandvliet DB (2007) Teacher education as or for social and ecological transformation: place-based reflections on local and global participatory methods and collaborative practices. J Educ Teach 33(2):207–223

Boström M, Andersson E, Berg M, Gustafsson K, Gustavsson E, Hysing E, Lidskog R, Löfmarck Ojala M, Olsson J, Singleton BE, Svenberg Uggla Y, Öhman J (2018) Conditions for transformative learning for sustainable development: a theoretical review and approach. Sustainability 10(12):4479. https://doi.org/10.3390/su10124479

Camillus JC (2008) Strategy as a wicked problem. Harvard Bus Rev 86(5):98

Coghlan D, Brannick T (2001) Doing action research in your own organization. Sage Publications, London, 133 p. ISBN 0761968873

Hampson GP, Rich-Tolsma M (2015) Transformative learning for climate change engagement: regenerating perspectives, principles, and practice. Integr Rev 11(3):171–190

Harris S (2020) Coronavirus: if we are in a war against COVID-19 then we need to know where the enemy is. The Conversation. https://theconversation.com/coronavirus-if-we-are-in-a-war-against-covid-19-then-we-need-to-know-where-the-enemy-is-135274. Accessed 25 May 2020

Hong CE, Lawrence SA (2011) Action research in teacher education: classroom inquiry, reflection, and data-driven decision making. J Inq Action Educ 4(2):1–17

Khalaim O, Tambovceva T (2018) ESD implementation in universities of Central and Eastern Europe: common barriers, challenges, and solutions for Kyiv and Riga. Stud Perieget J 3(23):77–92

Kostyuchenko N (2017a) IdeaLab as a place for creativity in Sumy State University, Ukraine. Learn Teach Mag 2:16–17

Kostyuchenko N (2017b) Teacher training in active learning techniques as a pace for changes at Sumy State University. Learn Teach Mag 4:19–21

Kostyuchenko N, Smolennikov D (2016) Active teaching methods in education for sustainability as applied in good practices of local communities. Stud Perieget 1(15):145–158

Kostyuchenko N, Smolennikov D (2018) Education for sustainable development through teacher training. Stud Perieget 3(23):11–21

Lotz-Sisitka H, Wals AEJ, Kronlid D, McGarry D (2015) Transformative, transgressive social learning: rethinking higher education pedagogy in times of systemic global dysfunction. Curr Opin Environ Sustain 16:73–80. https://doi.org/10.1016/j.cosust.2015.07.018

Mcmillin J, Dyball R (2009) Developing a whole-of-university approach to education for sustainability: linking curriculum, research and sustainable campus operations. J Educ Sustain Dev 3:55–64

Ojala M (2016) Facing anxiety in climate change education: from therapeutic practice to hopeful transgressive learning. Can J Environ Educ 21:41–56

RTU (2020) Riga Technical University. An official web site: https://www.rtu.lv/. Accessed 02 May 2020

Rumjaun AB, Urenje S (2017) Transforming teacher education for the 21st century: developing relevant competences for a changing. World J Mod Educ Rev 7(7):488–511

Sterling S (2004) The ecological and environmental dimensions of the holistic curriculum. In: Encyclopedia of life support systems (Theme 6.6.1 'Education for Sustainability'). EOLSS Publishers, UNESCO. http://www.eolss.net/Sample-Chapters/C11/E6-61-04-03.pdf. Accessed 25 May 2020

SSU (2020) Sumy State University. An official web site: https://sumdu.edu.ua/int/en/about-sumdu.html. Accessed 23 Apr 2020

Tambovceva T, Tereshina M, Samarina V (2019) Green innovations in regional economy. In: 18th international scientific conference engineering for rural development proceedings, Latvia, Jelgava, 22–24 May 2019. Latvia University of Life Sciences and Technologies, Jelgava, pp 1832–1839. https://doi.org/10.22616/ERDev2019.18.N357

Tambovceva T, Titko J (2020) Consumer perception of sharing economy: pilot survey in Latvia. Int J Econ Policy Emerg Econ (IJEPEE) 13(1):72–84. https://doi.org/10.1504/IJEPEE.2020.106681

UNESCO (2015a) Education 2030. Incheon declaration and framework for action: towards inclusive and equitable quality education and lifelong learning for all. Paris, 86 p. https://unesdoc.unesco.org/ark:/48223/pf0000245656. Accessed 25 May 2020
UNESCO (2015b) Rethinking education. Towards a global common good? http://unesdoc.unesco.org/images/0023/002325/232555e.Pdf. Accessed 10 Mar 2020
Uppsala University (2020) Department of Business Studies. An official web site: https://www.fek.uu.se/about-us/. Accessed 20 Apr 2020
Urenje S (2019) Confronting the climate change crisis by inspiring teachers to facilitate education for social transformation; lessons from the 'Sustainability Starts with Teachers' initiative. High Educ Res (in press)
Wals AEJ, Benavot A (2017) Can we meet the sustainability challenges? The role of education and lifelong learning. Eur J Educ 52:404–413. https://doi.org/10.1111/ejed.12250
Wals AEJ (2007) Learning in a changing world and changing in a learning world: reflexively fumbling towards sustainability. South Afr J Environ Educ 22:35–45
Wals AEJ (2010) Mirroring, gestaltswitching and transformative learning: stepping stones for developing sustainability competence. Int J Sustain High Educ 29:380–390
Wenger E, Trayner B, de Laat M (2011) Promoting and assessing value creation in communities and networks: a conceptual framework. Rapport 18, Ruud de Moor Centrum, Open University of the Netherlands, 60 p. ISBN: 978-90-358-1808-8
World Commission on Environment and Development (1987) Our common future. Oxford University Press, New York, p 300
Yin RK (1994) Case study research: design and methods, 2nd edn. Sage, Thousand Oaks, p 171

Oleksandra Khalaim, Ph.D. is graduated in Economics (2006), master's degree in Economics (2007), master's degree in Environmental Sciences (2010), and Ph.D. in Ecology and Environmental Sciences (2017). Post Doctorial Research at Uppsala University (2020). Her current research interests include education for sustainable development, climate change education and transformative learning methods in higher education, climate change adaptation and management of urban green areas.

Tatjana Tambovceva, Dr.oec. Professor at the Riga Technical University (Latvia). She has more than 25 years' experience in lecturing both at the undergraduate and graduate levels. Her research interests focus on sustainable development, circular economy, education for sustainable development, change of consumer behavior; corporate social responsibility; green production and consumption; project management and use of ICT in many areas. She is a member and an expert in a number of international Societies, Councils and Foundations. She is also the Managing Editor of the Scientific Journal "Economics and Business".

Lovísa Eiríksdóttir has a bachelor degree in Economics (2012) from university of Iceland, master's degree in Applied Ethics (2014) from University of Iceland, and a master's degree in Sustainable Management (2016) from Uppsala University. Now Lovísa is a Ph.D. student in the Department of Business Studies at Uppsala University, doing research on Sustainable Development in Business Schools, through the faculty of thought and judgement of business educators. Current research interests is Hannah Arendt's theorizing around thinking and responsibility, story-telling, narratives and higher education for sustainable development.

Shepherd Urenje, Ed.D received his doctoral degree in Development Education at the University College of London, Institute of Education (2012) and graduated with a master's degree in Environmental Education from Rhodes University (2004). He is an alumnus of the University of Zimbabwe (1983) where he graduated with a Bachelor in Education. He has worked as a teacher of environmental science and development education in Zimbabwe, a Regional Programme Manager for Education and Training at SADC, and Principal Examiner of Environmental Science in the UK. Currently Dr Shepherd Urenje is an ESD Programme Specialist at the Swedish International Centre of Education for Sustainable Development at Uppsala University (SWEDESD). His work supports ESD in Higher Education within and among countries in Scandinavia, Africa and Asia in responding to the twenty-first century sustainability challenges.

Case Studies on Sustainable Development at Universities

Education for Sustainable Development: A University Perspective

Maris Klavins

1 Introduction: Sustainable Development on the Political Agenda of Latvia

In Latvia, sustainable development (SD) was set as a political target for development following the Rio meeting in 1992. Later on, it was harmonised with the objective of sustainable development as laid out in the EU Lisbon Strategy, and in the Renewed EU Sustainable Development Strategy (CEU 2006). In keeping with these international and EU commitments, the political target of sustainable development and corresponding objectives must be declared in national-level planning documents (Saeima 2010). The responsibility for drawing up the National Development Plan and monitoring the implementation of the Sustainable Development Plan of Latvia until 2030 is laid in the Cross Sectorial Coordination Centre, subject to the supervision of the Prime Minister of the Republic of Latvia, with the planning and control process taking place in ministries. Thus, the governance of sustainable development, including its planning and implementation phases, is not implemented in a cross-cutting manner in Latvia. Admittedly, progress towards sustainability has largely been declarative, short of practical measures and outcomes. For example, impacts of restructuring of the economy and the recent recession arising from the economic crisis resulted not only in a major decrease in GDP but also in reductions of greenhouse gas emissions and use of fertilisers, as well as depopulation of the country due to massive emigration. These trends are misattributed as progress with respect to sustainable development.

At the same time, there have been marked achievements in the development governance of Latvia in several sectors. For instance, remarkable progress has been made in promoting a sustainable energy policy. Despite various challenges, a sustainable

M. Klavins (✉)
Department of Environmental Science, University of Latvia, 19 Raina Blvd, 1586 Riga, Latvia
e-mail: maris.klavins@lu.lv

W. Leal Filho et al. (eds.), *Universities, Sustainability and Society: Supporting the Implementation of the Sustainable Development Goals*, World Sustainability Series,
https://doi.org/10.1007/978-3-030-63399-8_18

restructuring of the energy system has occurred, as indicated by a significant reduction in energy use in the most consuming production branches. This was possible due to a major restructuring of the entire production system, major increase in the use of renewable energy sources (a target of 40% of renewable energy sources by 2020 is realistic), as well as improved energy production and saving technologies (Rasmussen 2003). Diversification of energy sources (previously relying largely on imported fossil fuels and imported electricity), attainment of energy independence and strengthening of the national economy by increasing the production of biomass energy and biofuels, and applying advanced energy production and saving technologies all play the key role in this process (Rasmussen 2003). In the energy sector, transformation of the production and consumption system is driven by national actors, considering both the national interests and international processes, most importantly—the requirement to implement Directive 2006/32/EC of the European Parliament and of the Council on energy end-use efficiency and energy services (EPC 2006). Concerted actions of all actors involved in the governance of the energy production and consumption system, with the public support, ensure transformation of the energy sector and the implementation of a sustainable energy policy at national and sub-sectoral levels (Blumberga et al. 2014).

On another note, there are evident problems in the governance of sustainable development in Latvia. One of the most problematic aspects, as indicated by several surveys, is a lack of knowledge about sustainable development both in society and among stakeholders (Klavins and Pelnena 2010). Since sustainable development is a general concept that must be adapted to specific local social, cultural and political conditions, and must be apprehensible and acceptable to a wider public, education on sustainability issues is one of the key factors in promoting integration of the concept into policies and everyday life. Although there is a political commitment to achieving the integration of Education for Sustainable Development (ESD) into the national educational system, the real progress is insignificant (Klavins and Pelnena 2010), and sustainable development issues are minimally considered in the education process. Nevertheless, integration of sustainability topics into the educational system can be considered to be a major direction of activity for the mainstreaming of sustainable development.

Among other factors hindering the integration of the concept of sustainable development into governance processes is the maturity of the governance system itself, and a number of other associated obstacles. This includes most importantly the cooperation of practices and political habits (mitigating tensions among groups with different political interests, political parties, and groups of stakeholders). Governance for sustainable development requires active cooperation among different sectors, and unwillingness to cooperate is the weakest element in the governance system (Zilans 2013).

On the local/municipal level, it has been demonstrated that (besides limited knowledge about sustainable development among municipal development planners and decision-makers) deficiencies in policy integration, intersectoral cooperation, municipality and stakeholder cooperation and urban management practices constitute the major reason for weak support of sustainable development in governance practice

(Zilans 2008). Considering the absence of reliable criteria for analysing the governance processes with respect to the mainstreaming of sustainable development, it is important to develop a methodology for assessing the sustainability of development policies and indicators for monitoring the policy outcomes and development trends with respect to sustainability (Zilans and Abolina 2009).

2 Sustainability and Universities

Universities play a key role in introducing the concept of sustainable development to society. Universities in Latvia are significant players in promoting sustainable development. The most important contribution of universities pertains to research on sustainable development at a national-level and studies into the development process conducted as part of sustainability reporting (Dale and Newman 2005). Several studies have analysed the sustainable development process in Latvia, revealing its achievements and failures (Zilans and Abolina 2009; Blumberga et al. 2014). However, the volume of analytical work dedicated to studying the development process is relatively small. This is especially true given that study of sustainable development is not recognised as a research priority, and analysis of the development process from a very classical perspective of the economy dominates in the existing studies. Notably, the main objective of the National Development Plan of Latvia 2014–2020 is a "Breakthrough in Production" (National Development Plan of Latvia 2014). Other major factors that hinder research on sustainability of the development of Latvia are the dominance of the public administration system in the national planning process, limited accessibility of data that characterise the development process, and obstacles for academics to participate in that process and discuss its results. Definitely, there is a large gap between the academic analytical process and the real planning process and evaluation of the development results (Bourn 2008). This situation is, for the most part, related to the national development plan being used as an instrument in deciding on the allocation of EU Structural Funds to Latvia as a "new" EU Member State. Thus, the development planning process, also including aspects of sustainability of development, in compliance with the 17 sustainable development goals and other aspects of sustainable development, was overwhelmed by economic and political interests of the national development planning process in Latvia, thereby preventing the involvement of academics in sustainability studies. Furthermore, the national sustainable development reporting process to evaluate compliance with the sustainable development goals has been taken over by the Cross-sectoral Coordination Centre, under the authority of the Prime Minister of the Republic of Latvia. The key impact of the existing impediments for academics to access the development process is a limitation of the possibilities to influence the planning process at a national level and to participate in discussions about this process and achieved results.

Sustainability aspects pertaining to universities are mostly those related to the Education for Sustainable Development. Initiatives related to the development and implementation of ESD represent a bottom-up approach, and emerge based on the results of local initiatives of academics and as result of active international cooperation, first of all in the Baltic Sea Region. Activities of the UNESCO Chair on Teacher Education and Continuing Education: Interplay of Tradition and Innovation in Education for Sustainable Development at Daugavpils University is a successful example of ESD development. The aim of the UNESCO Chair is to promote an integrated system of ESD research, training, information and documentation as well as to facilitate international cooperation. The Chair has initiated a national-level dialogue for reorientation of teacher education towards sustainable development in terms of analysis and exchange of results of scientific research and practical experience, as well as complementation of the existing teacher education curricula with relevant content of ESD. Significant achievements have been the emergence of an enabling institutional environment for the analysis of fundamental and applied research and practice, involvement of Latvian higher educational institutions, school teachers, students and NGOs in ESD action research initiatives, as well as the founding of the Baltic and Black Sea Circle Consortium in action research education. The best practice is shared in the internationally acknowledged *Journal of Teacher Education for Sustainability*.

Preparation of study materials on sustainable development basics and the implementation of a Sustainable Development course in the leading Latvian universities can be mentioned among other achievements of ESD in Latvia.

3 New Concepts—New Challenges

The new concept of Education for Environmental Citizenship presents a new challenge for ESD (Dobson, 2010). This concept is highly attractive, as it supports "pro-environmental behavior, in public and private, driven by a belief in fairness of the distribution of environmental goods, participation, and co-creation of sustainability policy". The concept covers several major aspects essential for the development of an educational system to improve citizens' public involvement (Schild 2016).

Education for Environmental Citizenship can be considered as a continuation of the efforts to implement Environmental Education (EE) and ESD in Latvia (Ryden 2009; Pelnena and Klavins 2009). The preparation of a highly educated workforce, investment in human resources, and social motivation for education are all deemed to be determining factors for the development of Latvia in the twenty-first century, ensuring its competitiveness at the European and global levels. Latvia's geopolitical situation and limited raw materials and energy resources create conditions where its main competitive advantage will be educated people and a qualified workforce. The system of general education institutions ensures that primary education can be acquired close to the place of residence, and parents and pupils can choose their general educational institution. National minorities are afforded equal opportunities for general education, while preserving and maintaining their national cultures. The implementation of educational programmes for national minorities is the precondition for social integration in Latvia. The financial allocations for the implementation of the general education policy, however, are insufficient. The available funds are used for maintaining the sector, but fail to support its development. The education system is characterised by a general shortage of a qualified workforce and an increasing mean age of teachers. Towns and rural areas differ in terms of access to and availability of high quality education, and there are regional differences in the available educational programmes. The physical infrastructure of many educational institutions of all levels and kinds has not been upgraded. This considerably decreases the quality education and makes the maintenance costs of such institutions more expensive. The structure of academic personnel is also disproportionate. The low wage standard makes it difficult to attract young academic staff and researchers. There is a wide offer of continuous education in the country at large, and it develops along with demand. It would be necessary to set up a system to assess the quality of the continuing education programmes. There is no regulatory framework governing the financing of continuous education, including shared financing of lifelong learning. Employers play a crucial role in providing apprenticeships. At present, the provision of internship opportunities is insufficient, requiring an economic stimulus for businesses to provide internship placements for students in vocational and higher education programmes. Neither is the apprenticeship system sufficiently developed in the country. Development of scientific research and innovations and intensive use of knowledge and high technologies is the main and only realistic path for the development of the Latvian economy in order to ensure a stable welfare standard. Essential means to this effect are Environmental Education and Education for Sustainable Development. The concept of Education for Environmental Citizenship is timely in the context of the transformation process of education. This process also includes the need to develop learners' competencies that are necessary for functioning in society and for successful learning, such as being able to manage information flows, acquiring skills for future use, and handling the rapid development of new methods to achieve education goals at all levels.

The results of the expert survey on Education for Environmental Citizenship indicate insufficient knowledge of the concept, even among experts in the field, not to speak of the general public and teachers at all levels. This aspect can be considered to be a significant weakness in the implementation of the Education for Environmental Citizenship, calling for actions to be taken. There are actors supporting and promoting EE and ESD. EE is supported by political parties (the Green Party) and NGOs at a national and international level. Therefore, it can rely on structural and institutional support. Numerous NGOs support some activities, from campaigns to the Ecoschool network, along with implementation of the Environmental Education concepts into the regular education system. ESD in Latvia is supported by the National Committee of UNESCO. Accordingly, its institutional support from ministries is secured. Major activities in the past that have supported the raising of ESD awareness at the UN level are reflected in the national mass media and are also on the discussion agenda. The existing efforts are integrated into legislation. Thus, the Environmental Protection Act, Article 42 (2), states: "The environmental protection course shall be included in the mandatory part of all study programmes of higher education institutions and colleges." Recent efforts have resulted in the creation of study materials and study programmes and in regular activities at schools. There have been significant endeavours to achieve the aims of the Environmental Education and Education for Sustainable Development (Klavins and Pelnena 2010). Another major threat which could compromise the progress towards the aims of Environmental Education and Education for Sustainable Development is related to the implementation of the third concept—Education for Environmental Citizenship. Since this concept covers the same major issues, the progress towards the achievement of the intended aims and targets may be hampered.

A major opportunity for Education for Environmental Citizenship in Latvia is its potential role in the reorganisation of the educational system, given that this highly attractive concept is aimed at the active involvement of the whole society. This opportunity relates to the still relevant aim to support the active involvement of all citizens in the political and social processes in the country, and to mobilise resources for solving actual problems. Questions can be raised about the capacity to increase the potential and significance of EE, ESD and Education for Environmental Citizenship, all of which have good prospects for finding their positions in the reorganisation of the educational system. Another opportunity is to develop study (teaching) materials in line with the latest achievements and methodologies, replacing outdated methods and materials.

Major threats to Education for Environmental Citizenship have been identified and are as follows: (1) a comprehensive Environmental Education system has not been established in Latvia; consequently, introduction of a new approach can attenuate the existing efforts; (2) insufficient resources (intellectual, material, financial) to assure Education for Environmental Citizenship; (3) shortage of study aids on the basic problems of Environmental Education to assure its representation at different levels; (4) shortage of knowledge, motivation and Environmental Education in the process of further and continuing education; (5) the content of education does not have sufficient, relevant and objective information on environmental problems, and this prevents an awareness of and responsibility for the interaction of humans with the environment; (6) the content of subjects is not harmonised with the guidelines for education and fails to assure succession at different levels of education; (7) an insufficient quantity of study and teaching aids for Environmental Education; (8) an insufficient number of qualified teachers; (9) a small number of students will acquire the knowledge of sustainable development and environmental protection during their studies at Latvian educational institutions; (10) shortage of the state system of syllabi in higher Environmental Education; (11) the country has no scientific research institutions, the activities of which would allow resolution of strategic and practical issues in environmental protection, as well assurance of a scientific basis for decision-making.

Education for Environmental Citizenship is a concept with a high potential for contributing to the reorganisation process of the Latvian educational system. In practice, however, major efforts are required in order to come to real implementation of the Education for Environmental Citizenship in the educational practice.

4 Conclusions

Considering the complexity and diversity of the concept of sustainable development, it is extremely important for universities to identify their priorities and aims, so that the contribution of academic stakeholders in promoting the ESD process could be optimised. The primary responsibilities of universities should concern the following: (1) development of the sustainability science; (2) integration of Education for Sustainable Development in the study process; and (3) promotion of an active stance towards problems related to the sustainability of national development.

For universities, the priority action must be to promote ESD is participation in the development of an ESD-supporting educational system—a task at which the contribution of university actors can be most efficient. Another priority task is the development of study approaches and study materials to support the implementation of ESD in all study programmes at a university level. A further challenge concerns the new initiative aimed at strengthening the links between ESD, Science Education and Citizenship Education in order to promote the Education for Environmental Citizenship.

Acknowledgements I acknowledge the support from the University of Latvia research grant for the "Climate change and its impacts on sustainability of natural resources" project.

References

Blumberga A, Blumberga D, Bazbauers G, Zogla G, Laicane I (2014) Sustainable development modelling for the energy sector. J Clean Prod 63:134–142

Bourn D (2008) Education for sustainable development in the UK: making the connections between the environment and development agendas. Theor Res Educ 6:193–206

CEU (2006) Council of the European Union. Renewed EU sustainable development strategy. 9 June 2006

Dale A, Newman L (2005) Sustainable development, education and literacy. Int J Sustain High Educ 6:351–362

Dobson A (2010) Environmental citizenship: rapid research and evidence review. https://www.sdresearch.org.uk/sites/default/files/publications/SDRN%20Environmental%20Citizenship%20and%20Pro-Environmental%20Full%20Report_0.pdf. Accessed 22 Dec 2019

EPC (2006) Directive 2006/32/EC of the European parliament and the council of 5 April 2006 on energy end-use efficiency and energy services and repealing council Directive 93/76/EEC. Off J Eur Union L114:64–85

Klavins M, Pelnena M (2010) Concepts and approaches for the implementation of education for sustainable development in the curricula of universities in Latvia. J Baltic Sci Educ 9:264–272
National development plan of Latvia (2014) https://www.pkc.gov.lv/images/NAP2020%20dokumenti/NDP2020_English_Final.pdf. Accessed 22 Jan 2020
Pelnena M, Klavins M (2009) The concept and evolution of education for sustainable development. In: Klavins M (ed) Environmental education at universities. Academic Publisher of University of Latvia, Riga, pp 69–88
Rasmussen LH (2003) A sustainable energy-system in Latvia. Appl Energy 76:1–8
Ryden L (2009) Society and environment—who takes responsibility. In: Klavins M (ed) Environmental education at universities. Academic Publisher of University of Latvia, Riga, pp 36–51
Saeima of the Republic of Latvia (2010) Sustainable development strategy of Latvia until 2030
Schild R (2016) Environmental citizenship: what can political theory contribute to environmental education practice? J Environ Educ 47:19–34
Zilans A (2008) Governance as a barrier to mainstreaming sustainable development in Riga, Latvia. Int J Environ Sust Dev 7:1–20
Zilans A (2013) Assessment of urban development policies in the context of sustainability in Latvia. WIT Trans Ecol Environ 175:135–147
Zilans A, Abolina K (2009) A methodology for assessing urban sustainability: Aalborg commitments baseline review for Riga, Latvia. Environ Dev Sustain 11:85–114

Professor Maris Klavins His major research fields include climate change, sustainable development, and innovation in the development of environmentally friendly processes and technologies. M. Klavins is currently a supervisor of five Ph.D. students, has supervised a total of 11 doctoral theses, is an author of numerous scientific articles (158 SCOPUS articles, $h = 18$). In the last few years, M. Klavins has been involved in the climate policy legislation development and adaptation in Latvia. M. Klavins has been a leader of several projects related to climate change aspects, for example, of the "Climate Change and Waters" National Research Programme (http://kalme.daba.lv/en/), the INTERREG project aimed at the development of adaptation strategies to climate change (https://www.eucc-d.de/past-projects/articles/climate-change-impacts-costs-and-adaptation-in-the-baltic-sea-region-782.html). He has supported development of study content and study materials on climate change (http://www.zalabriviba.lv/klimats/klimats-un-ilgtspejiga-attistiba/).

Students' Perception of Campus Sustainability in a Brazilian University

Claudio R. P. Vasconcelos, Paula Ferreira, Madalena Araújo, Davidson Cordeiro, and Shammio M. Dias Silva

1 Introduction

The sustainability issues in HEIs have been attracting a progressively increasing level of consideration from managers and scholars. Hundreds of applicational case studies of sustainable practices in HEI and dozens of sustainable assessment tools (SATs) have been created since the emergence of the environmental crisis reported in Stockholm, 1972.

Until now, much relevant knowledge has been generated on the topic of the sustainable system and SATs for HEIs. However, HEI are complex institutions composed of several interdependent subsystems, therefore sustainable improvement requires a holistic and integrated system and assessment measures to ensure its compliance with the established goals (Leal Filho et al. 2019a; Tim and Jutidamrongphan 2018). Despite the vast literature produced concerning sustainability in higher education, the need for the development of integrated and holistic systems to manage HEIs'

C. R. P. Vasconcelos (✉) · D. Cordeiro · S. M. Dias Silva
Laboratory of Sustainability Engineering and Consumption, Federal University of Paraíba, P.O. Box 5045, 58051-970 João Pessoa, PB, Brazil
e-mail: claudioruy@yahoo.com

D. Cordeiro
e-mail: davidsoncordeiro70@gmail.com

S. M. Dias Silva
e-mail: shammio.michel@gmail.com

C. R. P. Vasconcelos · P. Ferreira · M. Araújo
School of Engineering, Algoritmi Research Centre, University of Minho, 4800-058 Guimarães, Portugal
e-mail: paulaf@dps.uminho.pt

M. Araújo
e-mail: mmaraujo@dps.uminho.pt

W. Leal Filho et al. (eds.), *Universities, Sustainability and Society: Supporting the Implementation of the Sustainable Development Goals*, World Sustainability Series,
https://doi.org/10.1007/978-3-030-63399-8_19

efforts in assuming their role in implementing sustainability is still acknowledged. This claim is directed both to HEIs' internal routines and, from a broader perspective, to a global movement towards a more sustainable society.

In the existing literature, more attention has been given to the development of objective assessment tools, rather than human-centred ones which allow the generation of knowledge about the perception of individuals that make up an HEI, such as students, teachers, or staff. To date, no stakeholder perception study on the sustainability of HEIs has been identified that would consider the possibility of integrating their assessment tools with the sustainability dimensions already consolidated in the literature.

In order for HEIs to successfully achieve the sustainability goals, the cooperation and participation from all stakeholders are critical, which includes staff, faculty, students, funding bodies, government, employers, suppliers and community (Green 2013; Leal Filho et al. 2019b; Sammalisto et al. 2015). Among all of them, students appear as one of the key stakeholders in universities, not only for their much bigger number and HEIs' main target mission but also since there is empirical evidence that they have shown willingness to support and participate in university sustainable practices (Emanuel and Adams 2011). Many authors note the importance of placing students engaged in the university's sustainable practices as active agents of change, although they also recognise that there is still a dearth of previous studies about students' perception of sustainability in HEIs (Blanco-Portela et al. 2018).

Nejati and Nejati (2013) support that understanding how students evaluate the sustainability practices implemented by HEIs is crucial as it allows the decision-maker to become aware of the HEI performance from the perspective of one of their major groups of stakeholders. For these authors, "the study of students' perceptions towards sustainability remains under-researched and needs to be further explored" (Nejati and Nejati 2013, p. 102).

Concerning the gaps previously discussed, related to the shortage of tools to assess HEI sustainability relying on stakeholders' perceptions, and the absence of assessment tools that support its integration with the quantitative indicators of HEIs' sustainability established in the literature, this study has two main goals. The first intends to contribute to the literature by designing a sustainability assessment tool to assess the students' perception of the campus's sustainability, based on a Brazilian HEI case study. The second aims to analyse the adherence of the designed tool to assessing the key dimensions of sustainability proposed in the literature.

2 HEIs' Sustainability

Since Stockholm, 1972, the higher education institutions (HEIs) have been adapting themselves to assume their social role in supporting societies in the promotion of sustainable lifestyles. From 2015, since the development of the new sustainable agenda, the Sustainable Development Goals (SDGs) have been established—an expansion of the eight Millennium Development Goals (MDGs), compounded by a set of actions grouped into 17 goals which aim to end poverty in all its forms by 2030 (Leal Filho et al. 2019b). This new agenda turns the role of HEIs into a more meaningful and convoluted challenger in terms of conceiving more sustainable societies.

Bizerril et al. (2018) recognise HEIs as a strategic agents in promoting sustainability. This perspective takes into account different aspects, such as the fact that they are institutions that promote innovation (Lozano 2006b); play a relevant role in the education of leaders, teachers and professionals from different areas of society (Cortese 2003); have been considered responsible for ensuring to ensure that the curriculum taught prepares individuals for sustainability challenges; take a leading role in promoting regional sustainable development (Karatzoglou 2013).

Hopefully, the students will become individuals prepared to understand the complexities of sustainability and to convert the knowledge acquired into systemic, anticipatory, critical thinking and actions to implement environmental management systems that support the social change to a more sustainable living standard (Brandli et al. 2011; Sammalisto et al. 2015). To overcome their challenges in the promotion of sustainability, HEIs should develop skills to reduce the environmental impact of their activities (Alshuwaikhat and Abubakar 2008; Findler et al. 2019). Thus, according to Ceulemans et al. (2015), university institutions, due to their specificities and importance, should be considered differently from other public or corporate institutions. To meet the expectations set out in the 2030 New Global Environmental Agenda, a lot of universities have taken on the challenge of incorporating sustainable development practices into their education, research, internal management, and community engagement processes. Higher education institutions taking action in this direction are being usually designated as sustainable HEIs.

Conceptualising the Sustainable University designation is not an easy task due to the variety and diversity of activities commonly undertaken in a university campus. Velazquez et al. (2006, p. 812) compiled empirical data from sustainable programmes and actions carried out by about 80 universities around the world and defined a sustainable campus as "a higher education institution, as a whole or as a part, that addresses, involves and promotes, on a regional or a global level, the minimization of negative environmental, economic, societal, and health effects generated in the use of their resources in order to fulfill its functions of teaching, research, outreach and partnership, and stewardship in ways to help society make the transition to sustainable lifestyles". This definition is mainly restricted to the minimisation of negative impacts already happening. As such, it tends to shorten HEI scope focusing on concrete and limited aspects, not addressing the proactivity in anticipating other effects and concerns related to the HEI complexity, the novelty of sustainability in this kind of institution and finally the new challenges related to sustainability issues that keep coming daily.

A definition that has been recurrently used, from Sterling et al. (2013, p. 23), states that a sustainable university is one that "through its guiding ethos, outlook and aspirations, governance, research, curriculum, community links, campus management, monitoring, and modus operandi seeks explicitly to explore, develop, contribute to, embody and manifest—critically and reflexively—the kinds of values, concepts, and ideas, challenges and approaches that are emerging from the growing global sustainability discourse". This last definition might be more appropriate since the authors perceive HEI from a much broader perspective. The HEIs' sustainability practices often extrapolate the boundaries of their geographical area, bringing benefits to their local, regional and sometimes national environment.

At the beginning of this century, the work of Cortese (2003) stated that the achievement of HEI sustainability is attained by considering the following four dimensions: education, research, campus operations and reporting. Later, Lozano (2006a) and Lozano et al. (2015) complemented the model including three more dimensions: institutional framework, on-campus experience and outreach and, finally, uniting the existing dimension reporting with the assessment practices. Table 1 details briefly each dimension of HEIs' sustainability proposed by Lozano et al. (2015).

Table 1 Dimensions of HEIs' sustainability

Dimension	Description
Education	It includes proposals related to the presence of sustainability themes in the course curriculum, the development of skills and teacher training programmes. This dimension relates not only to the theme of sustainable HEIs but also to a much broader scope of knowledge which includes the central role that education plays in the science of sustainability and the promotion of SD
Research	It is related to the existence of structures and financial support for the production of knowledge, technology and innovations in sustainability
Campus operations	It addresses the presence of sustainability practices in the day-to-day management of HEIs, including resource efficiency and management of water, energy, waste and greenhouse gases, transport and accessibility, as well as access to good quality food
Institutional framework	It deals with the commitment of the higher management and the councils of the institution to sustainable development. It considers the presence of DS in policies, missions and other official institutional documents
On-campus experience	It considers that working groups and other sustainable practices among students, teachers and staff are indicators of the daily presence of sustainability concerns in the academic community
Outreach	It refers to actions related to the integration of the university with society, which includes other universities, governments, companies, schools, civil society organisations and the local community
Assessment and reporting	It involves the implementation of an integrated environmental management system to monitor and control the environmental impacts of campus operations, processes and routines, as well as the internal and external dissemination of the results of this monitoring and the adoption of continuous improvement principles

Adapted from Lozano et al. (2015)

The model of Lozano et al. (2015), presented in Table 1, has been broadly cited by many authors because it captures the core facets of HEI sustainability and, as shown later, its adherence to SATs was empirically tested.

3 Sustainability Assessment Tools in HEIs

To endorse the effectiveness of HEIs' sustainability practices, various Sustainability Assessment Tools (SATs) were created and are considered a crucial element to enable the path towards sustainability. They support the HEIs' decision-makers on the improvement of their plans and policies towards a sustainable higher education institution and make it possible to publish the sustainability reporting of HEIs (Berzosa et al. 2017).

The work of Lambrechts (2015), which provides an overview of existing sustainability assessment tools, identifies the SAT contribution to the HEIs' sustainability process as threefold. According to this work, SATs usually contribute to (1) policy development; (2) mainstreaming sustainable development in higher education, and finally, (3) improving transparency and communication.

Fischer et al. (2015) studied 12 sustainability assessment tools in HEIs to analyse the understandings of a sustainable university that are underpinning contemporary sustainability assessment tools. Their research findings showed that these SATs comprised at least three different monitoring purposes, from affording compliance to predetermined standards, to determine the state of internal processes, and to provide data for competitive performance comparisons. Several other authors carried out similar comparative SATs analysis.

Yarime and Tanaka (2012) used a mixed-method approach and analysed 12 SATs, and the results showed that most tools indicators were focused on operations (44%), governance (39%) and education (8%). Berzosa et al. (2017) simultaneously applied three SATs to compare on a real case study, namely to assess the sustainability of the Universidad Europea de Madrid (UEM) in Spain. The authors analysed advantages and differences between tools and concluded that it may be feasible to use more than one tool for diagnosis and planning. In another study by Asmuss and Kamal (2013) four tools were reviewed to select the best benchmarking tool for the purposes of the University of Saskatchewan (UofS) in Canada. This work considered the following five areas of campus life: education, operation, governance, research and community engagement. After analysing the strengths and weaknesses of each of the following tools: Sustainability Assessment Questionnaire (SAQ), the Campus Sustainability Assessment Framework (CSAF), the College of Sustainability Report Card (CSRC), and the Sustainability Tracking Assessment and Rating System (STARS), the last one was chosen. STARS was considered by the authors to be the best benchmarking tool to satisfy the UofS's needs for assessing sustainability in all designated areas of campus life—education, research, operations, governance and community engagement.

Finally, in the work carried out by Findler et al. (2018), the extent to which SATs are capable of measuring the impact that HEIs have on sustainable development was analysed. To achieve the purpose of their study, the authors performed the analysis of 19 SATs and 1134 indicators for sustainability assessment. According to the adopted methodology, each indicator was exclusively assigned to one of the Lozano et al. (2015) sustainable development dimensions. Those indicators related to administrative structure and broad-scale policies were assigned to the institutional framework, while indicators addressing assessment and reporting processes were categorised into the "Assessment and reporting" dimension. Further, indicators related to the HEIs on an institutional level were related to the new category "higher education institution (HEI)", such as demographic effects on the region through student in-migration. The column "not applicable (NA)" included those that did not fit in any of the other dimensions of the Lozano et al. (2015) model. Table 2 presents these SATs and their relation to the seven Lozano et al. (2015) sustainable development dimensions and, also the new categories, HEI and NA, as proposed by Findler et al. (2018).

According to Table 2, the dimension with the highest number of indicators is Campus operations, followed by Institutional framework (20.90%), Education (16.04%) and Research (7.85%). The results are in line with the works of Fischer et al. (2015). The study of Findler et al. (2018) is particularly relevant because it highlights the possibility of a relationship between the sustainability indicators of the 19 studied SATs with the key dimensions of the sustainability proposed in the Lozano et al. (2015) model (Table 1).

Much of the attention of sustainability research focusing on HEIs has been directed towards the dimensions of education and research. Moreover, considerable attention has been given to isolated aspects of the campus sustainability operations dimension, such as green building (Hopkins 2016), waste (Zen et al. 2016) and carbon emission (Altan 2010; Larsen et al. 2013; Ramos et al. 2015). Nejati and Nejati (2013, p. 102) assert that "sustainability practices within the academic setting need to be understood and practiced by all members of the organization at various levels. Only then can a collective force for achieving the sustainability mission be mobilized successfully".

While literature concerned with Sustainable Assessment Tools (SATs) in HEIs recognises that most of those are focused on inside impacts, authors like Findler et al. (2019) and Beynaghi et al. (2016) have noticed that HEI sustainable development efforts have an effect that reflects beyond its organisational boundaries. The HEI SD impacts might emerge from a variety of contrasting areas such as economy, societal challenges, natural environment, policy making, culture, and demographics (Findler et al. 2018).

4 Students' Perception of Sustainable HEIs

Although tools that use perception measurement to assess sustainability in HEIs differ from traditional ones, which often use objective measurement variables such as energy consumption in kWh, water consumption in m^3, and tons of selective waste collection, among others, they contribute to a better understanding of HEIs' sustainability. The SATs performed by the subjective approach of assessing service users'

Table 2 Adherence between Lozano et al. (2015) sustainability dimensions and SATs

Sustainability Assessment Tools (SATs)	Institutional framework	Education	Research	Outreach	Campus operations	Campus experiences	Assessment and reporting	HEI	NA
AISHE[a], AMAS[b], BSIS[c], CSA[d], CSAF[e], D-SiM[f], DUK[g], GASU[h], GMID[i], P&P[j], PENN[k], SAQ[l], SCE[m], SPT[n], STARS[o], STAUNCH®[p], TUR[q], UIGM[r], USAT[s]	20.9	16.04	7.85	5.20	34.48	2.56	2.91	5.03	5.03

Source Adapted from Findler et al. (2018)
[a]Auditing Instrument for Sustainability in Higher Education; [b]Adaptable Model for Assessing Sustainability in Higher Education; [c]Business School Impact System; [d]Campus Sustainability Assessment Review Project; [e]Campus Sustainability Assessment Framework; [f]Driving force-pressure-state-exposure-effect-action; [g]Deutsche UNESCO Kommission; [h]Graphical Assessment of Sustainability in Universities;[i]Graz Model of Integrative Development; [j]People and Planet's University League; [k]Penn State Indicators Report; [l]Sustainability Assessment Questionnaire; [m]State of the Campus Environment; [n]Sustainable Pathways Toolkit; [o]Sustainability Tracking, Assessment and Rating System; [p]Sustainability Tool for Auditing for University Curricula in Higher-Education; [q]Three-dimensional University Ranking; [r]UI GreenMetric World University Ranking; [s]Unit-based Sustainability Assessment Tool

perceptions of their sustainability effectiveness may complement a more holistic perspective, by bringing new insights to the assessment process. As a complementary approach, it would concur to a better understanding of HEIs' effort to become more sustainable.

The use of subjective tools to measure sustainability in HEIs may induce an improvement of social control in managing the HEIs' system, which is a gain in terms of governance, and, in addition, would improve the students' engagement, as key stakeholders, in achieving the institution's sustainability goals.

5 Description of the Study Areas and Methodology

5.1 Case Study

The Federal University of Paraíba is a national public university located in the north east of Brazil. It is the biggest of the Paraíba State. It has 127 undergraduate and 111 postgraduate courses that enrol 38,880 students. For this study, the sample was composed of students from two of the 16 HEI study centres, namely the Technology Centre and the Renewable Energy Centre. These two centres comprise most of the engineering courses offered by the HEI.

5.2 Research Methods—Scale Design

In line with Malhotra et al.'s (2018, p. 378) methodological proposal, a new multi-item quantitative tool was designed to measure students' perception of HEIs, following the steps shown in Fig. 1.

The first step encompassed an extensive literature review to identify the main dimensions of sustainability in higher education institutions. As presented in the

Step 1: Extensive literature review →	Identify the main dimensions of HEIs' sustainability; Item generation (43 items).
Step 2: Survey design and pre-validation →	Final questionnaire survey (43 items and 8 variables); Pilot test with a reduced sample of undergraduate students.
Step 3: Data collection →	Survey administration to a sample of technology undergraduate students (207 questionnaires were administered).
Step 4: Data analysis →	Item extraction through principal component analysis; Reliability analysis through Cronbach's alpha and validity analysis (composite reliability, convergent and discriminant validity); Gender, income and education impact on sustainability perception through T-test/Anova.

Fig. 1 Questionnaire development and validation process

introduction session, the sustainability of HEIs is composed of the seven following dimensions: (1) Education; (2) Research; (3) Campus operations; (4) Institutional framework; (5) On-campus experience; (6) Outreach; (7) Assessment and reporting. 43 items were generated through literature review, based on the works of Findler et al. (2019), Emanuel and Adams (2011), Lozano (2006a), Lozano and Young (2013), Nejati and Nejati (2013), Luiz et al. (2015), Savelyeva and Douglas (2017) and Thomashow (2014).

Once the set of items derived from the literature review was identified, the next step was the design of a final questionnaire comprising 51 items of which five were demographic (course, age, gender, income, education); three dichotomous questions to measure student connection with sustainability in the course; and the 43 items identified in Step 1. For the 43 multi-item scale, responses were provided using a five-point Likert-type scale from (1) strongly disagree to (5) strongly agree with a (3) neutral response option. Thus, the survey was refined through a pilot test, applying it to a reduced sample of 12 individuals to evaluate the following criteria: (a) assess respondent's reaction and understanding of the items and variables allocated on the questionnaire; (b) obtain feedback with regard to content, length, arrangement, wording accuracy and relevance. As a result of this phase, two items were rewritten to improve wording accuracy.

In Step 3, the final questionnaire was administered by a structured and assisted survey to a sample of 207 undergraduate students of 12 courses of engineering provided by the case study Brazilian university (Table 3).

After data collection, a statistical analysis of the results was conducted in Step 4. In this phase, as will be demonstrated in the results section, other items were discarded due to their lack of statistical adherence to the proposed tool. Finally, the methods adopted presented some limitations. For instance, the sample size and composition, although allowing the analyses performed, limited the possibility of generalising the results to other HEIs and to students other than engineering; the survey was designed to be comprehensive for the majority of respondents, however it may be challenging for some respondents to have enough knowledge about all topics addressed in the survey.

6 Results and Discussion

6.1 Principal Component Analysis

For grouping the items into their specific dimensions, the principal component analysis (PCA) was performed on the 43 items of the scale. To assess the factorability of the data and to ensure the adequacy of the sampling, Bartlett's Test of Sphericity and the Kaiser-Meyer-Olkin (KMO) Measure of Sampling Adequacy were applied. The Bartlett's Test of Sphericity analyses whether the correlation matrix has significant correlations among at least some of the variables and it is significant ($p < 0.05$) for

Table 3 Demographic profile of respondents

Demographics	Frequency	Percentage (%)
Gender (valid N = 207)		
Male	138	66.70
Female	69	33.30
Age (N = 207)		
Below 20 years old	29	14.00
20–22 years old	77	37.20
23–25 years old	71	34.30
26–28 years old	18	8.70
Over 28 years old	12	5.80
Monthly familiar income (N = 184)		
Lowest thru 500€	90	43.50
500–999€	50	24.20
1000–1499€	30	14.50
1500–2000€	14	6.80
Over 2000€	23	11.10
Course (valid N = 202)		
Industrial Mechanical Engineering	27	13.40
Mechanical Engineering	30	14.90
Renewable Energy Engineering	35	17.30
Environmental Engineering	34	16.80
Civil Engineering	20	9.90
Industrial Engineering	19	9.40
Industrial Chemistry	5	2.50
Chemical Engineering	15	7.40
Electrical Engineering	14	6.90
Food Engineering	2	1.00
Materials Engineering	1	0.50

the PCA to be considered appropriate (Field 2009; Hair et al. 2014; Nejati and Nejati 2013). The KMO corresponds to a measure of sampling adequacy (MSA) that looks not only at the correlations but also at patterns between variables. It ranges from 0 to 1 and its accepted values are equal to or above 0.6 (Hair et al. 2014). Further, the component loadings were analysed. Based on sample size, a loading of 0.6 or greater on one component was considered significant (Hair et al. 2014). The values ranging from 0.609 to 0.850, as shown in the fourth column of Table 5, were considered as achieving the accepted threshold. To solve the cross-loading issues, the criteria adopted by Nejati and Nejati (2013) were used, whereby items having a loading

Table 4 Rotated component loading matrix (VARIMAX)

	C1	C2	C3	C4	C5	C6	C7	C8
Sum of squares (eigenvalues)	3.776	3.060	2.776	2.721	2.675	2.445	2.015	1.779
% of trace	12.149	9.871	8.954	8.777	8.628	7.887	6.500	5.738
Cumulative % of trace	12.149	22.020	30.973	39.751	48.379	56.266	62.766	68.504

difference across components less than 0.10 were suppressed. Applying the criteria described above, 11 items were removed from the model.

The final model was composed of 31 items, grouped into eight components with eigenvalues higher than 1, explaining 68.504% of the variance. The 31 items model obtained a significant Bartlett's Test of Sphericity ($p \approx 0.000$) and also collectively meets the necessary threshold of sampling adequacy, measured through KMO, with an MSA value of 0.860. The individual MSA of each item was also measured and these ranged from 0.709 to 0.933.

The items included in each component were considered, relating to the literature, and labelled as: 1—Waste (6 items), 2—Emissions/Procurement (3 items), 3—Energy (4 items), 4—Quality of Life in the Workplace (4 items), 5—Fauna and Flora (4 items), 6—Institutional framework (4 items), 7—Education/Research (3 items) and, 8—Water (2 items). The eigenvalue percentage of the trace of each component is presented in Table 4.

6.2 Reliability and Validity Analysis

To assess reliability, Cronbach's alpha was computed for each subscale. A commonly accepted rule of thumb for describing the internal consistency calculated by Cronbach's alpha is as follows: $\alpha \geq 0.9$: Excellent; $0.7 \leq \alpha < 0.9$: Good; $0.6 \leq \alpha < 0.7$: Acceptable; $0.5 \leq \alpha < 0.6$: Poor; $\alpha < 0.5$: Unacceptable (Hair et al. 2014; Jorge et al. 2015). As shown in the last column of Table 5, values of Cronbach's alpha (α) for each component range between acceptable and good.

Finally, to ensure the quality of measurement, the composite reliability, convergent validity and discriminant validity were also tested. Composite reliability (CR) is a robust measure of internal consistency in scale items (Byrne 2016). Fornell and Larcker (1981) found thresholds for composite reliability to be above 0.60. The values of CR, shown in Table 5, exceed the limits established in the literature. The average variance extracted (AVE) for each component surpasses the recommended level of 0.5 (Hair et al. 2014); thus, it is possible to conclude that convergent validity was achieved.

Table 6 summarises the measured coefficients for discriminant validity. The diagonal elements, in bold, are the square root of the average variance extracted (AVE).

Table 5 Descriptive statistics, loadings (VARIMAX), reliability and validity tests of constructs

Item	Mean	SD[a]	Load	Reliability validity
Component 1: waste	2.52	0.808		
The implemented composting system is efficient	2.77	0.942	0.769	CR[b] 0.85 AVE[c] 0.49 α[d] 0.871
UFPB encourages, through campaigns, the correct disposal of its waste	2.39	1.117	0.733	
UFPB has an efficient selective waste collection program	2.26	1.043	0.726	
UFPB performs proper disposal of its chemical waste	2.57	0.962	0.666	
UFPB promotes reverse logistics of cartridges and toners used by the Institution	2.70	0.928	0.662	
Recycling bins scattered around campus motivate students to discard waste properly	2.43	1.200	0.649	
Component 2: emissions/procurement	2.26	0.823		
UFPB prioritises the use of biofuels in its vehicle fleet	2.19	0.944	0.847	CR[b] 0.86 AVE[c] 0.61 α[d] 0.878
UFPB monitors greenhouse gas emissions from its fleet	2.03	0.975	0.820	
UFPB has procedures to optimise the use of its vehicle fleet	2.37	0.946	0.761	
UFPB cleaning, safety and telephone contracts take into account sustainability issues	2.43	0.895	0.683	
Component 3: energy	2.28	0.869		
UFPB invests in renewable energy generation strategies	2.25	1.059	0.753	CR[b] 0.83 AVE[c] 0.54 α[d] 0.834
UFPB embraces energy efficiency principles by replacing LED lighting	2.23	1.049	0.747	
UFPB adopts practices committed to reducing non-renewable energy use	2.27	1.002	0.731	
UFPB promotes campaigns to rationalise the use of electricity	2.37	1.137	0.719	
Component 4: quality of life in the workplace	2.86	0.894		
UFPB encourages respectful treatment among students	2.94	1.087	0.812	CR[b] 0.81 AVE[c] 0.52 α[d] 0.829
UFPB encourages respectful treatment between students and lecturers	2.96	1.112	0.809	
The UFPB workload required for course activities is adequate	2.67	1.128	0.623	
Student rights are respected	2.86	1.070	0.609	
Component 5: fauna and flora (ff)	3.03	0.817		

(continued)

Table 5 (continued)

Item	Mean	SD[a]	Load	Reliability validity
UFPB performs proper wildlife management on its campuses	3.14	1.143	0.799	CR[b] 0.80 AVE[c] 0.51 α[d] 0.757
UFPB takes care of its forest areas	3.44	1.073	0.729	
UFPB performs the correct management of domestic fauna on its campuses	2.43	1.205	0.681	
The institution complies with environmental legislation	3.12	0.842	0.627	
Component 6: institutional framework	2.82	0.767		
UFPB's portal and social media detail the institution's sustainability initiatives	2.77	0.983	0.755	CR[b] 0.78 AVE[c] 0.47 α[d] 0.778
UFPB has a specific sector to address the environmental issues of its campuses	3.21	1.067	0.739	
Overall, sustainability issues are adequately addressed at UFPB	2.65	0.948	0.636	
Important decisions related to campus sustainability are made in a participatory manner on university councils	2.65	0.958	0.613	
Component 7: education/research	3.30	0.855		
The course offers institutional research and extension programmes with themes related to sustainability	3.51	0.994	0.795	CR[b] 0.80 AVE[c] 0.57 α[d] 0.687
The institution's postgraduate programmes related to students' field of study offer sustainability-themed lines of research	3.19	1.107	0.780	
The course offers enough sustainability subjects for students' education	3.19	1.161	0.686	
Component 8: water	2.27	0.950		
The drinking water distributed by UFPB is of high quality	2.04	1.001	0.850	CR[b] 0.80 AVE[c] 0.68 α[d] 0.804
UFPB has a good drinking water supply	2.50	1.074	0.797	
Overall score for Student Perception of HEI's Sustainability (SPHEIS)[e]	2.67	0.560		

[a]SD: standard deviation
[b]CR: composite reliability
[c]AVE: average variance extracted
[d]α: Cronbach's alpha
[e]Average value obtained from scores of the items included in each category

Table 6 Discriminant validity coefficients

Component	C1	C2	C3	C4	C5	C6	C7	C8
C1—waste	**0.700**							
C2—emissions/procurement	0.325	**0.781**						
C3—energy	0.409	0.350	**0.735**					
C4—quality of life in the workplace	0.385	0.390	0.280	**0.721**				
C5—fauna and flora	0.358	0.233	0.383	0.367	**0.714**			
C6—institutional framework	0.375	0.329	0.379	0.463	0.345	**0.686**		
C7—education/research	0.163	0.226	0.075	0.272	0.156	0.302	**0.755**	.
C8—water	0.270	0.363	0.266	0.357	0.214	0.327	0.176	**0.825**

Off-diagonal elements are the correlation among components. To examine discriminant validity, diagonal elements should be larger than off-diagonal elements (Nejati and Nejati 2013).

6.3 Gender, Secondary Education and Income Analysis

Despite the gender difference between male (66.7%) and female (33.3%) in the number of students questioned, the t-test results show that there isn't statistical evidence to confirm gender influence on the perception of campus sustainability ($t(205) = 0.297$; $p = 0.767$), with male mean equal to 2.67 and female 2.65. This result is in line with the work carried out by Meek and Sullivan (2018) which developed a new measure of sustainability orientation among entrepreneurs. Further, the gender result is similar to the study carried out by Dagiliūtė et al. (2018) that compared students' attitudes towards sustainability in two Lithuanian universities. Although the study of Zhang et al. (2017) found that gender is influential on sustainability perception, using a sample of 509 undergraduate students from 10 university campuses in Beijing, China, we could not find evidence of gender differences on perception towards sustainability in the case of these Brazilian students.

Similarly to the gender result, no statistical evidence was found to assert that secondary education in public (mean 2.71) or private (mean 2.62) schools influences the perception of sustainability of the surveyed sample ($t(203) = -0.240$; $p = 0.216$). Lastly, there was also no statistically significant difference regarding income as a predictor of the perception of sustainability ($F(3;180) = -0.127$; $p = 0.944$). This result regarding income is congruent with the work of Bosona and Gebresenbet (2018).

6.4 Model Performance and Sustainability Perception for the Case Study

As a result of the principal component analysis, the 31 remaining items of the final model (Table 5) are related to five of the eight dimensions of the higher education sustainability model designed by Lozano et al. (2015), described in Table 1.

The components C1—Waste, C2—Emissions/Procurement, C3—Energy, C5—Fauna and Flora and C8—Water, are congruent with the dimension Campus operation. The component C4—Quality of Life in the Workplace has similarities with the dimension On-campus Experience; likewise the component C6—Institutional Framework is consistent with the analogous Lozano dimension. Component 7 (Education/Research) has items compatible with the dimensions Education and Research on Lozano's model. The items related to the dimensions Outreach and Assessment and report were removed in the refinement phase of principal component analysis. The adherence between the performance of the Student Perception of HEI's Sustainability (coined now as SPHEIS), which is the proposed model, and the one designed by Lozano et al. (2015) is illustrated in Fig. 2.

Results of the analysis indicate the following: firstly, five components obtained average scores above the midpoint of the scale (mean = 2.5). These are: C1 – Waste (mean = 2.52); C4—Quality of Life in the Workplace (QLW) (mean = 2.86); C5—Fauna and Flora (mean = 2.82); C6—Institutional Framework (mean = 2.82); and C7—Education/Research, which had the highest score, obtaining mean equal to 3.30. In contrast, the three following components achieved scores below the midpoint of the scale: C2—Emissions/Procurement, which obtained the lowest score, with mean equal to 2.26; C3—Energy (mean = 2.28); followed by C8—Water (mean =

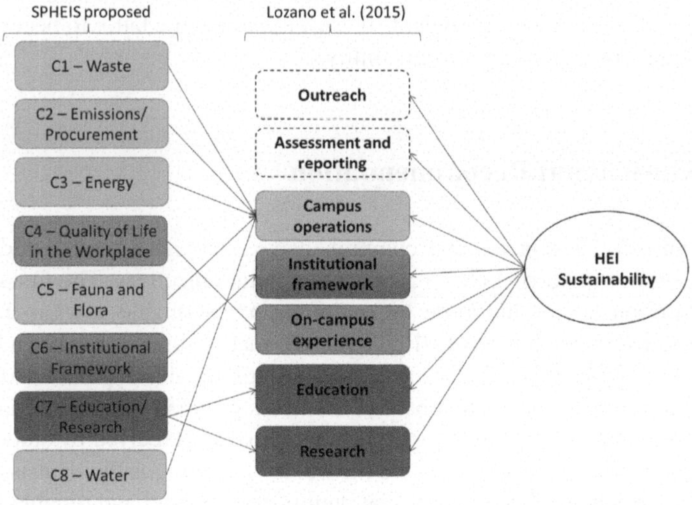

Fig. 2 Adherence between the proposed model SPHEIS and Lozano et al. (2015)'s model

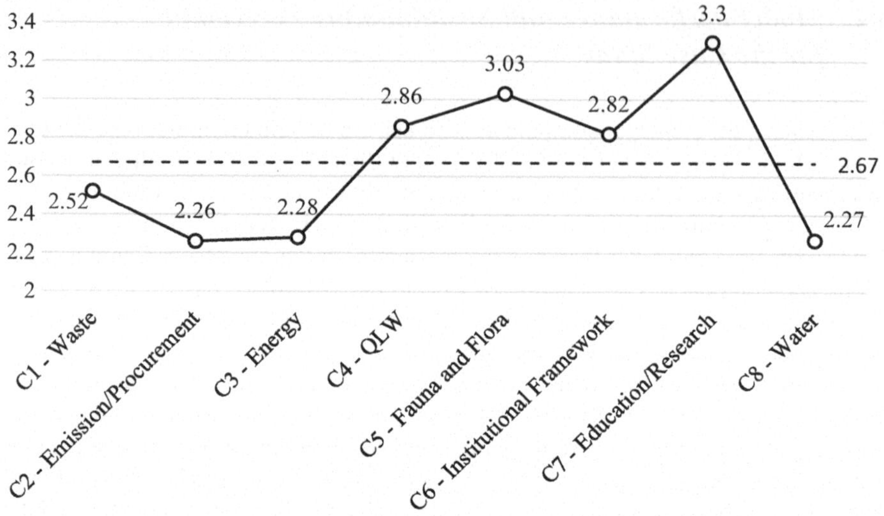

Fig. 3 Score of the Student Perception of HEI's Sustainability (SPHEIS) in each component

2.27). Secondly, students' overall perception of campus sustainability was weak to moderate, with a score of 2.67 (SD = 0.56), as shown in Fig. 3.

This low score obtained in the assessment of the sustainability performance perceived by the surveyed students needs to be considered through the analysis of each of the items that make up the developed scale, by those responsible for the implementation of sustainability practices. In this way, it will be possible to conclude whether the results are due to low investment in some sustainability practices implemented by the institution, or if it is due to a lack of communication channels between the institution and its students that would allow the latter to become aware of institutional efforts towards sustainability.

7 Conclusion and Recommendation

This section offers a concise and comprehensive conclusion of the study's findings. This study was carried out to achieve two main objectives. The first objective was to contribute to the literature by designing a sustainable assessment tool to assess the students' perception of HEI campuses' sustainability, based on a Brazilian HEI case study. This objective was achieved through the development and application of a multi-scale survey, composed of 31 items grouped into eight components that cover the main aspects of campus sustainability as perceived by students. The validation procedure adopted statistical measures to ensure results' consistency and therefore acceptable statistics scores that demonstrate a valid and reliable tool. The proposed instrument would work as a complementary tool to assess HEIs' sustainability performance and assist managers in improving their efforts to increase the students' commitment to building a sustainable HEI which is able to face and accomplish the new society requirements towards sustainable development. Taking into account the performance obtained through the collected data, a gap may exist between the implementation of sustainable practices and their perception by the students. This gap may be overcome with an effort to improve communication towards sustainability practices by using the available social media channels to provide information about achievements pertinent to sustainable development on campus.

The second objective of the study intended to evaluate the adherence of the proposed tool, Student Perception of HEI's Sustainability (SPHEIS), in relation to the dimensions of HEI sustainability designed by Lozano et al. (2015). The proposed tool was partially adherent to the Lozano et al. (2015) model once the eight components of the SPHEIS were related to five dimensions of the Lozano et al. (2015) model. In fact, five of the eight components were found adherent to the Campus operations dimension which are in line with the work of Findler et al. (2018), who analysed 19 SATs and concluded also that the analysed tools include more items focused on Campus operations. Table 2 showed that the highest amount of the analysed items, 34.48%, were grouped into this dimension. A justification for the dimensions Outreach and Assessment and reporting not being considered in the proposed SPHEIS model would be that these two dimensions tend to be less easily perceived by the students, since usually students are more focused on activities related to campus operation, education and research.

Considering the results, implications and recommendations could be designed for university planners and decision-makers to increase sustainability in HEIs and correlated institutions. As an illustration, a few of them are presented below. (1) Future studies may expand the sample and include more items, such as those related to Assessment and reporting as well as Outreach, in order to comply with all dimensions of HEIs' sustainability and to provide results that are more representative. (2) Future research may examine the validity of the introduced assessment tool in another regional context. (3) The assessment tool may be adapted to other correlated institutions, like hospitals or secondary schools, to measure customer/user perception of sustainability. (4) Besides, due to time limitations, this study was applied with a cross-sectional approach, therefore it is suggested for future research to adopt a longitudinal approach as a way to control the sustainability performance and implement the principles of continuous improvement.

References

Alshuwaikhat HM, Abubakar I (2008) An integrated approach to achieving campus sustainability: assessment of the current campus environmental management practices. J Clean Prod 16(16):1777–1785. https://doi.org/10.1016/j.jclepro.2007.12.002

Altan H (2010) Energy efficiency interventions in UK higher education institutions. Energy Policy 38(12):7722–7731. https://doi.org/10.1016/j.enpol.2010.08.024

Asmuss M, Kamal ASM (2013) Benchmarking tools for assessing and tracking sustainability in higher educational institutions Identifying an effective tool for the. Int J Sustain High Educ 14(4):449–465. https://doi.org/10.1108/IJSHE-08-2011-0052

Berzosa A, Bernaldo MO, Fernández-Sanchez G (2017) Sustainability assessment tools for higher education: an empirical comparative analysis. J Clean Prod 161:812–820. https://doi.org/10.1016/j.jclepro.2017.05.194

Beynaghi A, Trencher G, Moztarzadeh F, Mozafari M, Maknoon R, Filho WL (2016) Future sustainability scenarios for universities: moving beyond the United Nations Decade of Education for Sustainable Development. J Clean Prod 112(Part):3464–3478. https://doi.org/10.1016/j.jclepro.2015.10.117

Bizerril M, Rosa MJ, Carvalho T, Pedrosa J (2018) Sustainability in higher education: A review of contributions from Portuguese Speaking Countries. J Clean Prod 171:600–612. https://doi.org/10.1016/j.jclepro.2017.10.048

Blanco-Portela N, R-Pertierra L, Benayas J, Lozano R (2018) Sustainability leaders' perceptions on the drivers for and the barriers to the integration of sustainability in Latin American Higher Education Institutions. Sustainability (Switzerland) 10(8). https://doi.org/10.3390/su10082954

Bosona T, Gebresenbet G (2018) Swedish consumers' perception of food quality and sustainability in relation to organic food production. Foods 7(4). https://doi.org/10.3390/foods7040054

Brandli LL, Frandoloso MAL, Tauchen J (2011) Improving the environmental work at University of Passo Fundo, Brazil—towards an environmental management system. Braz J Oper Prod Manag 8(1):31–54. https://doi.org/10.4322/bjopm.2011.002

Byrne BM (2016) Structural equation modeling with AMOS: basic concepts, applications, and programming, 2nd edn. Routledge, New York

Ceulemans K, Molderez I, Van Liedekerke L (2015) Sustainability reporting in higher education: a comprehensive review of the recent literature and paths for further research. J Clean Prod 106:127–143. https://doi.org/10.1016/j.jclepro.2014.09.052

Cortese AD (2003) The critical role of higher education in creating a sustainable future. Plan High Educ 31(3):15–22

Dagiliūtė R, Liobikienė G, Minelgaitė A (2018) Sustainability at universities: students' perceptions from Green and Non-Green universities. J Clean Prod 181:473–482. https://doi.org/10.1016/j.jclepro.2018.01.213

Emanuel R, Adams JN (2011) College students' perceptions of campus sustainability. Int J Sustain High Educ 12(1):79–92. https://doi.org/10.1108/14676371111098320

Field A (2009) Discovering statistics using SPSS. Statistics, vol 58. https://doi.org/10.1016/j.landurbplan.2008.06.008

Findler F, Schönherr N, Lozano R, Reider D, Martinuzzi A (2019) The impacts of higher education institutions on sustainable development. Int J Sustain High Educ 20(1):23–38. https://doi.org/10.1108/IJSHE-07-2017-0114

Findler F, Schönherr N, Lozano R, Stacherl B (2018) Assessing the impacts of higher education institutions on sustainable development—an analysis of tools and indicators. Sustainability (Switzerland) 11(1). https://doi.org/10.3390/su11010059

Fischer D, Jenssen S, Tappeser V (2015) Getting an empirical hold of the sustainable university: a comparative analysis of evaluation frameworks across 12 contemporary sustainability assessment tools. Assess Eval High Educ 40(6):785–800. https://doi.org/10.1080/02602938.2015.1043234

Fornell C, Larcker D (1981) Structural equation models with unobservable variables and measurement error. J Mark Res 18(1):39–50

Green TL (2013) Teaching (un)sustainability? University sustainability commitments and student experiences of introductory economics. Ecol Econ 94:135–142. https://doi.org/10.1016/j.ecolecon.2013.08.003

Hair JF, Black WC, Babin BJ, Anderson RE (2014) Multivariate data analysis, 7th edn. Pearson Education, Essex

Hopkins EA (2016) Barriers to adoption of campus green building policies. Smart Sustain Built Environ 5(4):340–351. https://doi.org/10.1108/SASBE-07-2016-0016
Jorge ML, Madueño JH, Cejas MYC, Peña FJA (2015) An approach to the implementation of sustainability practices in Spanish universities. J Clean Prod 106:34–44. https://doi.org/10.1016/j.jclepro.2014.07.035
Karatzoglou B (2013) An in-depth literature review of the evolving roles and contributions of universities to Education for Sustainable Development. J Clean Prod. https://doi.org/10.1016/j.jclepro.2012.07.043
Lambrechts W (2015) The contribution of sustainability assessment to policy development in higher education. Assess Eval High Educ 40(6):801–816. https://doi.org/10.1080/02602938.2015.1040719
Larsen HN, Pettersen J, Solli C, Hertwich EG (2013) Investigating the carbon footprint of a university—the case of NTNU. J Clean Prod 48:39–47. https://doi.org/10.1016/j.jclepro.2011.10.007
Leal Filho W, Doni F, Vargas VR, Wall T, Hindley A, Rayman-Bacchus L, Emblen-Perry K, Boddy J, Avila LV (2019a) The integration of social responsibility and sustainability in practice: exploring attitudes and practices in Higher Education Institutions. J Clean Prod. https://doi.org/10.1016/j.jclepro.2019.02.139
Leal Filho W, Shiel C, Paço A, Mifsud M, Ávila LV, Brandli LL, Molthan-Hill P, Pace P, Azeiteiro UM, Vargas VR, Caeiro S (2019b) Sustainable Development Goals and sustainability teaching at universities: falling behind or getting ahead of the pack? J Clean Prod. https://doi.org/10.1016/j.jclepro.2019.05.309
Lozano R (2006a) A tool for a Graphical Assessment of Sustainability in Universities (GASU). J Clean Prod 14(9–11):963–972. https://doi.org/10.1016/j.jclepro.2005.11.041
Lozano R (2006b) Incorporation and institutionalization of SD into universities: breaking through barriers to change. J Clean Prod 14(9–11):787–796. https://doi.org/10.1016/j.jclepro.2005.12.010
Lozano R, Ceulemans K, Alonso-Almeida M, Huisingh D, Lozano FJ, Waas T, Lambrechts W, Lukman R, Hugé J (2015) A review of commitment and implementation of sustainable development in higher education: results from a worldwide survey. J Clean Prod 108:1–18. https://doi.org/10.1016/j.jclepro.2014.09.048
Lozano R, Young W (2013) Assessing sustainability in university curricula: exploring the influence of student numbers and course credits. J Clean Prod 49:134–141. https://doi.org/10.1016/j.jclepro.2012.07.032
Luiz LC, Pfitscher ED, da Rosa FS (2015) Plano de Gestão de loGística Sustentável: ProPosição de ações e Indicadores socioambientais Para avaliar o desemPenho nos órGãos Públicos federais. Rev Admin UFSM 8:8. https://doi.org/10.5902/1983465917696
Malhotra NK, Nunan D, Birks DF (2018) Marketing research: an applied approach, 5th edn. Pearson Education Limited, Edinburgh
Meek WR, Sullivan DM (2018) The influence of gender, self-identity and organizational tenure on environmental sustainability orientation. J Dev Entrep 23(03):1850018. https://doi.org/10.1142/S1084946718500188
Nejati M, Nejati M (2013) Assessment of sustainable university factors from the perspective of university students. J Clean Prod 48:101–107. https://doi.org/10.1016/j.jclepro.2012.09.006
Ramos TB, Caeiro S, van Hoof B, Lozano R, Huisingh D, Ceulemans K (2015) Experiences from the implementation of sustainable development in higher education institutions: Environmental Management for Sustainable Universities. J Clean Prod 106:3–10. https://doi.org/10.1016/j.jclepro.2015.05.110
Sammalisto K, Sundström A, Holm T (2015) Implementation of sustainability in universities as perceived by faculty and staff—a model from a Swedish university. J Clean Prod 106:45–54. https://doi.org/10.1016/j.jclepro.2014.10.015
Savelyeva T, Douglas W (2017) Global consciousness and pillars of sustainable development: a study on self-perceptions of the first-year university students. Int J Sustain High Educ 18(2):218–241. https://doi.org/10.1108/IJSHE-04-2016-0063

Sterling S, Maxey L, Luna H (2013) The sustainable university: progress and prospects, 1st edn. Routledge, Abingdon, UK
Thomashow M (2014) The nine elements of a sustainable campus. The MIT Press
Tim S, Jutidamrongphan W (2018) Life cycle cost analysis and energy performance of president's office, prince of Songkla university, Thailand. Songklanakarin J Sci Technol 40(2):439–447. https://doi.org/10.14456/sjst-psu.2018.51
Velazquez L, Munguia N, Platt A, Taddei J (2006) Sustainable university: what can be the matter? J Clean Prod 14(9–11):810–819. https://doi.org/10.1016/j.jclepro.2005.12.008
Yarime M, Tanaka Y (2012) The issues and methodologies in sustainability assessment tools for higher education institutions: a review of recent trends and future challenges. J Educ Sustain Dev 6(1):63–77. https://doi.org/10.1177/097340821100600113
Zen IS, Subramaniam D, Sulaiman H, Saleh AL, Omar W, Salim MR (2016) Institutionalize waste minimization governance towards campus sustainability: a case study of Green Office initiatives in Universiti Teknologi Malaysia. J Clean Prod 135:1407–1422. https://doi.org/10.1016/j.jclepro.2016.07.053
Zhang H, Liu J, Wen Zg, Chen YX (2017) College students' municipal solid waste source separation behavior and its influential factors: a case study in Beijing, China. J Clean Prod 164:444–454. https://doi.org/10.1016/j.jclepro.2017.06.224

Sustainability Concept (Whole-Institution Approach) of the Catholic University of Eichstätt-Ingolstadt (Germany)

Ingrid Hemmer and Anne-Kathrin Lindau

1 Introduction

German universities today face major and diverse challenges in meeting the current and future expectations of society (Barlett and Chase 2013). On the one hand, students have to fulfill the requirements of current labor markets, on the other hand, the university itself as an educational institution and company should act as a role model.

The concept of sustainable development and education for sustainable development currently represents an approach that universities around the world and especially in Germany are increasingly adopting. In Germany, some universities have already made progress in the process of adopting the whole-institution approach and can serve as a model for other universities on their way to becoming more sustainable.

The aim of this article is to present the KU as a positive example of a German university that has been on the path towards becoming a more sustainable university for ten years and thus, according to Leal Filho (2009), belongs to the group of universities that have made a long-term commitment to sustainability. It should be emphasized that, in addition to research and teaching, the focus is on four further fields of action in connection with sustainability (governancc, student initiatives and commitment, campus management and operations, and transfer) for implementation of a whole-institution approach. This stringent approach is still too rare in Germany in this scope and complexity (HRK 2017).

I. Hemmer (✉) · A.-K. Lindau
Department of Geography, Catholic University of Eichstätt-Ingolstadt, Ostenstraße 18, 85072 Eichstätt, Germany
e-mail: ingrid.hemmer@ku.de

A.-K. Lindau
e-mail: anne.lindau@ku.de

W. Leal Filho et al. (eds.), *Universities, Sustainability and Society: Supporting the Implementation of the Sustainable Development Goals*, World Sustainability Series,
https://doi.org/10.1007/978-3-030-63399-8_20

With these remarks, other universities can be given an idea of a sustainable university and are encouraged to start implementing Education for Sustainable Development (ESD), to continue their efforts and to make the entire university more sustainable. This example can also be used to help document the paths taken by universities towards becoming more sustainable.

To a large extent, the explanations refer to the documentation of the annually published sustainability reports (https://www.ku.de/unileben/nachhaltige-ku/nachhaltigkeitsbericht).

2 Sustainable Development and Whole-Institution Approach at Universities

Within the current sustainability debate, ESD is said to be of utmost importance (UNESCO 2014). The Global Action Programme (GAP) on "Education for Sustainable Development" (2015–2019) has defined these top five priority action areas that focus on the following aspects:

> "Action Area 1: Advancing policy,
> Action Area 2: Transforming learning and training environments,
> Action Area 3: Building capacities of educators and trainers,
> Action Area 4: Empowering and mobilizing youth, and
> Action Area 5: Accelerating sustainable solutions at local level" (DUK 2014, p. 15).

The National Action Plan (Nationale Plattform Bildung für nachhaltige Entwicklung 2017) for the implementation of the GAP in Germany underlines the outstanding position of universities and their responsibility within the educational landscape when it comes to implementing sustainable development measures. Emphasis is placed on the formation of a whole-institution approach by carving out five interdependent fields of action for the higher education sector:

> "Action Area I: Aligning higher education funding and incentive systems with substantive and structural sustainability and ESD" (p. 52),
> "Action Area II: Systematically linking research and ESD using quality criteria" (p. 54),
> "Action Area III: Supporting a diversified higher education system with a range of ESD paths, ESD pioneers and second followers" (p. 59),
> "Action Area IV: Encouraging, supporting and enabling real participation of students and graduates as key architects of sustainable development" (p. 62) and
> "Action Area V: Developing transformative metaphors and narratives" (p. 66).

The aim is to integrate ESD in all educational areas (Michelsen and Fischer 2015; Rieckmann 2016). The effectiveness of ESD is closely connected to the transformation of the entire educational institution. This holistic transformational concept (whole-institution approach) aims to integrate sustainability and sustainable development in all areas of the university by incorporating and combining aspects such as teaching practice, operations and organizational cultures, student participation, management level, ties with the respective city administration and research orientation. An important objective of the higher education institution must be to become a role model for students and the region (UNESCO 2017). "Universities are institutions of society and, as the core of the scientific system, are responsible for contributing to the forward-looking development of society." (HRK and DUK 2010).

In connection with the Earth Summit in Rio de Janeiro (1992), a number of charters and declarations (e.g. Copernicus Charter 1994; Thessaloniki Declaration 1997; Lüneburg Declaration 2001; Lucerne Declaration on Geographical Education for Sustainable Development 2007) were developed especially for the higher education sector that additionally underlined and substantiated this responsibility. The Copernicus Charter (1994) set out ten principles of action for a sustainable higher education development: institutional commitment, environmental ethics, education of university employees, programs in environmental education, interdisciplinary approaches, dissemination of knowledge, networking, partnerships, life-long learning programs, and technology transfer (Hemmer and Bagoly-Simó 2016).

Also in Germany, the German Rectors' Conference (HRK) underlined the large scope of influence and the significance of higher education institutions and universities within the educational landscape and their important role in connection with the sustainability transformation process (HRK and DUK 2010). With the HRK declaration "Towards a culture of sustainability", German universities have committed to act in accordance with the guiding sustainability principle and ensure the implementation of structures for a holistic sustainable development of educational institutions (HRK 2017). According to Schneidewind (2014), there are two elements that are of key significance within the intended holistic transformation processes at higher education institutions: on the one hand, there is the "deliberate orientation of the research and teaching practice towards central social challenges" and, on the other hand, the "involvement of social actors in the definition of and work on scientific questions from the very beginning" (p. 2). The involvement of a large number of actors is of crucial importance in connection with the successful implementation of a whole-institution approach at universities (Brinkhurst et al. 2011).

Within the "Bavarian Network for Sustainability in Higher Education", the project KriNaHoBay has drafted a scientifically substantiated criteria catalog for a systematic, standardized and whole-institutional evaluation regarding sustainability aspects at Bavarian universities which is, amongst others, based on the STARS criteria developed by AASHE (2012, 2015). A total of 45 criteria was set down for the categories of understanding sustainability, research, teaching, operations, governance, transfer and students. The whole-institution approach at universities is especially supported by structural anchoring of the topic of sustainability within the university, the support provided by the university management and committed groups of persons (e.g. students) – so-called 'change agents'. Inhibiting factors for sustainability development at higher education institutions are a lack of understanding for the topic of sustainability as well as the lack of (or insufficient) support by the university management and politics (KriNaHoBay 2017; Fig. 1). The two-way top-down and bottom-up process was determined to be a crucial, expedient factor for successful implementation of a sustainability concept at higher education institutions (Gräsel and Parchmann 2004; Fig. 1). For documenting the sustainability implementation process at higher education institutions, a sustainability report which is issued in regular intervals and can also be understood as an evaluation tool has proven to be very useful (Sassen et al. 2014). For the analysis, researchers compared criteria such as content, structure and approaches of the sustainability reports (Shriberg 2002; Ceulemans et al. 2015; Fischer et al. 2015).

Along with just a few examples of sustainability-oriented universities in Germany, such as the Leuphana Lüneburg, Vechta University, the University of Applied Sciences Eberswalde, and the University of Hamburg, the Catholic University of Eichstätt-Ingolstadt is an institution with a well advanced and far-reaching overall sustainability concept that has stipulated six action areas in which different actors collaborate in synergy. In the following, it will be illustrated how the whole-institution sustainability approach is realized, controlled and implemented using the example of the Catholic University of Eichstätt-Ingolstadt.

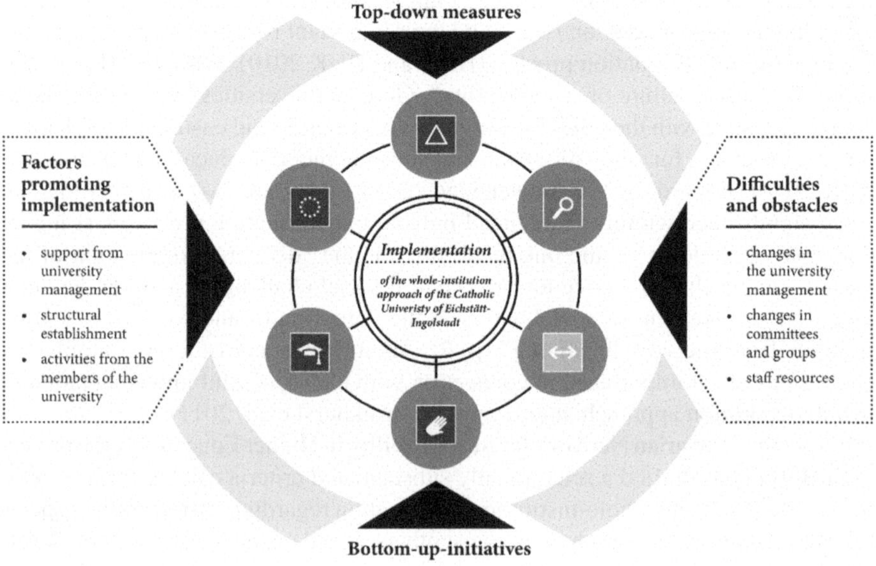

Fig. 1 Factors promoting implementation and challenges (Hemmer and Bagoly-Simó 2016, modified, illustration: C. Pietsch)

3 The Whole-Institution Sustainability Approach at the Catholic University of Eichstätt-Ingolstadt (Germany)

3.1 The Idea of a Sustainability Concept and Its Ten-Year History

The Catholic University of Eichstätt-Ingolstadt (KU) was founded in 1980. It is located in Bavaria approx. 120 km north of Munich. Currently, just under 5000 students are enrolled at the eight faculties of the KU. While the university places its focus on social sciences and the humanities, it also represents a natural sciences component with its Faculty of Mathematics and Geography. Already after the conference in Rio in 1992, KU students started initial efforts for making the university more sustainable (Fig. 2).

In 2010, the KU University Management at the time decided to introduce an overall sustainability concept by way of a whole-institution approach as a reaction to a joint declaration made by the German UNESCO Commission and the German Rectors' Conference (HRK and DUK 2010) and a conference of the national working group

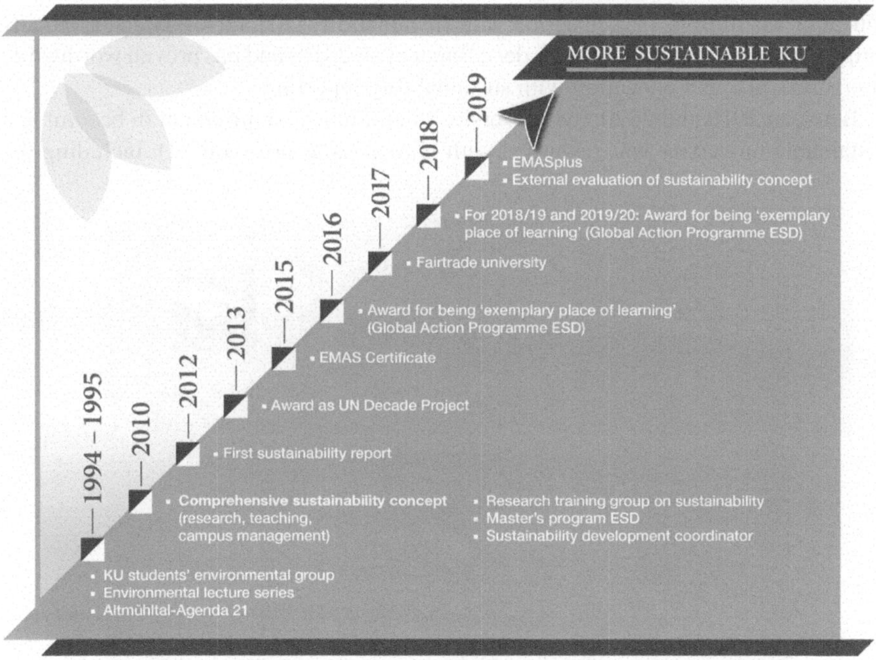

Fig. 2 The Catholic University of Eichstätt-Ingolstadt on its way to a more sustainable university (Hemmer and Limmer) (Katholische Universität Eichstätt-Ingolstadt 2019, p. 10, illustration: C. Pietsch)

on universities and sustainability ("AG Hochschule und Nachhaltigkeit") of the UN Decade ESD. This concept included the three action areas research, teaching and environmental campus management that were all grounded in the guiding principle of sustainable development. Objectives as well as short, medium and long-term measures were formulated for each of the three action areas. In addition, the concept also included a corresponding implementation strategy. The starting conditions for the concept were favorable, as there were already different existing activities in the fields of research and teaching and beyond. A very important step was to firmly anchor the topic of sustainability in the KU's organizational structure with the help of this concept. At the same time, a sustainable development coordinator was entrusted with the coordination of the implementation process of the sustainability concept (carried out as secondary function in addition to the position as Professor of Didactics of Geography) with the support of a steering group. Since 2012, the KU has issued an annual sustainability report (hhttps://www.ku.de/unileben/nachhaltige-ku/nachhaltigkeitsbericht). In subsequent years, the KU was able to achieve almost all objectives as stipulated in the concept and implement nearly all proposed measures. Several years ago, the university has extended the number of action areas by adding three additional fields: Governance, transfer and student initiatives & commitment were added as additional action fields besides the existing areas of research, teaching and campus management (Fig. 3). The decision to expand the focus to these six fields of action is based on the scientific debate dealing with the transformation of higher education institutions in connection with sustainability (KriNaHoBay 2017; HochN (https://www.hochn.uni-hamburg.de/); Bauer et al. 2018) and has proven worthwhile in practice, also in connection with sustainability reporting.

Between 2010 and 2020, the university's continuous commitment in becoming a sustainable university was rewarded with numerous awards (Fig. 2), including the

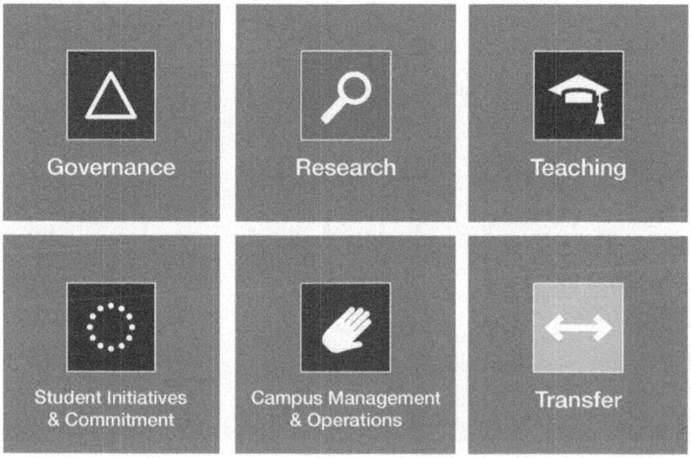

Fig. 3 The sustainability action areas at the Catholic University of Eichstätt-Ingolstadt (illustration: C. Pietsch)

title as UN Decade Project in 2013 and as exemplary place of learning for ESD in 2016, 2018 and 2019 in connection with the UN Global Action Programme ESD. Furthermore, the KU bears the title 'Fairtrade University' since 2017. Between 2011 and 2014, the institution implemented an environmental management system [eco management and audit scheme (EMAS)] and was awarded the EMAS certificate and later on, in 2019, also the EMASplus certificate for its efforts. In 2019, the KU's overall sustainability concept was evaluated by external experts. The evaluation confirmed that the KU was among the few higher education institutions in Germany that served as a role model for consistently pursuing an overall sustainability concept by way of a whole-institution approach. In Germany, the KU has the leading position in the fields of EMASplus and sustainability reporting. Together with the University of Applied Sciences Munich, the KU has founded a Bavarian university network in 2012 which has the aim of supporting other higher education institutions on their way towards becoming more sustainable.

3.2 *The Six Sustainability Action Areas at the KU*

3.2.1 Action Area 'Governance'

In order to be able to implement the topic of sustainability at a higher education institution, different activities and initiatives must be carefully coordinated. The task of coordinating these processes in a structured way is referred to as 'governance' (Bormann et al. 2018). This definition includes both organizational structures and controlling complex processes. All groups within a university with their very different skill profiles, perspectives and interests play an important role for the governance of a sustainability process: university management, students, professors, employees in research and teaching and administrative and technical staff. Often, these different players have very different perspectives on the relevance of sustainable development, the detailed objectives and the means that are pursued or are to be implemented in this context or in view of the relevant anchoring of responsibilities in connection with sustainability. This is why the coordination of all these positions is one of the central tasks of such governance. Communication and consideration processes as well as active participation are also important pillars in connection with governance. It is of utmost importance to create necessary structures and allocate responsibilities in order to effectively support the sustainability agents at a university.

At the KU, the above-mentioned whole-institution sustainability approach, the appointment of a sustainable development coordinator and the establishment of a steering group created an important framework for coordinating sustainability activities. The steering group is made up of members from all KU faculties, administrative departments as well as students. Since 2012, the Chancellor is the official contact person for all sustainability matters within the university management. In 2014, the topic of sustainability was included in the development plan and later on also in the KU's guiding principles. The university management supports the process by providing staff resources: In 2012, a part-time position for campus management was established. In 2016, the university created a full-time position for the assistance of the sustainable development coordinator, who was already granted a teaching load reduction of 1 h in 2015. Furthermore, the sustainability initiatives are supported financially in order to enable implementation of measures from the different action areas. Currently, a 'Green Office' is being established following the proposal of students in order to support student involvement and increase the visibility of the sustainability topic on campus.

3.2.2 Action Area 'Research'

In the relevant literature, the action area of research is divided into three categories (LeNa 2014; Leuphana Universität Lüneburg and Ludwig-Maximilians-Universität München 2018):

(1) Research as a social responsibility
Research that reflects its social and ecological impact with regard to its conceptualization, implementation and results.
(2) Research for sustainable development
Research that can contribute to sustainable development.
(3) Sustainability research
Research that is oriented towards sustainability issues with regard to its choice of topics, conceptualization, implementation and expected outcomes and that explicitly seeks to contribute to solving sustainability problems.

At the KU, research category (1), which makes the research process itself the subject of ethical and sustainable criteria in order to make these processes more environmentally friendly and socially acceptable, is to date only rarely represented. Research categories (2) and (3) are predominant. There are quite a few projects, including some large-scale third-party funded projects, in category (2) that for example deal with climate and climate impact research, river restoration and dynamization of floodplains as well as with migration issues and justice and thus contribute to the field of sustainable development. Furthermore, there are also projects that address sustainability issues directly, such as, for example, research on sustainable financial systems, sustainable land management, sustainable tourism, sustainable entrepreneurship, sustainable lifestyles and ESD. A number of these projects have an interdisciplinary and transdisciplinary orientation with project partners from different fields, such as municipalities, water resources authorities, nature conservation authorities, tourism enterprises and schools. The majority of research projects on sustainability at the KU is carried out by geographers. However, there are also researchers at all other KU faculties who work in the field of sustainability. On average, 10% of KU research projects are explicit sustainability projects and 30% contribute to a sustainable development. The university is currently developing indicators for the field of research (projects, publications).

From 2010 to 2015, there was an interdisciplinary research training group on sustainability for environment, humans and society. The research results of the group were published in a final volume (Altmeppen et al. 2017). The KU Center for Flight and Migration and the KU Center for Latin American Studies dedicate their research, teaching and transfer practice to sustainable development. Currently, the KU plans the establishment of a research-oriented center for sustainable development.

3.2.3 Action Area 'Teaching'

Based on the UNESCO Global Action Programme ESD, the Federal Ministry of Education and Research (BMBF) and the German UNESCO Commission passed the National Action Plan ESD (Nationale Plattform Bildung für nachhaltige Entwicklung 2017) in 2017. This paper set down action areas, objectives and measures for all fields of education including the higher educational sector. The aim is to contribute to achieving the SDG 4 quality education, and in particular 4.7 ESD. Today's students will be the decision makers and change agents of the future—which is why it is especially important to raise their awareness for the challenges connected to sustainable development and enable them to actively participate in this development. In this context, a firm, structural implementation of ESD in higher education institutions is paramount.

At the KU, the interdisciplinary Master's degree program ESD was conceptualized as early as in 2010. The program is unique, at least in the German-speaking world, and trains future change agents who are enabled to conceptualize, implement and evaluate educational processes in ESD and carry out research in the field (https://www.ku.de/studienangebot/geo-bne-ma). The interdisciplinary lecture series on sustainable development, which came into being through dialog between the members of the above-mentioned research training group, is offered at the KU since 2011. During the last few years, the KU has added further modules and study specializations on sustainability to its curriculum, e.g. in the subjects of Geography, Politics, Social Work and Business and Economics. The degree program in Tourism and (Sustainable) Regional Development has strengthened its focus on sustainability in 2019. Project modules enable students to transfer sustainable innovations into the region. In field trips and internships that are either completed in sustainable places of learning and sustainable companies or are dealing with the topic, KU students get the chance to broaden their horizon with regard to sustainability aspects. Furthermore, the KU conceptualized a sustainability certificate worth 20 ECTS credits in 2019. It gives all KU students the possibility to approach questions connected to sustainability more intensively and was very well received from the outset. In recent years, the KU has been very active in the field of continuing education and professional development and offers training especially in the field of university teaching methodology and teacher training.

3.2.4 Action Area 'Student Initiatives and Commitment'

Active involvement and student commitment is a driving force for transformation on a university's path to more sustainability. This becomes evident in initiatives such as the international Green Office movement (https://www.greenofficemovement.org/de/) or the corresponding German-speaking network 'Netzwerk n' (https://netzwerk-n.org/). Student initiatives can initiate and significantly advance transformation processes. Interestingly, they are only rarely mentioned as an action area at universities in relevant literature. The Bavarian project KriNaHoBay is the only one to explicitly mention this action area (KriNaHoBay 2017, https://www.nachhaltigehochschule.de/projekte/).

At the KU, the "Environmental Department" ("Umweltreferat") was already founded around 1990 as an independent sub-group of student representatives at the KU. Its members actively worked towards the KU becoming a more sustainable university in the years following the Rio Conference of 1992. With a large-scale project, they also contributed to increasing the awareness for sustainability within the region. In the end, they did not succeed in transforming the university. However, student initiatives dealing with the topic of sustainability remained and started to pick up pace again since 2008. In summary, it can be said that the students also played an important role in connection with the KU agreeing on a sustainability concept (see Sect. 3.1). They participated in drafting the concept and have been members to the sustainability steering group ever since. The "Environmental Department" carries out a number of own sustainability activities every year and participates in other initiatives, such as lecture series with representatives from professional practice, clothes swap events, dance theaters on sustainable topics with actors from the global south, sustainable Christmas markets etc. (Schöps and Hemmer 2018).

In 2010, students founded the association "ThinkSustainably!" ("DenkNachhaltig!") to address topics such as sustainability, social innovation and entrepreneurship. The association is a platform enabling exchange between students and representatives from society and the industry. It offers presentations and talks, discussion rounds and workshops in regular intervals. Past workshop topics included 'Agile Management', 'Design Thinking' and 'Scrum'. A format on sustainable business ideas is currently being planned.

In addition to these two initiatives, there are meanwhile five other student initiatives that are committed to the concept of sustainability and support corresponding activities. In 2019, students founded the initiative "Students for future" and also the respective student umbrella organization has intensified its activities in this field.

3.2.5 Action Area 'Campus Management and Operations'

The topic of sustainability became increasingly relevant within a higher education context approx. since the 1990s, when legal regulations were introduced that required universities to fulfill certain employment law and environmental criteria in its operational practice. Since 1999, validated environmental management systems such as the Eco-Management and Audit Scheme (EMAS) gained increasing relevance and determined responsibilities and an allocation of tasks as well as a controlling system for sustainable processes. Currently, only around 5% of German universities have been awarded the EMAS certificate. The implementation of sustainability and integration into operational processes at a university comprises many different fields that are tracked by an environmental management system: Procurement, waste management, mobility, facility and energy management, controlling, research practice, event management, employment relationships and communication (Technische Universität Dresden, Hochschule Zittau/Görlitz 2018).

The overall sustainability concept of the KU that was drafted in 2010 already set out the objective of being awarded the EMAS certificate. The certificate requires comprehensive preparatory work and staff and financial resources that were provided by the university management. The KU was able to successfully implement the process and was awarded its first EMAS certificate in January 2015. In January 2019, the Catholic University of Eichstätt-Ingolstadt was the only university in Germany to be awarded the EMASplus certificate. This award also takes into account social and economic aspects in addition to ecological criteria that play a major role in the EMAS certification. The EMAS process, which required drafting an annual sustainability program with approx. 15 objectives, significantly contributed to the improvement of sustainability at the KU. For example, the institution was able to save large amounts of energy, all printing and copying stations now use recycled paper, waste separation was optimized and the procurement process was adapted to take better account of ecological and social criteria. Needless to say, not all fields have been optimized yet: the university is currently reviewing the sustainability status of its canteen and mobility processes. It took a lot of effort of a continuously dedicated environmental management officer on campus to achieve the above-mentioned successes. In addition, the willingness of different groups within the university to actively participate and contribute their ideas was and still is of utmost importance for the process.

3.2.6 Action Area 'Transfer'

The requirements of a higher education institution change over time along with changes in society that take place very fast. In addition to their core responsibilities of research and teaching, universities also increasingly assume additional tasks as requested by society (Hochschule für nachhaltige Entwicklung Eberswalde 2018). These activities are referred to as transfer activities and include interactions between universities and actors from other fields of society, such as the industry, education or municipalities. Transfer challenges scientific thinking and offers the opportunity to advance and deepen science and research. This is a win-win situation because science and academia can have a stronger impact on society and simultaneously gain new impulses from practical fields. Transfer can take many different shapes.

At the KU, there has always been a substantial number of researchers who were successfully engaged in transdisciplinary projects because of their subject specification, amongst others from the fields of Geography, Applied Mathematics, Social Work, Religious Education and Business and Economics. The projects increasingly involved representatives from external professional practice in all stages of the projects from planning to implementation and evaluation. Furthermore, the KU offered teacher training courses in regular intervals. Since 2018, the field of transfer was strengthened considerably when the university successfully acquired the third-party funded project "Mensch in Bewegung" (https://mensch-in-bewegung.info/). Over a period of five years, the project especially aims to develop ideas for improving living conditions in fields such as sustainability, social engagement, mobility and digitalization within the broader region together with the Technische Hochschule Ingolstadt and in continuous exchange with the general public. The project team has already hosted a number of interesting events, workshops, social innovation camps and so-called 'future workshops' for different target groups. Furthermore, the KU supports initiatives in neighboring municipalities that seek to increase their efforts to become more sustainable.

3.3 A Look Ahead—The New Sustainability Concept

Since 2018, the university's sustainability team has been revising and updating the KU's overall sustainability concept that was published in 2010. The three original action areas (research, teaching, campus management and operations) have meanwhile been extended by another three fields (governance, transfer and student initiatives and commitment). The new concept was jointly drafted by all members of the sustainability steering group in which all groups of the university are represented (see Sect. 3.1). In 2019, the new draft concept was reviewed by external evaluators. Their feedback and recommendations were subsequently included in the new concept which is expected to be published in 2020. Main objectives of the new concept are, amongst others, strengthening sustainability research and applying a more extensive implementation of sustainability in the teaching practice in as many subject areas as possible. In 2020, the KU will set up a Green Office that will support student initiatives and commitment. Last but not least, the KU has the objective of becoming a climate-neutral university within the next couple of years.

3.4 The University's Contribution to Transformation at Higher Education Institutions

3.4.1 Promoting and Inhibiting Factors for Transformation at Higher Education Institutions

The developments at the KU offer a lot of potential for being transferred to other universities. Drawing on the example of the KU, advantages and disadvantages can be identified that can be said to promote or inhibit the transformation of universities in general. For example, a very favorable factor is the KU's size. Being a small university with approx. 5000 students and 120 professors, the KU has relatively lean administrative structures and regular exchange between University Management and KU members is the norm. Another beneficial factor is that the KU has decided to involve the Chancellor in the coordination of sustainability matters, providing for a direct connection to the University Management. The commitment of the University Management also plays a decisive role in connection with how fast, intensively and comprehensively the sustainability process is being advanced at the KU. This also becomes evident in the results of the project KriNaHoBay (2017). Furthermore, the KU is a successful example of efficient implementation of a university-wide overall sustainability concept, which had the aim of continuously promoting sustainable development by defining objectives and measures for the relevant action areas for ten years already. Particularly noteworthy in this context is the consistent creation of participatory structures (see Sect. 3.2.1) and an increasing degree of professionalization. The fact that the KU is a Catholic university is also beneficial, especially in view of the Encyclical Laudato Si, although it should be pointed out that the sustainability activities were not initiated by the subject of Theology, but primarily by the subject of Geography.

Inhibiting factors in the sustainability development process at the KU were the changes within the University Management over the last ten years and other staff turnover in important and responsible positions. As regards the fields of research and teaching, it can be stated that processes of change in these areas are subject to a certain inertia and thus require great commitment, initiative and persistence of responsible players in these fields.

Without a doubt, the KU also serves as an illustrative example of how well a jointly implemented top-down bottom-up process supports targeted implementation of such concepts at higher education institutions (see Sect. 2; Gräsel and Parchmann 2004; Fig. 1).

3.4.2 Work for the Network of Bavarian Higher Education Institutions

In 2012, the KU and the University of Applied Sciences Munich founded "The Bavarian Network for Sustainability in Higher Education" (http://www.nachhaltigehochschule.de) which emerged from the national working group "University and Sustainability" ("Hochschule und Nachhaltigkeit"), an initiative of the UN Decade "Education for Sustainable Development". The aim of the network is to expand and tighten the network of Bavarian universities and higher education institutions in the field of education for sustainable development. Since its year of foundation, the network hosts meetings two times a year and welcomes representatives from university managements, researchers, students from all disciplines and administrative staff from almost all Bavarian universities. Representatives from ministries, politics, the industry and civil society are also invited to attend the conferences. These biannual network meetings are held at different universities in Bavaria and address different sustainability topics that are relevant in a higher educational context, such as ESD, digitalization or university teaching methodology.

The network was already successful in reaching some political objectives, for example the anchoring of the topic of sustainability in the higher education act and the examination regulations for teaching degree programs. All Bavarian universities benefit from the project KriNaHoBay (see Sect. 2) that was supported financially by the Ministry for the Environment. Universities that have already gained experience in the implementation of sustainability measures assist other universities that have just started out on their paths towards becoming more sustainable. Since 2018, the Ministry for the Environment has been subsidizing a project that concentrates on training university lecturers with a particular focus on teacher training programs ('teach the teacher'). The effects of the project are monitored scientifically. Just over 50% of Bavarian universities have concluded a memorandum of understanding in 2019 in which they declared their willingness to increase their focus on sustainability.

The above development clearly shows that German, and in particular also Bavarian, higher education institutions are making enormous progress when it comes to sustainable development at universities. Universities have a special responsibility in connection with transformation processes leading to a more sustainable society. The example of the KU encourages other universities to persistently follow their path towards more sustainability.

4 Conclusion

Due to their great social significance as educational institutions, universities play an important pioneering role in the implementation of sustainability. Drawing on the KU as an example, the previous remarks show a way towards becoming a more sustainable university. Essential conclusions for a successful sustainable orientation of a university are multi-perspective approaches to cope with the interrelated complex challenges. These include top-down measures and bottom-up initiatives on the one hand and simultaneous processing of several fields of action (governance, research, teaching, student initiatives and commitment, campus management and operations, transfer) on the other hand, in order to achieve a whole-institution approach.

These steps must be consistently pursued in the years to come. Progress is documented in detail in the annual sustainability reports. In the future, it is planned to accompany the path towards becoming a more sustainable university from a research perspective and to compare the implementation successes with those achieved at other universities. In addition, continuously updated recommendations and challenges will be formulated.

References

AASHE (2012) framing campus sustainability, STARS quarterly review. The Association for the Advancement of Sustainability in Higher Education. https://eric.ed.gov/?id=ED538365. Accessed 18 Jan 2020

AASHE (2015) STARS Technical Manual. The Association for the Advancement of Sustainability in Higher Education. https://stars.aashe.org/pages/about/technical-manual.html. Accessed 18 Jan 2020

Altmeppen K-D, Zschaler F, Zademach H-M, Böttigheimer Ch, Müller M (eds) (2017) Nachhaltigkeit in Umwelt, Wirtschaft und Gesellschaft. Interdisziplinäre Perspektiven. Springer, Wiesbaden, p 285

Barlett PF, Chase GW (2013) Sustainability in higher education: stories and strategies for transformation. The MIT Press, Cambridge, p 316

Bauer M, Bormann I, Kummer B, Niedlich S, Rieckmann M (2018) Sustainability governance at universities: using a governance equalizer as a research heuristic. High Educ Policy 31(4):491–511. https://doi.org/10.1057/s41307-018-0104-x

Bormann I, Rieckmann M, Bauer M, Kummer B, Niedlich S (2018) Nachhaltigkeitsgovernance an Hochschulen (Betaversion). BMBF-Projekt „Nachhaltigkeit an Hochschulen: entwickeln—vernetzen—berichten (HOCHN). https://www.hochn.uni-hamburg.de/-downloads/handlungsfelder/governance/hoch-n-leitfaden-nachhaltigkeitsgovernance-an-hochschulen.pdf. Accessed 18 Jan 2020

Brinkhurst M, Rose P, Maurice G, Ackerman JD (2011) Achieving campus sustainability: top-down, bottom-up, or neither? Int J Sustain High Educ 12(4):338–354

Ceulemans K, Molderez I, Van Liedekerke L (2015) Sustainability reporting in higher education: a comprehensive review of the recent literature and paths for further research. J Clean Prod 106:127–143

DUK—Deutsche UNESCO-Kommission e.V. (2014) UNESCO Roadmap zur Umsetzung des Weltaktionsprogramms „Bildung für nachhaltige Entwicklung“. https://www.bmbf.de/files/2015_Roadmap_deutsch.pdf. Accessed 18 Jan 2020

Fischer D, Jenssen S, Tappeser V (2015) Getting an empirical hold of the sustainable university: a comparative analysis of evaluation frameworks across 12 contemporary sustainability assessment tools. Assess Eval High Educ 40(6):785–800

Gräsel C, Parchmann I (2004) Implementationsforschung—oder: der steinige Weg. Unterricht zu verändern. Unterrichtswissenschaft 32(3):196–214

Hemmer I, Bagoly-Simó P (2016) Making the campus sustainable. The example of the Catholic University Eichstätt-Ingolstadt. In: Albiez M, Banse G, Lindemann KC, Quint A (eds) Designing sustainable urban futures. KIT Scientific Publishing, Karlsruhe, pp 129–140

Hochschule für nachhaltige Entwicklung Eberswalde (2018) Nachhaltigkeit an Hochschulen: entwickeln—vernetzen—berichten (HOCHN)—Transfer für nachhaltige Entwicklung an Hochschulen. https://www.hochn.uni-hamburg.de/-downloads/handlungsfelder/transfer/hoch-n-leitfaden-transfer-fuer-nachhaltige-entwicklung-an-hochschulen.pdf. Accessed 18 Jan 2020

HRK—Hochschulrektorenkonferenz (2017) Für eine Kultur der Nachhaltigkeit. Empfehlung der 25. Mitgliederversammlung der HRK am 06. November 2018 in Lüneburg. https://www.hrk.de/fileadmin/redaktion/hrk/02-Dokumente/02-01-Beschluesse/HRK_MV_Empfehlung_Nachhal tigkeit_06112018.pdf. Accessed 18 Jan 2020

HRK, DUK (2010) Erklärung der Hochschulrektorenkonferenz (HRK) und der Deutschen UNESCO-Kommission (DUK) zur Hochschulbildung für nachhaltige Entwicklung. https://www.hrk.de/fileadmin/migrated/content_uploads/Hochschulen_und_Nachhaltigkeit_HRK_DUK.pdf. Accessed 18 Jan 2020

Katholische Universität Eichstätt-Ingolstadt (Hrsg.) (2019) Siebter Nachhaltigkeitsbericht Katholische Universität Eichstätt-Ingolstadt 2018|19. https://www.ku.de/fileadmin/190811/6_N achhaltigkeit_Berichte/1_NHB_PDFs/NHB_18-19_HP_final.pdf. Accessed 18 Jan 2020

KriNaHoBay (2017) F + E-Projekt des StMUV „Nachhaltige Hochschule: Kriterien für eine Bestandsaufnahme". https://www.nachhaltigehochschule.de/projekte/. Accessed 18 Jan 2020

LeNa (2014) Teacher education for a sustainable development from pilot projects and initiatives to new structures. A memorandum on reorienting teacher education in Germany, Austria and Switzerland. https://netzwerklena.files.wordpress.com/2020/07/memorandum_lena_english_stand_august_15.pdf. Accessed 18 Jan 2020

Leuphana Universität Lüneburg, Ludwig-Maximilians-Universität München (2018) Nachhaltigkeit in der Hochschulforschung (Betaversion). BMBF-Projekt „Nachhaltigkeit an Hochschulen: entwickeln—vernetzen—berichten (HOCHN)". https://www.hochn.uni-hamburg.de/-downlo ads/handlungsfelder/forschung/hoch-n-leitfaden-nachhaltigkeit-in-der-hochschulforschung.pdf. Accessed 18 Jan 2020

Michelsen G, Fischer D (2015) Bildung für nachhaltige Entwicklung. Hessische Landeszentrale für politische Bildung, Wiesbaden

Nationale Plattform Bildung für nachhaltige Entwicklung (2017) Nationaler Aktionsplan Bildung für nachhaltige Entwicklung Der deutsche Beitrag zum UNESCO-Weltaktionsprogramm. https://www.bne-portal.de/files/Nationaler_Aktionsplan_Bildung_f%c3%bcr_nachhaltige_Ent wicklung_neu.pdf. Accessed 18 Jan 2020

Rieckmann M (2016) Bildung für nachhaltige Entwicklung—Konzeptionelle Grundlagen und Stand der Implementierung. In: Schweer M (ed) Bildung für nachhaltige Entwicklung in pädagogischen Handlungsfeldern—Grundlagen, Verankerung und Methodik in ausgewählten Lehr-Lern-Kontexten. Peter Lang Verlag, Frankfurt am Main, pp 11–32

Sassen R, Dienes D, Beth C (2014) Nachhaltigkeitsberichterstattung deutscher Hochschulen. Beiträge zur rechts-, wirtschafts- und sozialwissenschaftlichen Umweltforschung. Zeitschrift für Umweltpolitik & Umweltrecht: ZfU 37(3):258–277

Schneidewind U (2014) Von der nachhaltigen zur transformativen Hochschule—Perspektiven einer „True University Sustainability". http://netzwerk-n.org/wp-content/uploads/2017/07/Schneidew ind-2014-Transformative-Hochschule.pdf. Accessed 18 Jan 2020

Schöps A, Hemmer I (2018) Participation of student authors in reports on sustainability. Int J Sustain High Educ 19(2):249–265. https://doi.org/10.1108/IJSHE-08-2016-0155

Shriberg M (2002) Institutional assessment tools for sustainability in higher education: strengths, weaknesses, and implications for practice and theory. Int J Sustain High Educ 3(3):153–167

Technische Universität Dresden, Hochschule Zittau/Görlitz (2018) Nachhaltigkeit an Hochschulen: entwickeln—vernetzen—berichten (HOCH N)—Nachhaltigkeit im Hochschulbetrieb. https://www.hochn.uni-hamburg.de/-downloads/handlungsfelder/betrieb/hoch-n-leitfaden-nachhalti ger-hochschulbetrieb.pdf. Accessed 18 Jan 2020

UNESCO (ed) (2014) Roadmap for implementing the global action programme on education for sustainable development. UNESCO, Paris
UNESCO (ed) (2017) Education for sustainable development goals. Learning objectives. UNESCO, Paris. http://unesdoc.unesco.org/images/0024/002474/247444e.pdf. Accessed 18 Jan 2020

Professor Ingrid Hemmer has been involved in geography didactic research and teaching for almost forty years. For about 15 years she has been primarily concerned with ESD in both areas. Ingrid Hemmer earned her doctorate in economic geography at the University of Münster and habilitated at the University of Augsburg in geography didactics for the scientific orientation of geography teaching. In teaching, she introduces ESD into the teacher training courses in geography, has been leading an interdisciplinary Master's programme in ESD since 2010 and a Certificate in ESD since 2019. In research, she focuses on the areas of implementation of ESD in universities and schools and competence development for ESD. Ingrid Hemmer has been sustainability officer of the KU from 2010 to 2020 and has coordinated an overall institutional approach to sustainability there since 2010. Since 2012 she has been spokesperson of the Network University and Sustainability Bavaria and is active in numerous other committees on SD and ESD.

Professor Anne-Kathrin Lindau has been active in research and teaching in the geography education since 2002. For about ten years now, she has been increasingly concerned with the field of ESD in schools and universities. Anne-Kathrin Lindau studied teaching Geography and German at grammar schools. After her studies she completed her teacher training at a grammar school. Afterwards she did her doctorate at the Martin-Luther-University Halle-Wittenberg interdisciplinary in Geoecology and Geography Education. In teaching, she is committed to the increased implementation of ESD in schools as well as in teacher training courses, especially in Geography. Her research interests include teacher professionalisation, excursions in geography and wilderness education with a special focus on ESD. Since December 2019 Anne-Kathrin Lindau, the Professor of Didactics of Geography and Education for Sustainable Development at the Catholic University of Eichstätt-Ingolstadt, has been working in the field of ESD and since 2020 she has been the coordinator of sustainability at the KU.

Perspectives of Climate Change Adaptation in Organizations—A University' Example

Hardy Pundt, Martin Scheinert, and Andrea Heilmann

1 Introduction

The societal impacts of climate change are diverse. Depending on which IPCC scenario is taken as a basis, climate change will have repercussions that require the development of sustainable measures to mitigate harmful incidents such as droughts, floods, heavy storms, heatwaves, and other weather phenomena (Nakicenovic and Swart 2000). The explanation why global warming occurs and intensifies and what the consequences of those rising temperatures are crucial in numerous fields of society.

Agriculture, forestry, water management, soil protection, urban and regional planning, energy provision, desertification, heat management in rural and urban areas, melting of ice masses, the invasion of new species, traffic coordination, leisure—there is hardly any societal nor economic sector that is not concerned with climate change in a certain way. Therefore, it is essential to discuss climate change in politics, society, and sciences and to bring ideas, expertise and the latest insights of such different groups together.

Climate change is a challenge for mankind, and therefore the main institutions which teach young people facts, concepts and strategies to improve living conditions on Earth, namely schools and universities must address this challenge. Universities have a public responsibility to inform society about facts and data on climate change with reference to the Sustainable Development Goal (SDG) 13 (SDG Knowledge Platform 2020). Furthermore, they should contribute to developing concepts, strategies and concrete measures to deal with the problems arising from new climate phenomena. Sustainable measures, however, can only be achieved if the relevant stakeholders are involved, thus avoiding that solutions end up in the drawer.

H. Pundt (✉) · M. Scheinert · A. Heilmann
Department of Automation and Computer Sciences, Harz University of Applied Sciences, Friedrichstr. 57-59, 38855 Wernigerode, Germany
e-mail: hpundt@hs-harz.de

W. Leal Filho et al. (eds.), *Universities, Sustainability and Society: Supporting the Implementation of the Sustainable Development Goals*, World Sustainability Series,
https://doi.org/10.1007/978-3-030-63399-8_21

A second aspect of universities' responsibilities concerning society is that they themselves must turn into sustainable organizations. "A sustainable university is an educational institution that trains global citizens in sustainable development, offers relevant insights into societal challenges and reduces the environmental and social footprints of its campus operations, empowers students and staff to act sustainably, and makes sustainability a central priority" (GreenOfficeMouvement 2019).

One of the central mandates of universities is to educate students. Increasingly, curricula are provided that concern climate protection, and/or climate change adaptation in particular. Additionally, universities have a mandate concerning research. They can initiate and participate in projects that are related to (a) scientific foundations of climate change, or (b) the development of concrete measures to support climate change adaptation. Especially (b) concerns the ability of universities to participate in both inter- and transdisciplinary projects.

This paper picks up both perspectives, universities as educational and as research organizations, but will focus on the second aspect. An applied research project serves as an example to explain universities' role in coordinating a transdisciplinary team of scientists and local authorities, towns, forest and agricultural administration, districts, larger administrative bodies, in order to make suggestions on sustainable and societally accepted climate change adaptation measures in a collaborative way.

2 Four General Aspects of Universities' Engagement in Climate Change Adaptation

Universities can provide significant contributions to the mitigation of negative, if not harmful, consequences of climate change. Universities bundle expertise and knowledge. Both have to be communicated to society to support an objective and scientifically sound debate about climate change. In contrast to enterprises and political parties or organizations, universities are open and independent organizations. This makes them ideal partners in participating not only in pure scientific projects, but also in those that aim at integrating science and "the real world outside university". Climate change adaptation is a field that requires a multi-perspective view of possible threats and risks. In many cases it is not sufficient to look only at one single sector. Instead, different sectors and their specific expertise are of vital interest. The integration of a range of perspectives, opinions, and goal-settings of various stakeholders can ideally be managed by universities. Their interest to produce recommendations and solutions that are independent from economic and financial interests, is a profound advantage.

2.1 *Climate Change Adaptation as Educational Mandate*

Even though the next section will discuss a project-based approach to climate change adaptation, the fundamental mandate of universities to educate young people should be stressed before. Education must include both teaching of scientific facts about climate change, as well as intensive discussion of risks and threats resulting from it. This must be done on a broad and scientifically sound basis in lectures and seminars. Looking at the European universities' landscape this can be observed. More and more, complete programs on climate protection, and climate change adaptation are provided. Table 1 gives only some selected examples of Master's curricula that are provided at European Universities (Masterstudies 2019; Finadamasters 2020; Master and More 2020; PdT 2020).

Apart from curricula that are specifically dedicated to climate and climate change adaptation subjects concerning climate change are increasingly integrated into "classic" programs as well. Natural, technical, IT- and even societal and philosophical sciences take up climate change adaptation as a subject in lectures, seminars and research. They all aim at developing offers and solutions to mitigate the threats and

Table 1 Universities as organizations that have an educational mandate in terms of climate change adaptation—examples of curricula with relationships to "climate" and "climate change"

Curriculum	University	Country
Master in Marine Science and Climate Change	University of Gibraltar	Gibraltar
Master in Geographic Information and Climate Change	Swansea University	United Kingdom
Master in Urban Environment, Sustainability and Climate Change	University of Rotterdam	The Netherlands
Master in Climate Change	University of Barcelona	Spain
Master in Disaster Risk Management and Climate Change Adaptation	University of Lund	Sweden
Integrated Climate Systems Sciences	University of Hamburg	Germany
Master of Climate Change	University of Copenhagen	Denmark
Master in Climate Change: Adaptation and Mitigation Solutions	Polytechnic University of Torino	Italy
Master Degree in Adaptation to Climate Change	Paris-Saclay University	France
Climate Change: Science and Impacts	University of Dublin	Ireland
Hydrologia, meteorologia i climatologia	University of Poznan	Poland
Energy and Climate	University of Antwerp	Belgium
Klimawissenschaften (Climate Science)	University of Bern	Switzerland

risks on the one hand, and to prepare society, on the other hand, thus emphasizing that adaptation is a fundamental challenge.

2.2 *Climate Change Adaptation as an Organizational Goal of Universities' Policies*

Universities have a mandate to adapt themselves to climate change. Lawrence Bacow, president of the Harvard University, emphasizes this in a general matter that should impact all universities as organizations: "At Harvard, we are acting on climate change through research that occurs across disciplines and throughout the world; through teaching and learning, by providing our students with the tools to confront this issue for generations to come; and on our campus, by modeling an institutional pathway to a healthier, more sustainable community (Bacow 2020). In such a sense, universities have to turn themselves into climate friendly organizations, e.g. they should search for sustainable ways to support the reduction of waste and emissions. The consequence for employees and students is in the long or even short run a more climate friendly behavior. Universities can, for instance, implement environmental management systems (EMS). University EMS can help to avoid waste, to protect water, to support the use of sustainable products and the provision of food resulting from sustainable agriculture. Universities can do more in this field and try to reduce carbon emissions and ensure that students, regardless of what they are studying, understand sustainable development and the need for climate change adaptation. Moving away from investments in fossil fuels, reinvesting in renewables, planting trees on campuses are concrete measures that can be taken by universities (Mayo 2019).

2.3 *Climate Change Adaptation as a Purely Scientific Research Task*

Another important field is the universities' contributions to research in climate change adaptation. Such research can occur in two formats. One opportunity is to set up consortia that focus exclusively on science and scientific partners, e.g. universities and research institutes outside universities. Such projects contribute to the enhancement of theories and expertise in climate change adaptation. Such projects can be distinguished from activities that follow the goal of cooperating with concrete stakeholders.

2.4 *Climate Change Adaptation as Subject for Applied Research*

The close cooperation of universities with stakeholders in society can be summarized under the term "applied research". In such projects, scientists collaborate with administrations, organizations, and/or enterprises to develop problem solutions that concern different levels. Climate change adaptation can occur in a multifaceted way in such projects. For example, universities can offer scientific expertise through information about their latest findings. They can also play a role as coordinator or moderator of teams that involve different partners. This can be done, for instance, in the form of living labs. Such labs are "A kind of giant sandbox in which there is the freedom to explore—creatively and collaboratively—the technological, environmental, economic and societal aspects of sustainability" (UBC 2018). In such a sense, a living lab is an ideal environment to discuss, develop and implement climate change adaptation measures under comprehensive consideration of the goals and needs of different societal stakeholders.

3 The University as Scientific Partner in Real-Life Planning Procedures Which Aim at Climate Change Adaptation

3.1 *Project Background: The University as Coordinator and Facilitator*

Within the framework of the "BebeR-project", a pilot project to develop and assess the vulnerability of a specific mountainous area with regard to erosion, various partners collaborated to discuss problems that might arise from a variety of climate change scenarios. The area that was investigated for the development of vulnerability assessment methods is the administrative district of Mansfeld-Südharz, situated in the south of the German federal state Saxony-Anhalt. The area is dominated by agriculture and forestry dotted with some small towns and villages.

3.1.1 Setting up a Transdisciplinary Geographical Database

Within "BebeR", the set-up of a comprehensive spatial database was a crucial step to enable scientists and various administrative partners to use the data from not only one, but all relevant sectors. The database was established using Geographical Information Services. They enable users to pile a great variety of spatial data layers upon each other. These data come from all kinds of sources, e.g. agriculture, forestry, water management, weather services, and more. Using the data, it was e.g. possible to integrate data in such a way that areas vulnerable to soil erosion were identified.

These areas are especially threatened by heavy rainfalls. The latter are predicted to increase due to climate change in future. The data were provided by a scope of stakeholders. The analysis of such data layers led to maps that were provided by an online and interactive map service developed during the project. The participants always had the opportunity to look at the maps, to integrate datasets from different sources, and to make comments so as to improve the maps. Furthermore, they were enabled to produce new maps, using the numerous data layers. The maps, however, were soon accepted as an ideal basis for discussion of problems considering varying perspectives and scenarios. This way, the maps were appreciated as decision support tools, on the one hand, and as a way to inform the public, on the other.

3.1.2 Defining Different Adaptation Options Based on the Geographical Database

During workshops, the maps and different alternative measures to adapt to climate change were discussed intensively. A prioritization of potential measures had to be carried out, so as to find the most important and feasible measures that should be realized in the pilot region within the next few months or years. The following list mentions some of the options discussed:

- Reduction of soil erosion in agriculturally used areas.
- Change of the management of (small) rivers.
- Concept to deal with invasive neophytes.
- Adaptation of forest management.
- Improvement of disaster management (especially in conjunction with extreme weather events).
- Change of administrative methods to include climate change adaptation in urban planning.
- Adaptation of water provision and wastewater removal.

More and more, the focus of discussion was on problems in the field of riverbed erosion and water management, coupled with the consideration of special interests of farmers who cultivate land along the river. Buildings close to rivers, as well as traffic (roads crossing rivers) close to the river can be endangered by erosion as well. Therefore, two river catchments have been selected as exemplary within the "BebeR"-project. The catchment area of the "Regenbeek" represents an area with partly high riverbed erosion, which is caused by strong currents after snow melting periods and/or heavy rainfall. Figure 2 presents the catchment area including a 3D-digitale elevation model. Riverbed erosion can lead to the destruction of the riverbanks and an unusual deepening of the riverbed. Mud flows can occur (as this already happened some time ago, see Fig. 1), thus polluting villages. Furthermore, infrastructure (e.g. roads, pipes) or buildings are threatened by the mud flows. Heavy rainfall can lead to anomalous bank erosion. This is accompanied by extensive erosion of surrounding arable land.

Fig. 1 Due to heavy rainfall, surface erosion was reinforced. It resulted in mud flows which pollute roads in a small town in the "BebeR" pilot area, the Regenbeek catchment area

Fig. 2 Topographic map and 3D digital elevation model of river catchment "Regenbeek"

3.1.3 Universities' Roles

To develop solutions for erosion problems, the "BebeR"-project established a platform for all relevant stakeholders. People from local and regional administrations, farmers organizations, and politicians took part in workshops with the goal to find a consensus on measures to mitigate the threats. The university had three roles within this project:

- as project coordinator,
- as moderator,

- as developer and provider of methodologies and software tools.

The role of facilitator was quite important as some stakeholders had conflicts with others about the "best" adaptation measures. However, most participants appreciated the role of the university as a helping hand to achieve hands-on results. An important conclusion is that the participation of the relevant actors from different sectors and administrative levels is a prerequisite to take actual decisions about the implementation of adaptation measures. The entire process from problem analysis to solution and suggestions of measures needs an independent moderator.

The methodology developed during the "BebeR"-project promotes a multidisciplinary, cross-sector, and transparent approach, supported by the analytical capabilities of Geoinformation services (Pundt et al. 2017). A further step in the methodology is the evaluation and prioritization of adaptation measures. Such an assessment must be based on a sound multicriteria evaluation method. In many cases not only one decision on a specific adaptation measure has to be taken, because alternatives exist. Deciding which option is best requires a prioritization of the alternatives that occur. A multi-criteria evaluation of the different outcomes can contribute to generally accepted measures.

3.2 Inter-sectoral, Transdisciplinary Decision-Making

Within "BebeR", two approaches were chosen to support the development of climate change adaptation measures. First, spatial data and geographical information services were used to integrate spatial data from various sectors to identify vulnerabilities, patterns and new insights concerning the suitability of specific sites and areas for adaptation measures (as shown before). Second, a multi-criteria analysis was applied to help in the prioritization of measures. The university provided methodological and technical support in these fields.

3.2.1 Methodological Approach

As mentioned previously, climate change adaptation requires proper consideration of the data, goals and specific views of different sectors. Integrating data, goals and views can be a quite difficult task. A multi-criteria analysis is a tool to support such complex, sometimes ill-structured decision-making processes:

> In a situation where multiple criteria are involved, confusion can arise if a logical, well-structured decision-making process is not followed. Another difficulty in decision making is that reaching a general consensus in a multidisciplinary team can be very difficult to achieve. By using MCA the members don't have to agree on the relative importance of the Criteria or the rankings of the alternatives. Each member enters his or her own judgements, and makes a distinct, identifiable contribution to a jointly reached conclusion. (LDACMP 2020)

Figure 3 shows the methodological approach that was chosen within the BebeR-

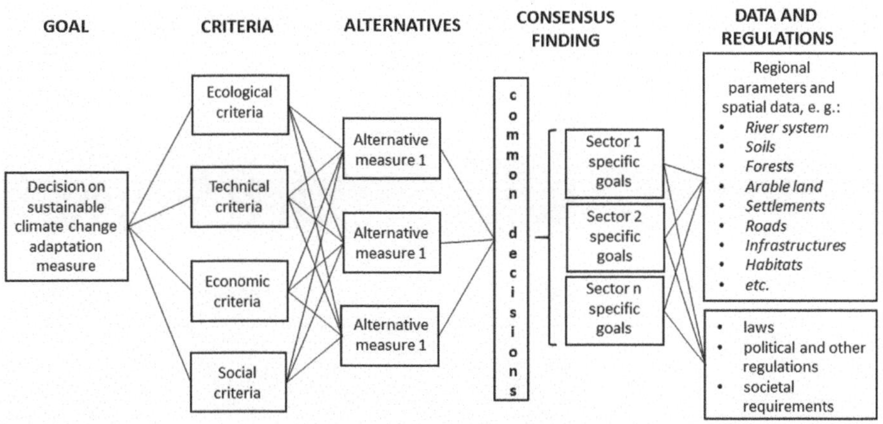

Fig. 3 From data to prioritized climate change adaptation measures using multicriteria analysis

project. The basic idea of this figure is from (IZNE 2018), but it has been extended and modified in view of the BebeR requirements. Based on the mentioned database and the online mapping system, the relevant sectors defined their specific goals in terms of the climate change adaptation measures. In a next step, the sector-specific goals were discussed comprehensively based on the vulnerability maps. Pros and cons were highlighted. The discussion resulted in a multitude of measures that would help to mitigate climate change threats. However, it was necessary to find consensus about the measures that had the highest priority. To support prioritization of measures, a multi-criteria analysis was proposed by the project coordinator. Such an analysis helps to deepen the insights into necessities and limitations of every participant on the one hand, and to prioritize measures based on a transparent method, on the other. Measures to mitigate soil erosion in the Regenbeek-catchment were evaluated using a specific software tool, called PRIMATE.

3.2.2 GIS-Based Simulation

Finding adequate sites for infrastructure or for buildings is a typical application field of Geographical Information Systems (GIS). Within BebeR, the university's role was to set up the aforementioned online GIS tool to enable stakeholders to explore data provided by the different sectors and to carry out spatial analyses aiming at producing new maps of the investigation area. Among other aims, erosion rates were simulated to identify areas that are explicitly endangered by soil erosion. This required the usage of a specific model that was integrated as a GIS tool. One of the measures to mitigate soil erosion identified by the project partners was a rain storage reservoir. Project partners agreed that the *detention of surface runoff* in case of heavy rainfalls would simultaneously reduce surface and riverbed erosion. To find out, which areas were suitable for building such a reservoir, a GIS analysis was conducted to rule

Fig. 4 Based on a GIS analysis, large areas in the Regenbeek catchment were excluded for building a rain storage reservoir

out unsuitable parts. Figure 4 shows areas that were excluded because they were identified as being inappropriate for various reasons.

3.2.3 Multicriteria Analysis to Support the Prioritization of Climate Change Adaptation Measures

A further step was the application of a multicriteria assessment tool to evaluate potential measures to mitigate erosion. With this assessment tool a prioritization of measures becomes possible. This tool is not presented here in its entirety due to its complexity. However, the following matrix in Fig. 5 shows results of the use of a software tool ("PRIMATE") that considers criteria that were defined by the project partners (Heilmann and Reinhold 2017). It determines ranks with regard to criteria and their importance. These were discussed comprehensively during workshops by all project partners. As a confirmation of the preliminary discussions on potential measures and the findings achieved using the GIS service, the multicriteria analysis resulted in the highest prioritization for the rain storage reservoir. In Fig. 4 the "F+"-value in the first line of the right column (bold framed, the 5.264 marks the option "rain storage reservoir") hints at this result of the multicriteria analysis.

Taking both steps (GIS- and multicriteria analysis) into account, the next step would be the implementation of the rain storage reservoir. The example shows that the expertise in terms of GIS analyses and technology, as well as the multicriteria analysis, came from the university. This expertise supported the process of finding

Total
InvestKost
LaufKost
zeitUmse
Sichl eh

	Regenr	Mäande	Kaskad	Rückha	Entsieg	Zisterne	Dachbe	Uml_Ka	Uferbe_	Uferbe_	Gewäss	Schlam	F+
Regenr	0	0.534	0.450	0.488	0.416	0.488	0.416	0.522	0.522	0.450	0.450	0.522	5.264
Mäande	0.071	0	0.522	0.327	0.424	0.327	0.327	0.488	0.522	0.424	0.255	0.203	3.897
Kaskad	0.155	0.083	0	0.071	0.255	0.071	0	0.232	0.522	0	0.083	0.190	1.666
Rückha	0.118	0.278	0.534	0	0.436	0.319	0.353	0.266	0.522	0.471	0.374	0.287	3.962
Entsieg	0.190	0.181	0.350	0.169	0	0.134	0	0.266	0.522	0.118	0.118	0.350	2.402
Zisterne	0.118	0.215	0.534	0.287	0.436	0	0.436	0.350	0.606	0.215	0.215	0.287	3.703
Dachbe	0.190	0.181	0.606	0.253	0.606	0.169	0	0.350	0.606	0.278	0.181	0.350	3.773
Uml_Ka	0.083	0.118	0.374	0	0.339	0.255	0.255	0	0.290	0.339	0.339	0.118	2.514
Uferbe_	0.083	0.083	0.083	0.083	0.083	0	0	0.083	0	0.083	0.083	0.083	0.751
Uferbe_	0.155	0.181	0.522	0.134	0.424	0.134	0.327	0.232	0.522	0	0.083	0.350	3.070
Gewäss	0.155	0.350	0.522	0.232	0.488	0.390	0.424	0.232	0.522	0.488	0	0.266	4.074
Schlam	0.083	0.402	0.416	0.255	0.255	0.319	0.255	0.488	0.488	0.255	0.339	0	3.560
F-	1.404	2.610	4.918	2.304	4.167	2.611	2.797	3.513	5.650	3.126	2.525	3.010	

Fig. 5 Result of a multicriteria analysis of potential measures to mitigate erosion in a small river catchment

a consensus and simultaneously documents the prioritization in a transparent and comprehensible manner. This was appreciated enormously by all project partners. It can be interpreted as an important contribution of a university to achieve sustainability of climate change adaptation measures, a field of high practical and societal significance.

4 Conclusions

Climate change adaptation is a very complex and multi-faceted issue that is appreciated highly in society. Universities as organizations can get involved in the development of climate change adaptation measures in different ways. The goal, however, is to integrate "science", on the one hand, and stakeholders outside university on the other, e.g. in so-called living labs, in order to develop sustainable concepts and solutions. On the whole universities have (a) an educational mandate, they have to (b) develop themselves into sustainable organizations and they must (c) develop new expertise and methods for society outside universities to adapt to climate change in a sustainable way. Universities should take up all these different roles. They should actively advertise their competence not only in education, but especially in society at large and make known that they are willing to facilitate transdisciplinary projects with stakeholders in local communities, towns, forest and agricultural administrations, districts, larger administrative fields etc.

Acknowledgements Thank go to the Federal Ministry of Environment, Nature Conservation and Nuclear Safety of Germany who funded the BebeR project. Special thanks goes to Maria Lobe van Camp for supporting in English translation.

References

Bacow L (2020) Tackling climate change. https://www.harvard.edu/tackling-climate-change. Accessed 5 Feb 2020

Findamasters (2020) Master degrees in climate and climate change. https://www.findamasters.com/masters-degrees/climate-and-climate-change/?30gjUh1

Green Office Mouvement (2019) What is a sustainable university? https://www.greenofficemovement.org/sustainable-university/. Accessed 30 Dec 2019

Heilmann A, Reinhold S (2017) Evaluation of a transdisciplinary research project for a sustainable development. In: Filho WL, Skanavis C, do Paço A, Rogers J, (Hrsg.) (eds) Handbook of theory and practice of sustainable development in higher education, vol 2. pp 201–214

IZNE (2018) Interdisziplinäres Zentrum für Nachhaltige Entwicklung (IZNE) -Teilprojekt A 1: Multikriterielle Mehrzielentscheidungsunterstützung für Biomassenutzungskonzepte. http://www.bioenergie.uni-goettingen.de/?id=8. Accessed 9 Jan 2020

LDACMP (2020) Local people devolution & adaptaive collaboration management programme. https://www.cifor.org/acm/methods/mca.html. Accessed 22 Jan 2020

Masterstudies (2019) 10 MSc programs in climate change. https://www.masterstudies.com/MSc/Climate-Change/. Accessed 2 Feb 2020

Master and More (2020) Das Master-Portal Deutschlands. https://www.master-and-more.de/nc/mastersuche/suchen/5/15/0//%5BKlimawissenschaften%5D/

Mayo N (2019) How green is my university? https://www.timeshighereducation.com/features/how-green-my-university#survey-answer. Accessed 13 Feb 2020

Nakicenovic N, Swart R (Eds) (2000) Emission scenarios, IPCC 2000, Cambridge University Press, UK. Cambridge CB2 2RU, England

PdT (2020) 2nd level master. Polytecnico do Torino. http://www.diati.polito.it/en/focus/climate_change/2nd_level_master

Pundt H, Heilmann A, Scheinert M (2017) Assessing vulnerabilities as a step toward climate change induced hazard preparedness. Int J Safety Secur Eng 7(2):137–146

SDG Knowledge Platform (2020) Sustainable development knowledge platform-sustainable development goal 13, take urgent action to combat climate change and its impacts. https://sustainabledevelopment.un.org/sdg13. Accessed 29 Mar 2020

UBC (2018) Campus as living laboratory. University of British Columbia. https://sustain.ubc.ca/our-commitment/campus-living-lab. Accessed 29 Jan 2020

Exploring New Technology's Meaning for a Sustainable Future via Collaborative Science-Fiction Prototyping: A Novel Method for the Engineering Curriculum

Johanna Kleinen and Lisa Kurz

1 Introduction

Our society is increasingly facing the challenges of a volatile, uncertain, complex and ambiguous world (Mack et al. Mack and Khare 2016) with wicked problems (Rittel and Webber 1973). This new reality is characterized by economic, ecological and social interdependencies, oftentimes making it hard to see how one's own actions can make an impact towards a desirable future. Issues like climate change seem too complicated, the individual too powerless and knowledge barriers too high for any single person's engagement to make a difference.

To adapt and progress as a society, we often rely on new technology, without realizing that the technology available to us is not fixed to a default by nature, but a product of human engineers. It is up to them to devise effective solutions to the right problems without creating new ones. This increasingly necessitates multi-dimensional solutions, designed to fit the systems and human needs they interface. To affect a future our society can thrive in, these also need to be sustainable and work in alignment with the UN's Sustainable Development Goals.

Universities today educate the engineers who will develop the technologies available to build our future. How can they teach the future-oriented, collaborative and interconnected way of thinking and working needed to develop sustainable solutions? How can they equip these individuals with tools and skills tailored to the entangled

J. Kleinen (✉)
Mercedes-Benz AG, ARENA2036 Research Campus, Pfaffenwaldring 19, 70569 Stuttgart, Germany
e-mail: johanna_maria.kleinen@daimler.com

L. Kurz
Institute for Human Factors and Technology Management IAT, Fraunhofer Institute for Industrial Engineering IAO, University of Stuttgart, Nobelstr. 12, 70569 Stuttgart, Germany
e-mail: lisa.kurz@iat.uni-stuttgart.de; lisa.kurz@iao.fraunhofer.de

W. Leal Filho et al. (eds.), *Universities, Sustainability and Society: Supporting the Implementation of the Sustainable Development Goals*, World Sustainability Series,
https://doi.org/10.1007/978-3-030-63399-8_22

nature of sustainability challenges and necessary to reach out and create a desirable future for our society?

There is a wealth of knowledge on making university operations sustainable (Leal 2009) and on integrating knowledge on sustainability topics into the curriculum (Leal 2018). Our research, however, focusses on building a bridge between the student's identity and practices as budding professionals as well as conscious and active citizens shaping the future of our society.

Science-Fiction Prototyping is a method used to generate research ideas for new technology while considering their implications for people and society (Johnson 2011). It does so by guiding its users in writing and reflecting upon a "futuristic" short story. This allows participants to jump to novel ideas, identifying new research directions as well as interdependencies.

Originally, Science-Fiction Prototyping was "out-sourced" to professional science-fiction writers or used by individual scientists to add to their own research practice. It has also been turned into different formats, for groups of subject matter experts (Kymäläinen 2015) and for Business Innovation (Wu 2013). However, Wu notes that there is need for more educational material on the method as well as for a refinement of structures and rules.

We developed a game-based intervention to address this gap: "Collaborative Science-Fiction Prototyping" (CSFP) workshops about "the sustainable factory of the future", previously implemented for cross-organizational groups of researchers and developers in an open innovation network organization. CSFP provides a recipe for how to turn the telling of a short story into an easy, reliable and replicable process and a truly collaborative effort.

Serious games' primary purpose is not to entertain (Abt 1987), but to harness the benefits which games and play provide, e.g. higher motivation, engagement (Rigby 2014) and creativity; like gamification, "the use of game elements in non-game contexts" (Deterding et al. 2011). Note that games are formalized systems with a goal, while play "is not tied to objects but brought by people to the complex interrelations with and between things that form daily life" (Sicart 2014). The concept of meaningful gamification reconciles this by using "gameful and playful layers to help a user find personal connections that motivate engagement with a specific context for long-term change" (Nicholson 2015).

Our CSFP concept builds on this, aiming to create play through a storytelling game. It adds to the existing knowledge by detailing a successful implementation of this concept, which combines the motivational benefits of game-based methods with future-oriented ideation and reflection on the connection between society and technology in one offering. For the application at universities, we adapted our original format to fit a larger student group by substituting a card game for the facilitator-led game. This study's contribution is therefore twofold: The re-design of the SCFP method, and the analysis of its test as an offering to introduce the topic of sustainability into an engineering and technology management curriculum. Our aim is to foster the intrinsic motivation, perspectives and skills enabling our future engineers to develop the technology and systems for society's sustainable future.

2 Study Design

The CSFP method combines the SFP approach with game mechanics inspired by the game "For the Queen" by Roberts (2019). It is one of the most prominent games leading players in a collaborative storytelling process: Every player in turn has to answer a question and the next player has to build on the previous contributions with their answer. In this way, a shared understanding of imaginary places, people and problems is created, taking on a dynamic of its own.

In "For the Queen", the questions are printed on cards and drawn randomly. For our purposes, it was important for a clear story to emerge, so questions based on those used by Wu (2013) were sorted into 8 phases ("The Seven Point Plot" by Algis Budrys (Johnson 2011) plus an extra step focusing on technology). In addition to this, questions with the function of increasing the immersion into the imagined world were added.

In our previous work, the questions were posed and order was enforced by a facilitator ("game master"). This smoothed and optimized the story's flow, allowing us to fluidly tailor the questions to the emergent input from the participants as well as to set the atmosphere. The biggest change in this study was the substitution of this person with question cards and written instructions. This was done in order to allow hosting several 5-person groups in parallel to fit the offering into the university's operating mode, with one to two instructors per course offering. Also, further exercises to bridge the gap between science-fiction story and real-world realizations were added.

This offering was tested with 13 master's students of technology management and engineering specializations at Stuttgart University. They must take part in 3–6 practical courses as part of their studies. These usually last two hours and are intended to connect their theoretical studies to real-world applications. We offered the workshop in November 2019 as one of these, however, scheduled for seven hours. 12 students chose this offering and another one was allocated to it. They ranged from 21 to 27 years of age, 85% of them being male and 15% female. The majority was studying technology management (77%), but mechanical engineering (15%) and energy technology (8%) were also represented. Because technology management students have to complete more of these courses, this cannot be said to be indicative of different amounts of interest in the offering.

Data was collected using mixed methods. In order to capture the information-dense storytelling process, it was recorded on audio, transcribed and analyzed using inductively generated categories. This approach was chosen because this was the first, explorative instance of using this format. To evaluate how the students received the offering and perceived its connections to sustainability, a quantitative survey was conducted at the end of the workshop. Here, participants rated how given statements applied to them on a scale on one ("applies completely") to six ("definitely does not apply"), with the additional option of "I do not know".

3 Results and Discussion

The outcomes of this study can be divided into four areas: The adapted design of the method, the effect of the design changes (as analysis of the audio tapes), the evaluation of the format's reception by the students (as analysis of the survey) and the resulting guidelines for practitioners interested in using the method.

3.1 Workshop Design

The workshop's structure is depicted in Table 1. Its main features are the skill-building, what-if, storytelling and implications blocks.

Skill-building serves the purpose of familiarizing the participants with the storytelling game mechanic[1]: One person answers one question, the next person the next question while building on the answer of the first, and so on. It also familiarizes the participants with one another, building trust.

Once the participants know how to tell a story, they determine its subject. For this, they develop a "what if"-scenario for the sustainable factory of the future. Selecting from a wide array of image and text prompts (including the UN Sustainable Development Goals), they identify shared interests and discuss possible "what if"-questions as the basis of a future world. Framing the task in this way encourages curious, imaginative and playful exploration.

During the storytelling, the questions are divided into eight phases:

- "In a future world" builds a shared vision of a world, characterized by the chosen "what if"
- "Influenced by a technology" dives deeper into how the sustainable technology works and impacts the people and society
- "Lives a person" introduces a protagonist through whose eyes the rest of the story is viewed
- "With a problem" poses that the "what if" creates not only good but also a challenge in the protagonist's life
- "He/she tries solving the problem intelligently and fails" leads to a first proposed solution
- "The situations gets worse" explores the consequences of that failed first attempt
- "And reaches a climax" pushes the story to an extreme, exploring both positive and negative outcomes before deciding on a final course of action
- "At the end, stand the consequences" steps further into the future, reflecting on how that situation played out.

Each question is printed on a separate card, which the students draw in a fixed order.

[1]Similar to how the LEGO© Serious Play© method uses skill-building.

Table 1 Workshop structure

Step	Content
Introduction	• Introduce instructor(s) • Explain your "why" • Goal of the workshop offering in the context of engineering studies • Core idea behind science-fiction prototyping as a method • Reasoning behind using it to develop ideas for the sustainable factory of the future • Divide students into groups of 4–5 • Explain that groups will be led through the workshop mainly by following printed instructions but that the instructor(s) will be available for questions throughout and check in between steps • Hand out first instructions: To (in their groups) state their subject of study and how well they know each other
Skill-building	• Hand out cards for skill-building exercises to groups • Skill-building l: "Erik" • Goal: To practice the storytelling game mechanic • Instructions explain the mechanic and cards facilitate the telling of one day in the life of an imaginary fellow student called "Erik". They build on common experiences of students and should stay realistic • Skill-building 2: "James Bond" • Goal: To practice stretching your imagination in a story and the building of a mental image of new technology • The cards ask questions following the scheme of abasic "James Bond" story, focusing on the villain's and spy technology
What if?	• One wall of the room serves as an "inspiration wall" (prepared in advance): Here, you po st printouts of: • News or journal articles, tweets, etc. concerning futuristic technologies/ latest breakthroughs and concepts which relate to the areas of: • Sustainable production and economy • Novel factory and production techno logy • Human factors in factory work and engineering o f the future • Futuristic megatrends • Articles about Science-Fiction genres, particularly those with implications on sustainable approaches (e g Solarpunk, Hopepunk, Cli-Fi, Thingpunk) • Concept art from Science-Fiction worlds respect copyright) • Every participant browses wall and takes two of these elements from the wall, which most appeal to them, then presents them to their group • Based on this input, the group discusses which premise to build the future world for their story on. Not everything needs to be incorporated, it does not need to be directly related to the elements, often the >3rd idea is the most interesting keep it as specific as possible formulate it as a "What if"-statement • Facilitators check the "What-if statements to see if they are specific enough
Storytelling	• Hand out storytelling cards to each group: the participants answer the questions

(continued)

Table 1 (continued)

Step	Content
Implications	• Still in their groups, the participants silently write research questions of interest based on the story they explored onto past- its and share them afterwards in the group • Each participant develops one project idea from these, which they would most like to start now
Sharing	• The groups briefly present their stories to each other
Stations	• Optional: Take break and recommence with energizer • The groups are dissolved and four new groups are randomly formed, each one as signed to one table • At the tables, different stations are set up with these tasks: • Sketching the blueprint of a piece of technology from their story (individually) • Writing the diary entry of a person working in/for their sustainable factory of the future, starting with "18th November. So, today something happened at work, that I didn't expect… "(individually) • Mapping out a timeline of how one of their future worlds came to be (as a group, using post-its) • Discussing how to fix what was wrong with one of their future worlds (as a group)
Reflection	• Facilitators guide discussion between whole group regarding their work with the method
Wrap-up	• Feedback round, good-bye and evaluation survey

After the storytelling, participants are asked to consider which aspects of their imaginary world would be interesting to study and to write down these "research questions". After sharing them with each other, everyone develops one final idea for a research project they would most like to engage in. This connects the science-fiction storytelling to their current life as an aspiring engineer or technology manager.

3.2 *Effects of Design Changes*

Aside from its application to the engineering curriculum at a university, the novelty of this offering lies in the redesign as a card game and the addition of further exercises in the last workshop part.

The premise was that no facilitator is needed to pose the questions to the group. Analysis of the audio tapes confirmed that the groups were indeed able to understand and follow the written instructions without such guidance.

Removing the facilitator and the strict upholding of the game mechanic allows for dynamic exchanges, similar to improvisational theater's "yes, and…" method. This allowed one group to spontaneously take their world to a place the participants would not have reached without their interaction, e.g.:

Person 5: "If the communication center breaks down, we'll have a real problem…"

Person 6: "But just as in old movies, someone will have an old satellite phone or use a phone booth."
Person 7: "And then your giraffe[2] can call you."
Person 5: "Yes, because it's from the first generation, they still have that functionality."
Person 4: "And they are keeping them, because they are so cute, you don't throw them away."
Person 7: "And for sustainability reasons! You don't throw them away, you use them. When they're broken, you recycle them and take every part out."
Person 6: "And cobbled together like patchwork ... and you can see it's first generation from the yellow color. Because most are yellow or blue, because they are cobbled together from different robots."

This type of emergent storytelling energizes a group and unlocks the group's creativity. Because it is unpredictable, however, the game's mechanic was implemented to formalize such interactions.

Group A completely abandoned the rule that each question is answered by one designated participant with the power to decide. Instead, they collectively discussed each question. This led to the common pattern of behavior in which a few group members talk the most and make the decisions. This group stayed at a superficial and cerebral level. They related the prompts from the question to previous knowledge more than any other group, both regarding Science-Fiction, current discourse regarding sustainability issues and content of their university courses (e.g. "Now we're at Blade Runner status.", "I know that's already possible, my friend has a PhD in biochemistry.").

More than the other groups, group A talked in first person plural (e.g. "if we had the solution, we could make real money"). When the prompts instructed them to choose a protagonist, they did, but never considered that person's perspective, e.g.:

- Person 1, reading the question from the card: "When does he/she notice for the first time, that this represents a problem for them?"
- Person 2: "When did we first notice this?"
- Person 3: "... we already noticed it."

As a result, collaborative storytelling did not take place. Group A still developed a vision for a sustainable factory of the future which they got invested in, and considered its implications. They discovered unexpected societal consequences of their technology, like groups B and C. Therefore, personal learning has taken place. Yet they did not equally tap the brainpower of all team members nor experience the effect sustainability-related challenges can have on individuals: "These stupid person-questions, for real. This is about the sustainable *factory*".

An interpretation of why this happened is connected to the very nature of play and games: The students in group A performed a task they were given as a university assignment: to answer a set of questions on cards. This was not voluntary and they

[2] A giraffe-shaped robot from a visual prompt that was picked from the inspiration wall.

completed it as quickly as possible. Games as systems of rules can be executed regardless of whether the participants want to do it or not. Play, though, can only occur voluntarily (Carse 1986; Huizinga 1949). Deterding (2012) described play as "the open-ended recombination of behaviors and meanings in a safe space of 'as-if' and 'what if'" and this is what the method attempts to channel. However, these play-centric benefits are only reaped if "play" takes place. The experience of this group implied that they did not "play". In contrast, group C, which spent the most time immersed in storytelling, actually referred to the task as "a game", perceiving themselves to be playing.

Groups B and C also -at times- deviated from the strict storytelling structure and branched out into discussions. In contrast to group A though, they always returned to the structure after one aspect was discussed and they upheld the decision-making power of the person whose question it was. For example, either the person with the question would ask for the others' input (e.g. "Do you have an idea who could also be on the losing side?") or the others would address clarifying questions to them: "Will these also be constructed by the AI?". The answers given by the one in charge of the question were accepted as "real" and fixed, reacting to it with statements like "Oh, it's getting worse and worse."

Moreover, groups B and C referred to the imaginary world they were creating as though it was real: " As soon as he leaves the factory, there is a big park and, a little to the right, the factory has a larger lake. There is a bench, which he actually likes to sit on quite a bit." They made little use of the subjunctive, in contrast to group A.

Both groups spent more time on the storytelling than group A and added more depth to the storytelling than specified in the instructions. They chose to name their protagonist of their own accord. Group C even requested additional material: "We are starting to need a map of the galaxy" and they considered multi-sensory aspects of their world: "What does it sound like, when an AI laughs?" and reported experiencing losing track of time, a testament to their engagement with the task.

To aid with this kind of immersion into the world of the fiction, a facilitator serves two important purposes: Enforcing the storytelling mechanic, and setting the foundation for the desired atmosphere by tailoring the questions and how they are framed to the individual group dynamics, which stabilizes the method.

Most participants are not used to being allowed to weave an imaginative story. This is something they address among themselves: "We need to do this seriously.[3]" Group C explicitly gave themselves the permission to follow their impulses several times:

- Person 4: "I think we're answering the questions a little too excessively, well, more excessively than it is."
- Person 5: "Probably."
- Person 6: "Probably it's not bad, then it's becoming a story."

They also stated: "we can think a little more sci-fi" and "well, it's science-fiction…". A facilitator can reassure and push them here.

[3]German: "Seriös", meaning seriously with strong connotations of respectability.

The last part of the workshop consisted of four extra tasks:

- Sketching blueprints of the sustainable factory technology aimed at encouraging the participants to think about how the futuristic technology might actually work. It predominantly resulted in sketches of the core process of the factory, not the technology itself. Every sketch contained labels providing explanations but the design and workings of the technological components were not detailed further, only their function.
- Writing diary entries explored what would result from changing the participants' perspective from third person to first person, i.e. stepping into the shoes of affected people and seeing the science-fiction world through their eyes. The outcomes included:
  - Describing the main plot challenge (before and after its resolution) through the eyes of the affected (54%)
  - A previously undiscussed solution for the main challenge (8%)
  - A technology-focused production challenge connected to the main plot that was not mentioned previously (8%)
  - New social aspects resulting from the sustainable technology not mentioned before (8%)
  - Exploring a side story (23%)
  - 62% contained expressions of hopes and feelings.
- Developing a time-line for the realization of their sustainable factories of the future was aimed at getting the students to think about how to make their future world possible. Overall, the participants stayed on a high level, yet all cases included one extra step with details not included previously.
- Discussing how to fix the main problem inherent in the imagined sustainable factories did not yield new insights and was perceived as repetitive.

3.3 Evaluation: Results of the Survey

To assess how the participants received the offering, a survey was conducted at the end of the workshop. The students rated whether they agreed with a range of statements on a six-point Likert-type scale in which the one was "applies completely" and 6 "absolutely does not apply". This survey analysis is divided in four parts: The first, regarding its aspects as a university teaching method; the second, its perception regarding sustainability-related aspects; the third, its effects as a collaborative method and, finally, its effects as an innovation method.

The CSFP offering was well received as a teaching method: When asked whether they would recommend the format, 62% scored it a one (agrees completely) and the rest either a two (23%) or three (15%). 65% agreed completely that they would participate again (with a different selection of topics, if necessary), with the rest divided between a two (31%) and a three (15%).

Furthermore, 85% of the participants completely agreed that they could actively participate, and 15% answered with a three. 92% completely agreed that their previous knowledge sufficed to actively participate and also that the requirements of the format were achievable for them, while the remaining 8% scored it a two. This shows that the instructions were clearly understandable and followable, as well as that the format was active instead of passive. Participants also stated that the format was more fun than other teaching methods (92% completely agree) and they felt that the majority of participants were engaged in the work (46% for each—one and two—; 8% chose three).

In contrast to this, only 54% of the respondents agreed completely that the subject of the format interested them, with 31% following at a two and 15% at a three. Their answers regarding sustainability-related aspects were also mixed: The participants also positioned themselves between a one and three with regard to the question "The format sensitizes one for the subject of sustainability". 46% completely agreed that the format sensitizes you for the subject. "Through the format, I gained new insights into the topic of sustainability" and "The format activates me to engage with the topic of sustainability beyond this format" received replies ranging from one to four, with no clear preference.

62% of the participants completely agreed that the format enables interdisciplinary cooperation (with 15% scoring a two and 23% a three). On top of this, 54% completely agreed that the format creates a frame within which one can talk about societal challenges (with 38% scoring it a two and 8% a three). This is important, because sustainable challenges are interconnected and thus one needs to create interdisciplinary and even shared technological and societal solutions for them.

Nevertheless, only 23% of the participants completely agreed that the methods employed within the format would be fruitful for their (future) work. Still, the majority of 62% scored it a two, and only 8% agreed with three. This might be related to their rating of the question whether the format expanded their professional skills, from which no clear picture emerges: Here, the responses ranged from a one (8%) to a five (23%), two and four both received 15% and the majority of 38% chose an agreement of three out of six. The participants seem to be unsure whether using CSFP to create visions for the sustainable factory of the future would be relevant for their future work and skillset. Verbally, the students expressed that they did not believe that the topic of sustainability was relevant for their studies *currently* but, in their opinion, definitely should be. Therefore, there is a discrepancy between the skills they think they need in order to be successful at university or to find employment, and the skills they think they need in the world of the future, where sustainability will be highly important.

To evaluate the collaborative effect of the method, the students were asked how much they agree with three statements: Whether the format enables making new contacts, whether the group grows together during the format and whether a feeling of community develops during the format. All participants either completely agreed or scored their agreement a two. Naturally, successful collaboration forges groups together. Still, not every in-class work group enjoys their cooperation. This decidedly positive result is in line with the known community building effect of games and play (Bowman 2010; Caillois 1958).

Participants also made new contacts and said that the skill-building exercises acted like "teambuilding, but not awkward", because they very naturally built on universal shared experiences of students. The participants filled them with fragments of their own experiences in which the others could recognize themselves.

Lastly, CSFP also attempts to foster innovation, because it considers sustainable *science-fiction* technology. One prerequisite for collaborative innovation to take place is to exchange ideas. All participants completely agreed that they were able to share ideas openly and without reservation. This high level of trust is a sign that the format managed to establish a safe space.

The next important aspect is that new ideas often emerge when previously disconnected knowledge collides; therefore, being exposed to a variety of new and different ideas is conducive to innovation. When asked whether the format provided insights into manifold ideas and thoughts, 46% agreed completely and 54% agreed with a two out of six. Furthermore, 23% completely agreed that the format offers them inspiration through the ideas of others, with the majority of 69% agreeing with a two and only 8% with a three, suggesting that the format provided an environment for innovation.

In addition to this, innovators need tools and structures to develop their ideas. We asked the students whether our methods provided an inspiring possibility to think about things in a new way. The response was predominantly positive, as 31% completely agreed, the majority of 46% agreed with a two and 23% with a three. Finally, the students assessed whether the format enables them to develop particularly creative ideas, which would not have been created in prevalent teaching formats: 62% completely agreed, 31% agreed at a two and 8% at a three.

Summing up, the evaluation of the format showed that card-based SCFP is suitable as a teaching format, as well as a collaborative innovation method. Its application to introduce sustainability-related topics into the engineering curriculum needs to be discussed in more detail: It sensitizes participants to the topic and creates engagement with it through the groups investing time and energy into their own idea and seeing it play out in a story. It also allows the students to generate tangible ideas for research projects they could tackle in the present in order to create a sustainable future. However, the survey results and verbal feedback showed that such an isolated offering is hard to fit into the mental framework of their studies, i.e. how does the topic relate to the rest of their curriculum and to the challenges they will face in the work force?

3.4 Guidelines for Practitioners

We derived a set of guidelines for practitioners interested in incorporating the method into their own work, based on the above analyses of both audio recordings and survey results.

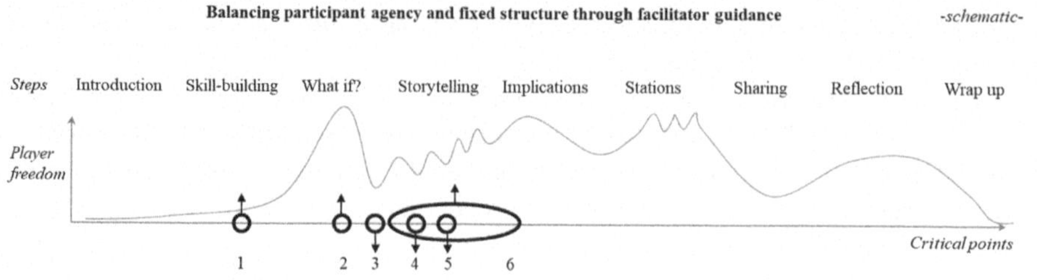

Fig. 1 The critical points for steering the participants' freedom

CSFP's main challenge is to balance the participants' agency with achieving a reliable result. Giving the students freedom in how to tell the story opens up space for surprising discoveries. These will feel more profound and owned by their creators. However, without sufficient structure, the method will not work reliably, and is in danger of yielding empty talk. The largest part of the structure is upheld through the inspiration wall, the question cards and the instructions. Still, the instructor's job is to monitor the groups' progress and intervene if necessary. If there are multiple groups of five, the instructor must keep an overview and be prepared to intervene at the following points (see Fig. 1):

1. Encouraging the participants to think "science-fiction", stretching their imagination, after skill-building one
2. Letting them explore several different options for a "what if"-statement, involving all group members
3. Keeping the "what if"-statement narrow
4. Making sure that the groups take the perspective of their protagonist
5. Checking that the protagonist's problem is relevant for the future world
6. Reminding the participants that stretching their imagination and considering human factors is the goal; also, enforcing the storytelling mechanic in case discussions start to dominate.

4 Conclusion and Outlook

Overall, the CSFP method provides a suitable first point of contact to introduce sustainability into the engineering and technology management curriculum. It enables students with little previous knowledge to open their minds, to create and explore a collaboratively generated science-fiction story revolving around sustainable factories and to reflect the thought experiment. We did not evaluate the creativity of the generated ideas. Instead, we prioritized learning about interconnected nature of sustainability challenges and tapping into the participant's intrinsic motivation to engage with the topic as its output.

The participants accepted the CSFP method well, both as an offering in their curriculum and as a collaborative innovation method. They identified it as an active and inter-disciplinary format and the majority would attend it again.

However, it was difficult for the participants to connect the sustainability content with the rest of their studies. We therefore recommend embedding CSFP into a larger context: It can easily be expanded by adding a phase for literature/inspiration research, reflection of how the chosen what-if would influence each of the Sustainable Development Goals, role-play scenarios and physical prototyping activities. One could also repeat the steps with different content; by varying the input via the inspiration wall, different sessions could be held, centering on different UN Sustainable Development Goals, technology fields, roles of the protagonist or domains within the Sustainability Innovation Cube (Hansen et al. 2009). Augmenting the sketching and timelines can be coupled with classes from construction design and project management, or business model generation. Therefore, it would be an ideal tool to include in project-based learning approaches. Expanding the format beyond the boundaries of the engineering faculty, it could be used to bring together students from all disciplines to work on multiple aspects of one future world; inviting participants from outside the university could connect the students with players from industry and politics.

CSFP enables the participants to engage with the topic of sustainability in a playful and creative way, while still yielding usable insights. This is an important way of changing the perception of sustainability-oriented technology development in our future engineers' minds: To experience it as opening up new creative opportunities to shape the world according to a new paradigm, instead of viewing it a barrier imposed by policy makers to hinder innovation. The students learn to create mental pictures of possible futures they want to live in. And only what you see, can you create.

References

Abt CC (1987) Serious games. University Press of America, Lanham

Bowman SL (2010) The functions of role-playing games: how participants create community, solve problems and explore identity. McFarland & Company, Jefferson, NC

Caillois R (1958) Man, play and games. University of Illinois Press, Chicago

Carse JP (1986) Finite and infinite games. Free Press, New York

Deterding S, Dixon D, Khaled R, Nacke L (2011) From game design elements to gamefulness: defining "gamification". In: MindTrek '11. Proceedings of the 15th international academic mindtrek conference: envisioning future media environments. ACM Press, New York, pp 9–15

Deterding S (2012) Playful technologies. In: Wiedermann C, Zehle S (eds) Depletion design: a glossary of network ecologies. Institute of Network Cultures, Amsterdam, 123p

Hansen E, Grosse-Dunker F, Reichwald R (2009) Sustainability innovation cube - a framework to evaluate sustainability-oriented innovations. Int J Innov Manag 13:683–713

Huizinga J (1949) Homo ludens: a study of the play-element in culture. Routledge and Kegan Paul, London

Johnson BD (2011) Science fiction for prototyping: designing the future with science fiction. Morgan and Claypool Publishers, San Rafael

Kymäläinen T (2015) Science fiction prototypes as design outcome of research. Ph.D, Aalto University
Leal Filho W (2009) Sustainability at universities: opportunities, challenges and trends. Peter Lang Scientific Publishers, Frankfurt, p 340p
Leal W (2018) Implementing sustainability in the curriculum of universities. Springer, Berlin
Mack, O., and Khare, A. (2016). "Perspectives on a VUCA World" in Mack O., Khare, A., Krämer A., and Burgartz, T. (eds.) *Managing in a VUCA World*. Springer, Switzerland. 3p
Nicholson S (2015) A RECIPE for meaningful gamification. In: Reiners T, Wood L (eds) Gamification in education and business. Springer, Berlin, pp 1–20
Rigby CS (2014) Gamification and motivation. In: Deterding S, Walz ST (eds) The gameful world. MIT Press, Cambridge, pp 113–138p
Rittel HWJ, Webber MM (1973) Dilemmas in a general theory of planning. Policy Sci 4(2):155–169
Roberts Alex (2019) For the queen [Card game]. Evil Hat Productions, Silver Spring
Sicart M (2014) Play matters. MIT Press, Cambridge, p 2p
Wu H (2013) Imagination workshops: an empirical exploration of SFP for technology-based business innovation. Futures 50:44–55p

Johanna Kleinen (M.Sc.) develops tools to empower researchers and developers from industry, start-ups and academia to collaboratively create sustainability-oriented innovations for the production of the future. After studying physics, she went into innovation consulting before turning to research. As part of the Research Campus initiative of the German Ministry of Education and Research, she is now studying Open Innovation methods and how to innovate for sustainability at Mercedes-Benz AG. Her sphere of action is ARENA2036, a Research Campus dedicated to mobility and production. Here, she leads the Innovation Forge—connecting the cross-organizational researchers working on innovation through collaboration and on human/organizational factors in the factories of tomorrow—and takes part in the project Agile Innovation Hub.

Lisa Kurz (M.A.) does research in the fields of organizational and corporate culture, working environments, sustainability and transformation processes. After studying culture and media education and empirical cultural studies, she started her scientific activity at the Institute for Human Factors and Technology Management IAT at the University of Stuttgart and the Fraunhofer Institute for Industrial Engineering IAO. Her work focuses on organizational ethnographic case studies in the context of the Research Campus ARENA2036 and on the development of innovative forms of cooperation and collaboration tools for industry and science. She leads the subproject "Cooperative Innovation Culture" in the project Agile Innovation Hub, which is funded by the German Federal Ministry of Education and Research.

"The Value of Mixed Teaching Methods in the Context of the Sustainable Development Goals—An Evaluation of a Compulsory Course Using the Example of a Large Group of Students"

Nadine Husenbeth and Merle Katrin Tegeler

1 The Sustainable Development Goals

The first earth summit was held in June 1992 with 178 participating countries in the city of Rio de Janeiro (Brazil). The countries came together to develop the Agenda 21 which had the goal to produce a global cooperation for sustainable development in humans and climate (United Nations Department of Public Information 2019). For example, the agenda included sections to save global resources and the strengthening of disadvantaged human groups (Mccammon 1992). In 2000 the millennium summit decided unanimously on the millennium agreement. This agreement includes eight different millennium development goals (MDG) to reduce extreme poverty by 2015. The main goal was to minimize the group of people who live below an average of \$1.25 on a daily basis (by half). This was achieved before the deadline of 2015, even if the agreement did not have legally binding mechanisms for regulations. The economic growth before 2015 in countries like china, who have a high population density, could be the reason why they reached the goal earlier. In 2012 the united nation conference of sustainable development decided the start of a new duty to achieve a bigger frame of sustainable development. They called it the sustainable development goals (SDG) which should be built upon the outcomes of the Agenda 21 and the MDG's. In 2013 they built a task force of 30 members for the development of these goals. 2015, after a three years' preparation process the sustainable development goals were implemented in the Agenda 2030. This agenda included 17 main goals and 169 lower goals, which cover factors like the reduction of world hunger and

N. Husenbeth · M. K. Tegeler (✉)
University of Bremen, Enrique-Schmidt-Str. 1, Postfach 330 440, 28359 Bremen, Germany
e-mail: tegeler@uni-bremen.de

N. Husenbeth
e-mail: nadine.husenbeth@uni-bremen.de

W. Leal Filho et al. (eds.), *Universities, Sustainability and Society: Supporting the Implementation of the Sustainable Development Goals*, World Sustainability Series,
https://doi.org/10.1007/978-3-030-63399-8_23

optimization of climate politics (Federal Government 2018). The sustainable development goals cover five key messages, which were agreed upon between Germany and the European Union. Germany focuses on humans, the planet, prosperity, peace and cooperation. To this day the SDG's are supervised as well as advanced by the "High-level Political Forum on Sustainable Development" and the "Division for Sustainable Development Goals" (DSDG).

1.1 Key Factors of the SDG's

Some key factors need to be achieved before the implementation of the SDG's can be successful. For the start, people need to know that the SDG's show overlapping between each other (Gaffney 2014). An example would be economic growth and the focus on new innovations. Both SDG's have a different definition but the investment in one of the SDG's also grants progress in the other. Another key factor is the deadline of the goals. The SDG's are included in the Agenda 21, in which the lower goals are meant to be achieved till 2030 (Griggs et al. 2013). Besides that, people need to watch the increasing population in the world. By 2050 there will be nine billion humans living on our planet. Such a population needs to be regulated in order to save the planet. For this reason, we need to see the SDG's not only as a finish line but also as an urgency (Robert et al. 2005). Furthermore, the people and the politics need to change their actions and behaviors. Also, the global markets need to work on the goals of the SDG's. Success will be in cooperation instead of competition. To compare and save the results of the efforts from every single country, the High-level Political Forum on Sustainable Development collects and evaluates the given data (Federal Government 2018). Countries are not allowed to falsify or beautify their results and need to pay attention to the regulations which were agreed upon in the Agenda. Most importantly the lower goals need to be clearly defined. Gaffney (2014) stated in his article some options of boundaries for lower goals. For the SDG named "clean water and sanitation" it could be an option to reduce global water drain to 4000 cubic kilometers in a year. As a final remark it should be noted that every country needs to engage in cooperation to develop a sustainable future for our world population and the planet.

1.2 German View on the SDG's

After the federal government decided in 2015 on the SDG's in Germany, they published 2017 the pamphlet "German Sustainable Development Strategy" as a first action plan (Federal Government 2018). This pamphlet, updated 2018, works as a summary on how to act on a regional basis across the country. The next German report will be published in 2021. Next to this report the High-level Political Forum on Sustainable Development holds a conference every single year. In the report 2021,

Germany wants to show their progress concerning the SDG's and what the next goals are for the future. Besides the country's internal strategies, the federal government also engages on an international level in the progress of the SDG's. The Ministry of environmental issues and the ministry of developmental issues started the initiative "Partners for Review" to cooperate with international representatives of the science community as well as the economic community and many more. The initiative should ensure the implantation and monitoring mechanisms of the SDG's. For this year the focus of Germany was on the SDG of equal treatment of the genders and the inclusion of every human being on national and international level (Federal Ministry for Economic Cooperation and Development 2019).

The SDGs thus form a core element of international and national policy against climate change and in any case need to be integrated into academic teaching in Germany. In order to achieve this, different teaching methods of the good practice event were established. In the following, these four fields will be explained in more detail.

Field 1: Basics (disciplinary—receptive)

One of the earlier approaches to learning theories is John B. Carroll's (1989) model of school-based learning, which assumes that knowledge is taught in a disciplinary and receptive way. In this model he distinguishes five different variables which are necessary for success at school. The first three refer to the time a learner needs to learn knowledge, the remaining two describe variables related to learning success. Carroll refers to the time it takes a learner to learn a particular course, to the extent that it is sufficient to master it successfully, as the "aptitude". A high "aptitude" assumes that under the best conditions (motivation, explanation, etc.), the learner needs very little time to master a particular topic. A low aptitude assumes that the learner needs more time than the average under best conditions. "Opportunity to learn" is the second variable in Carroll's model and refers to the time a learner has available to learn a particular subject. If this variable is related to today's university life, it can be assumed that students have too little time to learn and therefore the stress level of these students increases (Töpper 2016). "Perseverance" describes the motivation variable in Carroll's model and refers to the length of time a learner is willing to learn. It is the shortest of the three time variables and refers to a learner's ability to learn, minus the time during which he is unwilling to learn. The variable is based on the theory of operant conditioning by B. F. Skinner (Iversen 1992). Skinner describes as a supporter of behaviourism that living beings acquire behaviour better if they are either rewarded or punished for it. If this is now applied to the present day, it can be said that students have a higher motivation to learn if they can get a good grade for it or if they want to avoid a bad one. The fourth and fifth variables in Carroll's model are "quality of instruction" and "ability to understand this instruction". The two variables are difficult to separate in this model because Carroll does not really describe how he defines the quality of the statement. However, the example of a lecture can be used to explain this very well. Carroll describes that the quality depends on the clarity of the task. If the task of the respective lecture is clear to the student and he can work on the required material for preparation, he needs less time to understand the learned material. However, if the quality of the materials or that of the lecture is poor, the student needs longer to learn the knowledge. Carroll emphasized in his model that the student must be explicitly told what he has to learn and how he has to structure the workload so that he has optimal conditions to acquire the knowledge. Nowadays it is often discussed whether lectures are still a good way to pass on knowledge. It is often said that the more complex the lecture is, the less students understand the material to be learned (Bender 2019). In addition, the question arises as to whether, in the course of digitisation, the presence of students on site is still relevant or can be newly implemented through online courses.

Field 2: Consolidation (disciplinary—reflexive)
This field is mainly concerned with the teaching method in seminars. The structure of a seminar is often similar to school lessons, therefore the Offer- and Use-Model by Helmke (2008) can be used as an explanation. According to Helmke, the offer is defined in such a way that knowledge is presented to the student in class. This knowledge is processed and reflected upon together with the lecturer. The lecturer defines goals and demands on the student, but also gives the student a framework in which he or she wants to use the material. The variable of benefit is described in such a way that the student can expand and reflect on his knowledge in the seminar as well as outside of the seminar, if they want to. Students should learn better in this way, as they have to teach and develop knowledge themselves. The benefit variable of the offer and use model is also supported by psychological motivation theories. The fact that students can choose what they want to use for learning and what they do not want to use increases their intrinsic motivation (Deci and Ryan 1993). It is also important to note here that students' self-determined learning is enhanced by learning methods such as the seminar. Self-determined learning is an important intrinsic motivation for successful knowledge absorption (Ryan et al. 1990).

Field 3: The extension to inter- and transdisciplinary knowledge Examples of this would be courses of study that have different subject combinations, such as business psychology or business informatics. In this way, students from two different basic courses of study come together to acquire a greater wealth of knowledge through the exchange of knowledge. Psychological theories speak here of knowledge expansion through training of neural networks (Braun 2013). The basic knowledge that the students have acquired in their basic courses of study is passed on and at the same time new knowledge is acquired, which leads to an exchange of experts. In the theory of neural networks it is described that knowledge is best stored when it is linked to existing knowledge (Ritter et al. 1991). Our brain works even more efficiently if old networks can be eliminated. This is known in science as synapse elimination and describes the destruction of old knowledge stores to create new ones (Frey 2017). Students in courses of study that are combined from transdisciplinary areas often do not need some sub-areas from their basic studies any further and can thus clear out existing knowledge in the brain to make room for new neural connections. Eric Kandel (Kandel and Bischoff 2012) describes this form of knowledge optimization in his book. He assumes that by reducing neural networks, new ones can be created.

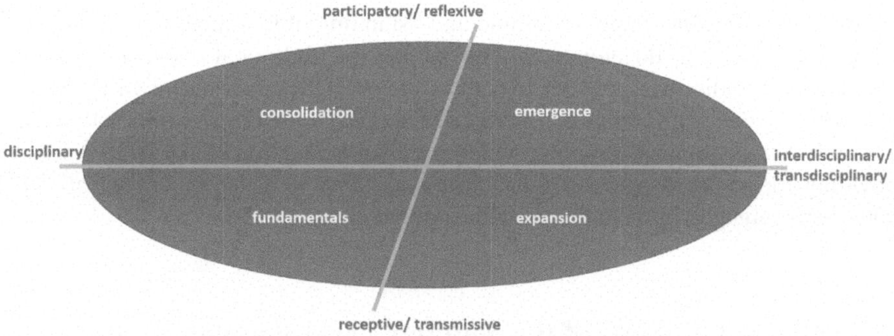

Fig. 1 Four dimensions of academic teaching (Source: Bellina et al. 2018)

Field 4: The Emergence Room as a Source of Innovation

This field is mainly concerned with the reflection of jointly acquired knowledge. Once again, the focus is on students from business psychology or business informatics. Then it becomes apparent that these students gather their knowledge from different basic studies. This collected knowledge then forms a new knowledge base and should be reflected. Reflection is the most important term here. Only by talking and discussing about existing knowledge can people build up a new treasure trove of knowledge (Sosa 2009). In many fields of research this is already done with "think tanks" and "knowledge hubs" in order to find new problems and to solve them (Bellina et al. 2018). However, such measures not only stimulate knowledge transfer, but also creativity and problem analysis (Medvetz 2008). This helps students to develop competencies in and alongside teaching (Kuhn et al. 2016). In this field, the focus is not only on the pure extension of basic knowledge, but also on the improvement of fluid and crystalline intelligence. These two types of intelligence are based on Raymond Cattell's two-component theory (Kleinheinz et al. 2019) and describe factual intelligence (crystalline) and fluid intelligence (fluid). Crystalline represents the pure knowledge that a student will acquire about life, and the fluid properties in turn are mutable (e.g. flexibility, creativity) (Fig. 1).

2 Good Practise: The Seminar *Sustainable Management* at the University of Bremen

Since 2001, the seminar *Sustainable Management* has been firmly anchored in the curriculum of the Bachelor of Economics at the Bremen University. Generally, the objective of this course, which is mandatory for all students, is that students are being offered theoretical rationales for describing, understanding and analysing corporate actions and communication on the sustainability topic. For that purpose, the course provides the historical background of sustainability as well as its different meanings and justifications for it such as a societal as well as a corporate understanding of sustainability. Building on an economic understanding of sustainability, a resource-oriented framework is developed which contrasts the traditional efficiency-oriented model. As contradictions arise due to contrasting efficiency and sustainability, strategies on how to deal with contradictions are developed. Due to the high political, social, economic and ecological actuality, which means the adaption of the 2030 Agenda, providing for a shared global vision towards sustainable development for all. The SDGs were engaged in the main topic in summer semester 2018.

3 Learning Outcome of the Seminar "Sustainable Management"

Figure 2 shows the way in which the learning outcomes of the seminar "*Sustainable Management*" are structured.

The seminar aims to impart specialist knowledge to the students on the one hand and to open up space for personality and competence development on the other. For the transfer of knowledge, approaches of Research-Based Learning are used, which essentially influence the tasks of the Working Packages. In the previously introduced Fig. 1, this type of learning is located in the "Field 1: Basics" and "Field 2: Consolidation" and "Field 3: The extension to inter- and transdisciplinary knowledge". So, students should acquire specific basic knowledge about sustainability issues, especially about the SDGs, and they should be enabled to identify, evaluate and classify current social developments in relation to the Agenda 2030.

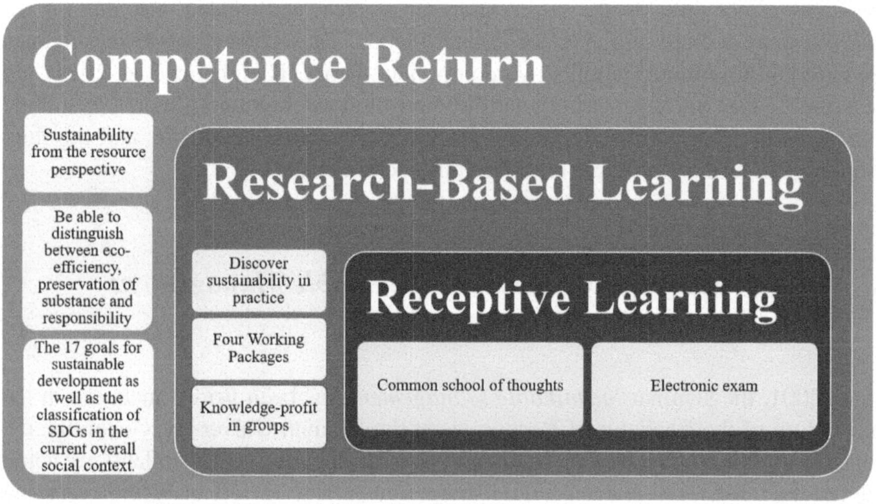

Fig. 2 Learning outcome of the seminar "Sustainable Management"

In addition, the required basic knowledge on sustainability issues is tested in form of an electronic exam. This can be essentially described as learning by reproduction. This type of academic teaching offered one major advantage for the Seminar "*Sustainable Management*": in the past, it has repeatedly occured out that working in groups confronts students with a high degree of coordination, which often leads to problems of cooperation. This can have an impact on the final grade. For this reason, there is a Working Package with an individual achievement in form of an electronic exam that can influence the group result both positively and negatively.

4 Structure of the Seminar "Sustainable Management"

A total number of 334 students attended the course and completed it successfully. Even though there was a large number of participants, the aim was to use a mix of different teaching methods to enable the students to develop a deep, comprehensive and sustainable understanding and knowledge of the SDGs. To meet this requirement, the course was structured as followed (Fig. 3).

During the event, students were invited to attend blended learning and present seminars. The student's total examination performance consisted of four Working Packages during the semester. In three out of four Working Packages the students

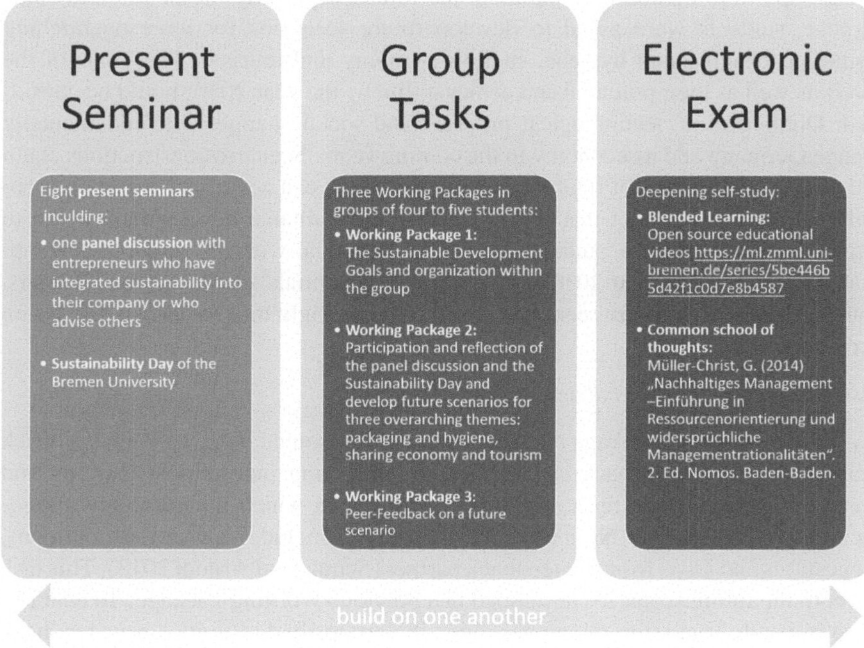

Fig. 3 Structure of the seminar "Sustainable Management"

were asked to work in groups of four to five students, while another Working Package was an individual achievement in form of a mandatory electronic exam.

The event was coordinated by the lecturers via a special learning environment provided and developed by the "Virtuelle Akademie Nachhaltigkeit", a third-party funded project of the Chair of Sustainable Management at the Bremen University. This learning platform made it possible to apply different teaching methods in the Working Packages, in which the students worked in groups, and thus give the students different and innovative tasks.

Task Example 1: Explanatory Video and scientific poster

The groups were asked to create an explanatory video and scientific poster for a special SDG. In order to be able to work on this task, the students had to deal intensively with the SDGs and the German Sustainability Strategy in order to understand the individual goals and their context. The resulting explanatory videos and scientific posters were of exceptionally high quality in terms of content and form. For this reason, individual videos and scientific posters were exhibited and presented by the students during the "Sustainability Day of the Bremen University" in 2018. A stand was set up for each SDG and supervised by a group of students. This allowed interested people to exchange information with the students and discuss the specific SDGs.

Task Example 2: Thinking out of the box-creating future scenarios

The groups were invited to think about the future, especially life in 2030. For this purpose, students were asked to develop future scenarios for three overarching themes: packaging and hygiene, sharing economy and tourism. The goals of the SDGs as well as their potential and achievability by the year 2030 should be considered. Digitalization, technological progress and social change will fundamentally change Germany and its economy in the coming years. Scenario constructions claim to describe possibilities of future development, which can occur with a certain probability. Based on the fact that the 17 goals for sustainable development claim to actively shape the future, students should actively deal with the possible development trends up to the year 2030 and identify the potentials, challenges and barriers. The development of future scenarios seems to be the right tool for such a discussion of content.

Task Example 3: Peer-Feedback

In a further Working Package the students were confronted with the Peer-Feedback method. The Peer-Feedback is a method of supporting independent learning and dealing with one's own texts and those of others, in which a student gives feedback to a fellow student. Normally a Peer-Feedback includes corrections, opinions, suggestions and ideas from the feedback partner (Schulz and Arnodl 2013). This task refers to the future scenarios developed in a previous Working Package. To enable a more in-depth discussion of another topic, the groups had to write a peer feedback on a topic that was new to them.

Table 2 Overview of grade distribution seminar "Sustainable Management"

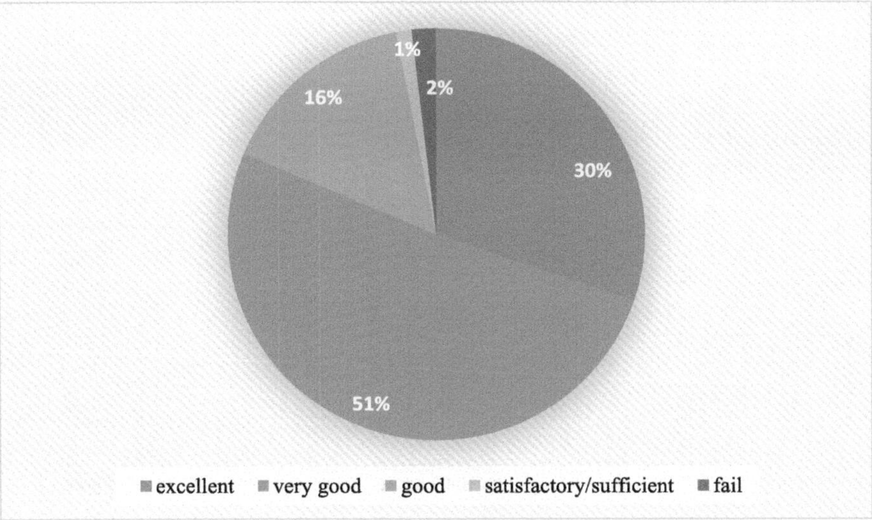

Table 3 Working on practical examples of SDGs has put me in a position to relate the theoretical principles of sustainability

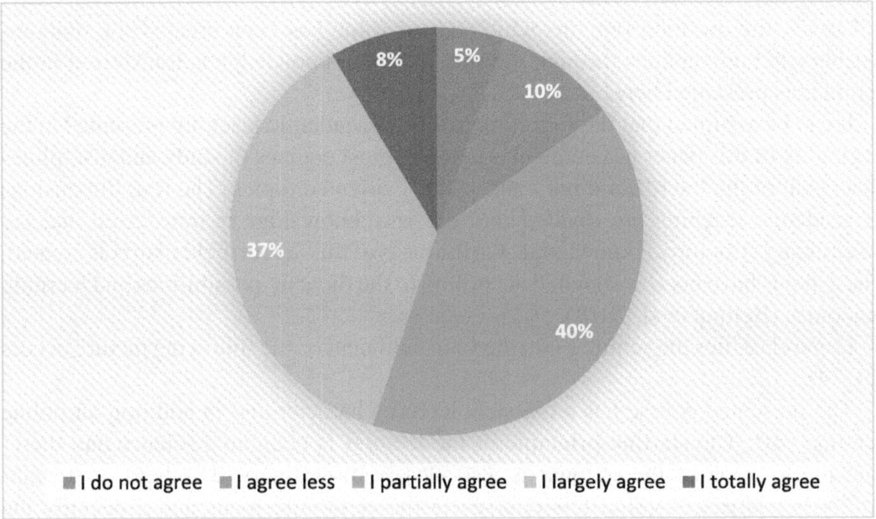

Table 4 The reflection on the individual progress of knowledge was helpful

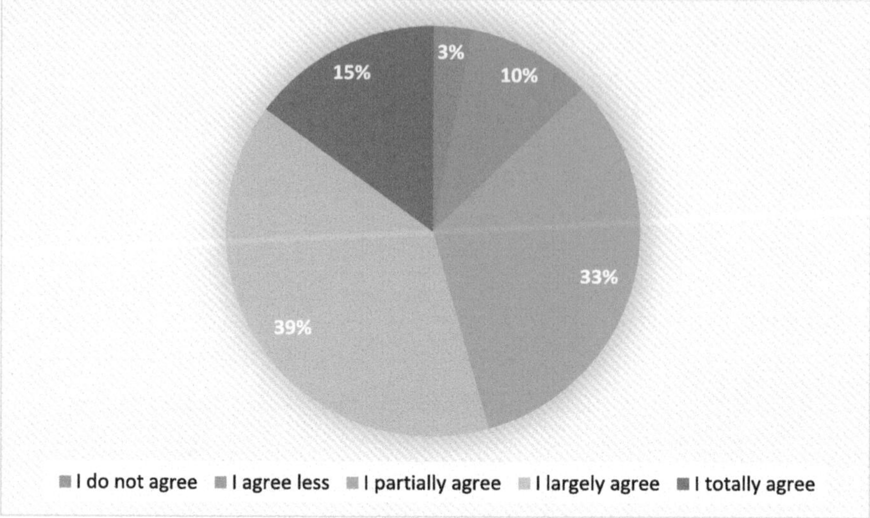

6 Conclusion

All in all, the question remains which added value has been created for students, teachers and universities through the way in which knowledge is transferred in the seminar as presented here.

It can be assumed that the four dimensions of academic teaching presented at the beginning of this paper has come into being in most courses of study and disciplines as a result of the interdisciplinary and socially oriented topics. The four dimensions of academic teaching are divided into the way knowledge is transferred and the disciplinary structure of knowledge. On the basis of this, a four-field matrix is created. These fields have been designated according to the didactic possibilities and learning outcomes (Bellina et al. 2018).

If one classifies the seminar into the four-field matrix, the following picture occurs (Fig. 4):

The presented course has a classical lecture character and in addition an online seminar part. The starting principle of the lecturer is economic science and therefore the location of the seminar is also clearly disciplinary. The field of sustainability is addressed within this course via the economic focus and especially the SDGs. The learning taxonomy is clearly focused on the basics with a receptive/transmissive teaching and learning format. The seminar offers transdisciplinary approaches through the structure, the work with and about SDGs and innovative task formats. Through this transdisciplinary exchange, which has a receptive character, new and innovative thought patterns can emerge among students and relevant sustainability competencies can be developed.

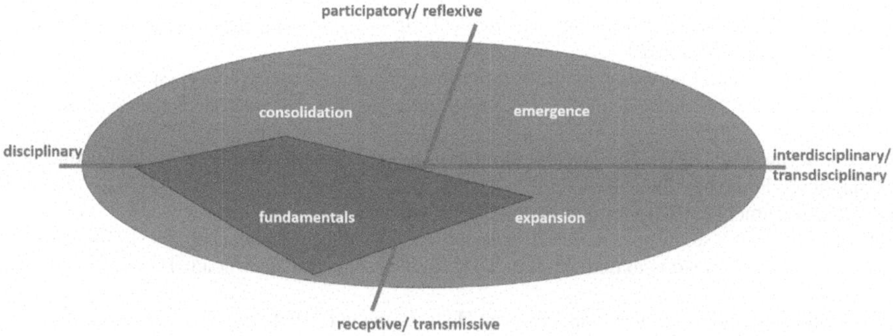

Fig. 4 Structural picture of the seminar "Sustainable Management" in the four-field matrix

The momentum for ESD has never been stronger. Global issues require a shift in lifestyle and a transformation of the way people around the world act and think. Education systems, as well as university teaching, must respond to this need by defining relevant learning objectives and learning contents. (Federal Ministry of Education and Research 2017) For this reason, it seems particularly useful for students to deal with the SDGs as a practical and comprehensible framework for the topic of sustainability. Through their structure, SDGs offer students good access to sustainability issues and enable larger cohorts to work on this topic as a whole. All in all, it is not only a matter of merely conveying theoretical facts, but also of enabling students to deal with social developments independently and in a reflective manner.

The implementation of such a teaching format can be regarded as ambitious. Essentially, it will probably require a fixed anchoring both in the curriculum of the corresponding program of study and a permanent personal responsibility. Overall, it is possible to bring the entire cohort of a mass study programme into contact with the topic of sustainability.

This article shows that it is possible to implement education for sustainable development and the SDGs in academic teaching. In addition, the different teaching formats in the event allow a multifaceted examination of the SDGs. The practical relevance of SDGs in particular has shown that this knowledge is deeply rooted among students (Table 3). Building on such a basic course, the four dimensions of academic teaching can be further filled with courses, providing students with comprehensive specialist knowledge of the SDGs and promoting personal competence development.

References

Bellina L, Tegeler MK, Müller-Christ G, Potthast T (2018) Education for sustainable development (ESD) in university teaching. BMBF project "Sustainability at universities". develop - network - report (HochN)", Bremen and Tübingen

Bender C (2019) The lecture: a discontinued model? https://www.academics.de/ratgeber/vorlesung. Last Accessed 15 Mar 2020
Braun H (2013) Neural networks: optimization through learning and evolution. Springer, Berlin
Carroll JB (1989) The Carroll model: a 25-year retrospective and prospective view. Educ Res 18(1):26–31
Deci EL, Ryan RM (1993) The self-determination theory of motivation and its significance for pedagogy. J Pedagogy 39(2):223–238
Federal Government (2018) German sustainability strategy federal cabinet decision 7 Nov 2018. https://www.bundesregierung.de/resource/blob/975274/1588964/1b24acbed2b731744c2ffa4ca9f3a6fc/2019-03-13-dns-aktualisierung-2018-englisch-data.pdf?download=1. Accessed 10 Mar 2020
Federal Ministry for Economic Cooperation and Development (2019) Targets for sustainable development worldwide. Retrieved from https://www.bundesregierung.de/breg-de/themen/nachhaltigkeitspolitik/ziele-fuer-eine-nachhaltige-entwicklung-weltweit-355966. Accessed 10 Mar 2020
Federal Ministry of Education and Research (2017) National action plan on education for sustainable development. The German contribution to the UNESCO Global Action Programme. https://www.bmbf.de/files/BMBF_NAP_BNE_EN_Screen.pdf. Accessed 25 June 2020
Frey A (2017) Clearing the brain: faster and better learning thanks to synapse elimination. https://onlinemarketing.de/jobs/artikel/schneller-und-besser-lernen-dank-synapseneliminierung. Accessed 25 June 2020
Gaffney O (2014) Sustainable development goals: improving human and planetary wellbeing. Glob Change 82:20–23
Griggs D, Stafford-Smith M, Gaffney O, Rockström J, Öhman MC, Shyamsundar P, Noble I (2013) Policy: sustainable development goals for people and planet. Nature 495(7441):305
Helmke A (2008) Classroom research. In: Tenorth H-E, Tippelt R (eds) Beltz- Lexikon Pädagogik. Beltz, Weinheim
Iversen IH (1992) Skinner's early research: from reflexology to operant conditioning. Am Psychol 47(11):1318
Kandel ER, Bischoff M (2012) Psychiatry, psychoanalysis and the new biology of the mind. Suhrkamp, Frankfurt am Main
Kleinheinz MM, Michael AHL, Heintze A, Hinkelmann DHJ (2019) Crystalline and fluid intelligence—do you know these forms of intelligence? https://open-mind-akademie.de/fluide-und-kristalline-intelligenz/. Accessed 25 June 2020
Kuhn C, Zlatkin-Troitschanskaia O, Pant HA, Hannover B (2016) Valid assessment of students' competences in higher education. J Educ Sci 19(2):275–298
Mccammon AL (1992) United Nations Conference on Environment and Development, held in Rio de Janeiro, Brazil, during 3–14 June 1992, and the 92 Global Forum, Rio de Janeiro, Brazil, 1–14 June 1992. Environ Conserv 19(4):372–373
Medvetz T (2008) Think tanks as an emergent field. Soc Sci Res Council, New York
Ritter H, Martinetz T, Schulten K (1991) Neural networks. Addison-Wesley, Munich
Robert KW, Parris TM, Leiserowitz AA (2005) What is sustainable development? Goals Ind Values Pract Environ Sci Pol Sustain Dev 47(3):8–21. https://doi.org/10.1080/00139157.2005.10524444. Accessed 20 Mar 2020
Ryan RM, Connell JP, Plant RW (1990) Emotions in non-directed text learning. Learn Individ Differ 2:1–17
Schulz F, Arnold R (2013) Peer feedback in der Hochschullehre hilfreich gestalten- Onlinegestütztes Peer-Feedback in der Lehrerbildung mit der Plattform PeerGynt. http://nbn-resolving.de/urn:nbn:de:hbz:386-kluedo-36292. Accessed 11 Mar 2020
Sosa E (2009) Reflective knowledge: apt belief and reflective knowledge, vol 2. University Press, Oxford
Töpper V (2016) Study: students are more stressed out than working people. https://www.spiegel.de/lebenundlernen/uni/aok-uni-stress-report-studenten-sind-gestresster-als-berufstaetige-a-1116064.html. Accessed 25 June 2020

UNESCO (2017) Education for sustainable development goals. learning objectives. https://www.unesco.de/sites/default/files/2018-08/unesco_education_for_sustainable_development_goals.pdf. Accessed 24 June 2020

"Evolution of Knowledge Sharing of Education for Sustainable Development in the Digital Space—Insights from Two Research Projects"

Jonas Joachims and Katharina Lingenau

1 Introduction

Society still faces a major challenge in terms of sustainable development. Numerous UNESCO programmes and the formulation of the Sustainable Development Goals have created a great awareness of global problems. Nevertheless, there is still little knowledge of solutions and changes in behaviour. Universities as research and educational institutions have the task of preparing society for the challenges of sustainable development and of creating knowledge with which these challenges can be met. Besides all the complexity of this task, there are other developments that are changing society, such as digitalization. This also confronts universities with new, complex challenges. How sustainability and digitalization can be brought together in university teaching, which challenges this poses to teaching and what digital ESD can look like, will be described in the following, by example of two case studies.

2 Overview Over the Embeddedness of ESD in German Higher Education

Since the declaration of the UNESCO Decade of Education for Sustainable Development in 2005, it has been a globally defined goal to anchor ESD in all areas of education. However, ten years were not enough to establish sustainability as an integral part of education. The Decade was followed by the World Action Programme

J. Joachims (✉) · K. Lingenau
Sustainable Management, University of Bremen, Enrique-Schmidt-Str.1, 28359 Bremen, Germany
e-mail: joachims@uni-bremen.de

K. Lingenau
e-mail: lingenau@uni-bremen.de

W. Leal Filho et al. (eds.), *Universities, Sustainability and Society: Supporting the Implementation of the Sustainable Development Goals*, World Sustainability Series,
https://doi.org/10.1007/978-3-030-63399-8_24

in 2015. With five concrete fields of action, the educational goal was to be further advanced, existing ESD projects consolidated and new ones created. As a result, exciting and diverse activities and concepts have been developed during this period, but the big goal—a systematic integration of ESD in education and training—has still not been achieved (Michelsen 2016). And so the World Action Programme 2019 ended with another follow-up programme: ESD for Sustainable Development Goals (SDG) (2020–2030).

But even if the big goal (ESD is integrated in every field of education) is still far away from being achieved, there are very big differences in the grade of the implementation of ESD in different educational sectors. For example, while great progress has been made in early childhood education in integrating ESD, German higher education is still lagging behind. In 2009, only about 2% of Germany's nearly 3 million students were reached by courses with a sustainability focus (DUK 2009). Although about 2000 different projects have been developed in Germany within the UNESCO Decade of ESD, the German higher education landscape is still far from becoming an ESD mainstream (Michelsen 2016). The same applies to the worldwide integration of ESD in curricula and research activities of universities.

Since 2015, the "Institut Futur" in Berlin has been conducting national monitorings of ESD in Germany. The overarching goal is to collect and describe a differentiated picture of the anchoring of ESD in Germany, the implementation of the UNESCO World Action Programme and the implementation of the German National Action Plan on Sustainable Development. The progress in the individual areas of education is regularly documented via document analysis (e.g. examining the existence of the topic of sustainability in university development plans or the integration of this topic in study programme plans) and expert interviews. For the higher education sector, the Institute comes to the conclusion:

> …that for the broad concepts of sustainability and sustainable development a significant increase can be recorded in the documents analysed, whereas pure educational concepts such as environmental education or education for sustainable development are only marginally found in the selected documents of the higher education sector. This means that further efforts are needed at all the levels examined in order to further advance the goal of comprehensively anchoring ESD in the higher education system. (Etzkorn and Singer-Brodowski 2017, p. 6)

These assessments were also made by the experts interviewed by the Institute. In addition to the issue of sustainability, universities are confronted with a variety of socially relevant topics such as diversity or digitalization. Moreover, approaches of ESD, such as interdisciplinarity, are confronted with the current characteristics of a course of study. All this means that the higher education system is moving only slowly towards sustainability, even though the pressure from politics, science and society is growing (Etzkorn 2017).

In addition to the current World Action Programme on ESD, the UN adopted the SDGs at the end of 2015. As part of Agenda 2030, 17 SDGs were drawn up and assigned with 169 sub-goals. The declared goal of the SDGs is to achieve sustainable development in developing and industrialised countries by 2030 (United Nations 2015). The Goals cover a wide range of complex social, ecological and economic challenges and address them in a solution-oriented manner with very concrete objectives. Here too, the relevance of ESD as part of the fourth objective "Qualitative Education" of the SDGs is emphasised. Compared to the concept of ESD, the SDGs are attributed several advantages by the expert community: For example, the SDGs serve as a common frame of reference for individual efforts to promote sustainable development (Ahel and Lingenau 2020). Additionally, the unique design (clear visualisation of the objectives in a striking tile form, catchy icons and different colouring) is also highlighted as particularly helpful. These forms of presentation offer a lot of orientation and at the same time create a feeling of difference and integration (Müller-Christ et al. 2017). Thanks to their unique role in society, universities are able to make an extremely important contribution in the sense of SDGs, just as in the UNESCO programmes on ESD. Through teaching and learning, students are given the skills they need to develop solutions to achieve the SDGs. At the same time, young people are encouraged, mobilized and networked to implement the developed solutions. And at the research level, universities also support the integration of SDGs into society by providing appropriate knowledge, methods, technologies and innovations and by researching solutions for concrete developmental problems. Published in 2017, the study on the integration of SDGs in the German education system conducted by Müller-Christ et al. (2017) comes to similar conclusions: SDGs are rarely integrated in universities. Innovative teaching methods and future-oriented content are often neglected in favour of attracting third-party funding and publications.

In summary, it can be seen that the transformation of the higher education system to ESD has not yet been achieved. Sustainability is often still seen as competition and not as a complement to other socially relevant issues. The complexity of the topic itself also makes integration in teaching, studies and organisational structures more difficult. The consideration of ESD and SDGs requires new ways of teaching and learning at universities and thus calls the system itself into question. Although the studies show that the German higher education system does not yet meet the goal of ESD, it is also clear that some progress has been made: More and more federal states are anchoring ESD in their education plans, new interdisciplinary study programmes are emerging and new, innovative teaching formats are being developed (Michelsen and Wels 2017).

3 Characteristics of ESD for Teaching in Higher Education Settings

The following section aims to outline the characteristics of ESD. Not only is ESD a wide ranging field of research, its special thematic orientation also impacts the way ESD has to be taught in higher education. The main characteristic of ESD is its diversity. Stemming from numerous—partially very different—fields of research, ESD has formed itself around the last turn of the millennium from, inter alia, ecological sciences, research concerning the global transition, socio-ecological studies and human ecology. The main element all of these former independent fields of research have in common is the long-term perspective on observed incidents (Vilsmaier and Maier 2017). Consequently, ESD-related topics are never going to be covered conclusively as Scott (2016) points out. They furthermore are characterised by social processes of learning, which generally follow the objective of improving the living conditions through an infinite joint negotiation between people. This social dimension is also stressed by Schneidewind and Singer-Brodowski (2015) who state that ESD's main concern is supposed to encourage learners to deal with societal questions and improve the personal ability to recognise and perceive important topics in this respect. Teaching ESD therefore is strongly concerned with the constant creation of spaces for simultaneously questioning and shaping aspects of the social, ecological and economical dimensions of everyday life. With respect to the cross-sectional nature of ESD, the mode of teaching in German higher education is strongly dependent on staff. This means that the scientific background of teaching staff strongly influences the thematic direction of lectures, research or student projects. This can be evaluated ambivalently, as the findings of a nationally funded project concerning sustainability and its integration into the German higher education system suggest:

1. a positive consequence is the creation of multiple thematic connection points with other scientific fields, bearing the potential for synergies and dissemination of sustainable principles, while
2. a negative consequence of a certain arbitrariness of the topic also has to be mentioned (Hoch N 2018).

One major challenge of including transdisciplinary perspectives in one single approach to ESD is according to Müller-Christ et al. (2015) the oftentimes dilemmatic relationship between decision principles. The challenge of dealing with contradictory success principles has to be overcome in order to be capable of acting in a sustainable manner—with this also being valid for researching or teaching ESD-related topics. A practical example from the context of businesses is the relationship between sustainability and efficiency, with the former limiting the latter and the latter being the, at least in past and present times, main driver for economic activity. Acting between conflicting poles induces researchers and practitioners to reinterpret the original meaning of terms, leading in above example to a concept like eco-efficiency, which streamlines sustainability to become an economic rationale of "ever-increasing profits while reducing resource consumption as best as possible" (ibid.).

Teaching in higher education has to face this challenge and find ways to incorporate the characteristics of ESD in approaches to knowledge transmission. Successful ESD enables learners—regardless of specific topics—to

1. identify dilemmatic relationships between concepts, terms or premises of action,
2. to cope with their contradictory nature, and
3. to find ways to integrate both poles on the same level of meaning (Müller-Christ 2014).

Table 1 Educational approaches towards ESD

Educational approach towards ESD	Degree of involvement of learners
Education *about* sustainability	Knowing facts and figures about certain topics of sustainability
Education *in* sustainability	Active involvement in understanding the scope and interdependencies of sustainability as a whole
Education *for* sustainability	Critical and reflective ability to discover opportunities for transformation towards greater sustainability, constant development of new capacities in relation to changing economical, ecological of social environments

Own illustration following Sterling (2001)

This ability to view questions from a multitude of perspectives, integrating these perspectives and finding ways to transform systems towards increased attention concerning sustainability is one main outcome of successful ESD, be it in higher education or other settings. In addition to above discussion, the thoughts of Sterling (2001) offer an insightful typology of teaching and learning ESD. The author distinguishes between education *about* sustainability, *in* sustainability and *for* sustainability. While education *about* sustainability only aims to transfer factual knowledge from lecturer to learner, education *in* sustainability reaches further and has the objective of more active learner-involvement in order to support greater overall understanding of sustainability. Education *for* sustainability however has an even wider, thus also more ambitious, transformative focus in encouraging the adoption of principles, ethical dimensions and values of sustainability. In this respect, successful ESD, as in above definition by Müller-Christ (2014), has to reach the level of educating for sustainability in order to enable students to take transformation-oriented perspectives in their respective fields of ESD. This requires teaching environments capable of developing critical and reflected capacity-building (Barth and Burandt 2013) (Table 1).

With these inherent contentual challenges of transmitting ESD in mind, the following part will take a closer look on additional obstacles of teaching sustainability-related topics in a purely digital setting.

4 Challenges of Transmitting ESD via Digital Methods

In trying to transmit the transformation-oriented and in its complexity oftentimes contradictory field of ESD, lecturers have to overcome a significant amount of obstacles solely from a contentual standpoint. As soon as teaching concepts are supposed to be digitised, a new set of challenges emerges. The following paragraph is going to showcase the most relevant of these.

First of all, the transfer of learning content from a classroom to a digital setting requires quite logically a few organisational and technological circumstances. In this context, this means for example a powerful technological setup, be it server capacities, hardware and appropriate software. Furthermore, teaching staff capable of and willing to deal with technologically driven questions is a crucial prerequisite (for a comprehensive overview over requirements for successful E-Learning see Davis 2004). These structural factors aside, the transfer into digital and media-based approaches to teaching pose other challenging questions to both lecturers and students involved.

Euler and Seufert (2005) speak of a shift of responsibility from lecturer to students, implying that in a setting where learning materials are provided electronically and students have the obligation learn autonomously, the lecturer clearly lacks control over the learning process and the intensity of engagement. In E-Learning, all activity is reliant on the individual in the first place. Even if techniques and forms for collaborative learning are designed and the technological requirements of conducting these are met, the individual learner still has to find and uphold motivation to start and stick to a certain course (Shephard 2004). Consequently, and due to the fact that learning in "e-contexts" is in many facets different from traditional learning in lecture contexts (Euler and Seufert 2005), students have to be familiarised with being self-reliant and –paced in their respective learning processes, with lecturers transitioning more and more to the role of learning coaches, who intervene in case of problems or moderate students through courses (Barth and Burandt 2013). According to practical experiences made by Israel (2018) who states that in a digital context, students have to repeatedly be reminded of the overall aims of courses, this can be achieved through a clear link between tasks and desired learning outcomes, or the usage of structured reflections, where students have the obligation to evaluate their own learning progression. Practical constraints like these, meaning that lecturers have to structure and scaffold E-learning courses quite differently than conventional lectures, equal additional challenges for people involved. A seriously planned E-learning course has to be clearly defined with respect to desired learning outcomes, the outline of the way students are supposed to reach these learning outcomes and measures for success monitoring and quality management prior to the start and without the possibility of adjustment on the fly (Dittler 2017).

Since 2011, the Virtual Academy of Sustainability (VAS) aims to face above challenges and transforms sustainability-related lecture contents to digital lectures with subsequent electronic assessments. Through an at the time state of the art approach to producing, designing and distributing online learning materials concerning a variety of sustainability-related topics in a purely digital manner, the project lead to numerous valuable experiences. The following chapter is going to outline the project's overall approach to transferring ESD in the digital context and aims to showcase major developments up until today.

5 Case Study I: Virtual Academy of Sustainability

The VAS was founded in 2011 with the aim to reach as many students as possible with basic knowledge about ESD. Based on experiences from the production of online courses and electronic exams at the University of Bremen, the Chair of Sustainable Management in cooperation with the Centre for Multimedia in Teaching developed the concept for an offer of online courses for all German universities.

5.1 Conceptual Approach of the Project

In order to offer creditable courses throughout Germany, which are used by universities and attractive for students, the VAS works in three main areas: (a) online courses, (b) learning platform and (c) electronic assessment (Fig. 1).

(a) In order to reach as many students with ESD as possible, the VAS produces and publishes online courses that are freely available and can be used by everyone. To ensure that students do not only voluntarily take these courses as an additional service and that universities also include them in their programmes, all course content is produced by university teachers. All university lecturers are experts in a specific sustainability topic. The production of a range of courses with varying experts is intended to create a wide range of sustainability topics. Students cannot only get credit for the courses they have taken for their own studies, but they can also take a course on a specific topic according to their own interests. Furthermore, the contents of all courses are designed for a basic level. On the one

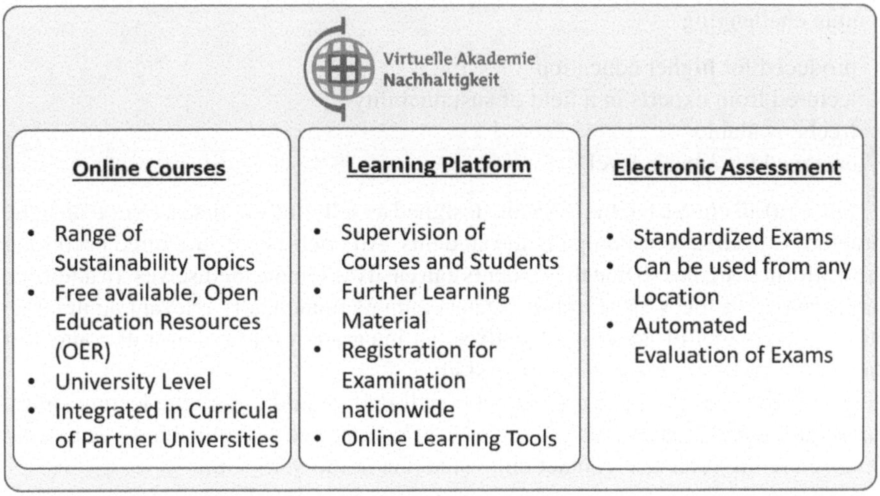

Fig. 1 Structure of the VAS. Own illustration

hand, this serves to ensure that all students regardless of previous knowledge can attend the courses (reaching as many of them as possible) and that universities can integrate this type of courses well into their own curricula in the form of elective courses without displacing their own teaching.

(b) The VAS operates its own learning platform to manage a large number of online courses and to ensure the supervision of these and the students. On this platform, all organisational aspects of each course are recorded and students receive further learning materials. The registration for the examination dates at the different locations is also managed completely online on this platform. In addition to administrative matters, the learning platform also provides various tools (e.g. blog posts, essay assignments, etc.) for the design of online courses.

(c) Supervising a nationwide course offering with a small team requires not only questions about the design of courses but also new ideas for the examination. These should also be able to be taken nationwide, have to meet the same standards at every location and while also being manageable in terms of effort. The VAS has opted for electronic examinations. By means of an examination software, these are available at any location, even if they are hosted and conducted from the University of Bremen. Closed questions and different variants enable an automated evaluation.

5.2 *The Didactical Concept*

Besides the conceptual requirements of the online courses of the VAS that are outlined above, the offering of the VAS is also based on a didactical concept for the (a) online courses, (b) the learning platform and (c) the assessment. As mentioned in some sections before, the teaching of ESD itself and especially the online teaching of ESD is quite challenging.

- produced for higher education
- lectured from experts in a field of sustainability
- freely available
- designed for a basic level

Since (a) all courses of the VAS are designed as self-study courses, a good didactic structure is needed to accompany the students. All courses are structured according to a uniform structure so that the students can easily orientate themselves. In addition, each episode of the VAS learning videos contains didactically relevant content: an outline, the listing of learning objectives, an integration into the overall context of the lecture and tasks for in-depth self-study.

(b) On the learning platform, students will find, in addition to the lecture slides, further learning materials such as quizzes, pre-structured learning diaries or a demo exam. Here they can also contact other students or the VAS team.

(c) The electronic exams also follow a didactic approach. For each exam question, students are shown which episode of learning videos the question refers to. Each question is accompanied by a short introduction for orientation. Since no specialist staff can be present at other locations for exams, there is also the possibility of commenting on questions within the software.

5.3 Competencies

In the guided self-study of the interdisciplinary courses, students of the VAS are acquiring shaping skills for sustainable development. Some of the 12 sub-competences which, according to de Haan et al. (2008), are important for shaping sustainable development are addressed by the VAS in particular: On the one hand, online learning promotes independence in defining, reflecting and evaluating learning and work processes and in shaping learning and work processes independently and sustainably. On the other hand, students are supposed to acquire the skills to search for new solutions for the topics of their own subject area with the help of the contents offered in the courses. This is made easier for them because they acquire relevant knowledge at the interfaces to other areas of knowledge. Shaping a more sustainable development means above all solving more complex problems in all areas of life. Complex sustainability problems also have characteristics which call for special skills. These include the competence to reflect on mission statements, the competence to deal with incomplete and over-complex information, the competence to cope with individual decision-making dilemmas and the competence to act morally.

5.4 Limits and Opportunities of the Concept

Today (early of 2020) the VAS offers 21 online courses on various sustainability topics. These are integrated into the curriculum of over 40 partner universities in Germany and Europe. In the meantime, more than 2500 students per semester take part in VAS examinations. The VAS is firmly integrated into the German university landscape.

However, the basic concept of the VAS also shows its limits: the teaching concept that is based on creating basic knowledge reaches many, but it does not go beyond certain competence levels (see Fig. 2). The concept of self-paced study, even if its guided, offers many advantages to the students, but there is little or no exchange between each other or no joint development of teaching content.

Being aware of these restrictions imposed by the course format, the VAS has continuously developed its teaching concept. Video formats have been adapted and a new teaching platform with possibilities for active participation and collaboration of students has been integrated. In order to make adjustments not only during operation, but also to enable major leaps in development, the VAS initiated the research project SCoRe at the end of 2018.

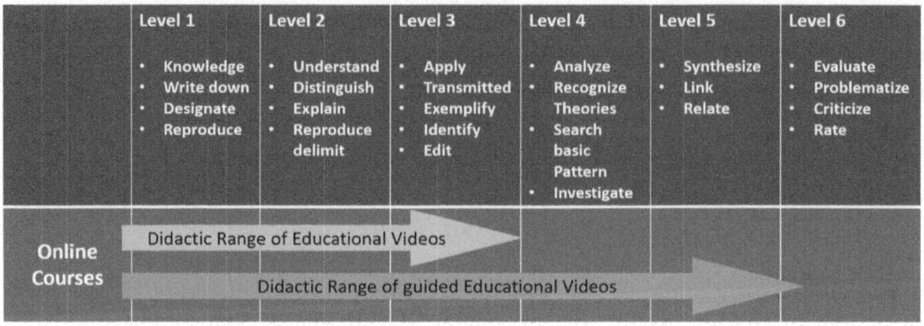

Fig. 2 Reachable competence levels in online education via video. Own illustration following DQR-Matrix, Arbeitskreis Deutscher Qualifikationsrahmen (2011)

6 Case Study II: Student Crowd Research

Since November 2018, a second third-party funded project is being conducted at the Chair of Sustainable Management of the University of Bremen. In cooperation with three higher education institutes and one software development company, Student Crowd Research (SCoRe) aims to (1) build a digital learning space for research-oriented activities concerning sustainability. Furthermore, the project seeks to (2) make use of innovative video techniques including, for example, collaborative annotation and analysis of video data. Finally, the, at least in German higher education, given (3) condition of ever-increasing cohorts of students is supposed to be utilised explicitly as a resource, instead of being seen as a threat for high-quality teaching.

Figure 3 illustrates the internal organisation of SCoRe. The process of and progress

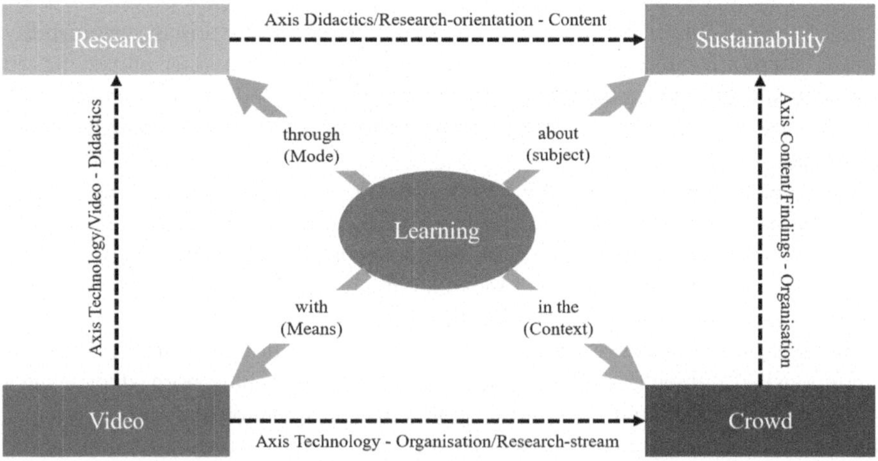

Fig. 3 Internal organisation of SCoRe (SCoRe-Consortium 2018)

in learning marks the center of all endeavours. The four main contentual, didactical and technological pillars influence the learning process and shape the distinct organisation of the project:

1. learning is supposed to happen through a research-oriented examination in the topic of sustainability (upper axis “didactics/research-orientation–content”),
2. didactical processes are supposed to be supported with innovative video techniques, e.g. collaborative annotation and analysis (left axis “technology/video–didactics”),
3. the condition of increasing cohorts of students is supposed to be utilised in order to constitute new didactical settings for crowd-learning (right axis “content/findings–organisation”), and
4. all resulting, mainly video-related artefacts, are supposed to be organised in a way that allows for iterative enhancements of overall quality user-generated content (lower axis “technology–organisation/research-stream”).

Each of the four pillars is related to one higher education institute, with the University of Bremen being responsible for ensuring the learning effectiveness of all developments with regards to the solely digital transmission of ESD. All conceptual considerations within SCoRe are being made after careful consultation of all respective partners and subsequently put into practice by the cooperating software development company.

6.1 Methodical Approach of the Project

The scientific and contentual approach of SCoRe is based on the principles of *design-based research.* In doing so, the project commits itself to an incremental approach of developing specific and tangible artefacts (*prototypes*) for students to engage with. The goal is to identify innovative approaches to current problems regarding the transmission of ESD in solely digital contexts. As mentioned above, the topic of ESD itself is characterised by certain difficulties regarding a holistic teaching outcome (“teaching *for* sustainability”). In combination with the setting of a truly digital implementation of teaching in the context of large and ever-growing cohorts, SCoRe has to face serious challenges—making it not entirely promising to solely explore and combine existing research findings. Rather, it seems senseful to create small entities (be they innovative digital task formats, video-techniques or experiments) that are capable of putting single facets of the overall research interest to test (Bakker 2019). Over time, these prototypes relate to more and more components of the overall problem-setting, which SCoRe has to find answers to. The design-based approach, switching between designing and evaluating resulting prototypes, is supposed to lead to the creation of an online learning platform, where large groups of students can engage with the topic of sustainability in a research-oriented way, while making use of innovative video-features.

The first approach to testing relevant features in order to gain insights with respect to the overall research interests of the project at the University of Bremen focused on a didactical design that incorporated as many of the online-based working features of the already existing learning platform as possible, which also is been used by the VAS. The aim was to change the mode of learning in direction of a research-minded attitude of students, challenging them to pose questions, answering these in a structured process, giving and incorporating feedback and communicating the results to the public (in this case the entirety of the course).

Figure 4 shows the didactical design of the course, which included 284 undergraduate students of the economics degree program and took place in the summer term of 2019. All described tasks have been conducted online and in working groups of four to five students. Group-based discussion was encouraged and happened online as well as offline. However, students were asked to use a variety of digital tools for gathering, structuring and communicating information on the learning platform, namely:

- identifying and discussing problems, developing these to research questions and documenting the whole process in a video,
- giving feedback via video comments and rating research results of peers via traffic light system, and
- structuring feedback on own research results in concept map and communicating final results in video.

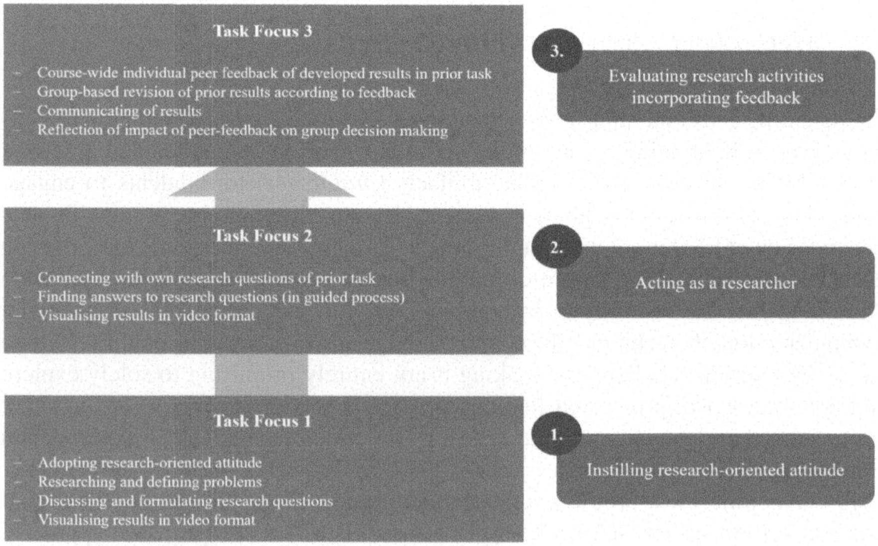

Fig. 4 Course design at University of Bremen. Own illustration

While the scope of this paper does not allow for further detailing on course contents, research activities or showcasing of student results, the following section will go into detail about the concluding evaluation and reflection.

6.2 Early Results of Evaluation

284 students completed the Sustainable Management module last semester. The aim of the evaluation is to condense the students' perspectives of experience by focusing on three different major issues, regarding (1) the general acceptance of the digital format of the course and the structuring of the module, (2) the quality of teaching in relation to the topic of sustainability and (3) the evaluation of the experienced intensity of the module. Based on a standardized questionnaire, the general topics were operationalized into several question units. Each question unit of the evaluation appeared as a five-level scale. In order to be able to include qualitative information in the evaluation in addition to the descriptive analysis of the general tendencies of the students, open fields of response were offered, in which criticism and suggestions on the three main topics were requested.

In the following, the descriptive analysis of the questionnaire is presented first. Afterwards, the qualitative statements are condensed into thematic areas in order to derive implications with semantic content for the further development of the project.

General acceptance of the course: In relation to the acceptance of the module design via a digital learning platform, 41% of all students state that the content introduced them "well" to the topic of sustainability. More than 60% of the students see a red thread, despite the structuring of the module via digital media combined with a focus on self-organized learning.

Quality of teaching Sustainability: With regard to the quality of teaching the topic sustainability, over 70% of the students assessed that they were able to gain "rather many" or "many" new perspectives on the topic of sustainability, as they continued to work with the results of others and exchanged views on the topic of sustainability in groups (see Fig. 5).

Accordingly, more than half of the students (56%) state that they have been able to benefit "very much" or "much" from the thoughts of the other students.

Experienced intensity of the workload: The relatively high evaluated benefit related to the quality of teaching, however, does not come without a price, as the students see an increased expenditure of time, which is caused by the use of new digital editing formats (e.g. videos, concept map, video commentary). Accordingly, over 70% of the students state that their workload has increased (40%) or greatly increased (35%) due to the use of digital formats and the self-organized teaching orientation.

Overall, the module sustainable management was evaluated by the students with a relatively high expenditure of time, compared to other completed modules with a scope of six Credit Points (CP), as Fig. 6 shows.

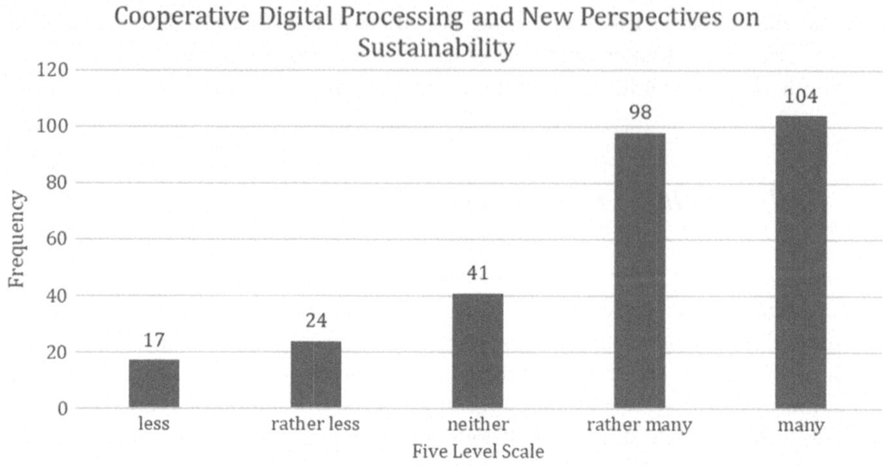

Fig. 5 Cooperative digital processing and new perspectives on sustainability. Own illustration

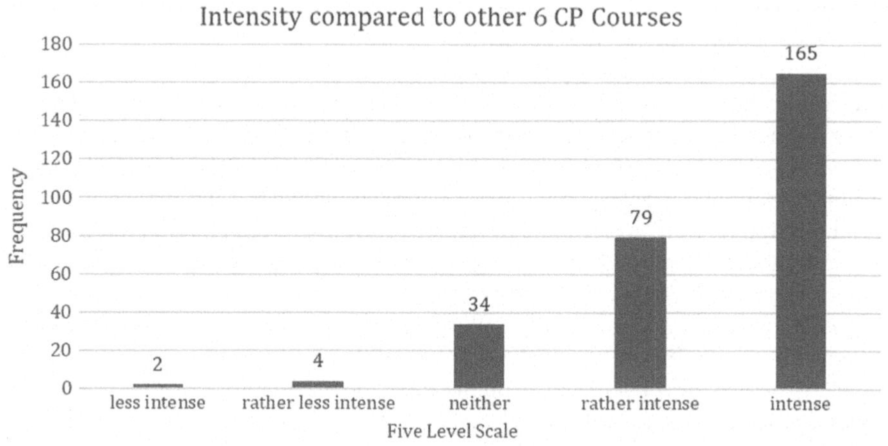

Fig. 6 Experienced intensity of the module sustainable management compared to other six CP courses. Own Illustration

(Credit points—also known as ECTS (European Credit Transfer System) points—were introduced in Germany as part of the Bologna Process, i.e. the changeover to Bachelor's and Master's programmes. They measure the amount of work ("workload") that an average student puts into his or her studies.)

Nevertheless, the students rated the use of digital processing methods on average ($M = 3.37$, $SD = 1.18$) as more conducive to conveying sustainability.

Qualitative analysis: In order to derive concrete implications for the further development of the project from the evaluation, the condensed results of the free response fields are presented below.

Positive highlights: The quality of teaching sustainability seems to be related with different learning achievements, frequently reported by the students. The students positively highlighted the intensive exchange and cooperation between students, creating a deeper understanding of sustainability, based on multiple perspectives. Furthermore, the students frequently reported that the intensive knowledge exchange, as well as the self-organized teaching orientation, invited the reflection of one's own action affecting the environment. It is also worth mentioning that the students emphasized the permanent availability of online material as positive.

Criticism: A frequently described point of criticism is aimed at the self-organized task processing, which is generated by a relatively high scope of interpretation of the presented tasks. Here, students relatively often wish for clearer instructions. As already shown in the evaluation of the questionnaire, many students rate the workload as relatively intense in relation to the scope of the course. In addition, some students mentioned the wish for support offers for multimedia formats. For example, tutorials for video and audio editing. Those offers would decrease the intense workload and increase the quality of expected outcomes.

6.3 Implications for Further Developments

One major point of interest for future adjustments of course design in order to transfer ESD in a purely digital setting is consequently going to be more effective instructions. Students seem to be overwhelmed by the situation of not being instructed by lecturers in lecture settings each week and therefore rate the effective workload as being higher than it actually is. Through placing more effort on showcasing successful examples, for example screencasts of previous student contributions, more transparency with respect to expectations towards the students is likely to be achieved. Reducing the perceived level of effort during online courses offers great potential for course designers to focus students' attention on engaging with the contents rather than comparing expected efforts with more common lectures.

7 Conclusion: Outlining the Trajectory of Transmission of ESD from 2011 to Today

Comparing the two case studies of this article offers valuable insights in the rapid change of approaching the transmission of ESD in digital settings. Early endeavours were in the first place dealing with the mere transportation from analogue lecture information to the digital space. Offering knowledge with maximum flexibility for students, therefore giving as many students as possible the possibility to engage with ESD, was a major driver for funding projects like the VAS. However, the specific characteristics of the topic, as well as changing demands from students and academics, led to a broader approach towards digital learning settings for ESD. Today's teaching concepts call for a different form of participation of students, who are not only supposed to consume and reproduce knowledge, but to actively engage with learning materials. This means the self-reliant identification of interesting topics, the instructed "expedition" of these and the production of own artefacts, which then in turn are possible learning materials for subsequent learners. The building of cross-institutional online environments, capable of facilitating such engagement of students from a technical, instructional and topical standpoint, remains one of the major challenges for the future, in order to ensure in-depth engagement (Education *for* ESD) of students in digital learning settings. Consequently, further research from both didactical and technological viewpoints regarding educational effectiveness of novel teaching settings in online learning environments is needed.

References

Ahel O, Lingenau K (2020) Digitale Ansätze zur Vermittlung der SDGs in der Hochschullehre im deutschsprachigen Raum. In: Leal (ed) Digitalisierung und Nachhaltigkeit: Chancen und Perspektiven für deutsche Hochschulen. Springer Verlag, München. (*in publications*)

Arbeitskreis Deutscher Qualifikationsrahmen (2011) Deutscher Qualifikationsrahmen für Lebenslanges Lernen. Verabschiedet vom Arbeitskreis Deutscher Qualifikationsrahmen (AK DQR) am 22.März 2011, DQR. https://www.dqr.de/media/content/Der_Deutsche_Qualifikationsrahmen_fue_lebenslanges_Lernen.pdf. Accessed 15 Jan 2020

Bakker A (2019) Design research in education—a practical guide for early career researchers. Routledge, London

Barth M, Burandt S (2013) Adding the "e-" to learning for sustainable development: challenges and innovation. Sustainability 5:2609–2622

Davis A (2004) Developing an infrastructure for online learning. In: Anderson T, Elloumi F (eds) Theory and practice of online learning. Athabasca University, Athabasca, pp 97–114

De Haan G, Kamp G, Lerch A, Martignon L, Müller-Christ G (2008) Nachhaltigkeit und Gerechtigkeit. Grundlagen und schulpraktische Konsequenzen, Springer, Heidelberg

Dittler U (2017) Ein kurzer historischer Rückblick auf die bisherigen drei Wellen des E-Learning. In: Dittler U (ed) E-Learning 4.0 - Mobile Learning, Lernen Mit Smart Devices Und Lernen in Sozialen Netzwerken. De Gruyter, Berlin, pp 5–42

DUK—Deutsche UNESCO-Kommission (2009) Bildung für nachhaltige Entwicklung. Tagungsbericht. http://www.ifa.de/fileadmin/pdf/abk/inter/unesco_esd09_rep.pdf

Euler D, Seufert S (2005) Change Management in der Hochschullehre: Die nachhaltige Implementierung von e-Learning-Innovationen. Zeitschrift für Hochschulentwicklung 3:3–15

Etzkorn N, Singer-Brodowski M (2017) Executive summary. Verankerung von Bildung für nachhaltige Entwicklung im Bildungsbereich Hochschule. FU Berlin. Institut Futur. https://www.bne-portal.de/sites/default/files/downloads/WAP_BNE_executive_summary_hochschule_0.pdf. Accessed 15 Jan 2020

Etzkorn N (2017) Executive summary. Hochschulen als Leuchttürme für Bildung für nachhaltige Entwicklung - Auf dem Transformationspfad die Strahlkraft erhöhen. FU Berlin. Institut Futur. https://www.bne-portal.de/sites/default/files/downloads/Experteninterviews_Executive_Summaries_Hochschulen_0.pdf. Accessed 15 Jan 2020

Hoch N (2018) Bildung für Nachhaltige Entwicklung (BNE) in der Hochschullehre (Betaversion). https://www.hochn.uni-hamburg.de/-downloads/handlungsfelder/lehre/hoch-n-leitfaden-

bne-in-der-hochschullehre.pdf. Consortium of the Hoch N Project, Hamburg. Accessed 20 Feb 2020
Israel D (2018) Alle Beteiligten auf dem Weg in die digitale Weiterbildung mitnehmen. In: Fischer H, Köbler T (eds) Postgraduale Bildung mit digitalen Medien. Problemlagen und Handlungsansätze aus Sicht der Beteiligten, Medien in der Wissenschaft. Waxmann, Münster, New York, pp 173–178
Michelsen G (2016) Policy, politics and polity in higher education for sustainable development. In: Barth M, Michelsen G, Rieckmann M, Thomas I (eds) Routledge handbook of higher education for sustainable development. New York, Routledge, pp 40–55
Michelsen G, Wels PJ (2017) A decade of progress on education of sustainable development. Reflections from the UNESCO Chairs Programme. UNESCO, Paris. https://unesdoc.unesco.org/ark:/48223/pf0000252319. Accessed 15 Jan 2020
Müller-Christ G (2014) Nachhaltiges Management: Einführung in Ressourcenorientierung und widersprüchliche Managementrationalitäten, 2nd edn. Nomos, Baden-Baden
Müller-Christ G, Liebscher AK, Hußmann G (2015) Nachhaltigkeit lernen durch Systemaufstellungen. Zeitschrift für Wirtschafts- und Unternehmensethik, zfwu 16:29–51
Müller-Christ G, Giesenbauer B, Tegeler MK (2017) Studie zur Umsetzung der SDG im deutschen Bildungssystem. Bremen: Universität Bremen. https://www.nachhaltigkeitsrat.de/wp-content/uploads/2017/11/Mueller-Christ_Giesenbauer_Tegeler_2017-10_Studie_zur_Umsetzung_der_SDG_im_deutschen_Bildungssystem.pdf. Accessed 15 Jan 2020
Schneidewind U, Singer-Brodowski M (2015) Vom experimentellen Lernen zum transformativen Experimentieren - Reallabore als Katalysator für eine lernende Gesellschaft auf dem Weg zu einer Nachhaltigen Entwicklung. Zeitschrift für Wirtschafts- und Unternehmensethik, zfwu 16:10–23
Scott W (2016) Sustainability and universities: an examination of conceptual frames, graduate attributes and learning outcomes. In: Barth M, Rieckmann M (eds) Empirische Forschung Für Nachhaltige Entwicklung – Themen, Methoden, Trends. Barbara Budrich, Opladen, pp 15–35
SCoRe-Consortium (2018) Projektantrag Verbundprojekt: Videobasiertes Lernen durch Forschung zur Nachhaltigkeit: Student Crowd Research
Shephard K (2004) The role of educational developers in the expansion of educational technology. Int J Acad Develop 9:67–83. https://doi.org/10.1080/1360144042000296062. Accessed 15 Jan 2020
Sterling S (2001) Sustainable education: re-visioning learning and change. Green Books Ltd., Foxhole
United Nations (2015) "A/RES/70/1". Transforming our world: the 2030 agenda for sustainable development. United Nations General Assembly, New York
Vilsmaier U, Maier E (2017) Forschendes Lernen in der Nachhaltigkeitswissenschaft. In: Mieg HA, Lehmann J (eds) Forschendes Lernen: Wie die Lehre in Universität und Fachhochschule erneuert werden kann. Campus, Frankfurt a. M., pp 357–366

Jonas Joachims has studied International Business and Management at the Bochum University of Applied Sciences and at the Manchester Metropolitan University. Afterwards he studied Business Psychology at the University of Bremen. After participation in a trainee program in the insurance sector he returned to the University of Bremen as a scientific associate in the field of Sustainable Management and digital learning. Since then, he worked for the project "Virtual Academy of Sustainability" and the research Project "Student Crowd Research (SCoRe). Besides this he is his writing doctoral thesis about the implications of digitalization on human resource development.

Katharina Lingenau has studied Economics at the University of Bremen, Germany and received her Diploma in economics/Sustainable Management in 2009. After working in the renewable energy sector, she returned to the University of Bremen as a scientific associate in the field of Sustainable Management and digital learning, especially for the project "Virtual Academy of Sustainability" in 2012. Since the beginning of 2017, she is responsible for the scientific coordination and the leadership of the Virtual Academy and the research project SCoRe.

Climate Change Education in Ukrainian Universities: Addressing Transformative Education for Sustainable Development

Oleksandra Khalaim and Shepherd Urenje

1 Introduction

As a society living in an emerging and highly dynamic global environment, we not only require future thinking but also effective enablers to transform our thinking practices. From a transformative perspective, the purpose of education for sustainable development (ESD) is to ensure that students are educated in a way that enables them to act for a sustainable future (Hopkins 2013). In many global reports focusing on the educational response to the issues, there is agreement that sustainability requires a reorientation of education and training, including higher education.

Thus, ESD focuses on rethinking how appropriate and successful educational practices may change and transform society. This social transformation often involves the emergence of new forms of human activities that sustain viable livelihoods. Biesta (2013) argues that, besides focusing on knowledge and socialization, one important goal of education is to focus on the emancipation of students from predetermined ways of thinking and being and the responsibility that follows with it and one pedagogical model that deals with this subjectification processes is transformative learning. That is why ESD allocates a wide range of participatory, experiential, critical and open-ended methodologies for teaching and learning. In this perspective, Climate Change Education for Sustainable Development (CCESD) ensures that education systems are prepared for current and emerging challenges, are responsive to them, serving at the same time as a proactive force in moving the sustainable development agenda forward (Mochizuki and Bryan 2015).

O. Khalaim · S. Urenje (✉)
Swedish International Centre of Education for Sustainable Development, Uppsala University—Campus Gotland, 621 67 Visby, Sweden
e-mail: shepherd.urenje@swedesd.uu.se

O. Khalaim
e-mail: oleksandra.khalaim@swedesd.uu.se

W. Leal Filho et al. (eds.), *Universities, Sustainability and Society: Supporting the Implementation of the Sustainable Development Goals*, World Sustainability Series,
https://doi.org/10.1007/978-3-030-63399-8_25

Our main contribution to CCESD related research is to expand the importance of higher education as an essential platform for introducing CCESD approach, using examples from Ukraine. We try to highlight possibilities for preparing university teachers and students who are more committed to engaging in personal and social transformation processes. In this paper, we discuss CCESD in three aspects:

1. As an environment that helps to extend transformative learning,
2. As a set of critical reflection processes to challenge existing beliefs, values and behavior,
3. As an enabler for co-creation of new, and more sustainable, ways of thinking and being.

The nature of current sustainability challenges demand that dominant pedagogies and forms of learning in higher education be reconsidered to enable students and staff to deal with accelerating change, increasing complexity, contested knowledge claims and inevitable uncertainty (Wals and Benavot 2017). Thus, the role of higher education to challenge learners with alternative ways of interpreting their experiences is an important part of transformative social learning, to disrupt unsustainable practices and create new, and more sustainable, ways of thinking and being (Wals 2007, 2010).

A UNEP Report suggests that while society has produced large amounts of knowledge to confront complex global environmental and climatic challenges, it is still lacking in responding effectively to these challenges. Due to insufficiency of knowledge of problematic issues in itself, UNEP suggests a range of action-oriented capabilities, including development of new modes of learning. In addition to acknowledging the need for adaptation and mitigation, there is also a need for transformative learning.

In this paper, we present transformative learning as a generic process that we believe is key to shaping the possible, probable, and preferred futures arising from the consequences of climate change. The paper aims investigating quality teaching and learning through transformational perspectives of CCESD in formal university education in Ukraine. The main objectives are to identify functional factors in the Ukrainian system of formal higher education that promote or hinder CCESD's embedment, and thus strengthen transformative ESD. In addition, we aim to deliberate on different transformative, flexible, locally applicable yet generic pedagogical and teaching methods related to CCESD, assessing the potential for effective teaching and learning opportunities in higher education.

2 Theoretical Background

2.1 Global Climate Change Phenomenon as a Specifically New Learning Objective

Nowadays, human-induced climate change impact is presented at the global level (IPCC 2018), affecting mostly all communities worldwide. At the same time, climate change can be regarded as a relatively new social-ecological phenomenon, raised in scientific discourses in the past 20–30 years only (Kronlid 2014). That is why education needs to develop capabilities to enable adaptation and mitigation, as well as reduce vulnerability. A necessity in a transformation of climate change related education occurred, caused by constant changes in paradigms about the nature of climate change adaptation, and its relation to social systems. Since climate change learning process is related to a challenge of "epistemic uncertainty" of climate change negative impacts, the focus should be given to "transformative approach to climate change learning as capability" (Kronlid 2014).

Historically, universities are being regarded as institutions responsible for forming the societal future by addressing the main ecological and socio-cultural challenges and enabling the development of key competences in their students (Rieckmann 2012; Kronlid 2014; Facer et al. 2019). Currently, this university function has been put in doubt considering the role of universities in responding to "one of the fundamental failures of contemporary anticipation: namely, the significant warming of the climate, and its consequent material, social and ecological disasters" (Facer et al. 2019).

Besides that, traditional university education tends to be "fact-based" one, losing the civic empowerment component. It is often focused on "acquisition metaphor" on learning, as one dealing with "knowledge, concept…, schema, fact, material, contents" (Sfard 1998). Considering the specificity of climate change phenomena, it could be important to introduce a "participation metaphor" as a more relevant view of university learning. It indicates key competencies as a clear way to action through "reflection, communication, … democratic competence and reflective knowing, … development through participation … and adult learning" (Sfard 1998).

A gap between climate change science and youth civic activism has been reported recently in US higher education as well. Universities do not provide comprehensive and uniform access to locally-produced climate change research resources (Goldman and Hyams 2019). Moreover, the climate change learning still is dominated by technical and scientific discourses, lacking educative analysis, social justice, and ethic research (Kronlid 2014). Hence, more effort on science communication and social participation is needed.

Climate change learning is directly related to key human capabilities of ensuring social security and well-being; climate change education can even serve as an institutional conversion factor. "Transformative climate change education … is important, as it might help turn adaptation resources into more viable functionings under the postnormal condition of climate change" (Kronlid 2014, p. 76). In other words, education could offer a sustainable source of local capacity and solutions due to knowledge transfer from learners to different social actors and future generations. As stated by Mochizuki and Bryan (2015), "once informed, citizens are more able to participate in civil society and to influence decision-making in areas that affect them, particularly at the local level."

Also, more personal authenticity in climate change learning methods should be added, regarding the specific context of the topic, which reflects a certain mixture of both global and local implications. According to Lundegård (2018), the type of students' authenticity created in the educational process highly depends on the level of students distance themselves from the subject matter (a climate change problem in our case). Hence, the learning questions should be formulated in a way to give the students "an opportunity to establish value-relations, or to dare to take a personal standpoint, as political subjects, to the issues" (Lundegård 2018).

2.2 ESD Incoherence in the Formal Education and Current Necessities of Methodological Transformations in ESD

The challenges of a certain incoherence and integrity violation of ESD in formal education are pointed out on the international level. As it is indicated in the post—Global Action Programme (GAP) position paper on the future of ESD, as an UNESCO analysis of the 5-year cycle of the GAP (2018), "ESD tends to be treated as a thematic topic… This has to change." Despite the fact that ESD has been positioned as a holistic approach, it still needs to be implemented in more systemic level. UNESCO emphasizes also an importance of a transformative action in ESD, pointing the ways of encouraging learners to undertake transformative actions for sustainability as a major preoccupation for ESD (A post—GAP position paper 2018).

It is also agreed (Kronlid 2014), that transformative rather than conservative climate change education could facilitate social and individual change processes, playing a key role in understanding and facilitating social change, both within and outside of the classroom. We understand transformative learning here as one that develops capacities, literacies and forms of citizenship based on sustainable values and principles. Learners are empowered with capabilities of altering their values and behavior, finding solutions for themselves, especially when the desirable course of action is unclear.

Included in the subject content, it can involve learning about the nature of people's relationship with each other and with the planet in order to enable citizens to tackle the root causes of environmental mismanagement. Emancipatory approaches should promote action-oriented, collaborative, participatory, and transformative learning (Wals and Benavot 2017). That is why a necessity in more flexible and universal climate change learning methods arises.

3 Contextual Profile

3.1 Quality Education and Climate Action in Ukraine

In September 2019, President of Ukraine Volodymyr Zelensky signed a decree to ensure compliance with 17 goals of sustainable development of Ukraine for the period until 2030 (Decree of the President of Ukraine on September 30 2019 № 722/2019). He instituted an expert group to analyze the compliance of state program documents to Ukrainian SDGs in two-month term, and to establish a working system of monitoring of SDGs' realization. At the same time, analyzing Ukrainian obligations on Agenda 2030, a formal rather than real baseline and benchmarks are indicated in "Sustainable Development Goals: Ukraine" national report provided by the Ministry of Economic Development and Trade of Ukraine in 2017. A closer look at the national targets related to SDG 4 (Quality Education), a strong focus on primary and secondary (school) education in set, leaving out of board the university level. For the national SDG target 4.4 "Improve the quality of tertiary education", a number of university towns is set as an indicator, which cannot be regarded as a qualitative one. Despite the fact that key recommendations for achieving the goal include a task to "update learning content by introducing curricula on sustainable development", ESD is not mentioned in the indicators and the report itself. Moreover, it is stated in the report, that "the level of school enrolment and the literacy rate of the adult population are traditionally high in Ukraine", which contradicts with the current analytical findings on the educational reality (Kahkonen 2018; OECD 2017; CEDOS 2015; Filiatreau 2011).

According to the World Bank data, in 2017 Ukraine spent 5.4% of GDP on education, which is one of the highest rates of public spending on education in the world, keeping the same tendency of previous years (CEDOS 2015). At the same time, according to an independent analytical review (CEDOS 2015), Ukrainian teachers "are paid several or even up to ten times less compared to teachers in Western Europe, as well as throughout the post-Soviet and post-socialist camp". Along with that, a high level of academic corruption is highlighted in the Ukrainian higher education recently (Filiatreau 2011; OECD 2017; Friedman and Trines 2019).

At the certain extent, it explains the situation with a rather low quality of the formal higher education in Ukraine, in terms of international integrity, transparency, teaching methodology, and content. "Both the approaches towards teaching and towards learning are out of date… Lessons are overburdened with theoretical knowledge and lacking in practical application, while textbooks are academically dry… dampening motivation and curiosity among students" (Kahkonen 2018). The last version of the Ukrainian law "On Education" tries to change the situation by prioritizing competencies (and the "environmental awareness" as one of them) rather than theoretical knowledge by means of lifelong learning under the "New Ukrainian School" concept—a key reform of the Ministry of Education and Science since 2017 (The New Ukrainian School 2016). However, ESD, as a core component of the 2030 Agenda for Sustainable Development, is not presented both in the Ukrainian national laws "On Education" and "On Higher Education" at all. There are no formal standards and detailed provisions of ESD, ecological education, and specifically climate change ESD as its part. Universities add climate change related topics to specific courses, hardly trying to link ESD to university and campus operations by adopting a 'whole-of-university' approach to sustainability. Thus, there is no evidence that formal education in Ukraine is responding to current climatic challenges, being not relevant to the actual situation with the climate change "out of classroom".

Our research could also make an input in covering the gap on climate change learning, not presented in SDG 13, formulated in "Sustainable Development Goals: Ukraine" national report (2017) as "Mitigate climate change impact"—with no word on adaptation. There is only one national target set there, which is "Limit greenhouse gas emissions in the economy", compared with 1990 level. One out of two key recommendations presented in this section emphasizes the necessity to "Inform the public about the consequences of climate change and mechanisms and practices to combat and adapt to them, and implement appropriate educational programs at all levels of formal and informal education."

Thus, there is a room left for initiatives related to CCESD in Ukraine. Few research questions could be posed at this stage: (1) what hindrances and opportunities concerning transformative ESD are expressed in interviews with Ukrainian teachers; (2) In what way does the respondents' information correspond with what is defined as key components in a transformative learning process?

4 Materials and Methods

In order to combine a necessity of certain scaling and randomization of field data with precise selection of the respondents that can highlight key problematic issues we focus on in this research, a downscaling design of the survey planning was introduced, as presented in Fig. 1.

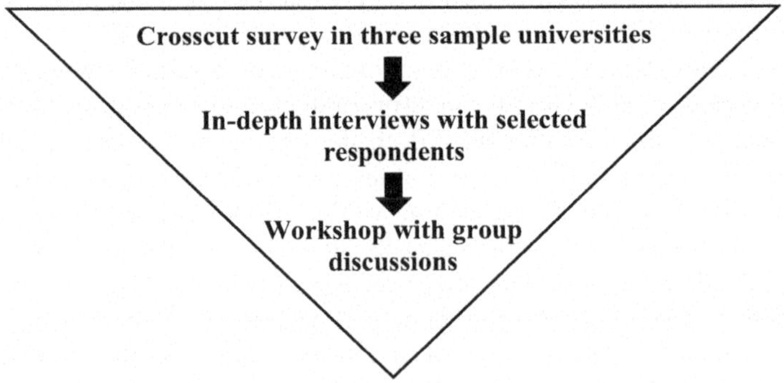

Fig. 1 Downscaling design of the survey planning. *Source* The authors

We started with a crosscut survey in three sample universities, disseminating the questionnaire related to climate change teaching through all university departments. Afterwards we selected a certain amount of interviewees among willing respondents for in-depth interviews, and a final face-to-face group discussion with research participants is planned in order to elaborate deeper on the problematic questions raised.

4.1 Crosscut Survey

Our case is defined as a phenomenon occurring in a bounded context of three universities in Ukraine. A purposive sampling method (Palinkas et al. 2015) was used to determine the unit of analysis because we wanted to access a particular subset of interviewees from three universities out of over 800 universities in the Ukraine that offer higher education. The criterion included universities in which we had collegial access, with an existing Uppsala University partnership, and with teachers who had previously taken part in Baltic University Programme (BUP) ESD courses. We then settled on a sample of three universities, Taras Chevchenko National University of Kyiv where there is a current Erasmus mobility partnership with Uppsala, and the other two, National University of Kyiv-Mohyla Academy and Sumy State University that are active members of the BUP. A detailed information about the surveyed institutions is presented in Table 1.

Considering the selection criteria used in our survey, we are aware that this might have left out credible respondents from who we could have generated quality contributions. Yin (2003) suggests placing boundaries on a case to prevent trying to work with a case, which is too broad. In binding the case (Yin 2003) we ensured the interviews which were to follow will remain reasonable in scope. Once we have decided on our case a survey was then conducted to determine the extent to which climate change education was included as described below.

In order to understand the actual level of presence of the climate change topic in university teaching through different disciplines, we designed a short online, open and cross discipline survey. We did not aim at developing a comprehensive database

Table 1 Surveyed institutions in Ukraine

	University	Abbreviation	Year of foundation	No of students	City
1	Taras Chevchenko National University of Kyiv	KNU	1834	26,000	Kyiv
2	National University of Kyiv-Mohyla Academy	KMA	1615	4422	Kyiv
3	Sumy State University	SSU	1948	16,363	Sumy

for any generalizations; rather we tried to get an initial picture, as well as to involve interested teachers for in-depths interviews at the next stage of the research.

A short google-form based on-line questionnaire was designed for the university teachers from all subjects, with the aim to highlight the university disciplines in which climate change is reflected or at least is being mentioned. The questionnaire was disseminated in December 2019 and January 2020 among teachers from KMA, SSU, and KNU, in Ukrainian. The form was sent by e-mail to heads of faculties and departments, as well as to some teachers, whose contacts were available at the university official web pages. Besides the contact information, the list of questions included requests on teacher's field of study and list of his/her disciplines for 2019/2020 academic year.

There was one "yes/no" type of question: "Is climate change mentioned in your discipline/courses?" In case of a positive answer, respondents were asked to specify their disciplines/courses that have climate change topic. Next, there was one multiple-choice question: "How often are climate change related issues being mentioned in your course?" Answer options were presented as follows:

1. More than two times during the lectures only;
2. More than two times, both during the lectures and seminars;
3. Once in the course, in brief/generally, during the lectures only;
4. Other (please elaborate).

Finally, the respondents were asked about the possibility and interest to participate in the in-depth interview session related to the climate change education afterwards. From the corresponding answers, we were able to select participants for the in-depth interviews.

4.2 *In-Depth Interviews*

In-depth interviews were designed according to methodological recommendations of Guest et al. (2013) and Boyce and Neale (2006). Interviews aimed exploring (1) some aspects of the CCESD presence in Ukrainian formal university education through different disciplines, learning methods, and content, as reflected by interviewees; and (2) an expressed necessity, barriers and possibilities to introduce transformative, flexible, locally applicable pedagogical methods related to CCESD in the teaching practice of interviewees.

Target audience included Ukrainian university teachers, who cover climate change related topics in their disciplines at any extent, and expressed their interest both in the given research and transformative learning methods in CCESD in general. Out of the 35 teachers who participated in the crosscut survey, 18 expressed their interest in participating in the in-depth interview. From 18 who expressed their interested we chose ten teachers of different fields of study and with different climate change teaching experience: both those who practice it intensively and those who do it in general during lectures from time to time. Four of them were from KNU, three from KMA, and three from SSU. It was a manageable number for interviewing, transcription purposes, and results processing considering the time available to us. We scheduled the interviews for January 2020; each interview average duration was about one hour. All interviews were video-recorded and transcribed in Ukrainian, with further partial English translation.

There were three blocks of questions designed in the interview protocol: (1) mechanics, processes, and methods of climate change teaching; (2) personal attitudes of teachers towards climate change problem and their observations at any climatic scientism in the university community; and (3) necessities, barriers, opportunities for better climate change teaching for respondents personally and in Ukrainian higher education generally. The detailed interview protocol is included as an Annex I to the paper.

In the selection process, we tried to access different specialties, as well as different level of engagement in the climate change teaching indicated by respondents in the first survey. Thus, among 10 selected interviewees, six answered to the question "How often are climate change related issues being mentioned in your course?" – "More than 2 times, both during the lectures and seminars", three answered "Once in the course, in brief/generally, during the lectures only", and one answered "other", meaning that her courses are directly related to climate change problem. Five fields of study were included as follows: five economists, one mathematician, one ecologist, two climatologists, and one phycologist. It just so happened that there were more participants from economics.

5 Results and Discussion

The problematic questions of this paper address hindrances, internal and systemic barriers, communicational and cultural challenges of educational environment that could inhibit a process of transformative learning methods' implementation into higher education. Here we highlight some challenges and cases of transformative learning in regard to CCESD necessities in Ukraine, emerging from the two-stage interviewing processes.

5.1 Crosscut Survey of Three Sample Universities in Ukraine

In December 2019–January 2020, we received 35 responses from three sample universities; the majority of respondents—fifteen—were from SSU following by thirteen from KNU, and with seven only from KMA (see Fig. 2). All respondents except two hold a PhD degree; four of them were working as heads of departments, and two—as deans.

Regarding the field of studies and disciplines, the biggest part of respondents teach Economics (related to Social Sciences according to UNESCO International

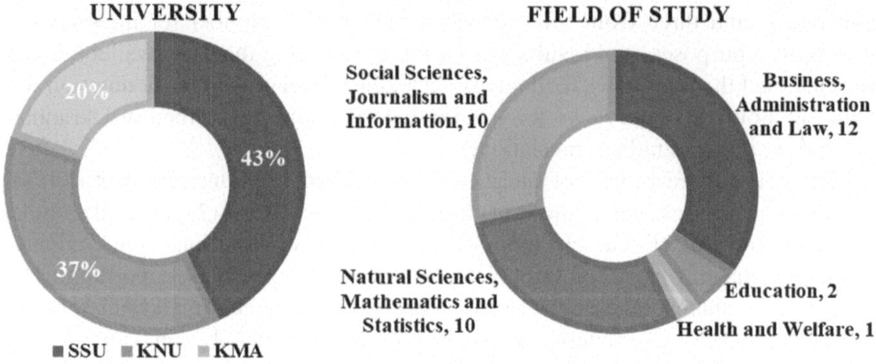

Fig. 2 The number of crosscut survey respondents of each sample university (%) and fields of study represented by them. *Source* The authors

Standard Classification of Education); there was a big number of Marketing disciplines named, and few cases of Mathematics, Ecology, Climatology, Pedagogy of geography, Psychology, Sociology, and Corporate Social Responsibility. The full list of disciplines indicated by respondents as that include climate change topic is available in the Table 2.

The disciplines where climate change was not mentioned, included Marketing of Innovations, Socio-Political Studies, Ukrainian Society in Sociological Dimensions, Social Structure and Inequalities in Society, Professional and Corporate Ethics, Methods of Data Analysis in Sociology, and Methodology of Sociological Research. Interestingly, some of disciplines that included climate change topic had almost the same titles as those not. It indicates a personal "goodwill" of a teacher in the decision of including the climate change topic, rather than any external factors, as well as a certain absence of program documents, recommendations, or policies on the faculty/university level regarding CCESD.

Under the conditions of equal informational access, a prevailing amount of economists participated in survey and interested in in-depth interview could be explained by higher level of their openness to any kind of international cooperation and external communication activeness, in comparison with teachers from natural sciences, who could be focused more on their current educational and research work.

Regarding the content, four respondents do not mention climate change topic in their disciplines at all: among them two sociologists, one marketing teacher, and one phycologist (see Fig. 3).

Eight teachers mention it "once in the course, in brief/generally, during the lectures only"; two mention it "more than 2 times during the lectures only", and 19 mention it "more than 2 times, both during the lectures and seminars". Undoubtedly, we cannot generalize this proportion even at the university scale; it rather means the participation in the survey mostly of those teachers who were interested in the topic, despite our attempt to emphasize that the survey is open for all teachers regardless their connection to climate change topic in teaching.

Table 2 The full list of disciplines indicated by respondents of the crosscut survey as those where climate change problem is mentioned* (translation from Ukrainian by authors)

Broad field	Narrow field	Disciplines
01 Education	011 Education 0111 Education science	Organization of teacher's work, pedagogical skills, extracurricular and extracurricular work in school, teaching methods of geography, research models and interpretations in social pedagogy
03 Social Sciences, Journalism and Information	031 Social and behavioral sciences 0311 Economics	International economic relations, world economy, economics of resource conservation, environmental economics, management decision-making methods, investment management, keystone, sustainable development studios, macroeconomics, social responsibility, history of economic thought, modern strategies of international civilizational development, international transfer of technologies, global economy
	0314 Sociology and cultural studies	Sociology, sociology of politics, sociology of morality, sociology of branding, methodology for the study of social inequalities, modern principles of consumer culture, quality management and social work, cross-cultural communication, cross-cultural competency
04 Business, Administration and Law	041 Business and administration 0413 Management and administration	Electronic governance, decision-making technologies in public administration, public procurement management, european integration, international public administration and security, crisis management, creative management, project management
	0414 Marketing and advertising	Marketing, basics of stock trading, green marketing, pharmaceutical marketing, marketing management
05 Natural Sciences, Mathematics and Statistics	052 Environment 0521 Environmental sciences	General ecology, ecology of terrestrial systems, environmental safety, environmental safety of human activities, occupational health and safety, soil studies

(continued)

Table 2 (continued)

Broad field	Narrow field	Disciplines
	053 Physical sciences 0532 Earth sciences	Fundamentals of biometeorology, urban meteorology, economic meteorology, meteorology and climatology, applied meteorology, geography of travel and tourism, geography of agro-recreation, geography of winter recreation, climate change and chemistry of atmosphere, climate of Ukraine, recreational climatology, fundamentals of meteorology, hydrometeorological risks and methods of their estimation, meteorological and hydrological studies, climatology with the basics of meteorology, fundamentals of earth sciences: Module 4—basics of meteorology, modeling in agrometeorology, diagnosis of geographical knowledge
	054 Mathematics and statistics 0541 Mathematics	Higher mathematics
	0542 Statistics	Statistical data processing and operations
09 Health and Welfare	092 Welfare 0923 Social work and counselling	Resocialization, social rehabilitation and social correction, International practice of social rehabilitation, quality management in social work, social phycology of trauma and traumatic actions

*According to UNESCO International Standard Classification of Education (ISCED F 2013)

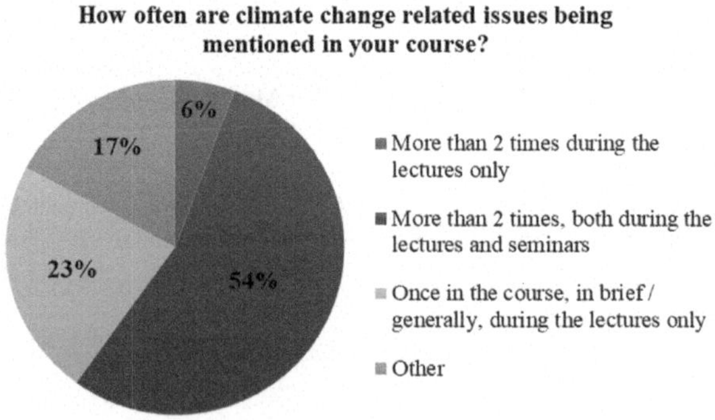

Fig. 3 The frequency and level of coverage of climate change related disciplines by respondents. *Source* The authors

5.2 In-Depth Interviews

The analysis of answers of the first block of questions, related to teaching methods and content (see Interview Protocol in Annex I), has shown a general tendency of non-distinguishing by respondents climate change problem from general environmental problems (waste management, air pollution), as well as from sustainable development concept and its other 16 goals. Teachers named a wide range of topics in connection to climate change: from psychological aspects of ecological awareness, influencing life quality, to mathematic modelling of industrial emissions depending on urban wind rose. The topics were clearly linked to teachers' field of studies and were mainly thematic applications of the climate change problem. Thus, economists mentioned climate change impact to supply and demand, business, agriculture, international economic situation, strategies of sustainable development. Mathematician used environmental data to practice statistical analysis and different types of modeling; psychologist was highlighting ecological impact on the cross-cultural communication. A number of teachers used climate change and sustainability related topics for student practical assignments and individual projects, such as drafting sustainable development strategies for a city, pollution modelling, business start-ups' development etc. Remarkably, none of respondents tried to link global climatic threat to students' lifestyle as individuals who bring an impact to climate as well. The problem of climate change was regarded by the majority of teachers from the academic perspective rather than a reason for any specific local ecological civil actions. This outcome is clearly connected with the teachers concern on students' demotivation in solving the ecological problems they claim in the next stages of the interview; we will return to this point later.

The amount of students covered by climate change related topics per one teacher widely varied, from 20 to 30 master students to more than 300 bachelor and master students during the academic year. It depended on the subject specialty (there were more students in economics than ecology), and on the size of a university.

Among pedagogic methods used in climate change teaching, respondents named mainly traditional ones, such as lectures, group work, individual assignments, case studies, sometimes debates. Few teachers use game method in course seminars, trying to invent their own games and visualization of course materials. Few teachers were involved previously in BUP ESD course, and used change project approach (Urenje 2019) in their academic work.

At the same time, the majority of teachers highlighted the necessity to organize lectures in more interactive way, to meet students' demand in new format of presenting the information. They include short videos describing the topic, initiate discussions during the lecture, and by default make Power Point presentations instead of talking or reading any printed materials in loud. Students are less eager to write down the biggest part of lecture materials; they do not perceive "long reads" and are less concentrated to one point. They are "short-memory generation", as one of respondents commented.

The assumption about a personal "goodwill" of a teacher in the decision of including the climate change topic in his/her discipline, preliminary indicated in the crosscut survey results, was clearly confirmed by answers under in-depth interview. All teachers include topics related to sustainability, environmental issues, and climate change to their disciplines as their own initiative and decision. Nobody claimed that it was an obligation of curricula or external requirement of the department or faculty. There were few courses mentioned as originally designed by teachers themselves; the majority of respondents expressed their satisfaction related to courses' content and explained it as an expression of their personal awareness, as well as an attempt to follow the current global challenges. Here it is important to recall the national policy documents analyzed in the "Contextual Profile" section above, that do not provide any formal basis for including CCESD in university disciplines. Since there are neither ministerial nor university policy recommendations on ESD and CCESD in particular, teachers have to make their own initiative based on their personal interest.

The second block of interview questions was related to personal attitudes of respondents towards climate change problem and observations on any climatic skepticism in the university community (see Annex I). Here all respondents agree that climate change is a relevant problem for Ukraine and feel its impact on themselves. The majority do not observe any climatic skepticism neither in terms of climate change denial, nor in its anthropogenic origin. Few teachers were complaining about the low level of students' interest in this topic, and somewhat practical, "cynic" attitude towards the educational process. "I have a feeling that they just need a university degree for a further carrier… They are not truly interested in the content and not motivated to really dive in the climate change problem"—claimed one of the respondents. This point turns us to a discussion on the purpose of education for sustainable development as that growing good citizens participating in civic activities and taking care of the future of society, rather than good specialists in their field. The students' attitude towards education described by respondent shows a misperception of the educational goals by students. On the one hand, it could be a result of general fatigue and helplessness caused by general negative pressure of mass media about the issue. On the other hand, it raises a question if the university community and the Ukrainian system of formal higher education efficiently introduce ESD to avoid these misperceptions. Additionally, teachers' personal example of ecological awareness and sustainable lifestyle can be questioned. Do they provide enough personal evidence for students to take action and be motivated in solving climate change induced threats? Alternatively, do we have a situation of a commonly inactive citizen position of the majority of teachers which results in the students' "eco-cynicism" described by respondents?

To solve the problem of students' lack of basic knowledge and motivation to interdisciplinary studies, some respondents proposed to set up an obligatory course of Ecology or Sustainable Development to all students regardless the specialty. However, none of them mentioned ESD framework as an approach to induce positive changes, to face lack of concern and lack of civic action. The promise of ESD is on rethinking how appropriate and successful practices may change and transform society by engaging with new forms of learning and human activities which sustain viable livelihoods. Four responses to the sustainability challenge were identified by

Sterling in 2004: (a) denial, (b) adding a 'green aspect' to a curriculum or programme, (c) integrating sustainability in all teaching and learning and (d) re-thinking (and re-orienting) the very foundations of what we currently do. In our experience, the first step usually begins with (b) and (c) with an ambition for a systems approach to institutionalization of ESD.

Finally, the third block of interview questions provided a possibility for the respondents to describe any barriers and necessities in their climate change teaching. Here many of them expressed the need in a common communicational interdisciplinary platform, to be able to share knowledge and expertise on climate change problem with colleagues from other fields of studies. Some teachers from Humanities described their demand for an expert support from colleagues of Natural Sciences, which had not been met due to working overload, lack of both corresponding procedures and any precedents of such interfaculty communication. "I could barely come for a consultation to my colleagues from Ecology department… but I do not know them, they are possibly busy"—says an Economy teacher. Another crucial challenge indicated by respondents concerning interdepartmental communicational gap was related to a traditional science division, which creates an additional barrier to sustainability teaching. "All our educational system is too regulated by the formal division of sciences… our university is not flexible for any interdisciplinary studies"—claims another teacher of Humanities. As possible solutions, respondents proposed to organize repeated common scientific seminars, which can set up a dialogue between academic staff in a systemic manner introducing a community of practice.

5.3 CCESD for Teachers

Apart from the abovementioned, a continuous professional development of university teachers could be helpful in ensuring the right working methods and socially proactive approach in CCESD. Baltic University Program (BUP) Teachers Course on Education for Sustainable Development in Higher Education can serve as a good example of initiative addressing this demand. The 7-month course focuses on the implementation of SDGs in the learning methods by a better integration of SDGs into teachers' own practical work, courses and study programmes. The core idea of the course is to introduce and practice the change project approach for bringing the interdisciplinarity to teaching practice. Course participants are expected to develop a change project based on their already existing courses/study programs in order to change the content for better ESD integration. Thus, planning and implementation of the change project is regarded as the main outcome of the course, with a peer review support by participants and course coordinators via the e-platform Moodle and during two field workshops (BUP official web site, 2020).

This format and example of mutual interdisciplinary collaboration and professional development of university teachers from different universities under the umbrella of the BUP ESD course can serve as a good starting solution to address challenges expressed by respondents in the block 3 of interview. Firstly, it forms a platform for interdisciplinary communication not only between teachers of different specialties, but also belonging to different universities of the Baltic region. Secondly, the course provides a clear roadmap to induce sustainability changes into teaching practice by developing and realization of a concrete small-scale individual change project. Finally, it brings a set of transformative learning methods to be practiced and afterwards introduced by course participants into their teaching work.

To enhance further the course achievements at the university level, a comprehensive and clear starting list of indicators to achieve basic CCESD knowledge and skills could be introduced. As proposed in UNESCO document "ESD Goals: Learning Objectives" (2017), three groups of learning objectives could be used: (1) cognitive learning objectives (related to the crucial understanding points); (2) socio-emotional learning objectives (related to capabilities to encourage, collaborate, and share the common understanding); and (3) behavioral learning objectives (related to ability to anticipate, estimate, and act). Here again we can observe a clear shift to the transformative approach to climate change learning as capability, more than steady knowledge on climate change-related issues.

Moreover, as suggested by Mochizuki and Bryan (2015), a number of key learning capabilities associated with CCESD could be introduced in a systemic manner. The list can include critical thinking, systems thinking and problem solving skills dealing with rapid change and uncertainties, assisting in analyzing, synthesizing and evaluating information, as well as management skills and life-long learning skills together with information, media and technology skills (Mochizuki and Bryan 2015).

In the twenty-first century, teaching and learning locations and processes are not merely about understanding social change; classrooms are catalytic incubators where personal and social transformation is generated and pursued. In order for these efforts to succeed, ESD has to consciously address and account to local context by initiating learning processes needed for achieving local sustainability challenges. The BUP ESD course provides a learning environment for experimenting on tools and approaches that university teachers can draw on to support students in developing the skills, capacity and capabilities needed to foster social transformation, becoming change agents in the face of local and global environmental challenges like climate change.

6 Conclusions and Recommendations

In 2020, the majority of universities worldwide are facing a transition period towards Education for Sustainable Development (ESD), in order to be capable to meet global threats and address the students' demand of being good citizens rather than barely good specialists in their field of study. Ukrainian universities are in the same process, as could be concluded by the given research in general.

On the one hand, there is an essential gap in the formal framework and regulations regarding ESD implementation in Ukrainian higher education. Since Education for Sustainable Development and Climate Change Education neither are clearly visible in official policy documents on higher education nor are reflected in the national targets of Agenda 2030, there is no platform for further university regulations' development towards ESD. The learning objectives for SDG 13 together with suggested topics and examples of learning methods derived from "Education for Sustainable Development Goals: Learning Objectives" (UNESCO 2017) could be taken currently as basic starting indicators for CCESD framework in Ukrainian higher education programs, along with the list of key learning capabilities indicated for CCESD by Mochizuki and Bryan (2015).

On the other hand, both the crosscut survey and in-depth interviews indicated not only challenges in climate change sustainability teaching that respondents face in their current work, but also a "grassroots" intension of university teachers to transform their educational practice towards ESD. The crosscut survey highlighted a personal teachers' "goodwill" in the decision of including climate change topic in their disciplines regardless the specialization and field of study. It was the primary reason with any influence of external factors, such as formal requirements of policy documents at the faculty/university/ministerial level. This observation was supported by results of in-depth interviews that lead us to conclusion about naturally induced individual action replacing the absence of regulatory framework.

The main insights derived from in-depth interviews allowed to draw the initial state-of-the-art, as well as challenges and necessities articulated by teachers together with their suggestions in order to change the existing situation for better.

Remarkably, the majority of respondents mixed climate change threat with other environmental problems and any sustainability issues in overall, at the same time hardly trying to link global climatic threat to their own and students' lifestyle as individuals who bring an impact to climate. In addition, teachers regarded the climate change problem purely from the academic perspective, not linking it to any local ecological civil actions. As a result, they were complaining to students' low interest in the topic and lack of motivation to dive into climatic research. It could be questioned, if they provide enough personal evidence for students to take action and be motivated in solving climate change induced threats. Should ESD be introduced more clearly in Ukrainian higher education to avoid such misconceptions?

In terms of methods, teachers highlighted the necessity to organize their lectures in more interactive way and to enlarge their methodological toolkit. Alongside, many of them expressed the need in a common communicational interdisciplinary platform, to be able to share methods and expertise on climate change problem with colleagues from other fields of studies. A working overload, lack of both corresponding internal procedures and any precedents of such interfaculty communication were named as main hindrances.

Hence, a number of working methods and cases could be highlighted to address these challenges and necessities described above. The format of BUP ESD Teachers Course based on the change project approach intends to address the lack of expertise and interdisciplinary communication. It could be empowered by an active campus network or a community of practice both on the inter-university and intra-university levels. The power of interdisciplinarity designed in a functioning way can bring faculties together for the common good.

What kind of criteria should we choose for methods and cases, applicable in CCESD, among a wide range of already existing ones? As already mentioned, in case of climate change learning we deal with a challenge of "epistemic uncertainty", constantly changing natural environment, and hence instability at all levels or learning stages—from a pre-knowledge level to final learning outcomes. That is why we regard three criteria as key and essential for CCESD methods: flexibility, universality, and applicability to the local context. In addition, five families of high-leverage approaches can serve as a methodological basis for climate change transformative learning: collaborative small-group learning, inquiry-based learning, experiential learning, service learning, and place-based learning (Nolet 2016). This enables social learning and further social transformation.

Therefore, a developed framework on methodological approach towards an effective CCESD in Ukraine can serve as a pilot case of transformative education implementation in Ukrainian system of formal education.

Acknowledgements Acknowledgements to the Swedish Institute (SI) for the financial support for the postdoctoral research of Oleksandra Khalaim under the supervision of Dr. Shepherd Urenje at Uppsala University (Campus Gotland).

Annex I

Climate Change Education in Ukrainian Universities: Addressing Transformative Education for Sustainable Development. Structure and content of in-depth interviews.

Concept, goals and objectives: The overall research is devoted to investigating quality teaching and learning through transformational perspectives of Climate Change ESD (CCESD) in formal university education. The interview outputs are expected to highlight:

1. Some aspects of the CCESD presence in Ukrainian formal university education through different disciplines, learning methods, and content, as reflected by interviewees.
2. Expressed necessity, barriers and possibilities to introduce transformative, flexible, locally applicable pedagogical methods related to CCESD in the teaching practice of interviewees.

Target audience: Ukrainian university teachers, who cover climate change related topics in their disciplines at any extent, and expressed their interest both in the given research and transformative learning methods in CCESD in general.

Timing and scale: up to 10 interviews with teachers from 3 universities; up to 1.5 h per interview; interview period—January 20–31 2020.

Interview Protocol

(Originally used in Ukrainian)

1. **Introduction**
   - A short description of the research
   - A short explanation about the idea and mechanics of the in-depth interview method
   - Disclaimer on confidentiality of the interview process
2. **Interviewing (list of questions):**
   Block 1: mechanics, processes, and methods.
   - What kind of topics do you reflect and how often do you deal with them during the academic year? How many students are covered?
   - How do you teach CC related topics (what methods do you use)? Is it lectures/seminars/workshops? Any informal meetings/initiatives related to CC leaded by you?
   - Are you required to bring CC issues by curricula or is it your personal choice?
   Block 2: personal attitudes.
   - Do you think CC is an urgent issue for Ukraine? Do you feel its impact on yourself?
   - Do you observe any climate skepticism or indifference among your colleagues/students/university administration? Are students interested in what you are sharing with them?
   - What CC related topics are interesting for you personally to be presented?
   Block 3: necessities, barriers, opportunities.
   - Are you satisfied with your CC teaching? Do you feel a necessity in new teaching methods?
   - What would you have changed in methods/content/courses' structure? Can you manage it by yourself? What kind of support is relevant?
   - What are the barriers for you to make CC teaching ideal, if any?
   - Do you see any system errors in Ukrainian higher education regarding the issue?
3. **Concluding remarks and short logistical information about workshop in Kyiv**

References

Biesta GJJ (2013) The beautiful risk of education, 1st edn. Routledge Publishers, Abingdon, p 178

Boyce C, Neale P (2006) Conducting in-depth interviews: a guide for designing and conducting in-depth interviews for evaluation input. Pathfinder Int Tool Ser Monit Eval 2:16

CEDOS (2015) Education in Ukraine digest (ed) Stadny Y. p 84. Available from https://cedos.org.ua/data/pdfs/EDUCATION_CEDOS_2015_en_updated.pdf [accessed Jan 17 2020]

Decree of the President of Ukraine on 30 Sept 2019 № 722/2019 "Про Цілі сталого розвитку України на період до 2030 року" Available from the web-database "Laws of Ukraine" of the Verkhovna Rada of Ukraine: https://zakon.rada.gov.ua/laws/show/722/2019?lang=en. Accessed on 17 Jan 2020

Facer K, Peters A-K, Holmberg J, Barrineau S, Mendy L, Granjou C (2019) How should universities care for the future? Anticipation Conference 2019, AHO Oslo

Filiatreau S (2011) Ukraine's participation in the Bologna process: has it resulted in more transparency in Ukrainian Higher Education Institutions? Int Res Rev J Phi Beta Delta Honor Soc Int Scholars. 1(1):49–64

Friedman O, Trines S (2019) Education in Ukraine. World Education News and Reviews (WENR). Available from https://wenr.wes.org/2019/06/education-in-ukraine. Accessed 17 Jan 2020

Goldman EB, Hyams M (2019) Climate science research in the United States and U.S. Territories: survey of Scientific Publications from Selected Public Universities (2014–2018). Published by the National Council for Science and the Environment. Available from https://www.ru.ac.za/elrc/people/staff/distinguishedprofessorheilalotz-sisitka/. Accessed 15 Dec 2019

Guest G, Namey EE, Mitchell ML (2013) Collecting qualitative data: a field manual for applied research. SAGE Publications, Thousand Oaks, pp 113–168. ISBN: 9781506374680

Hopkins C (2013) Educating for sustainability: an emerging purpose of education. Kappa Delta Pi Rec 49(3):122–125. https://doi.org/10.1080/00228958.2013.819193

IPCC (2018) Summary for policymakers. In: Masson-Delmotte V, Zhai P, Pörtner H-O et al (eds) Global warming of 1.5 °C. An IPCC Special Report on the impacts of global warming of 1.5 °C above pre-industrial levels and related global greenhouse gas emission pathways, in the context of strengthening the global response to the threat of climate change, sustainable development, and efforts to eradicate poverty. World Meteorological Organization, Geneva, 32 pp

Kahkonen S (2018) Why Ukraine's education system is not sustainable. World Bank official web-site. https://www.worldbank.org/en/news/opinion/2018/09/12/why-ukraines-education-system-is-not-sustainable. Accessed 19 Nov 2018

Kronlid DO (2014) Climate change adaptation and human capabilities: justice and ethics in research and policy. Palgrave Macmillan, US, p 211. ISBN: 978-1-137-42804-2

Lundegård I (2018) Personal authenticity and political subjectivity in student deliberation in environmental and sustainability education. Environ Educ Res 24(4):581–592. https://doi.org/10.1080/13504622.2017.1321736

Mochizuki Y, Bryan A (2015) Climate change education in the context of education for sustainable development: rationale and principles. J Educ Sustain Dev 9(1):4–26. https://doi.org/10.1177/0973408215569109

Nolet V (2016) Educating for sustainability: principles and practices for teachers. Taylor & Francis Group, Routledge, New York, p 218. https://doi.org/10.4324/9781315867052

OECD (2017) OECD reviews of integrity in education: Ukraine 2017. ORCD publishing, Paris, p 190. https://dx.doi.org/10.1787/9789264270664-en

Palinkas LA, Horwitz SM, Green CA, Wisdom JP, Duan N, Hoagwood K (2015) Purposeful sampling for qualitative data collection and analysis in mixed method implementation research. Adm Policy Ment Health 42(5):533–544. https://doi.org/10.1007/s10488-013-0528-y

Post-GAP position paper on the future of Education for Sustainable Development (ESD): draft for consultation (as of 15 June 2018) UNESCO, p 10

Rieckmann M (2012) Future-oriented higher education: which key competencies should be fostered through university teaching and learning? Futures 44(2):127–135. Available from https://www.researchgate.net/publication/232364850_Future-oriented_higher_education_Which_key_competencies_should_be_fostered_through_university_teaching_and_learning. Accessed 15 Jan 2020

Sfard A (1998) On two metaphors for learning and the danger of choosing just one. Educ Researcher 27(2):4–13. https://doi.org/10.2307/1176193

Sterling S (2004) The ecological and environmental dimensions of the holistic curriculum. In: Encyclopaedia of life support systems (Theme 6.6.1 'Education for Sustainability'). EOLSS Publishers, UNESCO

The Ministry of Economic Development and Trade of Ukraine (2017) Sustainable development goals: Ukraine. National Baseline Report

The New Ukrainian School (2016) Conceptual Principles of Secondary School Reform. In: Gryshchenko M (ed) Ministry of Education and Science of Ukraine, p 34
The World Bank Group (2017) Government expenditure on education, total (% of GDP)—Ukraine. ID: SE.XPD.TOTL.GD.ZS. Source: UNESCO Institute for Statistics. Available from https://data.worldbank.org/indicator/SE.XPD.TOTL.GD.ZS?end=2017&locations=UA&name_desc=true&start=2017&view=bar. Accessed 17 Jan 2020
UNESCO (2017) Education for sustainable development goals: learning objectives. p 67. ISBN 978-92-3-100209-0
Urenje S (2019) Confronting the climate change crisis by inspiring teachers to facilitate education for social transformation; lessons from the 'Sustainability Starts with Teachers' initiative. Higher Education Research, Science publishing group (in press)
Wals AEJ (2007) Learning in a changing world and changing in a learning world: reflexively fumbling towards sustainability. South Afr J Environ Educ 22:35–45
Wals AEJ (2010) Mirroring, gestaltswitshing and transformative learning: stepping stones for developing sustainability competence. Int J Sustain High Educ 29:380–390
Wals AEJ, Benavot A (2017) Can we meet the sustainability challenges? The role of education and lifelong learning. Eur J Educ 52:404–413
Yin RK (2003) Case study research: design and methods, 3rd edn. Sage, Thousand Oaks, CA, p 240

Oleksandra Khalaim is graduated in Economics (2006), master's degree in Economics (2007), master's degree in Environmental Sciences (2010), and Ph.D. in Ecology and Environmental Sciences (2017). Post Doctorial Research at Uppsala University (2020). Her current research interests include education for sustainable development, climate change education and transformative learning methods in higher education, climate change adaptation and management of urban green areas.

Dr. Shepherd Urenje received his doctoral degree in Development Education at the University College of London, Institute of Education (2012) and graduated with a masters degree in Environmental Education from Rhodes University (2004). He is an alumnus of the University of Zimbabwe (1983) where he graduated with a Bachelor in Education. He has worked as a teacher of environmental science and development education in Zimbabwe, a Regional Programme Manager for Education and Training at SADC, and Principal Examiner of Environmental Science in the UK. Currently Dr Shepherd Urenje is an ESD Programme Specialist at the Swedish International Centre of Education for Sustainable Development at Uppsala University (SWEDESD). His work supports ESD in Higher Education within and among countries in Scandinavia, Africa and Asia in responding to the twenty-first century sustainability challenges.

University Outreach and Their Contribution for the Sustainable Development Goals: A Case of Study Focused on Smart and Learning Cities

Luciana Brandli, Janaina Mazutti, Liane Dalla Gasperina, Michele Rocha Reolão, and Roberto Dos Santos Rabello

1 Introduction

Universities have being seen as spaces for evolution and social transformation. When evolving along with the problems and limitations of society in each era, these spaces have been strengthening their bond with the community and raising awareness for sustainable development. Universities must offer molds in terms of education, social concepts and behaviors (Bratianu and Bolisani 2015; Hapenciuc et al. 2016; Bejinaru and Prelipcean 2017; Bejinaru et al. 2018).

The campuses where the universities are located operate like small towns and often act as living laboratories, generating prototypes for application in large centers providing quality of life for citizens and the environment. In line with the thinking of Wagner and Alencastro (2018) stating that the role of the university goes far beyond spreading knowledge and professional growth, this role includes evolving the nation towards Human and Sustainable Development.

L. Brandli (✉) · J. Mazutti · L. D. Gasperina · M. R. Reolão
Graduate Program in Civil and Environmental Engineering, University of Passo Fundo, Campus I, BR 285, Passo Fundo, Rio Grande do Sul 99052-900, Brazil
e-mail: brandli@upf.br

J. Mazutti
e-mail: janainamazutti@gmail.com

L. D. Gasperina
e-mail: lianedallagasperina@hotmail.com

M. R. Reolão
e-mail: michelerreolao@gmail.com

R. Dos Santos Rabello
Graduate Program in Applied Computing, University of Passo Fundo, Campus I, BR 285, Passo Fundo, Rio Grande do Sul 99052-900, Brazil
e-mail: rabello@upf.br

W. Leal Filho et al. (eds.), *Universities, Sustainability and Society: Supporting the Implementation of the Sustainable Development Goals*, World Sustainability Series,
https://doi.org/10.1007/978-3-030-63399-8_26

In Brazil, to the University's roles be developed successfully, it is necessary that the focus is directed to the principles described in the Brazilian Constitution—Art. 207, which states that "obey the principle of inseparability between Teaching, Research and Extension" (Brazil 1988). No less important is innovation, which is inserted in the principles of inseparability. According to Miranda et al. (2016) the university campus can be seen and used for transferring and application of the knowledge in favor of society. Bejinaru et al. (2018) state that universities need to take their role seriously, facilitating access to innovations, providing implementations and disseminating research results.

In the view of Ortega (2016), Teaching has an elementary nature and ensures the transmission of the knowledge produced and all the new results. Together, teaching and research combine theory and practice and thus, promote the exchange of information. With regard to the Research, this dimension allows the construction of critical thinking, new responses, learning and academic thinking (Ortega 2016).

As the third item of inseparability, the extension is defined by Gonçalves (2015) as the affirmation of responsibility and social commitment, belonging to academic training and knowledge generated by the University. Similar to this view, Chaves et al. (2019) state that the extension presents a commitment to transmitting and instrumentalizing the knowledge generated at the university.

In order to achieve the social commitment, extension projects must operate in transversal dimensions, combining different areas of knowledge (Silva et al. 2016). Thus, extension projects are linked to several social, environmental, sustainable and educational purposes. In this way they contribute to the achievement of the UN Sustainable Development Goals by promoting the articulation of the activities of extension projects with the proposed goals of the SDGs.

Recognizing the value of each axis for the university to remain in constant evolution and highlighting the extension with its projects, the aim of this paper is to present an outreach project developed by a Brazilian university, with focus on smart and learning cities and in parallel highlight the social impact of the project and its contribution to achieving the Sustainable Development Goals.

2 Methodology

The methodology of this study addresses the analysis of secondary data, provided from the project reports and institutional news. This paper presents a case of study describing the interventions promoted by an outreach project guided by the University of Passo Fundo—UPF, in southern Brazil.

The UPF main campus is located in the city of Passo Fundo—Rio Grande do Sul (RS), however, the institution has campuses in other 6 municipalities in the north of the state. Figure 1 presents UPF's location.

As a community university, besides its responsibility as a teaching and research center in the north of the state of Rio Grande do Sul—Brazil, UPF also provides a relevant social role through its extension projects.

Fig. 1 Passo Fundo location, southern Brazil

The extension project entitled *UniverCidade Educadora e Inteligente* (Smart and Learning UniverCity) aims to promote formal, non-formal and informal education with different actors from Passo Fundo/RS and the coverage region of UPF. Also, it aims to highlight the pedagogical potential of city spaces, guaranteeing several opportunities to teaching and learning in the territory (UPF 2020a).

2.1 Methodological Steps

The research was divided into three steps, as shown in Fig. 2.

Fig. 2 The main goal and the methodological steps

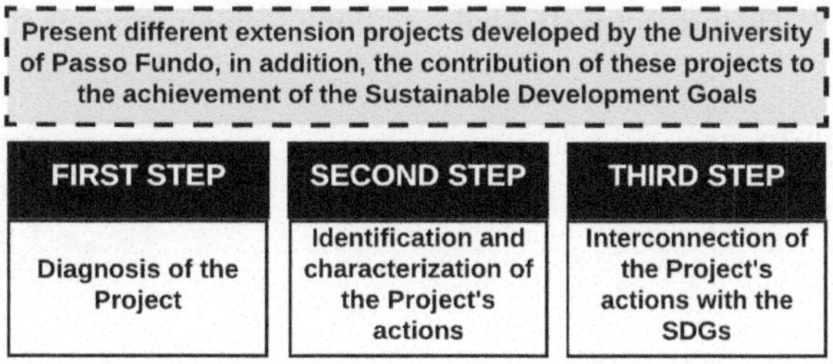

Step 1—Diagnosis of the project under study
This step aims to characterize and describe the project. The methodology adapted from Bardin (2011) was used, following the phases: analysis, exploration and treatment of the content. In this step, contact was made with the professors responsible for the project, and access to the project activity report was obtained.

Step 2—Identification and characterization of the project's actions
This step aimed to identify the actions carried out by the project and characterize them. A detailed analysis of the available documents and the collection of the necessary data were carried out. According to Bardin (2011), this is a relevant phase, as it consists in the choice of the data to be analyzed. First, it was searched by the actions promoted by he project so far. For this moment, the project's performance report was used as a source of information, describing the activities carried out. In parallel, a search was conducted on the UPF institutional website to identify news about the project's actions.

Step 3—Identification of the contribution to the SDGs
Finally, the project contribution to the consolidation of the SDGs was highlighted. The third phase covers the data treatment. The information was filtered and treated, as well analyzed regarding the contributions to the SDGs achievement. Each activity was analyzed from the perspective of the SDGs and their targets.

3 Results and Discussion

3.1 Diagnosis of the Project Under Study

The activities within the context of the "UniverCidade Educadora e Inteligente" Project started in 2011, but it was in 2016 that the project was officially launched and in 2017 it promoted its first event: "1st Meeting on Smart and Learning Cities". Among its purposes are the strengthening of social participation and the improvement and revitalization of public spaces (UPF 2020b).

The project is divided into three smaller projects related to each other (Fig. 3).

Fig. 3 The project division

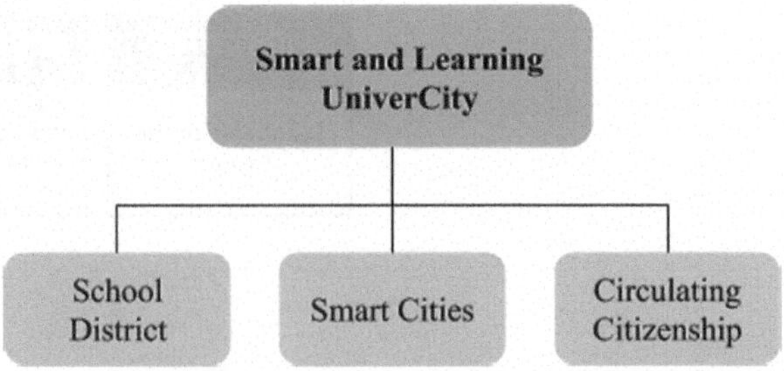

Table 1 Numbers of the project

Undergraduate courses involved	Cities involved
Law	Soledade/RS
Civil engineering	Marau/RS
Pedagogy	Camargo/RS
History	Passo Fundo/RS
Computer science	

School District: the sub-project is developed in partnership with the UPF's Faculty of Engineering and Architecture and works with two partner schools. It aims to promote the practice of walking in the city and motivate sustainable mobility.

Smart Cities: the sub-project works in partnership with the Master Program in Applied Computing, from the Institute of Exact Sciences and Geosciences of UPF. The sub-project intends to promote better spaces in the cities by using smart resources and services.

Circulating Citizenship: the sub-project is linked to the Faculty of Education of UPF, and works with schools focusing on mobility and sustainable cities. Also, it promotes activities such as courses for traffic instructors, in partnership with Detran (Rio Grande do Sul Traffic Department).

The common link that connects the three small projects is the city and, as an interdisciplinary topic, the projects that provide a rich and differentiated experience to all involved.

Currently, the project promotes the integration between 5 different undergraduate courses, involves 9 members (4 students and 5 teachers) and works with 4 cities in the north of the state of Rio Grande do Sul—Brazil. Table 1 presents that information.

3.2 *Identification and Characterization of the Projects' Actions*

In total, 29 actions promoted by the project with contribution to some UN SDGs were identified. These actions were divided into 4 categories according to their common characteristics: (1) *Promotion of activities*; (2) *Participation in events*; (3) *Guidance for city's certification* and (4) *Promotion of courses.*

Activities classified as *Promotion of activities* involved the organization of events about Smart and Learning Cities, as support for the *World Car Free Day* promoted in Passo Fundo, traffic education and accident prevention activities at schools, inclusive mobility activities, socio-environmental education workshops and activities to enhancing reading in schools.

The activities classified as *Participation in events* are described as moments in which teachers members of the project were invited to speak at events or meetings and expose the project's experiences.

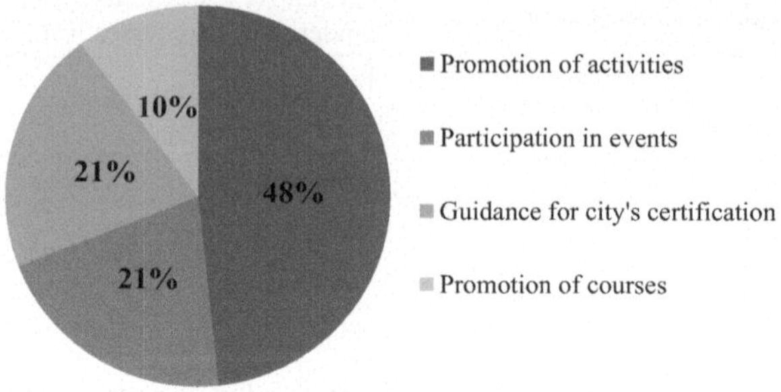

Fig. 4 Classification of promoted actions

The *Guidance for city's certification* refers to the assistance provided by the project so that the cities in the region can achieve the certification of recognition as a Learning city.

Finally, the *Promotion of courses* were moments in which the project provided technical training for traffic instructors. These activities represent an opportunity for young people and adults to get professional training and also provide public awareness regarding the importance of traffic care and accident reduction.

Figure 4 shows the result of this classification.

Initially, the most prominent action was the *Promotion of activities* equivalent to 48% of the total (14 actions). Within these activities are moments of awareness in traffic, promoted by the sub-project *Circulating Citizenship.* Most of the time, the activities were carried out in schools in the city of Passo Fundo, in partnership with the other sub-project *School District*. In other times, activities were promoted at events such as the *World Car Free Day*.

Still within the promotion of activities, is highlighted the promotion of events such as the "*Meeting of Learning and Smart Cities*" and the *World Car Free Day.* Events as mentioned provide important moments for good practices dissemination among cities, promoting dialogues and reflections of the role of the city in the improvement of society.

In general, the activities classified in this group involve the participation of professors and extension students, enabling those involved to have a human education that goes beyond what is proposed by academic curriculum.

Subsequently, *Participation in events* and *Guidance on cities certification* registered 21% of the actions (6 actions). The orientation for the certification of cities was successful in the northern region of the state, since in four months 3 cities were certify as Learning Cities and there is still other cities in progress.

Finally, 10% of the total (3 actions) represented the promotion of courses especially for traffic instructors.

Table 2 The activities contribution for the SDGs

Activities	SDGs related			Contribution
Promotion of activities	3 Good Health and Well-Being 4 Quality Education	5 Gender Equality 10 Reduced Inequalities	11 Sustainable Cities and Communities	Promotion of socio-environmental education, awareness of road traffic care, avoiding road accidents; promotion of events such as “Meeting of Learning and Smart Cities”
Participation in events	4 Quality Education	9 Industry, Innovation and Infrastructure		Sharing experiences and promoting dialogue to support technological development, research and innovation
Guidance for city’s certification	4 Quality Education 5 Gender Equality	9 Industry, Innovation and Infrastructure 10 Reduced Inequalities	11 Sustainable Cities and Communities	Promotion of lifelong learning spaces, including environmental questions. Encouraging the qualification of teachers and the improvement of school environment. Ensure an inclusive environment for all
Promotion of courses	4 Quality Education	8 Decent Work and Economic Growth		Technical learning and training opportunities for youth and adults

3.3 Contribution to the SDGs

The actions promoted were analyzed from the perspective of their contribution to the UN SDGs, taking into account how each activity could relate to each SDG individually. The results showed how different activities, classified in the same category, could contribute to the same SDGs. Table 2 shows the four categories of activities, the SDGs related and an explanation for their contribution.

In general, the activities promoted by the project were related to seven different SDGs: Good Health and Well-Being (SDG 3), Quality Education (SDG 4), Gender Equality (SDG 5), Decent Work and Economic Growth (SDG 8), Industry, Innovation and Infrastructure (SDG 9), Reduced Inequalities (SDG 10) and Sustainable Cities and Communities (SDG 11).

The SDGs Quality Education (SDG 4) and Sustainable Cities and Communities (SDG 11) stand out among all, since they are related to 25 and 16 activities, respectively. The goals Health and Well-Being (SDG 3) and Reduced Inequalities (SDG 10), still stand out with 9 and 10 related activities, respectively. Finally, although with less expressiveness, the goals of Gender Equality (SDG 5), Decent Work and Economic Growth (SDG 8) and Industry, Innovation and Infrastructure (SDG 9), show their relevance to the activities in which they were listed, 7, 4 and 5, respectively.

The *Participation in events* is related to only two different SDGs: Quality Education (SDG 4) and Industry, Innovation and Infrastructure (SDG 9), while the *Guidance for city's certification*, relates to the Learning and Smart Cities theme, addresses the largest number of different SDGs: 4, 5, 10 and 11. Finally, activities classified as *Promotion of courses* are related to the SDG 4 and 8.

The results showed how a single extension project can be related and contribute to seven different SDGs and integrate the community with the Sustainable Development Goals. Also, it demonstrates the relevance of extension projects at the university.

4 Conclusions

Through the discussions and reflections raised by the project, it was possible to consolidate the partnership between the University and the cities of the region, creating a unique opportunity to form Learning and Smart Cities cooperation network in the north of the state of Rio Grande do Sul, southern Brazil.

Therefore, the extension proved to be the channel whereby research and teaching promoted within a Higher Education Institution can reach society and drive transformations. Finally, it was possible to show that the dimension of lifelong learning at the university is also consolidated through extension.

This case of study reports how a practical experience of an extension project was able to bring the university closer to society and thus, promote mutual support in the implementation of the Sustainable Development Goals. Therefore, by documenting and sharing this experience it is expected to encourage future projects and partnerships between universities and cities, in different contexts.

Acknowledgements The authors acknowledge the CAPES (Coordination of Superior Level Staff Improvement—Brazil) for the financial support.

References

Bardin L (2011) Análise de conteúdo. Edições 70, Lisboa, Portugal, p 70
Bejinaru R, Prelipcean G (2017) Successful strategies to be learnt from world-class universities. In: Dima AM, Hadad S (ed) Proceedings of the international conference on business excellence, vol 11. De Gruyter and Stanford University Press cooperate, pp 350–358
Bejinaru R, Hapenciuc CV, Condratov I, Stanciu P (2018) The university role in developing the human capital for a sustainable bioeconomy. Amfiteatru Econ 20(49):583–598
Brazil (1988) Constituição da República Federativa do Brasil: promulgada em 5 de outubro de 1988
Bratianu C, Bolisani E (2015) Knowledge strategy: an integrated approach for managing uncertainty. In: Massaro M, Garlatti A (eds) Proceedings of the 16th European conference on knowledge management. Academic Conferences International Limited, pp 169–177
Chaves CJA, Oliveira EP, Romagnani P, Erbano CP (2019) Projetos de extensão universitária: um compromisso da Universidade com a inclusão social. HOLOS 2(1):1–17

Gonçalves NG (2015) Indissociabilidade entre Ensino, Pesquisa e Extensão: um princípio necessário. Perspectiva 33(3):229–256
Hapenciuc CV, Bejinaru R, Roman C, Neamtu DM (2016) The role of HES within the evolution of the business sector. In: Chova LG, Martínez AL, Torres IC (eds) EDULEARN—proceedings of the 8th annual international conference on education and new learning technologies Barcelona. IATED Academy, pp 5309–5317
Miranda JIR, Sidulovicz N, Machado DM (2016) O desafio da inovação tecnológica dentro da universidade. Revista de Desenvolvimento Econômico 2(34):389–406
Ortega LM (2016) Programa Empreendedorismo-Escola: influenciando a Universidade por meio do tripé ensino, pesquisa e extensão. Revista de Administração, Contabilidade Economia Fundace 7(1):118–132
Silva RR, Teixeira MRS, Rodrigues FTRL (2016) Uma análise da gestão de projetos de extensão de uma instituição federal de ensino. Revista Gestão Secretariado 7(3):150–171
UPF (2020a) Universidade de Passo Fundo. Extensão, Programas e projetos: Programa UniverCidade Educadora e Inteligente UPF. Available at https://www.upf.br/Iceg/curso/ciencia-da-computacao/extensao/programa-univercidade-educadora-e-inteligente-upf. Last Accessed 24 Jan 2020
UPF (2020b) Universidade de Passo Fundo. Programa UniverCidade Educadora e Inteligente: Apresentação. Available at https://www.upf.br/univercidadeeducadoraeinteligente/apresentacao. Last Accessed 03 March 2020
Wagner IC, Alencastro MSC (2018) Universidade, sociedade e urbano: reflexões sobre os modelos de universidade dentro do contexto urbano. Revista Tecnologia Sociedade 14(34):280–298

Prof. Luciana Brandli Ph.D. is an Associate Professor at the University of Passo Fundo, Brazil, and works in the Graduate Program in Civil and Environmental Engineering. Her current research interests include sustainability in higher education and green campus, management of urban infrastructure and sustainable cities, and the Agenda 2030 for sustainable development. She supervises a number of Master's and Doctoral students on engineering and environment and sustainability issues and has in excess of 300 publications, including books, book chapters, and papers in refereed journals.

Janaina Mazutti has a degree in Environmental Engineering (2018) and is currently a master student in Civil and Environmental Engineering at University of Passo Fundo, Brazil. Her interests are related to sustainable management projects, Sustainable Development Goals, climate change adaptation and sustainable and smart cities.

Liane Dalla Gasperina has a degree in Architecture and Urbanism (2016), and has a specialization in Architecture and Engineering Office Management (2017). She is currently a master student in Civil and Environmental Engineering at University of Passo Fundo, Brazil. Her interests are related to sustainable development goals, innovation of the environments including smart concepts.

Michele Rocha Reolão has a degree in Architecture and Urbanism (2014) and a master degree in Civil and Environmental Engineering (2019). She is currently a doctoral student in Civil and Environmental Engineering at University of Passo Fundo, Brazil. Her current interests include sustainable cities, Sustainable Development Goals, climate change adaptation, Urbanization and Green infrastructure.

Prof. Roberto Dos Santos Rabello has a Post-Doctorate in Strategic Digital City, he is currently professor at the University of Passo Fundo and works in the Graduate Program in Applied Computing and as extensionist professor. He is also the Rio Grande do Sul Director of RBCIH (Brazilian Network of Smart and Human Cities).

USP's Environmental Policy in the SDGs Approach

Tércio Ambrizzi, Tamara Maria Gomes, Fernanda da Rocha Brando, Flávio Pinheiro Martins, Tadeu Malheiros, Denise Crocce Romano Espinosa, and Paulo Santos de Almeida

1 Introduction

Sustainability has become a significant theme in the past decades. It has been raising the importance of promoting deep structural changes, economic efficiency and less environmental impact on feedstock equitable distribution. Therefore, universities play an important role as a mediator of local government and society by promoting discussion forums, and more importantly, as an agent of transformation in important changes. Hence, it becomes necessary the elaboration of a plan that guides the organization to achieve sustainability, and moreover, reinforce partnerships, and improve knowledge of what is needed to maintain the projects. Likewise, these actions will

T. Ambrizzi · T. M. Gomes (✉) · F. da Rocha Brando · F. P. Martins · T. Malheiros · D. C. R. Espinosa · P. S. de Almeida
Superintendence of Environmental Affairs—SGA/USP, Cidade Universitária, Rua da Praça do Relógio, 109, Bloco K- sala 309, São Paulo, SP ZIP05508-050, Brazil
e-mail: tamaragomes@usp.br

T. Ambrizzi
e-mail: tercio.ambrizzi@iag.usp.br

F. da Rocha Brando
e-mail: ferbrando@ffclrp.usp.br

F. P. Martins
e-mail: fpmartins@usp.br

T. Malheiros
e-mail: tmalheiros@usp.br

D. C. R. Espinosa
e-mail: espinosa@usp.br

P. S. de Almeida
e-mail: psalmeida@usp.br

W. Leal Filho et al. (eds.), *Universities, Sustainability and Society: Supporting the Implementation of the Sustainable Development Goals*, World Sustainability Series,
https://doi.org/10.1007/978-3-030-63399-8_27

promote a diagnosis establishment, actions, and, consequently short-, medium- and long-term goals (Lemos et al. 2018).

The University of São Paulo-USP/Brazil gathers a broad community (117,118.00 people) including professors, universities staff, as well as undergraduate and graduate students in a 76,437,742, 45 m^2 area on seven different *campuses* over the capital of São Paulo state, inland and coastal (USP Statistical Yearbook 2018).

Due to its many specificities and thinking about Universities sustainability, the elaboration of an Environmental Policy (EP) for USP had its onset in 2009 due to Working Groups (WG). The strengthening of this idea came with the creation of a Superintendence of Environmental Affairs (SGA) in 2012, along with the endorsement of the EP/USP publication in January 2018 by the 7465 resolution (USP 2018). Therefore, the EP main goal was preferably conducting and providing legitimacy to sustainability actions within the University, regarding preservation, conservation and the rational use of natural resources. The addressed topics were Green Areas and Ecological Reserves, Mobility, Greenhouse Gases and Pollutants Emission, Environmental Education, Fauna, Water and Effluents, Sustainable Buildings, Solid Waste, Energy, Territorial Use and Occupation, and Management (Lemos et al. 2019).

Since the EP publication, several actions have been developed on different *campuses* in order to implement their themes, considering the contextual alignment of their specificities. Among these actions was the Environmental Master Plans (EMPs) and Pilot Projects (PPs) elaboration, that are projects supported by SGA/USP resources to develop sustainable and replicable actions in different University contexts; participation in worldwide rankings such as Greenmetric (18th position in 2019, first place in Latin America), as well as taking part in different forums, as the Climate Change (COP), and the High-level Political Forum on Sustainable Development (HLPF) at the United Nations (Lemos et al. 2018).

USP's participation in international forums led the university to consider worldwide frameworks. The United Nations General Assembly established on September 25th, 2015 70/1 resolution a set of 17 global objectives denominated Sustainable Development Goals (SDGs), followed by 169 goals to be achieved by 2030, according to the Agenda 2030 (United Nations 2015).

The sustainable development proposed by SDGs englobes all the three Economic, Social and Environmental perspectives. The University of São Paulo has sought to bring its activities to meet the Agenda 2030, collaborating with the society to achieve the goals proposed by the United Nations. One of the main challenges when dealing with broad guidelines of international scope is an adequate local context, thus avoiding a thematic detachment.

The link between the University's actions and SDGs can be assessed through network analysis, a worldwide applied approach especially relevant when interrelationships between actors, organizations or themes are required (Batagelj 1998). With the advance on computer programs, the accessible Gephi Software appears allowing real-time associations visualization, interactive techniques, data filtering, and statistics application with a great visual result (Bastian et al. 2009).

The sustainability variables incorporation process developed by the university towards society is an important agent of transformation. Thus, the goal of this study is to correlate USP's sustainable actions adherence through its EP and PPs, with all SDGs, seeking examples developed in the university that might be replicable in the surrounding communities.

2 Literature Review

2.1 The Agenda 2030 Civilizational Aspirations

It has been observed that the planetary life support systems have been under stress in the last century (Rockström et al. 2009), beyond the safe boundaries needed for a truly civilizational human species advance. Among the many complex sustainable development challenges, such as plastic pollution ubiquity (Shen et al. 2020), biodiversity loss (Cardinale et al. 2012), and coral reefs bleaching (Beyer et al. 2018), the substantial synergic hazardous potential is the climate crisis (Hoegh-Guldberg et al. 2018; IPCC 2019). Recent studies, as the one on the Amazon rainforest deforestation (Lovejoy and Nobre 2018) has shown that many climate tipping points have already started (Lenton et al. 2019), and it is currently uncertain the effects it may cause, as well as the potential disruption, on the global climate.

Therefore, several initiatives and policies have risen as an international response to a problem that transcends the Nacional State barriers. It is important to highlight the United Nations Conference on Environment and Development (UNCED), also known as the Rio Summit or just ECO-92, and later on, Rio +10 and Rio +20. These conferences have leveled the sustainability debate to a political level never seen before, creating frameworks and directives for the civilizational agenda we are pursuing today.

Since 2012, the United Nations have been coordinating a widely participative process on Sustainable Development Goals (SDGs), and therefore, continuing the social advances (Mitra and Rodriguez-Fernandez 2010) obtained by the Millenium Development Goals (MDGs) to give the environmental crisis a higher attention profile.

The SDGs are the most advanced framework available for civilizational progress. In its 17 goals and 169 targets, most of the humanity challenges have been approached, from poverty eradication to gender equality, from biodiversity preservation to main political and cooperation goals.

The SDGs framework is complex and should be approached as one interconnected agenda with many systemic synergies (Nerini et al. 2018). Nevertheless, education plays an important role in turning this tide, and can be considered a focal point as well as an accelerator. The Higher Education Institutions have the responsibility and moral authority to improve their environmental resolutions by keeping a strong regulatory system towards efficiency, and at the same time, avoiding the eventual public policy regression, since they can improve common interest among social and academic communities.

USP has been careful towards environmental public policies non-regression and has been making an effort to the development of SDGs through its EP. The SDGs have been promoting non-regression principles, thus, emerging in states and at an international level applied in public policies, as included in "The Future We Want" (United Nations 2012).

Regression takes many forms. It is seldom explicit since governments do not dare to announce backtracking in environmental protection officially due to fear of an unfavorable public response from environmental and public policy. Note that this principle has been first established by a referendum in California (on November 2nd, 2010) when the majority of voters refused to suspend a climate change and greenhouse gas emissions reduction regulation, as requested by oil companies (Prieur 2012). It has been demonstrated that this principle may present significant implications on the environmental public policies with social interest.

3 Materials and Methods

To map the Agenda 2030 thematic context (United Nations 2015) at USP's environment, two thematic association stages were carried out using a consensus-based expert elicitation process with different backgrounds specialists. This approach has been inspired by the SDGs mappings developed by Nerini et al. (2018) and Vinuesa et al. (2020). It can potentially reduce uncertainty and also help to elaborate frameworks to support decision making. In this study, mainly institutional policy and project documents, also known as gray literature, were used. The specialists' board was created based on the relevance selection (Butler et al. 2015) to obtain a diverse and suitable set of expertise for the posit task.

The first association stage was based on the 11 USP's themes, which are the EP and its mirroring with the 17 SDGs (Table 2), while the second stage association evaluated PPs adherence developed on the university's *campuses* (Table 3) and their SDGs.

Table 1 Specialists' board, time of experience and background

	Years of experience	Background
Specialist 1	27	Climate changes
Specialist 2	19	Water resources
Specialist 3	15	Environmental education
Specialist 4	18	Sustainability indicators
Specialist 5	19	Solid waste treatment
Specialist 6	15	Environmental law

Table 2 USP's Environmental Policy (EP/USP) themes and Sustainable Development Goals (SDGs)

EP/USP
1. Green areas and ecological reserves
2. Mobility
3. Greenhouse gases and pollutants emissions
4. Environmental education
5. Fauna
6. Water and effluents
7. Sustainable buildings
8. Solid waste
9. Energy
10. Territorial use and occupation sustainability management
11. Sustainability in administration
Sustainable Development Goals (SDGs)
Goal 1. End poverty in all its forms everywhere
Goal 2. End hunger, achieve food security, improve nutrition and promote sustainable agriculture
Goal 3. Ensure healthy lives and promote well-being for all at all ages
Goal 4. Ensure inclusive and equitable quality education as well as promoting lifelong learning opportunities for all
Goal 5. Achieve gender equality and empower all women and girls
Goal 6. Ensure availability and sustainable water and sanitation management for all
Goal 7. Ensure access to affordable, reliable, sustainable and modern energy for all
Goal 8. Promote sustained, inclusive and sustainable economic growth, full and productive employment and decent work for all
Goal 9. Build resilient infrastructure, promote inclusive and sustainable industrialization and foster innovation
Goal 10. Reduce inequality within and among countries
Goal 11. Make cities and human settlements inclusive, safe, resilient and sustainable
Goal 12. Ensure sustainable consumption and production patterns
Goal 13. Take urgent action to combat climate change and its impacts
Goal 14. Conserve and sustainably use the oceans, seas and marine resources for sustainable development
Goal 15. Protect, restore and promote sustainable use of terrestrial ecosystems, sustainably manage forests, combat desertification, reverse land degradation and halt, along with halt biodiversity loss
Goal 16. Promote peaceful and inclusive societies for sustainable development, provide access to justice for all and build effective, accountable and inclusive institutions at all levels
Goal 17. Strengthen the means of implementation and revitalize the global partnership for sustainable development

The SGA participation in the project developed by the Research Center for Gas Innovation will allow research on sustainable mobility effectiveness throughout the creation of a public transport line powered by hydrogen buses, connecting São Paulo and São Bernardo do Campo cities.

The thematic adherence was evaluated and validated by the specialists' board; and the information was structured in the software network for future analysis. To create associations, we used the Gephi Software (Bastian et al. 2009) for graph modeling and network analysis. A graph is a connection network representation. The basic network composition includes "nodes" and "edges" that acquire meaning according to the thematic area and study goal (Newman 2010).

Table 3 The 17 Pilot Projects (PPs) developed by USP

Title	Description
PP1-living coverage, Faculty of Law in São Paulo	At USP law school in the annex building, a Pilot Project to revitalize the use of living coverage is under development, which consists of growing species in a garden that provides not only new possibilities for using the space but also inserts greener areas into the central area of São Paulo city
PP2-"Let's go by bike", on Pirassununga *campus*	Developed at USP Pirassununga, the project consists of encouraging sustainable movement and community bicycle use through the installation of bicycle posts that can be used with the USP portfolio
PP3-in vivo diversity bank on Ribeirão Preto *campus*	USP Ribeirão Preto covers soil is typical of seasonal semideciduous forest areas occupied in a diversified manner and intensely subjected to environmental impacts inherent to its location and occupation history. The project, in partnership with the Ribeirão Preto *campus* Administrative Building and the Department of Biology of the Faculty of Philosophy, Sciences, and Letters, aims to establish guidelines leading to the perpetuation of areas belonging to the USP-RP Ecological Reserves. The proposal to manage the seasonal semideciduous mesophyll forest in vivo genetic diversity bank involves academic-scientific activities and outreach activities to encourage scientific and didactic activities with the community inside and out the university
PP4-Raia olímpica, at CUASO	The Olympic Streak and SciTech Park projects are being developed in partnership with a multidisciplinary group, in which the State and the City of São Paulo, as well as USP, are taking part. The project aims Environmental Education activities outdoors using cutting-edge concepts, methodologies, and technologies toward mainly basic education at public schools

(continued)

Table 3 (continued)

Title	Description
PP5-"Bossa" Project, at CUASO	Firstly, the project seeks to assess how is the *campus* user behavior in administrative environments, regarding environmental comfort issues understood as the air conditioning and natural light use, among others. It addresses the sustainable building's theme, and mainly focuses on the user's behavior and how it can be extended to students in didactic environments
PP6-performance project, at CEPEUSP	The project consists of installing water-saving mechanisms at the Sports Practice Center of the University of São Paulo (CEPEUSP) located at the "Armando de Salles Oliveira"—University of São Paulo (CUASO-SP). The goal is not only reducing consumption but also develop a cost management project through the savings provided
PP7-shared bicycles, at CUASO	Developed at the "Armando de Salles Oliveira" University of São Paulo (CUASO-SP), the project consists in encouraging and facilitating access to the *campus* and the units using shared bicycles. In addition to the bike-sharing integration points on *campus*, an extra one integrating the Butantã terminal and its surroundings will be placed
PP8-community garden, at the Medical School in São Paulo	The project proposes the composting system expansion to meet the garden and vegetable garden needs of the Medical School (USP/SP). The community garden is a project in which several sustainable actions are developed in a daily practice, such as a rainwater harvesting system, composting, earthworm, Jataí bee breeding, and disposable containers reuse. It involves the internal and external community that frequents the place daily
PP9-biodigester for swine residues, on Pirassununga *campus*	In partnership with the Administrative building of Pirassununga *campus*, a biodigester is being implanted to treat pig farming residues, which also allows agricultural reuse through fertigation in pastures. The treatment will have an automated set for filtration and solids separation, an item financed by SGA/USP
PP10-chemical waste, on São Carlos *campus*	Project in partnership with the São Carlos *campus* Administrative Building. The chemical waste laboratory (LRQ), meets the demands of several teaching and research units, and also provides decontamination services as well as waste generated legal disposal on *campus*

(continued)

Table 3 (continued)

Title	Description
PP11-environmental restructuring in a community dentistry clinic	The project aims studies developing and local water measures, electricity, and gas management actions. SGA/USP has been offering since July/2017, a postdoctoral scholarship to produce a prototype for measurement and a homepage for information dissemination to instrumentalize the consumption resources monitoring (water and energy)
PP12-vegetation regeneration and coexistence, at EACH	The project aims the native plants and tree-shrub species growing in São Paulo city, considering the limiting physical aspects resulted from the old landfills areas at the School of Arts, Sciences and Humanities (EACH) *campus*. Bearing in mind that the *campus* is arid, sterile and a moisture-free terrain, afforestation will contribute to greater connectivity between the remaining vegetation in this region and its increase in the East Zone
PP13-digital radiography, at USP's School of Dentistry	An initiative of the School of Dentistry of Ribeirão Preto (FORP), São Paulo (FO), and Bauru (FOB) seeking to implement a digital radiography routine within the three units. Therefore, reducing the radiation exposure dose in patients, professors, technical and administrative staff, as well as undergraduate and graduate students, thus, avoiding radiographic films and radiographic chemical processing use. The project development implies teaching, research and outreach activities at the university, and can also suit as a model for other institutions, as well as to public policy proposals for the area
PP14-waste, at CUASO	The research project on Solid Waste Management Diagnosis at USP units had its onset at Poli and EACH. With the data acquired, it will be possible to come up with more efficient and sustainable management tools, using replicable methodologies among other *campuses* and units
PP15-small vertebrate fauna qualitative survey,7 kon Pirassununga *campus*	The project aims to implement wild species from Brazilian biomes, such as Cerrado and the Atlantic Forest, monitoring tools, gathering information on biodiversity and its areas of great occurrence; carrying out monitoring campaigns to verify the impacts caused by transport on the main roads and streets on *campus*, as well as in its surrounding areas and, finally to compare richness, abundance and diversity of species parameters in the vicinity

(continued)

Table 3 (continued)

Title	Description
PP16-mobility—hydrogen bus, at CUASO	Participation in the project developed by the Research Center for Gas Innovation that will allow research on sustainable mobility effectiveness throughout the creation of a public transport line of hydrogen powered buses, linking the cities of São Paulo and São Bernardo do Campo
PP17-sustainability indicator platform, on São Carlos *campus*	Research project for the development of a sustainability indicator applicable to USP *campuses*. The data analysis will help to build more efficient and sustainable management tools, and use of a replicable methodology among the other *campuses* and units

Source Lemos et al. (2018)

In this study, we have bipartite networks (Barber 2007), which means there are two types of non-connected "Nodes", and, therefore, they are represented by Type 1, configured as the 11 EP/USP thematic areas and the 17 PPs, and Type 2, represented by the 17 SDGs. The "Edges" represent a connection between Type 1 and Type 2 "Nodes", which was defined based on the thematic overlap between the EP/USP thematic areas and the listed SDGs.

4 Results and Discussion

The alignment proposed between EP/USP themes and the objectives guided by the Agenda 2030 was informed by the experts indicated consensus (Table 1), and it has shown that all 11 policy thematic areas present at least one thematic adherence with SDG (Fig. 1).

The results are shown in graphs, where the EP and the SDGs nodes are represented in Fig. 1, and the SDGs and PPs nodes are observed in Fig. 2. The nodes size represents the 'level of association' or the 'association number' the themes have with each of the SDGs, and the color represents the node type or node cluster.

The networks were subjected to modeling by the Hu (2005) and Force Atlas 2 (Jacomy et al. 2014) algorithms both available on the software, such algorithms act similarly to a gravitational system, in which the nodes are repelled or attracted by the strength of their connections (ties).The resulting graphs were adjusted (colored and resized) to improve visualization. The different colors represent the network bipartition, as the thematic association was made between SDGs and EP, and not between nodes of the same group.

It is noteworthy that the thematic areas with the greatest connection number are Sustainability in Administration, Fauna Management and Environmental Education (Fig. 1). Concerning Sustainability in Administration, the greatest alignment can be

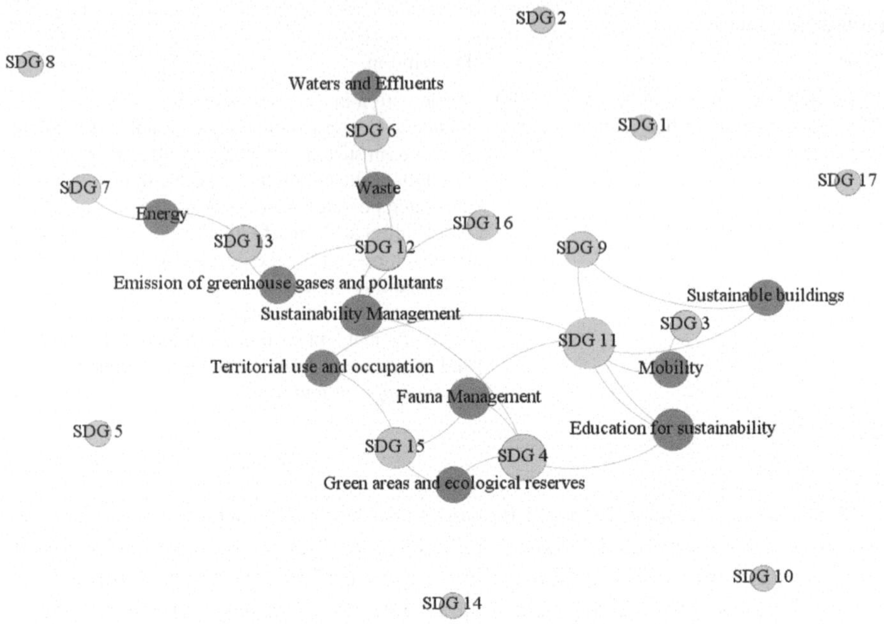

Fig. 1 USP Environmental Policy and the SDGs

explained in part by the relevance that the *campus* management pillar has in the university milieu. Similarly, Environmental Education is an eminently transversal theme, and it also represents the university mission itself.

A large number of connections with Fauna Management might be explained by considering the diversity of animal species existing on the *campuses* (Alexandrino et al. 2013). The highest number of connections is represented by SDG 11—Sustainable cities and communities, which is connected to 6 EP/USP themes. Likewise, SDGs 4—Quality Education and SDG 15—Life on land also present a significant number of connections, which suggest that most of the institution's policy themes are aligned with the needed response of growing urbanization challenges. *Campuses* are small cities, different in scale and activities, but still very similar in their administrative organization and structural arrangement.

The SDGs 11, 4 and 15 combination is configured as a summary of the main University of São Paulo *campuses* aspects, which are dispersed in seven different cities all over Brazil's economic pole state, acting as living laboratories for different city configurations (SDG 11). A significant amount of their *campuses* contains a cultural capital-rich heritage, and it is common for students to have classes in buildings dating from the beginning of the 20th century. Many of the administrative structures at USP Ribeirão Preto are set in buildings where once a coffee farm from 1920 was established. Such characteristics denote the importance of integrating the EP thematic areas, as "Territorial use and occupation" to pose a major challenge in

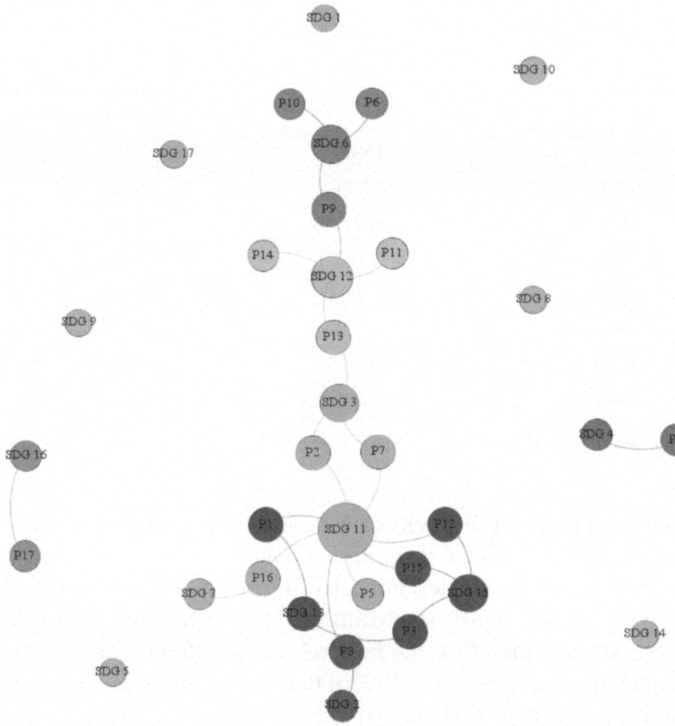

Fig. 2 SDGs and the Pilot Projects

advancing other thematic areas, such as "Sustainable buildings". Therefore, transforming the usual trade-off relationships with specific targets, such as the "SDG 11.4—Strengthen efforts to protect and safeguard the world's cultural and natural heritage" in synergistic relationships.

Another specific target aligned with the perspective of *campuses* as a model for cities is "SDG 11.2—By 2030, enhance sustainable urbanization and capacity for participatory, integrated and sustainable human settlement in all countries". The alignment of this specific objective with EP/USP indicates that the university has made efforts to reduce the gap between the university and its surrounding community. In order to achieve this goal, outreach activities acting as communication channels, as well as effective actions among students, researchers and the external community were performed. Likewise, another strategy was an open participatory process, in which society representatives were invited to the decision-making board, where they could share the city-*campus* impact on the surrounding community.

Considering now the education perspective, the "SDG 4.7—By 2030, ensure that all learners acquire the knowledge and skills needed to promote sustainable development" specific target is to promote education for sustainable development. When represented in the EP Thematic areas, it reveals an institutional commitment

Table 4 The seven USP *campuses* area

Campus	Ecological reserve area in ha	Total campus area in ha	% of ecological reserve considering the total area of each *campus*
Bauru	–	15.68	–
Lorena	8.16	37.34	21.85
Piracicaba	1111.88	3825.40	29.07
Ribeirão Preto	168.95	586.3	28.82
Pirassununga	881.62	2269.00	38.86
São Paulo	106.01	791.97	13.39
São Carlos	35.71	155.47	22.97
Total	2312.29	7681.16	30.10

Source USP Statistical Yearbook (2018)

to the necessary approach at different educational levels, such as the international framework adaptation to the local institutional policy.

SDG 15 is the one with the greatest impact on the institution's sustainability indicators. In 2012, with the SGA/USP establishment, 2165.98 ha were legally declared as Ecological Reserves and, therefore, has been receiving different status and protection from other university areas. Average, 30% of the seven campuses' total area consists of ecological reserves (Table 4) (USP 2017).

As such, the university significantly contributes to carbon capture on its *campuses* to the extent that forests in urban areas are effective mechanisms for carbon storage (Brack 2002; Nowak and Crane 2002). Usually, the *campuses* present a mixed composition of urban and rural areas. Another specific characteristic is that they are in a region where they can assist the Atlantic Forest preservation, which is a Brazilian biome with great biodiversity, and now with only 28% of its original coverage remaining (Rezende et al. 2018). The university perspective aligned with SDG 15 and 13, strongly contributed to it being classified as the most sustainable university in Latin America (GREENMETRIC 2019).

4.1 University as Test Bed for City Solutions

Analogously to the alignment with the EP, compliance with SDGs was also checked with the university's 17 PPs. The PPs are current initiatives on the seven university *campuses*, and are generally related to research, teaching or outreach teaching projects and permeate different thematic areas allied with the research diversity carried out at the institution (Lemos et al. 2018).

The projects had their alignment observed through Gephi Software treatment, with the same parameters used for the EP analysis. Similarly, it was observed that the highest connection number node was also the one represented by SDG 11—Sustainable cities and communities, followed by the SDGs 15, 13 and 12 nodes.

The modularity indicator with different color groups is a similarity indicator among the nodes and their connections. Besides, it has its main use on finding communities in large networks (Blondel et al. 2008), as well as it has a good use for the qualitative analysis carried out in this study.

The clustering formed with some PPs promotes joint efforts opportunities for networking, and also highlights how the projects can be converged in more complex programs or even new EP thematic areas.

The cluster formed by PP1, PP3, PP12 and, PP15 has shown how these projects can converge and work alongside, despite happening on different *campuses*. All projects involve biodiversity conservation and are also connected with the SDGs 15 and 13. Similarly, the cluster formed by PP2, PP5, PP7, and PP16, are associated with mobility innovations on *campus* and connect themselves to each other through the SDGs 11, 3 and 7, highlighting how mobility issues connect health and energy in a meaningful set of opportunities.

Figure 3 represents the Figs. 1 and 2 intersection, correlating the SDGs with EP and PP. The methodology applied to evaluate the adherence between the EP and PPs has not shown any correlation with the SDGs 1, 5, 8, 14, and 17, mainly because the environmental actions addressed throughout the themes do not target directly these goals.

However, it is important to stress that USP presents several actions that address to and with society: (i) SDG 1—public and tuition-free university able to offer several welfares and social services (housing, restaurants, nursery, among others) to support students in their studies, and besides, also enabling the under-represented groups to be admitted in all programs (at least 40% of the students are black, brown or indigenous, or from public schools); (ii) SDG 5—no gender differences in salary. A commission of only women was recently created to address and take action against all kinds of violence (physical, emotional, psychological, sexual, sexist, racial or LGBTphobic); (iii) SDG 8—due to its classification in different international rankings (QS—World University Rankings; U.S. News & World Report), USP plays an important role in developing high-quality professionals able to perform in several sectors in Brazil and the world; (iv) SDG 14—one of USP *campuses* is located in a coastal ecosystem protected area (CEBImar), in which a teaching and research center is promoting important knowledge development in aquatic life preservation; (v) SDG 17—the University has been broadening its partnerships and participation in the international scenario throughout university sustainability rankings (ISCN—International Sustainable Campus Network; IUSDRP—Inter-University Sustainable Development Research Programme; UI GreenMetric) with SGA/USP as mediator.

Highlighting the SDG 11 and concerning the local projects of each *campus*, it has been signaled that the EP/USP has been acting in a way to converge the thematic areas into practical and localized actions. The connection with the SDGs reinforces the need for a multilevel approach, which is especially relevant to sustainability challenges (Molthan-Hill et al. 2020), since the issues will not be solved if they are only fostered from the top or pursued at the individual level (Schwaninger 2015).

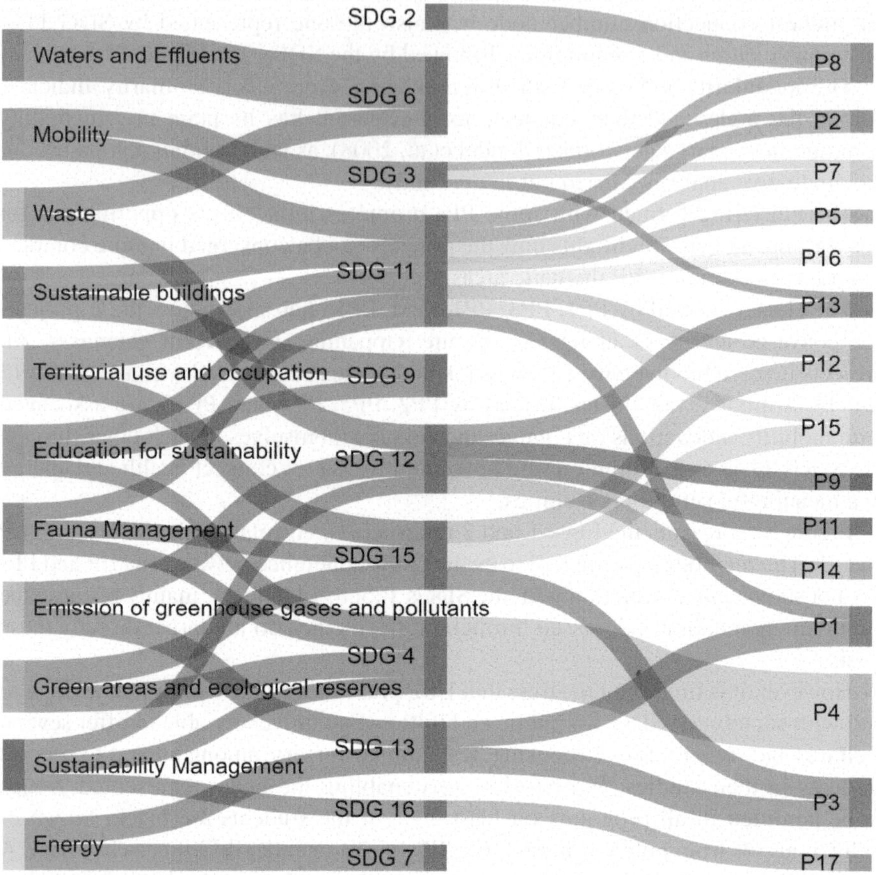

Fig. 3 Conceptual diagram: SDGs interface between Environmental Policy and Pilot Projects

University *campuses* emulate many common city features and also generate environmental, economic and social impacts on their communities (Verhoef et al. 2020). Such features make them able to act as interactive knowledge co-production models, also known as living labs (Evans et al. 2015) or test-bed (Robinson et al. 2013). Living Labs can be understood as an environment, infrastructure or space for experimentation, in which real-life contexts are shaped. It is a geographical or institutional limit where intentional experiments are carried out. It can also be used as a methodology, a systemic innovation, or a space used by researchers, citizens, companies and local governments from a user-centered perspective for experimentation (Mandai and Brando 2019). Therefore, USP, considering its *campuses* specificities, behaves as a Living Sustainable Laboratory, as they are potential environments for studying sustainability challenges, simulating small cities, and can also act as urban laboratories. These spaces include several teaching, research, and outreach activities, that are developed in the classroom, laboratories, or non-formal spaces, including the welfares and social services (housing, transportation, food, among others).

There is an urgent demand for cities to become more sustainable. Urban areas have expanded rapidly between 2000 and 2014. The area occupied by cities grew at a 1.28 rate, which is greater than their population growth. Therefore, there was a decrease in its density and an increase in the demand for urban growth management (United Nations 2019).

One of the main challenges relays on the agreement between policies and the international sustainable development agenda complexity at a local city scale (Mccarney 2019). Accordingly, universities can play an important role applying the SDGs, and can also use their framework as a guidance model for both diagnosis and prognosis for their *campus* sustainable solutions. That can be achieved by integrating university, the community itself and by fostering synergies among them.

5 Conclusion

The Gephi Software network analysis methodology was appropriated to verify the adherence of USP's environmental actions to the SDGs, but not to the ones with the themes not directly correlated to the EP and PPs, such as SDGs 1, 5, 8, 14 and 17.

EP/USP is strongly aligned with the Sustainability in Administration, Fauna Management and Environmental Education themes, SDGs 11, 15 and 4, respectively. The actions contemplated by the PPs on the different USP *campuses* have shown correlations with the SDGs, including Flora and Fauna Conservation (SDG 11 and 15); Waste Management (SDG 12) and Water Resources Conservation (SDG 6).

Although the SDGs 1, 5, 8, 14 and 17 have not shown adherence to the USP's environmental actions, the University has been providing important contributions in other segments strictly connected to these SDGs. We have as examples the following actions: public and tuition-free universities, gender salary equality, women commission against all kinds of violence, university for all (under-represented groups admission program), high-quality teaching and research to develop high-quality professionals, teaching and research center responsible for the aquatic life preservation, and, finally, partnerships with international institutions for sustainability experience exchange in Higher Education.

The data presented in this study confirm that the University of São Paulo develops its activities based on the management of green *campuses*, collaborating with the society on building sustainable communities.

Acknowledgements A special thanks to the Superintendence of Environmental Affairs (SGA/USP) interns, technical and administrative staff, as well as professors involved, for the support given in the activities and data gathering in this study.

References

Alexandrino ER, Bovo AAA, Luz DTA, Costa JC, Betin IGS, Ferraz KMP, Couto HTZ (2013) Aves do campus "Luiz de Queiroz" (Piracicaba, SP) da Universidade de São Paulo: mais de 10 anos de 158 observações neste ambiente antrópico. Atualidades ornitológicas (online). Ivaiporãn 173:40–52

Barber MJ (2007) Modularity and community detection in bipartite networks. Phys Rev E 76(6):066102

Bastian M et al (2009) Gephi: an open source software for exploring and manipulating networks. Icwsm 8:361–362

Batagelj M (1998) Pajek—program for large network analysis. Connections 21:47–57

Beyer HL, Kennedy EV, Beger M, Chen CA, Cinner JE, Darling ES, Obura DO (2018) Risk-sensitive planning for conserving coral reefs under rapid climate change. Conserv Lett 11(6):e12587

Blondel VD, Guillaume JL, Lambiotte R, Lefebvre E (2008) Fast unfolding of communities in large networks. J Stat Mech: Theory Exp 10:P10008

Brack CL (2002) Pollution mitigation and carbon sequestration by an urban forest. Environ Pollut 116:S195–S200

Butler AJ, Thomas MK, Pintar KD (2015) Systematic review of expert elicitation methods as a tool for source attribution of enteric illness. Foodborne Pathog Dis 12(5):367–382

Cardinale BJ, Duffy JE, Gonzalez A, Hooper DU, Perrings C, Venail P, Kinzig AP (2012) Biodiversity loss and its impact on humanity. Nature 486(7401):59–67

Evans J, Jones R, Karvonen A, Millard L, Wendler J (2015) Living labs and co-production: university campuses as platforms for sustainability science. Curr Opin Environ Sustain 16:1–6

GREENMETRIC (2019) Greenmetric world university rankings: rankings by Region 2019. Available at http://greenmetric.ui.ac.id/ranking-by-region-2019/>. Last Accessed 12 Feb 2020

Hoegh-Guldberg O, Jacob D, Taylor M, Bindi M, Brown S, Camilloni I, Guiot J (2018) Impacts of 1.5 C global warming on natural and human systems. In Global warming of 1.5 °C.: an IPCC Special Report, pp 175–311. IPCC Secretariat

Hu Y (2005) Efficient, high-quality force-directed graph drawing. Math J 10(1):37–71

IPCC (2019) IPCC Special Report on the Ocean and cryosphere in a changing climate

Jacomy M et al (2014) Force Atlas 2, a continuous graph layout algorithm for handy network visualization designed for the Gephi software. PLoS ONE 9(6):e98679

Lemos PFI, da Rocha Brando F, Gomes TM, Mulfart RCK (2018) Sustentabilidade na USP. SGA/USP, São Paulo, p 124

Lemos PFI, da Rocha Brando F, Gomes TM (2019) University of São Paulo environmental policy: master plan and pilot projects for pirassununga and Ribeirão Preto campuses. In: World sustainability series 1st edn., vol 1. Springer International Publishing, pp 73–90

Lenton TM, Rockström J, Gaffney O, Rahmstorf S, Richardson K, Steffen W, Schellnhuber HJ (2019) Climate tipping points—too risky to bet against

Lovejoy TE, Nobre C (2018) Amazon tipping point. Sci Adv 4:eaat 2340. Available at https://advances.sciencemag.org/content/4/2/eaat2340. Last Accessed 12 Feb 2020

Mandai SS, Brando FR (2019) Living labs for sustainability. In: Encyclopedia of sustainability in higher education, 1st edn., vol 1. Springer International Publishing, pp 1–8

Mccarney PL (2019) Cities leading: the pivotal role of local governance and planning for sustainable development. In: The Routledge companion to environmental planning. Routledge, pp 200–208

Mitra AK, Rodriguez-Fernandez G (2010) Latin America and the Caribbean: assessment of the advances in public health for the achievement of the Millennium Development Goals. Int J Environ Res Public Health 7(5):2238–2255

Molthan-Hill P et al (2020) Reducing carbon emissions in business through responsible management education: influence at the micro-, meso-and macro-levels. Int J Manag Educ 18(1):100328

Nerini FF et al (2018) Mapping synergies and trade-offs between energy and the Sustainable Development Goals. Nat Energy 3(1):10–15

Newman MEJ (2010) Networks: an introduction. Oxford University Press, Oxford, NY, USA, p 772
Nowak DJ, Crane DE (2002) Carbon storage and sequestration by urban trees in the USA. Environ Pollut 116(3):381–389
Prieur M (2012) Non-regression in environmental law, S.A.P.I.EN.S [Online], 5.2| 2012, Online since 12 August 2012, connection on 30 April 2019. Available at http://journals.openedition.org/sapiens/1405. Last Accessed 27 Feb 2020
Rezende CL, Scarano FR, Assad ED, Joly CA, Metzger JP, Strassburg BBN, Mittermeier RA (2018) From hotspot to hopespot: an opportunity for the Brazilian Atlantic Forest. Perspect Ecol Conserv 16(4):208–214
Robinson J, Berkhout T, Cayuela A, Campbell A (2013) Next generation sustainability at the University of British Columbia: the university as societal test-bed for sustainability. In: Regenerative sustainable development of universities and cities. Edward Elgar Publishing
Rockström J, Steffen W, Noone K, Persson Å, Chapin III FS, Lambin E, Nykvist B (2009) Planetary boundaries: exploring the safe operating space for humanity. Ecol Soc 14(2)
Schwaninger M (2015) Organizing for sustainability: an architecture for synergy
Shen M, Huang W, Chen M, Song B, Zeng G, Zhang Y (2020) (Micro) plastic crisis: un-ignorable contribution to global greenhouse gas emissions and climate change. J Cleaner Prod 120138
United Nations (2012) The future we want. Online. Available at https://sustainabledevelopment.un.org/content/documents/733FutureWeWant.pdf. Last Accessed 27 Feb 2020
United Nations (2015) Transforming our world: the 2030 agenda for sustainable development. RES/70/1. Available at https://sustainabledevelopment.un.org/post2015/transformingourworld. Last Accessed 12 Feb 2020
United Nations (2019) Special edition report: progress towards the sustainable development goals. Goal 11. Make cities and human settlements inclusive, safe, resilient and sustainable. Available at https://undocs.org/E/2019/68. Last Accessed 24 Dec 2020
USP (2017) USP, electronic journal. "Ecological reserves occupy 30% of USP's territory". Available at http://jornal.usp.br/universidade/reservas-ecologicas-ocupam-30-do-territorio-da-usp/. Last Accessed 12 Feb 2020
USP (2018) Resolução nº 7465, de 11 de janeiro de 2018. Available at http://www.leginf.usp.br/?resolucao=resolucao-no-7465-de-11-de-janeiro-de-2018. Last Accessed 12 Feb 2020
USP Statistical Yearbook (2018) Available at https://uspdigital.usp.br/anuario/AnuarioControle#. Last Accessed 12 Feb 2020
Verhoef LA, Bossert M, Newman J, Ferraz F, Robinson ZP, Agarwala Y, Hellinga C (2020) Towards a learning system for university campuses as living labs for sustainability. In: Universities as living labs for sustainable development. Springer, Cham, pp 135–149
Vinuesa R et al (2020) The role of artificial intelligence in achieving the Sustainable Development Goals. Nat Commun 11(1):1–10

Tércio Ambrizzi (Ph.D.) completed his Ph.D. in 1993 at University of Reading in the United Kingdom (UK). Currently, he is a Full Professor at Institute of Astronomy, Geophysics and Atmospheric Sciences (IAG), part of the Department of Atmospheric Sciences (DAS) at University of São Paulo (USP), in Brazil. He was also the head of DAS for 6 years (1997–1999 and 2001–2005), and since then he has been the Coordinator of the Climate Studies Group (GrEC) at IAG/USP and the General Coordinator of INCLINE (INter-disciplinary CLimate INvEstigation Center). From 2002 to 2006, he was Chief Editor of the Brazilian Journal of Meteorology, which was associated with the Brazilian Society of Meteorology. He has reviewed the IPCC's (Regional Impacts on Climate Changes) AR4 (2007) and AR5 (2013) reports and has over 100 publications in peer-reviewed journals and book chapters, including the publication of two books (in Portuguese). During his career, he has had over 200 extended abstracts published in national and international conferences/congresses/symposiums and has supervised several Master Dissertations, Ph.D. Theses, and Postdoctoral Assistantships in Geoscience. He is a Titular Member of the Brazilian Academy of Sciences, and he has also been nominated by the Federal Government (Science and Technology, and Environmental Ministries Interministerial Ordinance) as the Working Group 1 (WGI)—Climate Changes Physical Science Basis coordinator, and as part of the Scientific Committee of the Brazilian Panel on Climate Changes. Currently he is the Superintendent to the Superintendence of Environmental Affairs (SGA/USP). His research expertise is Atmospheric Sciences focused on Dynamic Meteorology, Atmospheric Numerical Modeling, and Climatology.

Tamara Maria Gomes (Ph.D.) completed her Ph.D. in Agronomy (Irrigation and Drainage) in 2001 at University of São Paulo (USP). She has developed significant research in "Agricultural use of treated sewage effluents" at USP's Postgraduate Research Program from 2007 to 2010. She was also an accredited auditor of the Rainforest Alliance Sustainable Agriculture Certification Program from 2009 to 2011. Currently, she is a Ph.D. Professor at College of Animal Science and Food Engineering (FZEA/USP), and a Technical Advisor to the Superintendence of Environmental Affairs (SGA/USP). Her areas of research include Irrigation and Drainage focusing on elaboration, implantation, and evaluation of irrigation projects, such as irrigation and fertigation management, water usage, and agricultural use of treated organic effluents.

Fernanda da Rocha Brando (Ph.D.) completed her Specialization in Environmental Management in 2001 at Federal University of São Carlos (UFSCar, 2001), Masters and Ph.D. in Science Education at State University of Júlio de Mesquita Filho (UNESP, 2005: 2010). Currently, she is an Associate Professor at University of São Paulo (USP)—Department of Biology of the Faculty of Philosophy, Science and Letters of Ribeirão Preto (FFCLRP-USP). She is associated to the Compared Biology Postgraduate Program (FFCLRP/USP), where she supervises several Masters and Ph.D. projects on "History, Philosophy, and Teaching of Ecology Evolution"; and also, to the Environmental Science Teaching Postgraduate Program (EESC-USP), supervising Masters projects on "Society and Environment". She has been a Technical Advisor to the Superintendence of Environmental Affairs (SGA/USP) since 2015, and Vice President of the Culture and Outreach Committee (FFCLRP-USP) since 2018, but she was the President from 2016 to 2018. Her research interest themes are Teaching in Science and Biology, Epistemology, Theoretical Biology, History of Ecology, Environmental History, Peirce's Philosophy, and Sustainability.

Flávio Pinheiro Martins (Postgraduate Student) completed his degree in Business Administration at School of Economics, Business Administration and Accounting at Ribeirao Preto (FEA/RP-USP). Currently, he is a Master Student of the Organizational Management Postgraduate Program (FEA/RP-USP) in Innovation and Sustainability. He works at Superintendence of Environmental Affairs (SGA/USP) and is a Principles for Responsible Management Education Committee (PRME) member (FEA/RP-USP). He attends the Working Groups (WG)—Sustainability in Management, Usage and Soil Occupation, and Environmental Education part of the Environmental Management Technical Committee at USP Ribeirão Preto. His expertise relies on auction and public contracts.

Tadeu Fabrício Malheiros (Ph.D.) completed his Masters in Resources Engineering at Universität Karlsruhe (1996), and his Ph.D. in Public Health at School of Public Health part of the University of São Paulo (USP) (2006). Currently, he is an Associate Professor at São Carlos School of Engineering (EESC-USP/SP). He is also the coordinator of the Professional Masters in Environmental Science Teaching Program and a member of the Sustainability Committee (EESC-USP/SP). He is a Technical Advisor to the Superintendence of Environmental Affair (SGA/USP). His areas of expertise are Environmental Engineering focusing on Environmental Health and Sustainability, especially on sustainability indicators, sustainable development, public health, solid waste, and environmental management.

Denise Crocce Romano Espinosa (Ph.D.) completed her Masters and Ph.D. in Metallurgic Engineering (1998 and 2002, respectively) at University of São Paulo (USP). She did a research internship at Massachussetts Institute of Technology (MIT-EUA) in 2001, and she was a visitor professor at University of British Columbia (Canada) from 2015 to 2016. She was an Associate Professor at Department of Metallurgic Engineering and Materials at the Polytechnic School of USP (EPUSP), where she was also a coordinator of the Postgraduate Program from 2012 to 2014, and an alternate coordinator of the Research Committee from 2009 to 2012. Currently, she is an Associated Professor at Department of Chemical Engineering at EPUSP, a titular member of the Chemical Engineering Postgraduate Program Coordination Committee at EPUSP (2014-actual) and Deputy Head of the Department of Chemical Engineering (2018–2020). Technical Advisor to the Superintendence of Environmental Affairs (SGA/USP) (2018-actual), and an USP's alternate representative at the State Environmental Council (2018-actual). She has significant experience in recycling, waste treatment, hydrometallurgy, extractive metallurgy, and electro-electronic equipment waste (EEEW) recycling.

Paulo Santos de Almeida (Ph.D.) completed his Ph.D. in Law and Social Relations at the Pontifical Catholic University of São Paulo (PUC-SP) in 2006. He is a Former Executive Director and President of the Conservation and Forest Production Trust of the State of São Paulo (Fundação Florestal). Currently, he is a Technical Advisor to the Superintendence of Environmental Affairs (SGA/USP), and a Ph.D. Professor at the School of Arts, Sciences and Humanities (EACH) at the University of São Paulo (USP), where he supervises in the Sustainability Postgraduate Program (PPG-S) and teaches at the Bachelor's Degree in Environmental Management Basic Cycle. He is a Lead Researcher of the Studies and Research in City, Sustainability and Environmental Management Group (CIDSGAM) certified by CNPq. Researcher at the Carbon Emissions Policy and Regulation Group (NUPPREC-IEE-USP) and Planning and Environmental Management Study Group (PLANGEA-USP). Lawyer and legal-environmental consultant in São Paulo. Effective member of the Sustainability and Environment Commission; Solid Waste Commission and consultant member of the Science and Technology Commission, all at the OABSP. His scientific production is based on sustainability, cultural environment, procedural environmental rights, waste, climate change, animal rights, and sustainable city and human rights.

Women's Empowerment Initiatives in Brazilian Universities: Cases of Extension Programs to Promote Sustainable Development

Carla Patricia Finatto, Camilla Gomes da Silva, Gabriela Carpejani, and José Baltazar Salgueirinho Osório de Andrade Guerra

1 Introduction

According to the United Nations (2019), women represent only 35% of students in the areas of Science, Technology, Engineering and Mathematics. According to data from the survey conducted by the Open Box of Science (2020), in Brazil, there are less than 77.8 million researchers in the five largest areas of knowledge. These areas being: linguistics, letters and arts; Engineering; applied social sciences; exact Sciences; and health sciences. Among the researchers studied, 59.69% are men and 40.3% are women. In the exact sciences, women represent only 31% of the total researchers with a doctorate. In engineering, representativeness is even lower: 26%. In the health area, women are the majority. However, in this area, surgeons are predominantly male.

According to data from the Brazilian Institute of Geography and Statistics (IBGE) (2018a), since the 1990s, women have predominated in participation in education at all levels. However, in 2018, only 16.9% of women aged 25 or over had a higher education degree, even higher than the percentage of men, this being 13.5%.

C. P. Finatto (✉) · C. G. da Silva · G. Carpejani · J. B. S. O. de Andrade Guerra
Center for Sustainable Development (Greens), University of Southern Santa Catarina (UNISUL), Araranguá, Brazil
e-mail: carlapatriciafinatto@gmail.com

C. G. da Silva
e-mail: camila2527@gmail.com

G. Carpejani
e-mail: gabrielacpj@yahoo.com.br

J. B. S. O. de Andrade Guerra
e-mail: baltazar.guerra@unisul.br

W. Leal Filho et al. (eds.), *Universities, Sustainability and Society: Supporting the Implementation of the Sustainable Development Goals*, World Sustainability Series,
https://doi.org/10.1007/978-3-030-63399-8_28

However, despite recent research indicating that women have been the majority at all levels of education in Brazil, including higher education, they receive a lower remuneration compared to men, even if they perform an identical activity (UNDP 2019).

In the last quarter of 2019, men earned, on average, 22% more than women. Among workers with higher education, the difference is even greater, reaching 38%. Between management and executive positions in private companies, the difference was 29%, totaling only 40% of leadership positions. However, the unemployment rate among women reached 13.1%, while among men it was 9.2%. This reality also occurs because women spend 95% more time than men on domestic tasks, sacrificing their academic and professional lives, often for maternal activity (IBGE 2018b).

Thus, even though there have already been significant cultural changes in society, especially regarding the way women see themselves and how they are seen, obstacles remain to achieve gender equality. For example, to be accepted and recognized in areas where historically men predominate, as in engineering and computing (Lima 2013). In addition, as explained, these obstacles can also be seen in medical areas. Although women now make up a large portion of the medical student body in industrialized countries, they are still underrepresented in various disciplines and at the highest levels of medicine. Besides, there are many possible obstacles to career goals, which presumably act synergistically, which include domestic responsibilities, rigidity in career structures and discrimination (Bonder 2015).

That said, the low percentage of women in key areas of science is worrying, a fact that requires initiatives from government, universities and society to minimize gender inequalities. In this regard, universities can contribute positively (Christie et al. 2017). Mainly through outreach programs that promote the United Nations Fourth and Fifth Sustainable Development Goals that involve quality education and gender equality, respectively.

The concepts of "quality education" (SDG4) and "gender equality" (SDG5) must be understood as interdependent issues. It is important to avoid considering them as isolated purposes, as the objectives can and should be interpreted as related to each other. Therefore, the idea of using university extension projects as a tool for the massification of quality education and the empowerment of women was adopted in the guidelines of Brazilian educational institutions. As women still suffer social inequalities, it is understood that, through education and empowerment, better jobs and political, economic and social representation can be guaranteed (McMahon 2010; Psacharopoulos and Patrinos 2018).

In this sense, the concept of "empowerment" of women can be understood as a definition of the development of their capacities, so that each one of them can, collectively and/or individually, have social, economic, political and cultural representation, self-sufficiency, and be able to establish their agendas and demands in society and the State (UNDP 2019). To promote this idea, the United Nations Entity for Gender Equality and Women's Empowerment—a commitment signed by the United Nations (UN-WOMEN 2019), with the support of 189 nations—highlighted equality as a fundamental element for the empowerment of women. The signatory nations have also pledged to effectively include the gender dimension in all their political institutions, advocacy activities, planning and mainstreaming.

This implies that education can act as a powerful mechanism to develop a more cohesive and egalitarian society (Groenez et al. 2007; Blossfeld et al. 2014). Currently, educators highlight the importance of developing gender equality (SDG 5), through initiatives that encourage more access for girls and women to research in areas such as Science, Technology, Engineering and Mathematics (STEM) (Panetta and Williams 2018). In addition, a highly qualified workforce is a proven motivator of decent work and strong economic growth (SDG 8) (Woessmann 2016; Frey 2017), which in turn contributes to the achievement of goals related to poverty (SDG 1) and hunger (SDG 2). Finally, quality education can also include growing skills about learning sustainability, and how to take care of our planet (SDG 12 and 13). It should also be noted that SDG 17 focuses on the role of partnerships to achieve the SDGs. It makes explicit reference to the need for governments to work together and cooperate with a wide range of stakeholders, in the private and non-profit sectors, which would tend to include more women.

It is emphasized that gender inequality is one of the challenges to overcome in order to achieve sustainable development. The empowerment of women, as well as the guarantee of the rights of all people, without exception, is fundamental elements for a legacy of a just society for future generations. Thus, the UN Sustainable Development Goals aligned with university extension programs are reflected in efforts to ensure more possibilities for women in society and the labor market. Nevertheless, recognition of their rights and access to their own income are key elements in building a feeling of empowerment.

Therefore, the role of universities is to mitigate the gender gap in leadership positions and intellectual circles, through educational empowerment that will generate financial independence. As a result, employment opportunities for men and women would increase, so that gender differences in the labor market become less pronounced at higher education levels. However, as will be demonstrated in this article, changes occur when universities connect the community with the academic environment, carry out studies, and when there are incentives for public policies and investments, and private partnerships within universities.

In this sense, this study aims to explore the role of Brazilian universities in contributing to the empowerment of women in science. This will be done through the analysis of two extension programs carried out by three Brazilian universities in southern Brazil, namely the Maternal-Child and Family Care Program (MCFCP) and the Girls in Science Program, through the qualitative approach and collection of information and bibliographic data. In summary, the cases show that the integration of the university with society is a crucial point to support the implementation of sustainable development, gender equality and, above all, access to education.

2 Women's Empowerment and Sustainable Development

Empowerment can be defined as an active multi-dimensional process which helps women to develop their identity and self-esteem in all spheres of life in due to greater access to knowledge and resources, more autonomy in decision making, greater ability to plan their lives, more control over the circumstances which influence lives, and freedom from customs, beliefs, and practices (Cornwall 2016).

Individual empowerment has, as indicators, self-esteem, self-confidence and self-affirmation, since the most successful actions are those that do not include only the community (Malhotra et al. 2002). Regarding gender equality, this involves elevating women through literacy, education, training and awareness (Alvarez 2013).

It is also necessary to establish the concept of gender equality. One can understand gender equality as a human right and essential for the achievement of peaceful societies and sustainable development (Shannon et al. 2019). It is also defined as a situation in which "all human beings are free to develop their personal skills and make choices without the limitations established by strict gender roles; that the different aspirations and needs of women and men are considered, valued and favored equally" (Holzner et al. 2010, p. 5).

Gender equality is included in the agenda of the Sustainable Development Goals as the fifth goal (UN 2010). To conceptualize, the most frequently cited definition is "the development that meets the needs of the present without compromising the ability of future generations to meet their own needs" (Brundtland 1985, p. 1). In other words, for Van Bellen (2004, p. 73), to live sustainably requires a standard of living "within the limits imposed by nature". It means society, in general, needs to act in favor of the interests of future generations while trying to satisfy the current needs to have a planet capable of supporting human activities. And Gender Equality is an essential goal to conquer a sustainable future for the planet since it collaborates for a better quality of life and social harmony (WEF 2016).

It is also noteworthy that sustainable development is not based on economic, social, ecological, or institutional dimensions in isolation of each other, but on the vision of the system as a whole. This is because the relationships between the systems must be reallocated in a logical structure to achieve coherent sustainability (Romeiro 2012).

Stevens (2010) states that gender disparity, unquestionably, is prevalent in all cultures in the world and that without serious measures to address it, sustainable development cannot be achieved. Also, the author emphasizes that "an increasing number of studies indicate that gender inequalities are extracting high economic costs and leading to social inequities and environmental degradation worldwide" (Stevens 2010, p. 3). To this point, "gender equality is a human rights issue and a precondition for establishing sustainable development" (Alvarez 2013, p. 85).

That said, UN-WOMEN (2019) believes that education is the most efficient way to achieve the fifth Sustainable Development Goal, thus establishing synergy between the fourth and fifth goals. However, it is worth mentioning that the search for gender equality and quality education depends on the interdependent interaction of actions based on the economic, social and environmental dimensions.

For such ideals to be attained, university extension presents itself as a satisfactory alternative to create the interaction between higher education institutions and vulnerable communities. This connection bridge between theory and practice promotes exchange with the environment in which both parts are inserted (Nogueira 2000).

Women represent half of the world's population and deserve equal access to health, education, economic participation, earning potential and political decision-making power (WEF 2016; UN-Women 2019). However, they face obstacles to empower themselves and reach decision-making positions due to the cultural, economic and social issues of the society that sees them as more fragile than men. Regardless of the historical model of the oppression towards women in society, social advancement and women's independence are slowly improving through education (Dalouh and Soriano 2017).

Quality education promotes a sense of protagonism, identity, leadership power and, consequently, better positions in the labor market for women (Shahtalebi et al. 2011; Yousefy and Baratali 2011). Higher education even has protective effects on health due to its impacts on the psychological and social processes of individuals (Chiu et al. 2019). And this transformative power of education, with wide-ranging effects across society, makes it the reason why it plays a significant role in gender equality.

3 Methodology

The nature of the research of this study is qualitative. According to Cooper and Schindler (2011), qualitative research includes a set of interpretative techniques that seek to understand a certain phenomenon through its description, decoding and translation.

Regarding the research approach, the case study was applied as it is an empirical investigation that investigates the contemporary phenomenon in depth. In other words, it investigates a contemporary phenomenon within its context, especially when the boundaries between the phenomenon and the context are not clearly defined. It is a comprehensive method that comprises the logic of the paper, data collection techniques and approaches to data analysis (Yin 2015).

Bhattacherjee (2012) defends that a case study is a method of intensive study of a phenomenon over time within its natural configuration. It consists of several methods of collecting secondary data and allows the researcher's inferences about the phenomenon, allowing a rich, contextualized and more authentic interpretation.

Also, Creswell (2007) asserts that when the researcher carries out the case study, he explores in depth a fact, an activity collecting detailed information using a variety of collection procedures. In this sense, this study chose three cases regarding two university extension programs in the south of Brazil, the Girls in Science program supported by the National Council for Scientific and Technological Development

(CNPq) executed by the University of the South of Santa Catarina (UNISUL) and the Federal University of Rio Grande do SUL (UFRGS); and the Maternal-Child and Family Care Program supported by the University of the Extreme South of Santa Catarina (UNESC), taking into account the action of these universities in relation to programs supported by governmental initiative and by the initiative of the educational institution itself. As the data collection technique, this study used sources such as "search sites" and "social networks" of the studied objects.

The analysis of the collected data used the typology defined by Yin (2015), the author suggests 5 analytical research techniques: (1) pattern combination; (2) construction of explanation; (3) time series analysis; (4) logical models; (5) cross-synthesis of data. For this purpose, a standard combination analysis will be carried out in which the subjects of the greatest congruence and relevance of the selected cases will be evaluated.

4 Results and Analysis

4.1 Gender Equality in Brazilian Universities

Brazil is a country with high levels of social inequality, and when filtered for gender issues, this development obstacle is still present. According to the UNDP report (2010), inequality between men and women causes a loss of 63% of human development potential in Brazil.

This is because, throughout the history of teaching in Brazil, universities and their role in society have undergone numerous transformations. This is linked to the transition in their functions, which then becomes a social, economic and cultural institution (Bartnik and Silva 2009). However, even though higher education in Brazil is at a slow pace, in terms of promoting the empowerment of women within and outside the community of higher education, it is currently possible to highlight some initiatives carried out by universities that seek to make higher education more accessible to be entered by women (Dalouh and Soriano 2017).

In this way, university extension programs can be conceptualized as a complement to traditional teaching, allowing students to expand their understanding during their academic journey, as well as involving the local community. In addition, it contributes to personal development and the formation of social values and critical emancipatory thinking. The aim is to empower the knowledge acquired reflecting in the adherence and articulation of public programs and policies that turn their attention to issues inherent, among others, to the empowerment of women (Casadei 2016).

In teaching, the extension allows learning beyond the classroom, since teaching is directed towards the construction of new knowledge that, as a rule, would not be taught in the classroom, especially given the participation of the community outside the university. At the same time, the extension allows the democratization of academic knowledge because, through it, the knowledge built with the community is produced in the dynamics between theoretical knowledge and practical constructions (Sousa 2010; Dias 2009).

In extension programs, the students present themselves as "subjects of change, capable of placing themselves in the world with an active and critical posture" (Castro 2004, p. 14). In addition to that, the extension, by covering the internal and external community of the university, allows all agents to be able to exchange knowledge widely, which would not happen only in the classroom.

Thus, deconstructing stereotypes and false dichotomies and moving towards equality of rights and gender equity are indispensable conditions for those who envision a democratic and citizen society (Prá and Epping 2012). It is in this context that this study utilizes two initiatives to demonstrate the role of universities in promoting sustainable development through the education and empowerment of women, which are presented and reflected on in the following two sections.

4.2 *The Maternal-Child and Family Care Program (MCFCP)*

As explicit above, there is a need for the empowerment of women in society at all levels to achieve greater access to equal rights, health, education, politics and job opportunities. In this context, the Maternal Child and Family Assistance Program was created. This extension program of the University of the Extreme South of Santa Catarina (UNESC) seeks to assist pregnant women and their families, within a social-community perspective, through group activities, in an interdisciplinary proposal (Amorim et al. 2018).

Among the actions of the program, it offers primary care to families through educational actions, support groups, physical activity (which includes aquatic physiotherapy and the use of the Pilates Method) and nutritional guidance through the "Healthy Kitchen" project. It also prepares women for childbirth and offers home visits to families. MCFCP stands out due to its interdisciplinary work proposal. The program includes professionals from the areas of Nursing, Physiotherapy, Nutrition and Psychology, with the possible participation of other areas of health knowledge such as Dentistry, Pharmacy and Medicine. The team is made up of 3 professors, 4 scholarship students and 9 UNESC volunteer scholars (UNESC 2019).

For the university extension programs to contribute to the training of more humane professionals, it is necessary to carry out actions and interventions on different fronts that contribute to the healthy development and adoption of positive practices. In the case of the MCFCP program, it promotes, mainly for pregnant women and their families, multidisciplinary care, within the biopsychosocial framework. The idea is to decentralize obligations, whether personal, psychological or educational, in the process of welcoming a child, shifting the responsibility from only the mother to the family (UNESC 2019).

The MCFCP program highlights that the main focus is low-income families due to the philanthropic nature of UNESC and the Organic Law on Social Assistance. Thus, meetings are organized every two weeks involving students, teachers and the community. In these meetings, activities are carried out according to the demand of the group, usually of about 20 pregnant women, with semi-annual registrations. In addition, several commemorative dates are encouraged, such as Women's Day, Mother's Day, National Mother's Reduction Week, Father's Day, National Breastfeeding Week, Children's Day and Adolescent Pregnancy Prevention Day, among others (Amorim et al. 2018).

Students, in the Psychology course, for example, benefit from the opportunity to use the MCFCP as a place for basic observation studies in the subjects of Developmental Psychology, Social Psychology and Group Dynamics; in the Physiotherapy course, they involve the subjects of Physiotherapy Practice II, Gynecological and Obstetric Physiotherapy, Hydrogymnastics Classes with Babies and Pregnant Women; in nutrition, the activities are carried out in the subject Integrator Seminar and, in physical education and nursing, they use it as a space of knowledge in the stages of academic studies (UNESC 2019).

The results expected by the MCFCP program include mainly the health of women, especially during pregnancy. In this way, the program encourages healthy pregnancies with various stimuli through the use of biopsychosocial health indicators or a positive capacity to react to adverse situations. Also, the program defends the preference of choosing normal delivery with the father's accompaniment in the procedure room and the family's harmony with the baby during the puerperium. It offers help when necessary about caring for the child, as well as working to prevent postpartum depression. It also emphasizes adherence to the practice of breastfeeding. It still promotes the affective bonds between family members and their children; citizenship and reaffirmation of the rights of pregnant women during childbirth, the puerperium and young children (UNESC 2019).

Initiatives like this raise the empowerment of women and the sense of independence. Especially in a country like Brazil, where the public health system fails to offer the minimum necessary support for pregnant women who are unable to afford the costs of care for their children. In addition, the students are rewarded with interdisciplinarity in academic performance. By MCFCP it is possible to contribute to the formation of future professionals who have a different and critical view regarding assistance to human beings. Moreover, the program offers, to these students, experiences to understand the benefits of teamwork and formulating strategies for the promotion of collective health through general intervention and in specific groups, contributing to the consolidation of more welcoming health in Brazil.

In this sense, it is possible to reinforce the thinking of Kumar (2017), Cornwall (2016) and Alvarez (2013), who state that empowerment refers to the intrapersonal factor and can include variables such as the self-esteem, quality education level and autonomy. As well as the representativeness in decision making positions at community and organizational activities. To that end, it is up to universities, to increasingly assume their role, aiming at sustainability and acting as key players in this new era. To offer solutions whose common objective is to involve the academy and the community, including empowering women, thus reducing gender inequality.

4.3 The Girls in Science Program

Given the gender gap in Brazil, especially when referring to education in vulnerable social classes, the National Council for Scientific and Technological Development (CNPQ) developed a program for the inclusion of girls and women in the scientific field. The program became known as "Girls in Science" and several universities in Brazil received funding to carry out actions for more equitable gender inclusion. Two of them are from the south of Brazil, such as the University of the South of Santa Catarina (UNISUL) and the Federal University of Rio Grande do Sul (UFRGS).

4.3.1 At the University of Southern Santa Catarina (UNISUL)

The University of Southern Santa Catarina is located in the South Region of Brazil, in the State of Santa Catarina. It is one of the best-rated community universities in the country. Its extension programs are geared to the University's commitment to connecting with the community. In this sense, the Girls in Science Program at UNISUL is a mechanism that integrates the teaching, learning and researching process to allow the students to experience the knowledge learned in the classroom (UNISUL 2020).

In this way, UNISUL aims to contribute through the Girls in Science Program to encourage girls to enter the universe of exact sciences, engineering, and computing. Thus, learning to discover, believe, invent and reinvent. This program seeks to modify the minority scenario of women in these scientific fields, currently represented by 30% of women (UNISUL 2020).

The Girls in Science program is essential because it effectively corroborates the fifth and seventh target of SDG 4, which maintains that by 2030, countries need to ensure the elimination of gender disparities in education and guarantee equal access to all levels of education and professional training for the most vulnerable. Also, to promote the knowledge and skills necessary to achieve sustainable development. Especially through education for sustainable lifestyles that respect the pillars of human rights, gender equality, promotion of a culture of peace and non-violence, global citizenship and diversity (UN 2020).

Thus, university extension programs that focus on scientific research enables an exchange of knowledge between students and the community since it influences the use of active teaching methodologies aimed at consolidating critical emancipatory thinking and empowering acquired knowledge. Besides, the programs can be tools of adherence and articulation of public health programs and policies that turn their attention to issues inherent, among others, to women's empowerment (Casadei 2016).

Moreover, with the benefits of university extension, an important role can be reconciled in achieving the targets of the SDG5 that seeks to guarantee the full and effective participation of women across all sectors and levels of decision-making, such as in the political, economical and the public life (UN 2020). Thus, it is understood that through sustainable education, the promotion of female empowerment becomes the accepted norm.

That said, in practice, the Girls in Science Program at Unisul involves five schools and offers 23 scholarships. Out of this total, 15 for high school students in public schools, 5 for teachers of exact sciences in public schools, 2 grants for engineering students and 1 for a cosmetology student at Unisul. As required activities, University scholarship students receive public school students monthly, stimulating them through workshops to enter engineering, computer or information systems courses and to acquire a real sense of the possibilities of a professional future in the technological area. The activities include soap workshops and calorie counting at the Pedra Branca Unit, in addition to activities with 3D printing, in the Fashion course at the Dib Mussi Unit. In addition, schools receive a visit from the "CiênciaMóvel" truck with the scholarship holders offering workshops on augmented reality (in the IT area) (Ximenes 2019).

The program members report some of their experiences on the official website of UNISUL, local newspapers and social networking pages of the Girls in Science Program. One of which is that after a few visits by the truck to schools in municipalities in the mountainous region of Greater Florianópolis, they realized that the reality of young women and adolescents in the schools selected to participate in the program was removed from the scientific world. In these schools, most of the students did not aim for higher education before the program, believing that it was an academic stage possible only for people with high purchasing power (Ximenes 2019).

Even though the Girls in Science program at Unisul does not provide statistical data or studies on the impact of the actions carried out, authors such as Sousa (2010) and Dias (2009) argue that the involvement of the university in communities, in this case at schools, can stimulate the entry of students in higher education courses, scientific initiation and the opening of the mind to the science as a whole. Therefore, by providing more access to education it is possible to mitigate several social issues like gender disparity.

This paper highlights that a sustainable society needs to empower people, respecting differences and diversities, including gender, race/ethnicity, social class, sexual orientation and culture. In this scenario, it is possible that the universities, as centers of higher education, assume their role as transformers and educators and implement initiatives that lead to reducing the social inequalities.

Thus, the Girls in Science program case is an example of university extension that allows community interaction and the role of the university in promoting sustainable development through the education and empowerment of women, seeking to eliminate gender disparities and promote knowledge and technical skills for young women, as well as encouraging access to quality higher education.

4.3.2 At the Federal University of Rio Grande do Sul (UFRGS)

Another initiative worth mentioning is the Girls in Science extension program at the Federal University of Rio Grande do Sul (UFRGS). It has been an extension action by the UFRGS Physics Institute since the end of 2013. In that university, the focus was to attract girls to Science and Technology (S&T) careers and to encourage women who have already chosen these careers to persist and become agents of change in the scientific and technological development of Brazil. The main goals are traced from the training of undergraduate students to disseminate Science and Technology through Astronomy, Physics and Robotics in public schools. The "Girls in Science" at UFRGS also aims to sensitize the academic community and the neediest communities about the role of women in society, contributing to the elimination of gender stereotypes (Brito et al. 2015).

The Girls in Science program shows relevance, considering that only 31% of women are researchers in technology and exact sciences in Brazil (Open box of Science 2020). In other words, women remain concentrated in some areas and almost expressionless in others. At least in Brazil, especially in the Humanities area, there are many more women than in the technological area, therefore, excluding them from leadership positions in the labor market (Barbosa and Lima 2013).

In addition to the low representation of women in the Exact Sciences, another factor that deserves to be highlighted is that as they progress in their careers, the percentage of women tends to decrease. This effect is called the Scissors Effect because it cuts women out of the career as it progresses (Agrello and Garg 2009).

Among the actions implemented some can be pointed out: (1) science workshops and gender discussions in public schools; (2) continuing education of undergraduate students in what concerns both the learning of Astronomy and Physics existing in the workshops and the awareness of gender issues; (3) training for teachers at the partner school; (4) a robotics course at a public school; (5) film production to disseminate the presence of women in S&T careers, creating female models in areas where women are underrepresented entitled "Lugar de Mulher" and (6) UFRGS campus opening and classroom activities for elementary and high school students (Brito et al. 2015).

The production of the films "Lugar de Mulher", mentioned above, has maintained a partnership with UFRGS-TV since 2014, promoting interviews with younger women, still in the beginning of their careers to encourage the areas of science and technology.

Still, primary schools are places where first expectations for the future and career planning are built. Thus, science and gender workshops in public schools are one of the main current activities of the program. The objective is to sensitize the whole society about gender issues and encourage students of both sexes to like science.

The Girls in Science program argues that the participation of women in research in exact science areas is important because expanding the debate on the participation of women in science also means thinking about a different science, inspired and renewed by life experiences historically excluded from scientific and technological production.

In this sense, for undergraduate students, the program invests in holding biweekly meetings to discuss issues related to gender disparity. In these meetings, roundtable meetings are held with teachers and students involved in the program. Each week, a group of students is responsible for conducting the meeting. The focus is on discussing the results of an existing workshop or designing more improved actions.

5 Conclusion

Higher education institutions have been developing strategies and actions to empower and include more women. However, inequalities persist. Gender inequality is still a social and historical issue that hinders access for women in many social spheres, especially in the labor market and high-quality education.

In this sense, a contemporary perspective of the image of women in society, in all its spheres, is being increasingly discussed. Issues related to equal rights, access to health and education are the main topics. And with this study, it can be understood that university extension programs have the potential to make a positive contribution through the ability to connect knowledge and people.

This paper presented a qualitative discursive analysis of two university extension programs with the potential to contribute to strategies for higher education institutions to collaborate with SDGs 4 and 5. The Brazilian programs called: Maternal-Child and Family Care Program and Girls in Science were held at three Brazilian universities in southern Brazil, in the state institution UNESC, in the federal institution UFRGS and in the private institution UNISUL. The programs focus on improving the goals of quality education and gender equality.

In this context, extension programs in science and technology have presented integrative and efficient initiatives to emphasize the socioeconomic and cultural role of women in society. In addition, the programs were representative of the empowerment of women by improving education, as a whole, when they connected the academic and scientific environment with girls and women with the potential to get into these areas.

The Maternal-Child and Family Assistance Program, applied by UNESC, showed how essential it is to offer quality-education opportunities for pregnant women. Pregnant women are the ones who have a greater stigma when entering the job market, which is why the project puts the whole family in the process of generating a child, shifting the cultural burden of the responsibility from only the mother to the family (woman). Thus, it favors the concept of women's empowerment through the feeling of independence.

The "Girls in Science" Program, in turn, proved to be efficient in the south of Brazil, hosted by two relevant institutions in the region, the Federal University of Rio Grande do Sul and the University of the South of Santa Catarina. The initiative sought to connect adolescent girls from public schools to undergraduate students, researchers and teachers. Thus, the project encourages leadership, knowledge, initiation and scientific representation; essential factors for women's empowerment and gender equality.

Thus, through this study, it was possible to understand that the presence of women in higher education institutions is increasing. However, institutions must encourage female participation in all areas of knowledge, especially in the areas of technology and exact sciences. In addition, it is necessary to encourage girls' access to study in these areas, through the connection between university, science and the community.

It is noteworthy that gender inequality is still one of the challenges to be overcome in order to achieve sustainable development. Therefore, the role of universities is to mitigate the gender gap in leadership positions and intellectual circles, through educational empowerment that will generate financial independence. As a result, employment opportunities for men and women would increase, so that gender differences in the labor market become less pronounced at higher education levels. However, as demonstrated in this article, changes occur when universities connect the community with academia and conduct studies. When there are incentives for public policies and investments, and private partnerships within universities.

Finally, it is worth to point out that this study had some limitations, such as the lack of presentation of quantitative data corresponding to the results of the projects, both for the community and for the university. Therefore, it would be necessary to carry out analyses of other initiatives in other contexts, including geographic and social contexts. Also, another suggestion would be for studies to verify the best approaches and the main difficulties of the projects. There is the possibility of research and interviews to be made with women who have already done and were favored by university extension programs, to verify the influence of these actions in their lives.

References

Agrello DA, Garg R (2009) Mulheres na física: poder e preconceito nos países em desenvolvimento. Rev Bras Ensino Fís 31(1):1305

Alvarez ML (2013) A case study of women's political empowerment in the Philippines. In: 12th National Convention on Statistics (NCS) EDSA, Mandaluyong City, pp 1–2

Amorim MP et al (2018) 4849477 PAMIF—Programa de Atenção Materno-Infantil e Familiar: O seu Papel na Extensão Universitária. Semana de Ciência e Tecnologia. Retrieved from http://periodicos.unesc.net/cienciaetecnologia/article/download/5169/4687

Barbosa MC, Lima BS (2013) Mulheres na Física do Brasil: Por que tão poucas? E por que tão devagar. Trabalhadoras: análise da feminização das profissões e ocupações. Editorial Abaré, Brasília

Bartnik FMP, Silva IMD (2009) Avaliação da ação extensionista em universidades católicas e comunitárias. Rev Avaliação Educação Super Campinas 14(2):453–469

Bhattacherjee A (2012) Social science research: principles, methods, and practices

Blossfeld H-P, Kilpi-Jakonen E, Vono de Vilhena D, Buchholz S (eds) (2014) Adult learning in modern societies: an international comparison from a life-course perspective. Edward Elgar, Cheltenham

Bonder G (2015) National assessments of gender, science, technology and innovation: Argentina. Women in global science and technology and the organization for women in science for the developing world: Brighton (Canada)

Brito C, Pavani D, Lima P Jr (2015) Meninas na Ciência: atraindo jovens mulheres para carreiras de Ciência e Tecnologia. Rev Gênero 16(1)

Brundtland GH (1985) World commission on environment and development. Environ Policy Law 14(1):26–30
Casadei EB (2016) A extensão universitária em comunicação para a formação da cidadania. Cultura Acadêmica, São Paulo, p 135
Castro LMC (2004) A universidade, a extensão universitária e a produção de conhecimentos emancipadores. Reunião Anual da ANPED, 27. ANPED, Textos Caxambu, pp 1–16
Chiu CT, Hayward MD, Chan A, Matchar DB (2019) Educational differences in the compression of disability incidence in the United States. SSM-Population Health 7:100347
Christie M et al (2017) Understanding why women are under-represented in Science, Technology, Engineering and Mathematics (STEM) within Higher Education: a regional case study. Production 27(spe):e20162205. https://doi.org/10.1590/0103-6513.220516
Cooper DR, Schindler PS (2011) Métodos de pesquisa em administração, 10th edn. Bookman, Porto Alegre
Cornwall A (2016) Women's empowerment: what works? J Int Dev 28(3):342–359
Creswell JW (2007) Investigação qualitativa e projeto de pesquisa: escolhendo entre cinco abordagens, 3rd edn. Penso, Porto Alegre
Dalouh R, Soriano E (2017) Second-generation moroccan women in europe: higher education and labour market positioning. Procedia-Soc Behav Sci 237:10–16
Dias AMI (2009) Discutindo caminhos para a indissociabilidade entre ensino, pesquisa e extensão. Rev Bras Docência, Ensino Pesquisa em Educação Física 1(1):37–52
Frey D (2017) Economic growth, full employment and decent work: the means and ends in SDG 8. Int J Hum Rights 21(8):1164–1184
Groenez S, Desmedt E, Nicaise I (2007) Participation in lifelong learning in the EU-15: the role of macro-level determinants. In: European Conference for Education Research (ECER), Ghent, Belgium
Holzner H, Neuhold B, Weiss-Gänger A (2010) Gender equality and Empowerment of Women Policy Federal Ministry for European and International Affairs, Directorate-General for Development Cooperation Austrian Development Agency. Gender Dev Unit Doc. Retrieved from https://www.entwicklung.at
IBGE (2018a) Estatísticas de Gênero Indicadores sociais das mulheres no Brasil. Retrieved from https://biblioteca.ibge.gov.br/visualizacao/livros/liv101551_informativo.pdf
IBGE (2018b) Indicadores IBGE. Pesquisa Nacional por Amostra de Domicílios Contínua Quarto Trimestre de 2019. Retrieved from https://biblioteca.ibge.gov.br/visualizacao/periodicos/2421/pnact_2019_4tri.pdf
Lima MP (2013) As mulheres na Ciência da Computação. Rev Estudos Feministas 21(3):793–816
Malhotra A, Schuler SR, Boender C (2002) Measuring women's empowerment as a variable in international development. In: Background paper prepared for the World Bank workshop on poverty and gender: new perspectives, vol 28
McMahon WW (2010) The external benefits of education. In: Peterson P, Baker E, McGaw B (eds) International encyclopedia of education. Elsevier, Amsterdam
Nogueira MDDP (2000) Extensão universitária: diretrizes conceituais e políticas: documentos básicos do Fórum Nacional de Pró-Reitores de Extensão das Universidades Públicas Brasileiras. Pró-Reitoria de Extensão da Universidade Federal de Minas Gerais, pp 1987–2000
Open box da Ciência (2020) Dados. Retrieved from http://www.openciencia.com.br/
Panetta K, Williams K (2018) Count girls in empowering girls to combine any interests with stem to open up a world of opportunity. Chicago Review Press, Chicago, IL
Prá JR, Epping L (2012) Citizenship and feminism in the recognition of women's human rights. Feminist Stud Mag 20(1):33–51
Psacharopoulos G, Patrinos H (2018) Returns to investment in education: a decennial review of the global literature. Educ Econ 26(5):445–458
Romeiro AR (2012) Sustainable development: an ecological economics perspective. Estudos Avançados 26(74):65–92

Shahtalebi S, Yarmohammadian MH, Ajami S (2011) Women's success factors from leadership in higher education. Procedia-Soc Behav Sci 15:3644–3647
Shannon G, Jansen M, Williams K, Cáceres C, Motta A, Odhiambo A, Mannell J (2019) Gender equality in science, medicine, and global health: where are we at and why does it matter? Lancet 393(10171):560–569
Sousa ALL (2010) A História da Extensão Universitária. Alínea, Campinas, SP
Stevens C (2010) Are women the key to sustainable development. Sustain Dev Insights 3(11):1–8
UN (2019) Desigualdades de gênero afastam mulheres e meninas da ciência, dizem especialistas. Retrieved from https://nacoesunidas.org/desigualdades-de-genero-afastam-mulheres-e-meninas-da-ciencia-dizem-especialistas/
UN (2020) About the Sustainable Development Goals. Retrieved from https://www.un.org/sustainabledevelopment/sustainable-development-goals/
UNDP (2019) United Nations Development Programme. Retrieved from http://www.undp.org/content/undp/es/home/sustainable-development-goals.html
UNESC (2019) Universidade do Extremo Sul de Santa Catarina. Retrieved from http://www.unesc.net/portal/capa/index/223/0/0/componente/projetoseacoes/verProjeto/10
UNISUL (2020) Projetos e Bolsas de Extensão da Unisul. Retrieved from http://www.unisul.br/wps/portal/home/extensao/projetos-e-bolsas-de-extensao
UN-WOMEN (2019) Paridade de gênero. Retrieved from http://www.onumulheres.org.br/planeta5050-2030/paridade/
Van Bellen HM (2004) Sustainable development: a description of the main assessment tools. Environ Soc 7(1):67–87
WEF—World Economic Forum (2016) Global gender gap report. Retrieved from http://www3.weforum.org/docs/GGGR16/WEF_Global_Gender_Gap_Report_2016.pdf
Woessmann L (2016) The economic case for education. Educ Econ 24(1):3–32
Ximenes M (2019) Project on wheels encourages teenagers to higher education in Greater Florianópolis. Retrieved from https://ndmais.com.br/noticias/projeto-sobre-rodas-incentiva-adolescentes-a-educacao-superior-na-grande-florianopolis
Yin RK (2015) Estudo de caso. Planejamento e métodos. In: Tradução de Anna Maria Vascocellos Thorell. Bookman, PortoAlegre, RS
Yousefy A, Baratali M (2011) Women, employment and higher education schoolings. Procedia Soc Behav Sci 15:3861–3869

Carla Patrícia Finatto is graduated in Law (2015), post graduate student in International Law and also is a researcher at Centre for Sustainable Development (GREENS) at University of Southern Santa Catarina (Unisul), Brazil.

Camilla Gomes da Silva is Master's student in Environmental Sciences and graduated in International Relations (2018) at University of Southern Santa Catarina (Unisul), Brazil. She is researcher at Centre for Sustainable Development (GREENS)

Gabriela Carpejani is graduated in Administration (2003). Actually master's degree student in Administration and student research in Centre for Sustainable Development (GREENS) at University of Southern Santa Catarina (Unisul), Brazil.

Professor José Baltazar Salgueirinho Osório de Andrade Guerra is a full Professor at the Graduate program in Management and Environmental Sciences at "Universidade do Sul de Santa Catarina" (Unisul, Brazil). Director and Founder from the Centre for Sustainable Development (Greens) at Unisul. Research Fellow from Cambridge Centre for Environment, Energy and Natural Resource Governance, University of Cambridge.

Implementing a Sustainability Plan on a Small, Young Campus in Brazil

Marcelo Bizerril, Leandro Oliveira Evangelista, Philippe Pomier Layrargues, Elaine Nolasco, Maria Cristina Oliveira, Luiz Henrique Oliveira, Rômulo Ribeiro, Antônia Roberto Sousa, and Luiz Felippe Salemi

1 Introduction: Sustainability in the Brazilian Public Administration

In Brazil, several instruments of command and control originated at the federal level, such as laws, rules and federal regulations, including criteria for the insertion of socio-environmental practices in public administration. A striking fact was the

M. Bizerril (✉) · L. O. Evangelista · P. P. Layrargues · E. Nolasco · M. C. Oliveira · L. H. Oliveira · R. Ribeiro · A. R. Sousa · L. F. Salemi
Faculty UnB Planaltina, University of Brasília, Área Universitária N.1, Vila Nossa Senhora de Fátima, Planaltina-DF 73.345-010, Brasília, Brazil
e-mail: bizerril@unb.br

L. O. Evangelista
e-mail: levangelista@unb.br

P. P. Layrargues
e-mail: philippe.layrargues@gmail.com

E. Nolasco
e-mail: enolasco@unb.br

M. C. Oliveira
e-mail: mcrisoliveira@unb.br

L. H. Oliveira
e-mail: embrageaunb@gmail.com

R. Ribeiro
e-mail: rjcribeiro@unb.br

A. R. Sousa
e-mail: embrageaunb@gmail.com

L. F. Salemi
e-mail: lfsalemi@unb.br

W. Leal Filho et al. (eds.), *Universities, Sustainability and Society: Supporting the Implementation of the Sustainable Development Goals*, World Sustainability Series,
https://doi.org/10.1007/978-3-030-63399-8_29

promulgation of Decree 7746 of 2012 that defines socioenvironmental guidelines to be implemented by federal institutions, and the resulting norms that establish minimum criteria and structures for the proposed instruments (Brasil 2012a). Due to this decree, the Interministerial Commission on Sustainability in Public Administration (CISAP) was created, which is the forum for discussion and proposals with a view to implement and regulate the actions that promote sustainability within the federal government (Brasil 2016).

The Decree institutionalized the Sustainable Logistics Management Plan (PLS), a planning tool to be developed by all the institutions of the Federal Public Administration (APF). The PLS works includes actions to promote sustainability and define respective goals, with deadlines, as well as monitoring and evaluation mechanisms. Thus, the PLS gives conditions for the organs and entities of the APF set their sustainability practices and rationalization of costs and processes.

In 2015, the Ministry of Planning, Budget and Management issued Ordinance 23/2015 that indicated good practices in the management of electricity and water, and their monitoring (Brasil 2015). In October of that year, three decrees were issued,[1] with a view to: (i) digitize all administrative records and information; (ii) rationalize the management and contracting of various services, from the rental of properties to the contracting of cleaning services, including dealing with surveillance and telecommunications contracts; and (iii) optimize carbon offsets for official vehicles and aerial (Brasil 2016).

In Brazil, federal public universities are considered as autarchies of the federal government, and, therefore, are subject to this legislation that becomes one of the factors that might mobilize universities for their transition to sustainability. Therefore, a movement towards sustainability is mandatory. However, federal public universities are slowly moving in such direction. With the expansion of the network of federal universities during 2003–2015 (Brasil 2010), a number of new campuses emerged which also had to follow the aforementioned regulations. In Brasília, the Planaltina campus is one of these campuses that are an extension of the University of Brasília (UnB), a federal university founded in 1960s. This young campus emerged in 2006 and consideration regarding sustainability arose in 2010 (see more details in next sections). Four undergraduate courses are offered: Environmental Management, Agribusiness Management, Natural Sciences and Rural Education. All of them with a clear relationship with environment and, therefore, with sustainability. New campuses such as this one are an opportunity to study and develop environmental practices towards the implementation of Sustainable Development Goals (SDG) at the university scale. The objective of this paper is to describe the process of implementing a sustainability plan in this small and young campus. Specific objectives were: (1) present its distinctions compared to the business sustainability model; (2) describe how such system is being developed in Planaltina campus and (3) presents the rich-learning opportunities for academic community during the whole process of understanding and developing such system.

[1]Decrees are: 8539, 8540 and 8541/2015.

2 Differences Between the Sustainability Approach at the University and Companies

In general, environmental management actions are guided by the logic of Eco-Efficiency, a perspective that was developed within the context of corporate environmentalism. Eco-Efficiency is an ecological imperative that emerged as a selection criterion amongst business environmentalism. The prevailing discourse at the time was under the influence of the ecological wave provided by Rio-92, wherein companies that did not conform to environmental constraints would perish. Since then, environmental considerations have changed from cost to investment, significantly changing corporate behavior in the face of new agendas within the business world. In view of the promises, green consumers are guiding the purchasing choices, investing in the environment has become a business opportunity, a market difference from the past (Layrargues 2000). It was in this scenario that business environmentalism and the respective green companies emerged to explore a new competitive advantage, also bringing its discursive logic in commercial advertising. From the perspective of advertising, what should be emphasized to attract the consumer is the final balance of the company's environmental investment. That is, seals of approval or certificates usually attest the commitment to results. It does not matter for the company to explain how or if it managed to produce in an ecologically sustainable manner, but only to signal that it was 'clean' production and guided by a commitment to sustainability. In other words, corporate environmental management carries a pragmatic approach and is mainly concerned with the dissemination of results.

Aligned with companies and other sectors, sustainability in Higher Education Institutions (HEIs) has become a worldwide concern for policy makers (Alshuwaikhat and Abubakar 2008). According to Kassaye (2018), universities have an ethical obligation to act responsibly towards the environment. They must be leaders in the movement for environmental protection and be protagonists in building a sustainable future, changing ways of thinking and fostering new skills, thus bringing social, economic and ecological changes. However, in the academic environment, this same environmental management process presents a different context than that of a business environment. It is precisely in this institutional distinction that marks the meaning of the academic world, that the differential of university environmental management is founded: the search for balance between the *process* and the *product*. In academia, the process covered by environmental management systems matter as much as the final product of institutional investment in the environmental cause. To the same extent that it is important to signal a commitment to sustainability through the presentation of statistics, indicators or environmental seals attesting to having achieved satisfactory results towards sustainability, it is also important for the university that the process is not reduced to a path bureaucratic and formally institutionalized, acting only in the engine room of university campuses, without appearing throughout the university community as an engaging opportunity for learning and, consequently, for changing habits. The university, as an educational space, cannot strictly follow the approach of corporate environmental management, at the risk of wasting the opportunity to be an educational structure that brings transformative learning during the implementation of sustainability actions. Environmental management in higher education stands out for its differential in being able to expand educational processes beyond the formal scope restricted to the classroom. In this sense, Petrelli and Colossi (2006, p. 71) pointed out that "Higher Education Institutions perform a significant social function: to provide higher education to people who can influence society's development process towards the improvement of life in planet".

To accomplish this mission, universities need to transcend the proposed models for the inclusion of sustainability into public administration as well as those related to environmental business management. Furthermore, in addition to the dimensions of teaching, research and extension, and especially university administrative management itself, emerges a privileged space for participatory environmental management with the horizon of bringing together the majority of the university community. Participation is the key idea contained in university environmental management.

3 Participatory Process of Fostering Sustainability at Faculty UnB Planaltina (FUP)

The FUP campus, located in Planaltina, a city in the northeast portion of the Federal District, Brazil, is one of the four campuses that make up the University of Brasília. With an interdisciplinary organization, FUP was created in 2006 and consists of 116 professors, 137 staff, and 1355 undergraduate and graduate students. The sustainability theme is related to the courses offered by the campus and the profile of the professors who research sustainability under different approaches. Both the four undergraduate courses offered (Degree in Natural Sciences, Degree in Rural Education, Bachelor in Environmental Management and Bachelor in Agribusiness Management) and the postgraduate programs in Environmental Sciences (PPGCA), Environment and Rural Development (PPGMADER) and Sustainability of Traditional Peoples and Lands (PPGMESPT) have close connection with the theme of sustainability (Bizerril 2018).

FUP has had influences from Environmental Education in the broadest sense since it opened. This influence was maintained by successive actions of projects organized by different professors without a formal continuity between projects, but in a complementary way. Thus, in 2007 the "Nosso Campus" (Our Campus) project was implemented, which used elements of communication and cultural promotion to develop a sense of belonging and care for the campus, with special attention to selective collection (Bizerril et al. 2009). It should be noted that the development of the sense of belonging and the culture of participation has been an important management mark of FUP since its inception (Bizerril 2015).

The idea of making the FUP campus an educational structure where all spaces and processes could become elements of living and learning about the sustainability culture gained strength with the creation of the "*Esperança Verde*" (Green

Hope) project in 2009, which coincided with the completion of "*Nosso Campus*" project, representing, therefore, a continuity, with greater theoretical strength and in the complexity of its proposals (Layrargues et al. 2011). The project generated two remarkable developments for the implementation of sustainability on campus: conducting research on the internalization of environmental issues into research and extension projects, and within the curricula of undergraduate courses at FUP (Layrargues and Dourado 2011), as well as the systematic inclusion of environmental coordination amongst the campus, which was formalized by a resolution of the Faculty Council (Resolution No. 3/2010). The project also made progress in training actions on the Solidary Selective Collection with cleaning servers and the elaboration of a new extension project, aimed at recycling cigarette butts.

In 2012, sustainability was formally associated with campus policies and missions since its inclusion in the Institutional Political Pedagogical Project (Universidade de Brasília 2012). In 2015, the "*Recicla FUP*"[2] project was created for the management of solid waste on the campus, carrying out qualitative and quantitative studies on waste, implementing and structuring the solidary selective collection, agreements with cooperatives of waste pickers in the region to receive the recyclable fraction and various actions of education and community awareness about the correct disposal of waste (Gonçalves et al. 2018; Silva et al. 2019). This same group has started to perform systematic monitoring of data related to waste production, water and energy consumption on campus (Ribeiro et al. 2019). In addition, from 2016, the "Composting" project, with the participation of students, staff and professors, implemented a small system for composting organic waste generated within the campus.

In 2019 the campus benefited from policies to improve energy efficiency at the University of Brasilia by receiving a Solar Photovoltaic Plant[3] with 132 plates and a capacity to generate 44 kWp (kilowatt-peak), representing savings of around R$ 4000/month in electricity bill of the unit - which corresponds to 12% of the bill, on average.

4 The Elaboration of FUP Sustainability Plan

In early 2019, a working group started to develop a sustainability plan for FUP. The plan included a proposal to have monitoring indicators and data collection for the first campus sustainability report. All faculty and staff were invited to a meeting of the working group and, in some cases, specific invitations were sent to those who were already developing research or extension activities aimed at promoting sustainability on campus. Nine people attended the meeting: two students from the Bachelor of Environmental Management who make up the Brazilian Junior Company of Environmental Management (Embragea), one technician, and five professors (one from

[2] www.reciclafup.com.br.

[3] https://noticias.unb.br/112-extensao-e-comunidade/2961-unb-ativa-sistema-de-geracao-de-energia-solar-fotovoltaica.

the Education and Languages area, one from Human and Social Sciences, one from Applied Social Sciences and Technology and two from Life and Earth Sciences). This group discussed the bases for building a sustainable campus from two sources: the recent literature on sustainability in higher education and the Brazilian government's guidelines for the implementation of sustainability in public administration (Brazil 2016). The group also outlined a set of indicators for monitoring the sustainability on campus and presented the main ideas to the academic community at the Campus General Assembly, as a part of the campus anniversary activities. On that occasion, the principles of sustainability and the set of indicators for monitoring it were validated, in addition to incorporating other suggestions.

The group started to work with 13 indicators, specifying the tools of measurement and analysis of each (Table 1), and initiating the data collection for the first campus sustainability report[4]. These indicators relate directly or indirectly to 13 of the 17 SDGs. It is worth noting, however, that the other four SDGs that are not related to the sustainability plan, are indirectly covered by the socio-economic profile of students enrolled on campus. More than half of the students have low income and receive financial support from the university.

In the legal context, the chosen indicators are aligned with Normative Instruction IN. 10 of 2012 of the Ministry of Planning, Budget and Management (MOP) (Brasil 2012b), which sets the rules for preparation of the Sustainable Logistics Plan (PLS). According to IN 10/2012, the PLS must contain, among other items, the practices of sustainability and rational use of materials and services as well as dissemination, awareness and training activities. In addition, according to the document, sustainability practices and the rational use of materials and services should cover the following topics: I—consumable material including printing paper, disposable cups and print cartridges; II—electrical energy; III—water and sewage; IV—selective collection; V—quality of life in the work environment; VI—sustainable purchases and contracts, comprising buildings, equipment, surveillance, cleaning, telephony, data processing, administrative support and building maintenance services; and VII—displacements, considering all means of transport, with a focus on reducing costs and emissions of polluting substances. The chosen indicators for monitoring sustainability at FUP, not only meet IN 10/2012, but go beyond this by including items 9, 10, 11 and 12 in Table 1, thus highlighting the need to adapt the instrument to the reality and vocation of a higher education institution with its peculiarities.

Reinforcing the presented legal instrument, there is also the voluntary adhesion program of the Environmental Agenda of Public Administration (A3P), created in 1999 by the Ministry of the Environment (MMA). A3P is an example of an environmental public management action by the Brazilian government, with the following thematic axes: rational use of resources, adequate waste management, quality of life at work, awareness and training of workers. Again, there is an alignment of sustainability indicators selected for the FUP with this program.

[4]http://fup.unb.br/wpcontent/uploads/2020/11/Relat%C3%B3rios-de-Sustentabilidade-da-FUP.pdf

Table 1 Indicators for monitoring sustainability on the FUP campus

	Indicator	Measurable items	Relationship with the SDGs[a]
1	Rational use of electricity	Consumption, per capita consumption, cost, cost per capita, cost by building area, awareness raising	7; 11; 12
2	Rational use of water	Monthly volume, volume per capita, monthly cost, cost per capita, awareness raising	6; 12; 14
3	Streamlining the use of paper and implementing an electronic process	Monthly paper consumption, per capita consumption, cost, disposal of toner recycling, existence of electronic process, awareness raising	12
4	Waste management and selective collection	Existence of selective collection, disposal, partnership with collectors union, awareness raising. Consumption of disposable cups	11
5	Compliance with accessibility requirements	Compliance with official standards (NBR 9050) of accessibility to buildings, furniture, equipment and urban spaces	5; 8; 9
6	Sustainable buildings	Lighting, comfort (thermal, visual, acoustic, olfactory), healthiness, materials, landscaping, security, water and energy saving structures	8; 11; 12; 13
7	Sustainable public procurement	Presence of sustainability in: contracts with outsourced companies and services, processes for purchasing consumable and permanent materials, building contracts	12; 13; 17
8	Mobility and greenhouse gases	Inventory of greenhouse gases, inventories of carbon sequestration actions (green areas of the campus, public transport use)	11; 13
9	Institutional Framework	Presence of Sustainability into the mission, organizational structure, budget and monitoring	11; 16

(continued)

Table 1 (continued)

	Indicator	Measurable items	Relationship with the SDGs[a]
10	Adherence to sustainability programs, networks and legislation	At the local, municipal, federal, international levels	17
11	Degree of internalization of environmental issues	In the curriculum, extension programs and research projects	4
12	Encouraging participation	Access to information, opinion, and decision-making processes	16; 17
13	Quality of life	Green area and constructed area ratio, green area per capita, self-community assessment	3

Management actions comprise items 1–10, and awareness and education actions include items 11–13

[a]SDGs: 1—No Poverty; 2—Zero Hunger; 3—Good Health and Well-being; 4—Quality Education; 5—Gender Equality; 6—Clean Water and Sanitation; 7—Affordable and Clean Energy; 8—Decent Work and Economic Growth; 9—Industry, Innovation and Infrastructure; 10—Reduced Inequalities; 11—Sustainable Cities and Communities; 12—Responsible Consumption and Production; 13—Climate Action; 14—Life below Water; 15—Life on Land; 16—Peace, Justice and Strong Institutions; 17—Partnerships for the Goals (UN 2020)

Rationality in the use of energy and water, as well as waste management and selective collection are consolidated items of all environmental management systems of public or private institutions, provided for in the sustainability guidelines for public administration bodies (Brazil 2016) and sustainability systems in higher education (Leal Filho 2010; Lozano et al. 2014). Accessibility and sustainable buildings have been being increasingly considered in urban planning, which results in an internal challenge for universities to apply these principles towards their own buildings. In the Brazilian case, this demand encounters barriers in the processes of public procurement of construction, as well as in the purchase, which drive the decisions towards the lowest price instead of including aspects related to sustainability. However, according to Lavor and Turatti (2018), the legal regulation for sustainable public procurement and contracting is extensive and complex. In order to make effective changes, the public manager must determine other ways of judging proposals that are not only based on price. Barki and Gonçalves-Dias (2014) affirmed that mechanisms of environmental education and awareness of civil servants must be implemented so that bidding becomes a legal instrument for the promotion of sustainability. Rec and Marini (2019) reinforce the issue of training and awareness of all those involved in the socio-environmental process of public management, so that the bureaucratic, cold and static aspects of the law and standards can be overcome and the proactive and committed posture of public managers prevails.

Brazilian university campuses tend to harbor large green areas, including parks and natural forest reserves and fragments for research and conservation. However, these areas are under pressure from being suppressed to make room for new buildings. The consideration of green areas as having sustainability value, including carbon sequestration, implies an increase in its visibility and appreciation. The formalization of institutional commitments to sustainability through its internal documents and adherence to international agreements has been considered as an important sustainability action by universities (Lozano et al. 2014) and may help to overcome the reduced commitment of Brazilian universities to the issue (Brandli et al. 2015).

The last three indicators show the greatest differential role of promoting sustainability at a university in relation to other institutions, which is the formative function, the university's primary mission. That is why it is essential that sustainability must be present in the learning processes while, at the same time, include formative actions of research and extension, which also have an impact on society. It is known that not only the students are educated in these processes but also staff, faculty and the community who interact with the university. This reinforces the fact that participatory processes are much more efficient in promoting sustainability in higher education, as already observed in others studies (Disterheft et al. 2014).

5 Discussion: Sustainability on a Small, Young Campus

In a study on the process of institutionalization of sustainability in a Portuguese university, Bizerril et al. (2018) indicated that the way sustainability is institutionalized in universities, including the dimensions and aspects that are most highlighted, would probably be related to the perception of the theme and the concept of sustainability adopted by the process coordinators. Thus, the process could be facilitated or hindered according to the structural and cultural circumstances of the university itself, as well as the institutionalization strategies assumed. In this scenario, the sustainability at the university will be more likely to be successful if it is encouraged, planned and carried out from the conditions and characteristics of each institution, rather than merely measured or quantified for purposes of evaluation and comparison.

In the case of FUP, size and age appear to be key elements facilitating the process, since it avoids the danger of "crystallization" of campus culture and dynamics. The small size, combined with the absence of departments, reduce the decision-making bodies providing greater monitoring and community control over the actions of the campus directors. It is also a facilitator of change and adjustment in management, as is the case of an implementation of a sustainability policy. In the case of the sustainability policy, the size reduces the distance between the community and the coordinating group of sustainability actions. Additionally, a young campus is more likely to deal positively with new management proposals than more traditional institutions.

Disterheft et al. (2012) analyzed European universities that had some kind of environmental management system and concluded that the participatory approach appears to be more comprehensive than a top-down process. This is because participatory approaches not only improve the environmental performance of the institution, but also more effectively incorporate sustainability in all the university's performance levels, especially in the training of students to address the issue in society. The need for an internal debate on sustainability also stood out in the opinions of the academic community in the study cited about the case of a Portuguese university (Bizerril et al. 2018).

Thus, although there are monitoring systems options applicable for universities, the discussion of the indicators to be adopted for the campus is itself a key stage of development of the understanding of the concept of sustainability to be built for the campus. Not least because the university needs to permanently exercise its role as a "laboratory of current reality", as suggested by González-Gaudiano et al. (2015, p. 71). Such is the case in rethinking the local application of concepts produced in the global context.

Even if participation is not widespread at first, it is important that it is qualified and representative so that bottom-up initiatives could be incorporated in the future, as noted in the present case. Indeed, pioneering sustainability projects on campus has a natural approach, and are essential to the establishment of an initial critical mass to conduct the work of the sustainability plan for the campus. However, the gradual engagement strategies of the academic community need to be addressed and strengthened. It should be expected that increased participation may imply changes in the indicators based on new suggestions, which is positive for the process, but care must be taken to respect the holistic/complex concept of sustainability assumed in initial stages.

Sustainability in higher education cannot be promoted only by top-down actions, as it shall not be an action of a management mandate but an institutional goal through internally approved legal provisions. For this reason, the FUP campus has advanced in the formalization of an environmental advisory, a sustainability policy, and now in the instrument for monitoring sustainability.

Another aspect already mentioned that seems to be strategic for sustainable universities is the sense of belonging of the academic community in relation to campus. That feeling certainly results in increased care with something that is not a private good. Rather, as in the case of Brazilian federal universities as public entities, it is a direct component in the formation of an environmental citizenship. The promotion of sense of belonging goes through various educational activities, but one aspect that has been strengthened in the FUP concerns the quality of the environment. Examples include: landscaping, the application of colors and drawings on the walls, the presence of animals (associated with responsible care projects for them), the expansion of conviviality and rest spaces. All of these are aspects that enhance the feeling of comfort and pleasure in being on campus, which, although there is a lack of empirical evidence, suggest establishing the conditions for the promotion of the sense of care and responsibility.

The existence of celebrations, strengthening democracy and happiness make up another aspect suggested by Disterheft et al. (2016) as determinants for sustainability in universities. In this sense, FUP has been practicing a schedule for years that includes regular general meetings with the academic community, and celebrations as the FUP's birthday, the June Festival (traditional Brazilian party), the University Week (with diversified activities for the whole academic and external community), as well as small gatherings dispersed throughout the year. It can be said that the small size of the campus intensifies the positive impact of these actions, bringing people together and creating a culture of engagement that can influence those who will integrate the campus over time.

These observations corroborate recent studies such as that of Caeiro et al. (2020), who analyzed the wide range of sustainability assessment tools available and, when studying cases from Portuguese and Spanish universities, conclude that it is necessary to emphasize the development of indicators that consider non-traditional aspects of sustainability and that represent little tangible aspects of society.

6 Conclusions

The Planaltina campus is developing a sustainability assessment tool that dialogs with the guidelines of the Brazilian public administration and recent debates on sustainability in higher education. This tool excels in considering the complexity of the understanding of sustainability, including the Sustainable Development Goals, but also the main responsibilities and possibilities of the university as an educational institution and a reference for society.

In the studied case, the academic profile, the size and age of the campus were key elements facilitating the process. In fact, given the differences between universities, we infer that it is not possible to standardize the sustainability actions that each institution is able to develop at first. Therefore, when creating the sustainability plan, the people profiles and the vocation of each campus must be considered.

Although there were options of environmental monitoring systems applicable to universities, the participatory discussion of the indicators that would be adopted on campus was, in itself, a fundamental step in developing an understanding of the concept of sustainability that would guide future actions on campus.

Acknowledgements The authors would like to thank Ryan Nehring for reviewing the English version

References

Alshuwaikhat HM, Abubakar I (2008) An integrated approach to achieving campus sustainability: assessment of the current campus environmental management practices. J Clean Prod 16(16):1777–1785

Barki TVP, Gonçalves-Dias SLF (2014) Licitações sustentáveis no Brasil: aspectos jurídicos e de gestão pública. In: VII Congresso CONSAD de Gestão Pública. Brasília, Brasil

Bizerril MXA, Andrade LG, Pereira IAFS, Souza DO, Coutinho ML, Rocha DMS (2009) Projeto Nosso Campus: uma experiência de implantação da educação ambiental em um campus universitário em formação. In: VI Congreso Iberoamericano de Educación Ambiental. San Clemente de Tuyú: PNUMA

Bizerril MXA (2015). Gestão participativa em uma equipe em formação: o caso do campus de Planaltina da Universidade de Brasília. In: Mano M (ed) Roteiro do Plane(j)amento Estratégico, Universidade de Coimbra, pp 488–493

Bizerril MXA, Rosa MJ, Carvalho T (2018) Construindo uma Universidade Sustentável: uma discussão baseada no caso de uma universidade portuguesa. Avaliação: Revista da Avaliação da Educação Superior 23:424–447

Bizerril M (2018) Universities in transition to sustainability: challenges and opportunities for the Campus of the University of Brasilia in Planaltina. In: Leal Filho W, Frankenberger F, Iglecias P, Mülfarth R (eds) World sustainability series, pp 361–370. Springer International Publishing

Brandli LL, Leal Filho W, Frandoloso MAL, Korf EP, Daris D (2015) The Environmental Sustainability of Brazilian Universities: Barriers and Pre-conditions. In Leal Filho W, Azeiteiro UM, Caieiro S, Alves F (eds) Integrating sustainability thinking in science and engineering Curricula, Springer, pp 63–74

Brasil (2010) Ministry of Education. REUNI—Programa de Reestruturação e Expansão das Universidades Federais. http://reuni.mec.gov.br. Last accessed 25 Mar 2020

Brasil (2012a) Decreto 7.746, de 5 de junho de 2012. Diário Oficial da União de 6 Jun 2012. http://www.planalto.gov.br/ccivil_03/_ato2011-2014/2012/decreto/D7746.htm. Last accessed 23 Mar 2017

Brasil (2012b) Ministério do Planejamento Orçamento e Gestão. Secretaria de Logística e Tecnologia da Informação. Instrução Normativa 10, de 12 de novembro de

Brasil (2015) Ministério do Planejamento Orçamento e Gestão. Portaria 23, de 12 de fevereiro de 2015. Diário Oficial da União de 13 fev. Brasília, DF, Brasil

Brasil (2016) Tribunal de Contas da União. Auditoria em Sustentabilidade na Administração Pública. https://portal.tcu.gov.br/biblioteca-digital/auditoria-operacional-em-sustentabilidade-na-administracao-publica-federal.htm. Last accessed 26 Mar 2020

Caeiro S, Hamón LAS, Martins R, Aldaz CEB (2020) Sustainanility assessment and benchmarking in higher education institutions—a critical reflection. Sustainability 12:543. https://doi.org/10.3390/su12020543

Disterheft A, Caeiro S, Ramos MR, Azeiteiro UM (2012) Environmental Management Systems (EMS) implementation processes and practices in European higher education institutions—top-down versus participatory approaches. J Clean Prod 31:80–90

Disterheft A, Caeiro S, Azeiteiro UM, Leal Filho W (2014) Sustainable universities—a study of critical success factors for participatory approaches. J Clean Prod. https://doi.org/10.1016/j.jclepro.2014.01.030. Last accessed 20 Feb 2015

Disterheft A, Caeiro SS, Leal Filho W, Azeiteiro UM (2016) The INDICARE-model—Measuring and caring about participation in higher education's sustainability assessment. Ecol Indic 63:172–186. https://doi.org/10.1016/j.ecolind.2015.11.057

Gonçalves JP, Oliveira MC, Abreu LM, Almeida AN, Ribeiro EN (2018) Avaliação do nível de conhecimento de uma comunidade acadêmica acerca da coleta seletiva de resíduos. In: Anais do 8º Encontro de Engenharia Sanitária e Ambiental (ENASB) e 18º Simpósio Luso Brasileiro de Engenharia Sanitária e Ambiental (SILUBESA), Porto, Portugal

González-Gaudiano E, Meira-Cartea PA, Martínez-Fernández CN (2015) Sustentabilidad y universidad: retos, ritos y posibles rutas. Revista de la Educación Superior 44(175):69–93

Kassaye AY (2018) Contemporary institutional solid waste management practices of Haramaya University, Eastern Ethiopia. Afr J Sci Technol Inovation Dev 10(2):219–238

Lavor AAA, Turatti L (2018) Contratações públicas sustentáveis no Brasil. Revista de Gestão e Sustentabilidade Ambiental 7(2):335–354

Layrargues PP (2000) Ideology and the environment: business leaders adopt a strategy of environmental discourse regarding ISO 14000. Ciência e Cultura 52(3):148–153

Layrargues PP, Dourado MF (2011) O grau de internalização da temática ambiental na Faculdade UnB Planaltina. In: Leme PCS, Pavesi A, Alba D, Gonzalez MJD (eds) Visões e Experiências Iberoamericanas de Sustentabilidade nas Universidades. Universidade de São Paulo, Universidad Autónoma de Madrid, Pontifícia Universidade Católica do Rio Grande do Sul, pp 235–240

Layrargues PP, Irineu Neto AA, Dourado BF, Andrade BR, Glória DS, Taulois IC, Rocha LFL, Silva SM, Nascimento WM (2011) Esperança Verde na Faculdade UnB Planaltina: um campus universitário modelo em gestão ambiental. Aprendizagens e perspectivas. In: Catalão VML, Layrargues PP, Zaneti ICBB (eds) Universidade para o Século XXI: educação e gestão ambiental na Universidade de Brasília. Cidade Gráfica e Editora, pp 289–296

Leal Filho W (2010) Sustainability at Universities—Opportunities, Challenges and Trends. Peter Lang Scientific Publishers, Frankfurt

Lozano R, Celeumans K, Alonso-Almeida M, Huisingh D, Lozano FJ, Waas T, Lambrechts W, Lukman R, Hugé J (2014) A review of commitment and implementation of sustainable development in higher education: results from a worldwide survey. J Clean Prod 108:1–18

Petrelli CM, Colossi N (2006) A quarta via das instituições de ensino superior: a responsabilidade social. Revista Catarinense da Ciência Contábil 5:71–78

Rek M, Marini MJ (2019) Gestão socioambiental na administração pública: uma análise de instrumentos legislativos aplicáveis ao âmbito institucional. Colóquio Revista do Desenvolvimento Regional 16:1

Ribeiro EN, Carneiro RL, Galdino OPS, Duraes PHV, Rocha DMS, Oliveira MC (2019) Diagnóstico ambiental de um câmpus universitário como estratégia para proposta de práticas sustentáveis. Revista Brasileira de Gestão Urbana 11:e20190029. https://doi.org/10.1590/2175-3369.011.e20190029

Silva LAX, Oliveira AKPS, Lima TJ, Brito AL, Costa DD, Ribeiro EN, Oliveira MC (2019) A coleta seletiva solidária no campus UnB Planaltina: o que sabem os calouros?. In: Atas da 9ª Conferência FORGES, Brasília: Editora IFB

United Nations (2020) Sustainable Development Goals. https://sustainabledevelopment.un.org. Last accessed 08 Apr 2020

Universidade de Brasília (2012) Faculdade UnB Planaltina. Projeto Político Pedagógico Institucional da Faculdade UnB Planaltina. http://www.fup.unb.br. Last accessed 20 Mar 2020

Marcelo Bizerril is Associate Professor at the University of Brasilia (UnB), with a doctorate in Ecology from the University of Brasilia, and a post-doctorate in Policy and Management of Higher Education from the University of Aveiro (Portugal). Over the last two decades, he has been developing a research program on sustainability in formal and non-formal Environmental Education Programs, Science Teaching, and Higher Education Administration.

Leandro Oliveira Evangelista has a degree in Veterinary Medicine, is part of the technical staff of the University of Brasília, and make up the working group for the sustainability plan of the Planaltina Campus.

Philippe Pomier Layrargues is Adjunt Professor at the University of Brasilia (UnB), with a doctorate in Sociology from the University of Campinas. Over the last three decades, he has been developing researches on the social field of Environmental Education.

Elaine Nolasco is Adjunt Professor at the University of Brasilia (UnB), with a doctorate in Industrial Biotechnology from the University of São Paulo (USP), and a post-doctorate in Infrastructure Engineering from the Aeronautical Technological Institute (ITA). Over the last two decades, she has been developing a research on industrial and domestic wastewater, urban solid waste management, and sustainability in public administration.

Maria Cristina de Olivera is Adjunt Professor at the University of Brasilia (UnB), with a doctorate in Forest Engineering and a post-doctorate in Dynamics of Plant Communities from the

University of Brasilia. Over the last two decades, she worked with vegetative propagation, structure and dynamics of vegetative communities, restoration of degraded areas and urban of solid waste management.

Luís Henrique M. Oliveira are undergraduate students in Environmental Management at the University of Brasilia, make up the Brazilian Junior Company of Environmental Management (Embragea), and are part of the working group for the sustainability plan of the Planaltina Campus.

Rômulo Ribeiro is Associate Professor at the University of Brasilia (UnB), with doctorate in spatial data analysis to aid in urban and environment planning, at the University of Brasilia. In the last two decades, he has been developing research on urban morphology, urban climate, environmental degradation, using spatial data, satellite images, UAV images.

Antônia R. Sousa are undergraduate students in Environmental Management at the University of Brasilia, make up the Brazilian Junior Company of Environmental Management (Embragea), and are part of the working group for the sustainability plan of the Planaltina Campus.

Luiz Felippe Salemi is Adjunt Professor at the University of Brasília. He holds a doctorate degree in Sciences (University of São Paulo). In the last decade, he has been developing research on catchment hydrology, biogeochemistry, management, recuperation and conservation. Furthermore, he is very interested in topics related to sustainable development in rural and urban areas.

Development and Delivery of a Sustainable Development Unit in UK Universities: A Higher Education's Guide for Future Sustainability Leaders

Eleni Michalopoulou, Aisling Tierney, Ed Atkins, Tabitha Stanmore, Lin Ma, Natalie Jester, and Chris Preist

1 Introduction

We are now in the era of the Anthropocene (Crutzen 2002) where humans are characterised as a "force significant enough to reshape the face of the Earth" (Clark et al. 2005). In 2015 the United Nations (UN), following a long history of discussions around Sustainable Development (SD), established the 17 Sustainable Development

E. Michalopoulou (✉) · E. Atkins
School of Geographical Sciences, University of Bristol, University Road, Bristol BS8 1SS, UK
e-mail: em15151@bristol.ac.uk

E. Atkins
e-mail: ed.atkins@bristol.ac.uk

A. Tierney
Education Services, University of Bristol, Great George Street, Bristol BS1 5QD, UK
e-mail: a.tierney@bristol.ac.uk

T. Stanmore
Department of Historical Studies, University of Bristol, Bristol BS8 1QU, UK
e-mail: Tabitha.stanmore@bristol.ac.uk

N. Jester
School of Sociology, Politics and International Studies, University of Bristol, Bristol BS8 1TU, UK
e-mail: natalie.jester@bristol.ac.uk

C. Preist
Department of Computer Science, University of Bristol, Woodland Road, Bristol BS8 1UB, UK
e-mail: chris.preist@bristol.ac.uk

L. Ma
SPAIS, School of Sociology, Politics, and International Relations, University of Bristol, Bristol, UK
e-mail: lin.ma@bristol.ac.uk

W. Leal Filho et al. (eds.), *Universities, Sustainability and Society: Supporting the Implementation of the Sustainable Development Goals*, World Sustainability Series,
https://doi.org/10.1007/978-3-030-63399-8_30

Goals (SDGs) (United Nations 2015; Sachs 2012). These goals reflect our societies' biggest concerns and aim to address humanity's biggest challenges such as poverty (SDG 1), inequalities (SDG 10) and climate change (SDG 13).

One key role that Higher Education Institutions (HEIs) have to play in this era of the SDGs is in relation to SDG 4 (Quality Education). As educators, we foster the leaders and SDG champions of tomorrow. As a result, it is imperative that these graduates have an excellent understanding of the complexity of sustainability, the urgency of global challenges and, most importantly, the impacts these challenges will have on different societies. One way to help address this is by increasing student engagement with sustainability within the HE curriculum and ESD. HE institutions must strengthen this link between Education in Sustainable Development (ESD), society and SDGs in order to ensure that graduates are appropriately equipped to tackle a variety of challenges, reflected in the different SDG targets.

With both student and institutional demand for more ESD-related courses and programs increasing, this paper presents, in detail, the development and delivery of an award winning, blended learning ESD unit at the University of Bristol: the Sustainable Development Unit (SDU). It provides information to encourage and enable other HEIs to replicate this unit, or develop their own. Such knowledge exchange between HEIs can be critical, as it can allow for less time spent in the development stage of such initiatives and more time on the delivery. This in turn can mean that more graduate cohorts are graduating with the skills needed to address societal challenges. Finally, this paper presents a detailed evaluation of the unit from the perspective of students and staff alike, which can help other educators to benefit from the learnings made during the development and delivery of three runs of the SDU.

The University of Bristol was one of the first UK universities to commit to the SDG accord on 9th September 2017 (SDG Accord 2020), and on 17th April 2019 it became the first UK university to declare a climate emergency (BBC 2019; Walsh et al. 2020). The Institution has committed to delivering high quality, research-led teaching that aims to provide students with the knowledge and skills that will enable them to become future problem-solvers of these global challenges (Tierney et al. 2015; Michalopoulou et al. 2019). However, SD is not only an "end result": it is also a means to enable society to address contemporary challenges. Therefore, any endeavour that sets out to explore the SDGs or local or global challenges must do so through the same frameworks it tries to explore. For this reason, the SDU was created with respect to, and in accordance with, the SDGs.

This paper will present the SDU developed as part of the *Bristol Futures: Sustainable Futures* pathway. It will discuss how this unit was developed, how it is delivered and which SDGs and theoretical concepts it aims to present to the students, continuing the analysis of previous work (Michalopoulou et al. 2019). Additionally, this paper will present which SDGs are used for the development and delivery of this unit.

1.1 Institutional Context

In 2005, the SDU was introduced to the optional curriculum at the University of Bristol and made available to students from many disciplines across the university. The unit was taught by a multi-disciplinary team and received a national teaching award in 2007 in recognition of its innovative nature (Hoare et al. 2008). Ten years later, as part of the wider *Bristol Futures* (BF) project to enhance the curriculum and student experience (Michalopoulou et al. 2019; Walsh et al. 2020), the unit was redesigned in a way which preserved the original interdisciplinary approach but increased accessibility to students by incorporating blended and problem-based learning approaches.

The SDU was (re-)designed with four core aims:

- To present the SDGs and, through them, develop a broad understanding of the challenges of SD and how different disciplines can provide insight into how they may be addressed.
- To invite comparisons and contrasts of experience across different SD challenges.
- To help students appreciate the problems and tensions in the application of SD ideals in practical circumstances.
- To introduce students to the experience of inter-disciplinary working focused on a specific challenge.

1.2 Pedagogy

From conception to delivery, the SDU is based on the recognition that both teaching staff and enrolled students do not just interact with and impact upon the peoples and environments in our immediate vicinity. Rather, they are connected to wider systems that affect the world as a whole. This is expressed through the interdisciplinary design of the SDU, which recognises that SD is multi-faceted, incorporating environmental, social and economic elements. This design allows for exploration of global challenges through a systems thinking approach. The design of the SDU is guided by a wider concept of global citizenship which refers to the "rights and responsibilities… duties and entitlements" (Davies 2006: 6) one has at a global level.

The SDU was designed to encourage students to take their criticality into their other units and everyday lives. The skill of criticality—and the related ability to approach and solve complex social and environment problems—is a key challenge to teach. In order to instil criticality, the course utilises the Vygotskyan principle of scaffolding, which refers to "the temporary assistance that teachers provide for their students to assist them to complete a task or develop new understandings, so that they will later be able to complete similar tasks alone" (Hammond and Gibbons 2019: 9). Employing scaffolding has been especially important as both staff and students enter each topic with different knowledge levels.

Scaffolding occurs at both the macro and micro level within the SDU and is incorporated into both the overall programme and individual topics. The flipped classroom model allows for the employment of specially designed introductory materials, followed by a lecture. The students apply the concepts in these materials to a particular scenario with support from the teaching team throughout a workshop: online or in person. Finally, students demonstrate their knowledge through discussion, or by presenting their ideas at the end of the class. Thus, the assistance provided is staggered, constantly adjusted in order to help students reach the minimum learning outcomes.

Sector research demonstrates that students respond positively to blended learning approaches (Sharpe et al. 2006). Research also suggests that students' learning improves in blended learning environments, compared to traditional lecture-style delivery (Gillette et al. 2018). "Blended learning" exists at the intersection between the Internet and the physical classroom, an intersection that can present a new range of possibilities for learning (Friesen 2012). While in-class hours may reduce in quantity, they increase in quality and students engage in more-informed discussion in the classroom (Smith 2014).

The flipped classroom approach expects the learner to engage with online material before face-to-face learning. At Bristol, digital learning takes place through the University's Blackboard Virtual Learning Environment (VLE) where resources such as discussion boards and instruction videos are hosted (Digital Education Office n.d.). Literature on the digital flipped classroom largely suggests a positive impact on learning. It can promote scaling-up of class sizes while retaining smaller discussion sizes and maintaining student satisfaction (Ryan and Reid 2016). Bergman and Sams (2012, pp. 6–7) highlight how the flipped classroom can also assist the "personalization" of education. In this way, students are catered for in a more accessible manner (Bergman and Sams 2012, p. 14). This approach allows students to conduct their learning at their own pace outside of the classroom, which allows them to manage their workloads alongside other commitments (Bergman and Sams 2012, pp. 21–23). This flexibility allows students who may have traditionally struggled with a non-flipped classroom to succeed. This can include students with caring responsibilities, or those with learning and physical disabilities. With concern for the environment at record high amongst young people in the UK in 2019 (with a YouGov poll finding that 45% of 18–24 year olds placed environmental issues as their second biggest policy concern) (Smith 2019), it is imperative that HEIs not only increase the availability of courses on environmental issues but prioritise the accessibility of such opportunities.

Students are given ownership of their learning through the flipped-classroom approach, harnessing and improving their personal motivation and engagement with course materials (Abeysekera and Dawson 2015) and their positivity towards their subject (Turra et al. 2019). Students who engage in the flipped classroom see an increase in cognitive learning outcomes and experience increased satisfaction from the process of learning (Sergis et al. 2018). This is true for students from a wide range of disciplinary backgrounds (Chen et al. 2018). Post-disciplinary education opens opportunities for these students to go beyond deepened disciplinary knowledge and bridge it into that of others for purposeful engagement with sustainability. This pedagogical edge enables a liaison for example between humanities and science students (Biagioli 2009). When disciplinary boundaries are broken from the teaching level, students begin to see their instructors less as information disseminators and more as facilitators of discussion with their peers: the true benefit of the flipped classroom approach (McLean and Attardi 2018).

2 Development

Several challenges emerge when working on cross-subject curriculum material in blended learning contexts. Challenges included the need for specialist staffing and filming abroad and domestically. At the core of these efforts was a need to foster and maintain excellent working relationships with academic, student and external collaborators. Venues, councils, and archives were also integral partners in the success and visual variety of the media content. Another challenge was the need for testing and delivery of innovative curriculum approaches for other types of digital assets. The project team supported the identification, acquisition, testing and implementation of tailored resources that balanced innovation with accessibility.

2.1 Collaboration

Development of the unit required collaboration with teams across and beyond the University. The Academic Quality and Policy Office (AQPO) was enlisted to provide guidance through the quality assurance process, while the timetabling team ensured that all unit-specific requirements were given institutional priority to allow as many students as possible to take the unit. The Digital Education Office (DEO) was a core point of contact to advise on Blackboard systems and blended learning support. The Secretary's office advised on legal requirements. Faculty administration teams supported small and large-scale activities, such as workload arrangements. The unit was hosted in the School of Geographical Sciences, which oversaw core curriculum actions such as arrangements for external examiners, registering students onto units, and collating assessment marks.

Community and international collaborations were also integral to creating the broad vision of the unit. The unit invited content contributions from the local city and international colleagues. Predominantly, this took the form of freeform video interviews and case studies, as detailed below. Furthermore, the unit worked with specialist groups, such as archives, councils, research and commercial organisations to ensure that multiple stakeholder perspectives to the SDGs were considered and integrated into the unit. Together, over two hundred people were involved in the development of the unit. In addition, interdisciplinary collaboration within the core teaching team also provided an important element of curriculum design and delivery, with colleagues drawn from the physical sciences (Chemistry), engineering (Computer Science), social sciences (Human Geography) and the humanities (Archaeology).

The student voice was considered particularly important in developing the unit. Between December 2018 and May 2019, as part of the *Bristol Futures* project, six workshops attended by over 100 students were held. Overviews of the planned units were shared with students across this consultation phase, and their comments and suggestions were adapted within the final design of the units.

2.2 Hourly Paid Teachers

Bristol Futures Hourly Paid Teachers (HPTs) are an essential component of the success of the SDU. Principles were established before HPTs were recruited, and all principles were adhered to throughout the project delivery:

- HPTs should be paid at same rate for all work.
- HPTs should be paid for all training, meetings, facilitation (online/offline), workshop delivery and workshop preparation.
- HPTs should tangibly benefit from the experience (e.g. experience serves as evidence for recognised teaching accreditation schemes).

The authors agree with the growing consensus that HPTs are an underappreciated afterthought of the realities of teaching in HE, and are often exploited (see, for example, Husbands and Davies 2000; Carroll 2003; Dematagoda 2016; Hall and Bowles 2016; Webster 2017). As such, it was an ethical principle to ensure that HPTs were suitably supported and financially remunerated for their work. A high rate of pay was selected for these roles to reflect the interdisciplinary complexities and pedagogical challenges that the HPTs would encounter while teaching on these units. The type of teaching on the optional units allowed the HPTs to expand their teaching experience beyond their own subject, which would appeal to those with experience in only one discipline, and especially those applying for teaching accreditation(s). Any HPT embarking on an accreditation scheme was supported by the project staff as required. Following the interdisciplinary character of the core teaching team, HPTs appointed were drawn from numerous disciplines—including English Literature, History, Politics and Sociology, Medical Science, and Engineering.

2.3 Digital Assets

The range of digital assets used on the unit includes Online Learning Environment software provided by the institution as standard (e.g. Blackboard discussion boards, interactive blogging platforms, multiple choice quiz templates) and non-standard software that is supplied through external sources (Padlet). Equivalent, institutional-level standard software (i.e. Online Learning Environments) and the use of such software is commonplace across institutions and will therefore be excluded from discussions. Due to the complex nature of groupwork design and digital components on the unit, a technical expert was needed to support the architecture design on Blackboard. A graphics assistant supported the resourcing, refinement, and accessibility of all non-standard software and bespoke learning documentation. The SDU uses Padlet (an easy-to-use, online bulletin board which allows instantly shareable images, video and text) in some sessions to collect comments, new content, and reactions from students during workshop sessions.

Unit directors generated in-house learning materials in the form of short articles and an editor was hired to both proofread and copy edit all assets. This ensured consistency in language used; accuracy of the writing; suitability for the learners' level; and that the tone, style and vocabulary were suitable for students from across subject disciplines. These documents were then reviewed by the graphics assistant to be standardised in terms of formatting, visual graphics, accessibility and checks on copyright to ensure all legal compliance. Some of the documents required the creation of new learning asset images for illustration and data presentation. All content materials, modes of teaching, and learning support were designed to both respond to the diverse learning needs of our students (i.e. in terms of disciplinary oversights) but also embrace the diversity of the cohort, thus drawing attention to the shared strength that can be found in such diversity. All materials were designed to be integrative, highlighting relationships and common themes (and differences) between concepts and allowing students to work through materials at their own pace.

The unit required bespoke video assets to create engaging and thought-provoking framing content in preparation for the face-to-face learning. Together, the project manager, film maker and project assistant worked as a team to create a streamlined service to the unit directors through the creative process. To begin the film-making process, academics completed a "creative brief" template. The team then provided an end-to-end service delivery encompassing storyboarding; conducting research for assets; sourcing and booking locations and contributors; scriptwriting; sourcing supporting intellectual property materials; and ensuring that all legal requirements and related paperwork were completed and recorded appropriately. The use of videos is understood to increase participation, emotional engagement, and overall course engagement by providing students with a framework to directly understand and apply the concepts discussed in previous content (Carmichael et al. 2018). Videos included interviews with scientific experts and local policymakers and included discussions of environmental policies (such as the plastic bag tax) in terms of personal anecdotes, foregrounding social sustainability content within individual experience. Numerous videos were purposefully shot with a tight frame, focused on the subject's head and shoulders – with no other materials or content presented. This is because research has highlighted that students find videos that include the teacher's images to be more engaging (Carmichael et al. 2018)

To ensure that the video assets met the highest quality standards, accessibility and representation were prioritised. This consideration was perceived as an ethical requirement that was embedded from the inception of each video asset, rather than an afterthought. All assets were designed with high-contrast visuals to ensure accessibility for those with visual impairment. Likewise, transcripts and timecoding subtitles were produced by default for each video. Throughout the development process, the project team worked with unit directors to ensure that all content prioritised gender, LGBTQ+ and BAME (Black, Asian and Minority Ethnic) representation.

3 Delivery

The unit is organised into an introductory workshop on the SDGs, followed by five interdisciplinary, themed blocks: science and understanding; behaviour and organisational change; economy, policy and law; equality and justice; and technology and innovation (Michalopoulou et al. 2019). Students gain insight into the concepts behind the challenges of SD and learn how different disciplines and interdisciplinary approaches are used. Through engagement across all teaching and learning approaches, students develop a broad understanding of the SDGs and learn how to work in an interdisciplinary team to analyse and critically evaluate challenges and potential ways forward. Each themed block presents real-world case studies with illustrated text documents and videos which include interviews and other dynamic content. Such content draws from institutional academic expertise and from professionals and volunteers in the commercial and charitable sectors.

The intended learning outcomes (ILOs) are broad in scope:

1. Be able to understand and define the key principles behind SD and relate these to the UN SDGs.
2. Be able to recognise ideas and concepts from their own discipline, along with those from others, in relation to different SD challenges, and critically reflect on their interplay.
3. Be able to analyse and critically discuss how contemporary SD challenges intersect with and influence our day-to-day lives.
4. Be able to work in an interdisciplinary team to analyse and critically evaluate challenges and potential ways forward in response to a specific SD case study, and present their findings.

Workshops are delivered both online and offline. This independent reading process forms the necessary background knowledge that prepares students for the offline, face-to-face workshops. Workshops begin with a mini-lecture to provide a link to the weekly content and show their relevance to SDGs. This is then followed by problem-solving activities where students work in small groups. Each group is named after an endangered animal, which serves to sharpen an awareness of Anthropocene and its impact. By pre-fixing group members, a familiarity is built, which allows groups to focus on understanding the given scenarios and solving particular barriers to achieving SDGs. Scenarios we have used in our workshops are shown in Table 1.

Table 1 Scenarios used in group activities in relation to the sustainable development goals (SDGs) that they cover

Link(s) to SDGs	Scenario
All SDGs	Mock international SDG summit whereby each group represents a country or organization; identifies SDG priorities and synergies; negotiates trade-offs
Variable focus, e.g. SDGs 7, 8, 9 and 10	Groups identify a suitable geographical area for a sustainability-related project, such as a wind farm or community well
SDG 12	Groups act as consultants to develop a strategy to reduce single-use plastic for one of four client organisations
SDGs 9, 12 and 16	Mock tribunal of a proposed airport expansion against a fictitious UK sustainable development law
SDGs 1, 8, 9, 10, 11 and 16	A participatory budgeting exercise set in one of six city-based environmental projects, considering the priorities of marginalized groups
Any SDG can serve as the focus	Developing a proposal for a mobile app to be used in a country in the Global South, and the presentation of a funding pitch based on this

The student groups also conduct a group project over the course of the teaching block. In this, they apply the ideas and concepts learnt online and the skills learnt in the workshops to study a specific sustainable development challenge and produce a brief for a fictitious client, critically exploring the underlying issues and potential strategies moving forward. They are expected to use ideas and concepts from at least three of the five interdisciplinary themed blocks. Students have a choice of four projects each time the unit runs and have some flexibility in how they interpret the brief. Example challenges used so far include:

- Approaches to decarbonising transport in a city by 2030.
- Approaches to reducing emissions associated with meat consumption in a given country.
- Advice to a large food manufacturing company subject to a boycott over palm oil use.
- Advice for a city or region in the Global South suffering from poor access to fresh water.

Challenges can be reused or changed each year, to match the interests of students and adapt to emerging issues.

3.1 Assessment

Students are assessed on their learning through the submission of two assignments: first, a group project (groups consist of three to five members) and second, a reflective portfolio that evaluates how and what they learned. These evaluate their understanding and critical thinking in relation to the SDGs and local or global challenges. The reflective report is made up of three elements: two workshop learning statements; reflections on the future; and a blog article on a SD theme of their choice.

The aim of the reflective part of the portfolio is to assess:

- How well students have understood and applied the key concepts taught on the unit, and whether they have drawn from several disciplines in their answers.
- Whether the student has engaged in constructive discussion with other students within the unit's workshops.
- How much students have understood and thought creatively about the potential applicability of the concepts covered in the unit to both their own future and the futures of those who study their discipline.

The aims of the group project element of the portfolio are to assess:

- The students' ability to apply the concepts from different disciplines introduced in the unit to a specific case study of a SD challenge.
- The ability to collaborate with others, particularly with students from other backgrounds, experiences, and educational disciplines.
- The ability to analyse and critically evaluate potential ways forward, drawing on resources external to the core materials of the course.

Each student submits an individual feedback form, where they outline what and how much they contributed to the report, as well as how much they thought was contributed by their peers. These feedback forms are useful for the teaching team while marking, as well as for the students: it encourages them to reflect critically on how individuals can work cohesively, what barriers there are to collaborative working and how individuals might be encouraged or held to account.

4 Evaluation

Evaluation of the SDU considered the student learner voice, HPTs who contributed to delivery on the unit, and internal reviews with the unit directors and support team. Both quantitative and qualitative questions were posed to students and HPTs. The results of all feedback were shared with unit directors and senior project staff and used to inform future improvements on the design of the units. The design of the end of unit feedback form was altered significantly from the first to the second run in order to accommodate new avenues of interrogation that emerged from continuous improvements and refinement of both content and teaching approaches.

Table 2 Percentage of student feedback responses as they occurred from the first two runs of the SDU

Run	Statement	Disagree strongly (%)	Disagree (%)	Neither agree nor disagree (%)	Agree (%)	Agree strongly (%)
Pilot	Staff have made the subject of this unit interesting	1	3	17	45	18
2nd run		3	3	6	47	41
Pilot	I have received sufficient advice and guidance with the work for this unit	1	8	20	37	18
2nd run		0	13	6	47	34
Pilot	The material covered in this unit is intellectually stimulating	9	10	20	34	19
2nd run		0	6	13	41	41
Pilot	The unit is well organised	5	10	19	31	19
2nd run		6	3	6	63	22
Pilot	The online content blocks (videos and written content on blackboard) supported my learning	0	10	16	46	25
2nd run		0	6	16	38	41

Anonymous student feedback was collected for the first two runs of the unit (Table 2). In total, 114 responses were collected and provide insights into student perspectives on the strengths and weaknesses of the approaches taken in the design of the unit. The pilot run of the unit captured feedback from 32% of the cohort during teaching on the latter half of the academic year. The second run received responses from 78% of the cohort from teaching taking place in the first half of the academic year.[1] Students on the pilot unit were, therefore, able to situate their reflections in the context of a greater level of exposure to their home discipline.

[1]This differentiated level of student response is rooted in the different sizes of the two cohorts. Whilst the pilot run involved over 260 students, the second run involved 40. At the time of writing, we are preparing for the third cohort—involving 300 students.

Feedback notes positive attitudes towards all areas queried across all questions duplicated in both runs of the unit (range: 67.5–75% positive alignment). The changes made between the first and second run of the unit were successful, as each criterion shows improvement (Table 2). The percentage of students who neither agreed nor disagreed either remained static or decreased. The criteria relating to online content showed the least variability, reflecting the small-scale changes made to this material between runs. Additional qualitative questions were posed to ensure that students could highlight areas not considered by the teaching team. The most frequent observed criteria related to feedback and the range of interactions with staff. As such, adaptations made to the unit for the second run were reflected in the additional feedback questions posed at the end of the unit (Table 3).

On the second run of the unit, high satisfaction is observed with 76% of students reporting that they were happy or very happy with the unit and 100% noting opportunities to work with other students. Other strong areas of positive responses concerned how staff explain the subject matter (82% positive), opportunities to explore ideas and concepts in depth (81% positive), opportunities to apply learning (91% positive), clarity of marking criteria (75% positive), timetabling efficiency (78% positive), communications (88% positive), and a sense of community (81% positive).

5 Conclusion

In early 2020 the eyes of the world's population focused on Australia, where record-breaking temperatures and a prolonged, severe drought have resulted in vast bush-fires across the south-eastern states of Victoria and New South Wales. As of 21 January 2020, these fires killed 30 people (including four firefighters) and burnt over 100,000 km^2 of forest, bush and property (BBC 2020). While attributing these fires to anthropogenic climate change is a challenging work in progress (Phillips and Nogrady 2020), the experience of Australians in January 2020 is widely asserted to be a sign of a new normal in a climate-changed world (McGrath 2020). It is within this context that governments and public institutions across the globe are declaring a climate emergency and a recalibration of society's behaviour, consumption and activity—and, with it, patterns of social-environmental interaction.

The University of Bristol declared a climate emergency in April 2019. In making such a declaration, the institution not only affirmed the urgency of addressing and mitigating climate change (by becoming carbon neutral by 2030) but also acknowledge the "deep concerns of our students, many of whom are worried about what the future holds in store for them" (University of Bristol 2019b). Sustainable is an institutional priority, evidenced in the University's Vision and Strategy, its commitment to the global Sustainable Development Goals Accord and its recent award for the publicly-accessible, online *Sustainable Futures* unit (University of Bristol 2019a). It is within this institutional context that the SDU has been redesigned, refocused and made available to students across the University of Bristol. The SDU is an optional, credit-bearing and open unit. From the outset, an interdisciplinary team

Table 3 Additional qualitative feedback questions from the second run of the SDU

Statements (second run only)	Disagree strongly (%)	Disagree (%)	Neither agree nor disagree (%)	Agree (%)	Agree strongly (%)	Not applicable (%)
Overall, I am happy with the quality of the unit	0	9	16	34	42	
Staff on this unit are good at explaining things	0	16	3	38	44	
This unit provided me with opportunities to explore ideas or concepts in depth	0	3	16	65	16	
This unit has provided me with opportunities to apply what I have learned	0	3	6	53	38	
The criteria used in marking were made clear in advance.	0	6	19	50	25	
Marking and assessment have been fair	0	3	23	6	0	68
Feedback on my work has been timely	0	3	13	16	0	69
I have received helpful comments on my work	0	3	6	31	0	59
I have been able to contact staff when I needed to	0	2	9	24	30	35
The timetable works efficiently on this unit	0	9	13	59	19	
Any changes in the unit have been communicated effectively	0	0	9	69	19	3
I feel part of a community of staff and students working on this unit	0	6	13	53	28	

(continued)

Table 3 (continued)

Statements (second run only)	Disagree strongly (%)	Disagree (%)	Neither agree nor disagree (%)	Agree (%)	Agree strongly (%)	Not applicable (%)
I have had opportunities to work with other students as part of this unit	0	0	0	25	75	

have committed to delivering content that both informs successive cohorts about contemporary sustainability challenges but also empowers them to develop skills of critical, imaginative, and innovative thinking about sustainable development in general and the SDGs in particular. Following scholarship elsewhere, we have found that this teaching is greatly assisted by the dedication to engaging students in critical discussions and for allowing reflection on the issues and contests that characterise the wider concept of sustainability (Kopnina 2018; Opoku and Guthrie 2018). Students are drawn from across the institution. Geographers learn alongside biologists, engineers alongside those from the arts and humanities. These students have different levels of knowledge and understanding of the subject matter discussed across the unit. As a result, there was a need to ensure that content was inclusive and open to all. With contemporary teaching across UK HEIs still highly siloed, this diversity of students created challenges to the objective to deliver cohesive, engaging content that challenges students as much as it informs them. In response to this student diversity, we embraced a post-disciplinary approach, understood as representing the incorporation of non-academic and both locally- and globally- focused perspectives. To do so, the curriculum and materials were designed to be integrative, highlighting relationships and common themes (and differences) between concepts and modes of research.

At the time of writing, much of the world is under lockdown. HEIs across the globe have—in the wake of the mass disruption caused by the Covid-19 pandemic—been forced to transfer *all* teaching online. As a result, there is renewed impetus for units like the one profiled above, which illuminates the role that blended learning can play in HE, both today and in the future. The central motivation for curriculum design was that the Unit was for it not to be a mere conduit for communicating facts. Instead, it was to both encourage personal, critical reflection and facilitate collaboration and exchange between students. This was in terms of the concepts taught, which moved from empirical discussions of renewable energy transitions to reflections on the role that different research methods (and their associated epistemological and ontological standpoints) could have in such change. This rationale for accessible, holistic content-delivery is also found in the way information was delivered. The first iteration of the unit involved over 260 students. This volume creates a danger of bottlenecks of teaching, assessment and marking and the creation of a passive learning environment that restricts student engagement (Exeter et al. 2010; Cash et al. 2017). To mitigate these issues, the unit adopted a pedagogical approach informed by the concepts of "blended learning" and "flipped classroom" to provide the necessary materials to many students, drawn from diverse disciplinary backgrounds.

Students can pause, revisit and review with ease, aided by the provision of high-quality closed captions. This is useful for students with accessibility needs, non-native English-speaking students, and those who need clarity on particular words or terminology used. Such an approach of blended learning allows for the creation of a multi-modal, post-disciplinary content that captures contributions from a variety of agents and constituencies present in discussions of SD and the SDGs.

The post-disciplinary character of this unit, its content and its delivery are evident in the disciplinary make-up of the teaching team. Chemists worked with historians, engineers with human geographers: all to demonstrate to students that voices from other disciplines can contribute to shared discussions on SDGs. The goal was to illuminate how no discipline has supremacy in understanding climate change, its drivers and potential solutions. Instead, a holistic vision is necessary: a vision that transcends and subverts traditional disciplines, epistemologies and ontologies, to provide students with a comprehensive view of what SD is that maintains relevance long after the assignments are submitted, and the marks bestowed. This post-disciplinary teaching did not take the form of disciplinary blindness – instead, students were encouraged to turn to different disciplines to understand the challenges faced (Stentoft 2017). On a personal note, we were particularly struck by how such a process forced us all to reflect on our respective disciplinary backgrounds, and its strengths and blindspots in teaching topics of environmental science and policy. It showed us that, more than ever, in teaching these topics HE must embrace not only the diversity of the classroom but also the difference within its teaching staff.

In an era where more and more HEIs across the globe are embedding climate change education into their curricula, we see the SDU as an exemplar of post-disciplinary engagement, creative and critical thinking, and the use of digital materials to increase accessibility and reach. However, we also see the unit as a work in progress. We will continue to respond to the feedback of both successive cohorts and groups of staff involved in delivery to improve this unit. We will continue to seek inspiration from all disciplines, and we will strive to provide students with a holistic vision of how to address challenges in an era of climate emergency.

References

Abeysekera L, Dawson P (2015) Motivation and cognitive load in the flipped classroom: definition, rationale and a call for research. High Educ Res Dev 34(1):1–14. https://doi.org/10.1080/07294360.2014.934336

BBC (2020) Australia fires: a visual guide to the bushfire crisis, BBC News, 21 Jan. https://www.bbc.co.uk/news/world-australia-50951043. Accessed 27 Jan 2020

BBC (2019) Bristol university declares climate emergency. https://www.bbc.co.uk/news/uk-england-bristol-47962554

Bergmann J, Sams A (2012) Flip your classroom: reach every student in every class every day. International Society for Technology in Education, Iste, Eugene and Washington; ASCD, Alexandria

Biagioli M (2009) Postdisciplinary liaisons: science studies and the humanities. Crit Inq 35(4):816–833

Carmichael M, Reid A, Karpicke JD (2018) Assessing the impact of educational video on student engagement, critical thinking and learning: the current state of play. Sage Publishing. https://us.sagepub.com/sites/default/files/hevideolearning.pdf

Carroll J (2003) We're exploited, not unqualified. The Chronicle of Higher Education, 23 (June). https://www.chronicle.com/article/Were-Exploited-Not/45253

Cash CB, Letargo J, Graether SP, Jacobs SR (2017) An analysis of the perceptions and resources of large university classes. Life Sci Educ 16(2). https://doi.org/10.1187/cbe.16-01-0004

Chen K, Monrouxe L, Lu Y, Jenq C, Chang Y, Chang Y, Chai PY (2018) Academic outcomes of flipped classroom learning: a meta-analysis. Med Educ 52(9). https://doi.org/10.1111/medu.13616

Clark WC, Crutzen PJ, Schellnhuber HJ (2005) Science for global sustainability: toward a new paradigm. SSRN. https://doi.org/10.2139/ssrn.702501

Crutzen PJ (2002) Geology of mankind. Nature 415(23). https://doi.org/10.1038/415023a

Davies L (2006) Global citizenship: abstraction or framework for action? Educ Rev 58(1):5–25. https://doi.org/10.1080/00131910500352523

Dematagoda U (2016) Academic staff: we graduate teaching assistants need you to fight for us. The Guardian (28 Sep 2016). https://www.theguardian.com/higher-education-network/2016/sep/28/academic-staff-we-graduate-teaching-assistants-need-you-to-fight-for-us

Digital education office, Blackboard. University of Bristol. http://www.bristol.ac.uk/digital-education/support/tools/Blackboard/. Accessed 10 Mar 2020

Exeter D, Ameratunga S, Ratima M, Morton S, Dickson M, Hsu D et al (2010) Student engagement in very large classes: the teachers' perspective. Stud High Educ 35(7):761–775

Friesen N (2012) Defining blended learning. Learning spaces. http://learningspaces.org/papers/Defining_Blended_Learning_NF.pdf

Gillette C, Rudolph M, Kimble C, Rockish-Winston N, Smith L, Broedel-Zaugg K (2018) A meta-analysis of outcomes comparing flipped classroom and lecture. Am J Pharm Educ 82(5). https://doi.org/10.5688/ajpe6898

Hall R, Bowles K (2016) Re-engineering higher education: the subsumption of academic labour and the exploitation of anxiety. Workplace 28:30–47. https://doi.org/10.14288/workplace.v0i28.186211

Hammond J, Gibbons P (2019) What is scaffolding? National centre for english language teaching and research, Macquarie University, Sydney, pp 8–16

Hoare A, Cornell S, Bertram C, Gallagher K, Heslop S, Lieven N, MacLeod C, Morgan J, Pickering A, Wells S, Willmore C (2008) Teaching against the grain: multi-disciplinary teamwork effectively delivers a successful undergraduate unit in sustainable development. Environ Educ Res 14(4):469–481. https://doi.org/10.1007/s10639-019-09886-3

Husbands CT, Davies A (2000) The teaching roles, institutional locations, and terms and conditions of employment of part-time teachers in UK higher education. J Further High Educ 24(3):337–362. https://doi.org/10.1080/030987700750022271

Kopnina H (2018) Teaching sustainable development goals in The Netherlands: a critical approach. Environ Educ Res Routledge 24(9):1268–1283. https://doi.org/10.1080/13504622.2017.1303819

McGrath M (2020) Climate change: Australia fires will be 'normal' in warmer world, BBC News, 14 Jan. https://www.bbc.co.uk/news/science-environment-51094919. Accessed 27 Jan 2020

McLean S, Attardi SM (2018) Sage or Guide? Student perceptions of the role of the instructor in a flipped classroom. Active Learn High Educ 1–13. https://doi.org/10.1177/1469787418793725

Michalopoulou E, Shallcross DE, Atkins E, Tierney A, Norman NC, Preist C, Ninos I (2019) The end of simple problems: repositioning chemistry in higher education and society using a systems

thinking approach and the United Nations' sustainable development goals as a framework. J Chem Educ. https://doi.org/10.1021/acs.jchemed.9b00270
Opoku A, Guthrie P (2018) Education for sustainable development in the built environment. Int J Constr Educ Res Routledge 14(1):1–3. https://doi.org/10.1080/15578771.2018.1418614
Phillips N, Nogrady B (2020) The race to decipher how climate change influenced Australia's record fires. Nature 577:610–612. https://doi.org/10.1038/d41586-020-00173-7
Ryan MD, Reid SA (2016) Impact of the flipped classroom on student performance and retention: a parallel controlled study in general chemistry. J Chem Educ 93(1):13–23. https://doi.org/10.1021/acs.jchemed.5b00717
Sachs JD (2012) From millennium development goals to sustainable development goals. Lancet 379(9832):2206–2211. https://doi.org/10.1016/s0140-6736(12)60685-0
SDG Accord (2020) The SDG accord. https://www.sdgaccord.org/. Accessed 17 Feb 2020
Sergis S, Sampson DG, Pelliccione L (2018) Investigating the impact of flipped classroom on students' learning experiences: a self-determination theory approach. Comput Hum Behav 78:368–378. https://doi.org/10.1016/j.chb.2017.08.011
Sharpe R, Benfield G, Roberts G, Francis R (2006) The undergraduate experience of blended e-learning: a review of UK literature and practice. High Educ Acad. https://www.heacademy.ac.uk/system/files/sharpe_benfield_roberts_francis_0.pdf
Smith M (2019) Concern for the environment at record highs. YouGov. https://yougov.co.uk/topics/politics/articles-reports/2019/06/05/concern-environment-record-highs. Accessed 27 Jan 2020
Smith V (2014) What is blended learning? University affairs. http://www.universityaffairs.ca/features/feature-article/blended-learning/
Stentoft D (2017) From saying to doing interdisciplinary learning: is problem-based learning the answer? Active Learn High Educ SAGE Publ 18(1):51–61. https://doi.org/10.1177/1469787417693510
Tierney A, Tweddell H, Willmore C (2015) Measuring education for sustainable development: experiences from the university of Bristol. Int J Sustain High Educ 16(4):507–522. https://doi.org/10.1108/IJSHE-07-2013-0083
Turra H, Carrasco V, González C, Sandoval V, Yáñez S (2019) Flipped classroom experiences and their impact on engineering students' attitudes towards university-level mathematics. High Educ Pedagogies 4(1):136–155. https://doi.org/10.1080/23752696.2019.1644963
United Nations (2015) The sustainable development goals. https://www.un.org/sustainabledevelopment/sustainable-development-goals/. Accessed 29 Mar 2019
University of Bristol (2019a) Sustainable futures online course wins national green gown award. University of Bristol, 27 Nov. https://www.bristol.ac.uk/cabot/news/2019/green-gown-award.html. Accessed 27 Jan 2020
University of Bristol (2019b) University of Bristol declares a climate emergency. University of Bristol, 17 Apr. http://www.bristol.ac.uk/biology/news/2019/university-of-bristol-declares-a-climate-emergency.html. Accessed 27 Jan 2020
Walsh A, Michalopoulou E, Tierney A, Tweddell H, Preist C, Willmore C (2020) Sustainability in higher education: beyond the green mirror. In: Universities as living labs for sustainable development, Springer, pp 183–191
Webster R (2017) Teaching assistants are being overstretched and exploited. TES. https://www.tes.com/news/teaching-assistants-are-being-overstretched-and-exploited

The "Sustainable Development Goals (SDG) Teaching Map" and Other Innovative Teaching Formats

Elisabeth Fröhlich, Marina A. Schmitz, and Silvia Damme

1 Introduction

Responsible management education (RME) is on everyone's lips, and especially whilst the world is grappling with the Coronavirus pandemic, the call for value-based management is getting louder. Global supply chains have completely collapsed in recent weeks, and the top management teams of most companies have underestimated the fragility of today's (business) system. In addition to the human rights and climate problems that have been discussed in the past years, there is now the demand for global hygiene regulations and handling the negative effects of continuous over-consumption in the textile sector on sewers and cotton farmers—to name just two of many challenges and affected target groups. The pessimistic news about not only the development of the world economy, but also private restrictions associated with it, make it necessary to approach a possible solution with optimism. The following quote from the poem "Be an optimist," by Wayne Visser (2020), seems very appropriate at this point: "I am optimistic, not because the future is bright but because bright people are working to make the future better; not because the news is good but because good people are showing that change is always possible; and not because the world is fair but because fair people are fighting for justice wherever it is needed."

E. Fröhlich (✉)
Strategic Supply Management, CBS International Business School, Hardefuststr. 1, 50677 Cologne, Germany
e-mail: e.froehlich@cbs.de

M. A. Schmitz · S. Damme
Center for Advanced Sustainable Management (CASM), CBS International Business School, Hardefuststr. 1, 50677 Cologne, Germany
e-mail: m.schmitz@cbs.de

S. Damme
e-mail: s.damme@cbs.de

W. Leal Filho et al. (eds.), *Universities, Sustainability and Society: Supporting the Implementation of the Sustainable Development Goals*, World Sustainability Series,
https://doi.org/10.1007/978-3-030-63399-8_31

Especially in times of crisis, the call for higher corporate social responsibility (CSR) standards shows that society no longer tolerates irresponsible behaviour, threatening the licence to operate of respective companies and organisations. Thus, business schools especially have to react by providing responsible education, thereby enabling future managers to make socially and ethically sound decisions in the respective organizations in which they will work (Cornelius et al. 2007). Future-fit leadership and leaders who are able to navigate through the challenges of our time are needed, thus creating added value for society (Jones Christensen et al. 2014). Although courses (and research) focusing on CSR, business and society, business ethics and the like seem to be en vogue, scarce attention has been paid in terms of pedagogical and didactic elements in relation to how they can best be integrated into (higher) education settings (Giacalone and Thompson 2006). With this in mind, CBS International Business School developed an SDG Teaching Map. It aims to show how business schools can contribute to value-oriented management, in order to enable companies to meet these social and environmental challenges of our common future (Schneidewind 2014) and showcase an outside-in perspective (Muff et al. 2013; Dyllick and Muff 2016; Muff et al. 2017).

The focus of this paper, however, does not rest on the presentation of the SDG Teaching Map but on its further development potential in terms of introducing new, innovative RME teaching formats. These are currently being developed in an EU Erasmus+ project under the title **I**nnovative **S**olutions for **SU**stainability in **E**ducation (ISSUE) (ISSUE 2019). Thus, this paper tries to provide a first impression of how best to reflect the SDGs in respective teaching formats.

2 Methodology: The Sustainable Development Goal (SDG) Teaching Map

The SDG Teaching Map is the result of a master thesis written at the CBS International Business School (Kul 2020), which is an internationally-oriented, state-recognised university and ranks among the best private business schools in Germany (Book et al. 2019; CBS International Business School 2019). The programme portfolio of the CBS covers all educational levels, from Bachelor to Executive Master of Business Administration. CBS degrees are obtained on either a full-time or a part-time basis, and the majority of the courses are taught in English.

2.1 Case CBS International Business School—SDGs and Teaching: How and Why?

In a first step, the research process for creating the SDG Teaching Map will be described (see Fig. 1). As a result of the internal curriculum analysis, a short overview

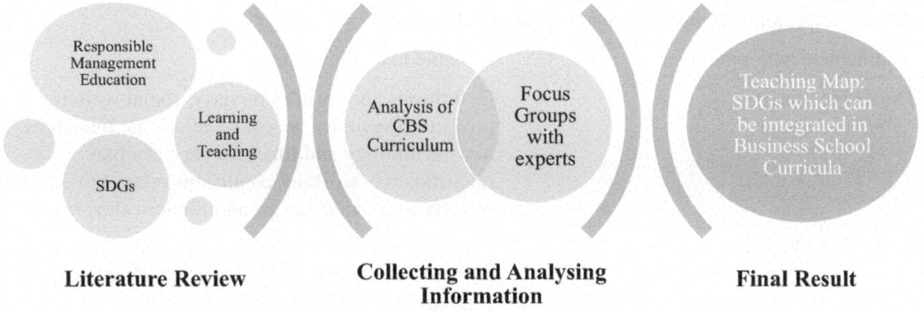

Fig. 1 Research process for establishing the SDG teaching Map. *Source* Own Illustration

of the SDGs and possible content in the respective teaching formats is provided (see Table 1). Figure 2 summarises the targets of the SDGs to be anchored in the CBS International Business School curriculum. The empirical insights derived from the six focus group expert interviews with international RME researchers belong to the PRME community (Kul 2020), which allowed initial statements about which sustainable teaching content can be supplemented or mapped by new formats.

The starting point of this research project can be characterised by the following three questions: (a) What content of the 17 SDG targets not covered so far can be integrated into the curriculum of a business school? (b) How can these contents be integrated (e.g. establish a new lecture or integrate them into existing lectures)? (c) What target content of the 17 SDGs cannot be represented by a business school? How can they be acquired?

Before deliberating how to integrate additional teaching content into the curriculum of a business school, an overview of possible teaching content regarding how to support the individual SDGs is required. Supporting the sustainable development of society is the major aim of RME and serves as a justification why the Sustainable Development Goals are at the forefront of our research interest. In essence,

> the 2030 Agenda for Sustainable Development, adopted by all United Nations Member States in 2015, provides a shared blueprint for peace and prosperity for people and the planet, now and into the future. At its heart are the 17 Sustainable Development Goals (SDGs), which are an urgent call for action by all countries—developed and developing—in a global partnership. They recognize that ending poverty and other deprivations must go hand-in-hand with strategies that improve health and education, reduce inequality, and spur economic growth—all while tackling climate change and working to preserve our oceans and forests (United Nations 2020).

These contents are considered essential for establishing a value-based management approach, and Rieckmann (2018) also recommends their integration into business school curricula.

SDG 2 "Zero Hunger" serves as an example to illustrate the connection made above. Hunger is on the rise, and malnutrition continues to harm millions of children around the world. Public investment in agriculture is declining worldwide, small

Table 1 SDGs and management teaching content

Sustainable development goals	Possible teaching content
SDG 1: No poverty	Definitions and form of poverty, social welfare protection systems, access to economic resources, new technology and financial services like microfinance, working conditions related to poverty (e.g. child labour and modem slavery)…
SDG 2: Zero hunger	Drivers and causes of hunger, relation between climate change and food security, trading systems concerning food, sustainable agriculture methods…
SDG 3: Good health and well-being	Direct and indirect strategies for health promotion, pollution of air, water and soil, communicable and non-communicable diseases…
SDG 5: Gender equality	Gender and labour, gender and education, gender and poverty…
SDG 6: Clean water and sanitation	The impacts of pollution on water quality, water scarcity, efficient water-use, recycling and reuse technologies…
SDG 7: Affordable and clean energy	different energy types, the environmental impacts of energy production, bridging technologies
SDG 8: Decent work and economic growth	Models of economic growth, financial systems and their relation to economic growth, entrepreneurship, social innovation, labour rights …
SDG 9: Industry, innovation and infrastructure	Information and communication technologies (ICT's) in supply chains, sustainable innovation and industrialization, sustainable infrastructure development…
SDG 10: Reduced inequalities	Fiscal, wage and social protection policies, global trade systems and regulations, international development aid …
SDG 11: Sustainable cities and communities	Sustainable energies and transportation, sustainable food, waste generation and management…
SDG 12: Responsible consumption and production	Environmental and social impacts of production and consumption, food production and consumption, concepts of green economy (e.g. circular economy approaches)…
SDG 13: Climate action	Greenhouse gases and their emissions: energy, agriculture and industry-related greenhouse gas emissions, impacts on eco systems…
SDG 14: Life below water	Oceans pollutants, the relationship of climate change and the sea, sustainable marine energy…

(continued)

Table 1 (continued)

Sustainable development goals	Possible teaching content
SDG 15: Life on land	Various threats to biodiversity, the extinction of species, desertification and deforestation, climate change and biodiversity…
SDG 16: Peace, justice and strong institution	Climate justice, child labour, corruption…
SDG 17: Partnership for the goals	Global systems and power structure, global trading systems, global governance and policies, international development aid…

Source Own Illustration based on UNESCO (2017)

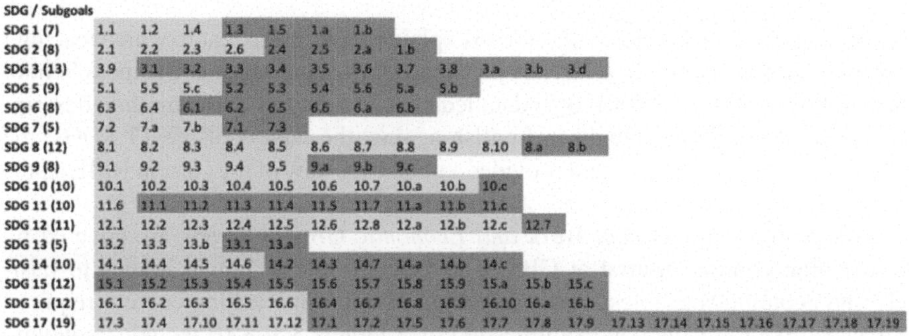

Fig. 2 SDGs and targets already embedded in the CBS curriculum. *Source* Own Illustration (Green = covered in teaching/red = not covered in teaching, yet)

food producers and family farms call for more support and increased investment in technology and infrastructure for a more sustainable agricultural approach is urgently required (United Nations 2019b). The SDG 2 therefore aims to stop hunger, to achieve food security and to promote forms of sustainable agriculture to improve nutrition. In terms of education in management, students have to understand the drivers of—and reasons behind—the worldwide phenomenon of hunger. Knowledge about sustainable farming methods is required to understand the relationship between hunger and climate change, or the effects of different types of trading systems on hunger (UNESCO 2017).

The importance of *SDG 10 "Reduced Inequalities"* (United Nations 2019a) is particularly clear as we face a global pandemic (Covid-19). Inequality in global supply chains is currently resulting in the impoverishment of entire regions, due to inequalities in income, access to labour markets and trade. For this reason, the call for a green economy and a new sustainable economic model is becoming ever louder. Management training must outline new approaches for the construction and regulation of global trading systems, wage and social protection and international development assistance (UNESCO 2017). This is the only way to ensure resilient supply chains as a prerequisite to reducing inequalities within and among countries.

These two examples should serve as a first indication to understand what teaching content in management might be relevant for the implementation of SDGs, as well as to consider how this content can be taught and embedded into different teaching formats. *SDG 4 "Quality Education"* in Table 1 does not have any content, as it is more about teaching formats. Some of these innovative teaching designs will be presented later in this paper as the preliminary output of the EU Erasmus+ project ISSUE.

2.2 Results: SDG Teaching Map: How to Improve a Business School Curriculum to Foster Value-Based Management

The following two examples will serve to explain briefly the extent to which teaching content related to the SDGs is already part of the CBS International Business School curriculum (see Fig. 2). Based on a detailed content analysis of the module descriptions of all programmes taught in English against the targets of the 17 SDGs, the following figure was created and provides a good overview of the current RME status of the university.

Almost all *SDG 8 Decent Work and Economic Growth* targets are covered by the teaching content offered at CBS. The fact that most content relates to business and economics enables a good coverage of relevant teaching content required to achieve SDG 8. Concrete examples to cover targets 8.1–8.10 include economic growth in emerging markets, technological modernisation and innovation, policies promoting entrepreneurship and innovation, full employment, youth unemployment, forced labour and modern slavery, labour rights and working environments, sustainable tourism and sustainable financial services for SMEs. Different lectures cover these topics, e.g. Human Resources Management, Managerial Finance, Advanced Economics or Supply Chain Management and Logistics. *SDG 15 Life on Land*, however, presents a completely different picture. An interdisciplinary approach would be necessary to address this SDG, since the respective teaching content is found mainly in the field of environmental and natural sciences, and CBS currently does not offer this content.

Even though CBS International Business School has already successfully implemented a so-called 'integrated sustainability curriculum' (Kolb et al. 2017; Raueiser and Kolb 2018), current research shows that great efforts are still required to enable business schools to offer an "SDG-relevant curriculum," in order to achieve Agenda 2030 from an educational perspective. As already mentioned, the experts taking part in the online focus groups (Kul 2020) were asked how missing SDG targets can be covered by the curriculum of a business school in the future. Although the answers received provide the first helpful insights, any piece of structured advice was missing. For this reason, it seems necessary to find a theoretical framework that will help tackle this lack of systematisation, and so we now dive further into the "Future Art Model" (Schneidewind 2018).

3 Theoretical Framework: The "Future Art Model"

At the centre of the "Future Art Model," or transformational literacy ("Zukunftskunst"), lies the question relating to the radical incremental change of the economic order (Schneidewind 2014, 2018; Schneidewind and Rehm 2019; Schneidewind and Singer-Brodowski 2013). According to Macy (2009), three main dimensions could initiate a great transformation: actions to slow the damage to Earth and its beings, analysis of structural causes and the creation of structural alternatives and a shift in consciousness. Society has to find its space on this continuum between raising its voice in protest (fighter) and reshaping and building a new future (artist) (Schneidewind 2019). Jointly, the four dimensions of transformational literacy (i.e. cultural, institutional, economic and technological) drive sustainability-oriented transformation.

Through our SDG Teaching Map, business schools should be enabled to fuel this change and cultural mindset shift. It is about reformation and not about revolution, and a "culture of experimentation" is necessary to create experimental spaces that show us the way to a more sustainable future. Business schools need to test new, innovative strategies and instruments to enable this necessary reformation of our economic thinking (Fig. 3). At the basis of this process lies a cultural shift leading to a broadening of perspectives, cooperation and a new mindset in relation to this necessary sustainable transformation. A new economic order also requires institutional reforms, to encourage new co-operation between state and markets, the private sector and society in the sense of a major socio-ecological transformation. Especially business schools should strengthen their interaction with institutional and

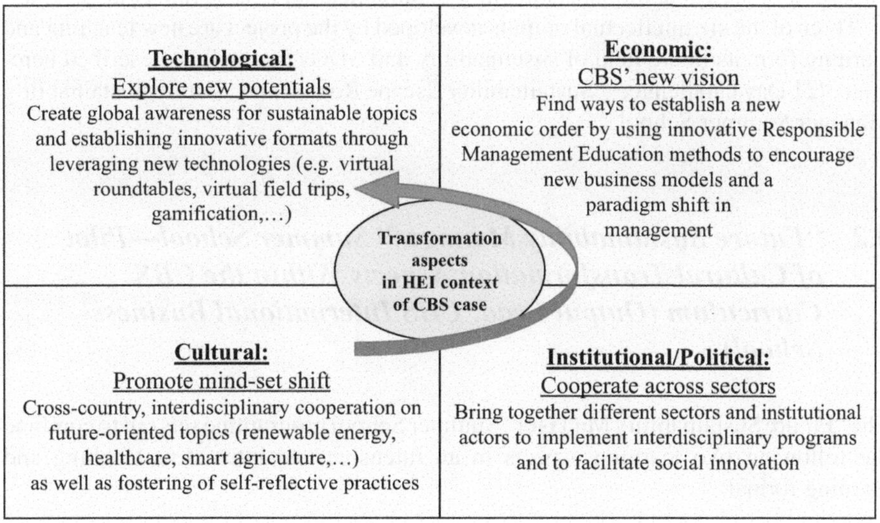

Fig. 3 Future Art/transformational literacy model in the context of the CBS curriculum. *Source* Own Illustration based on Schneidewind and Rehm (2019), p. 169

policy actors and try to integrate into their curricula fields relevant to sustainable development. The economic vision resulting from this notion underlines the need for RME to support new ways of doing business, aligned with a paradigm shift in management. This reformation path needs to be open to and leverage new technologies, since digital tools offer the potential to facilitate this new economic thinking, e.g. by realising a fully transparent global supply chain.

In the following, we introduce three teaching formats that CBS is currently developing within a European research project—and which can be matched to the respective clusters as practical examples.

3.1 Erasmus+ Transnational Partnership: Innovative Solutions for SUstainability in Education (ISSUE)

With the aim of developing in university teaching innovative teaching and learning concepts for the stronger integration of sustainability, the CBS International Business School has been a project partner of the ISSUE project (Innovative Solutions for SUstainability in Education) since December 2018 (www.issue-project.eu).

The eight European partner institutions will jointly develop a variety of intellectual outputs over the next two years directed at introducing new tools and teaching methods for sustainability within higher education. These will be piloted in the respective institutions, and resulting experiences will be incorporated into the final manuals and strategy recommendations. The project enables the development, testing and use of new content, teaching methods and transformative approaches that are important in the context of sustainability communication (Leal Filho et al. 2019).

Three of the six intellectual outputs developed by the project are new teaching and learning formats in the field of sustainability and SDGs that will be described hereafter: "21 Day Challenge," "Sustainability Escape Room" and "Future Sustainability Manager Summer School."

3.2 "Future Sustainability Manager" Summer School—Pilot of Cultural Transformation Aspects Within the CBS Curriculum (Output Lead: CBS International Business School)

The "Future Sustainability Manager" Summer School programme sets out to combine the following two learning aspects in an intensive and interactive teaching and learning format:

(1) Learning for Sustainable Development

Knowledge transfer and awareness-raising on sustainability issues, in particular in relation to sustainable economic transformation in the context of the SDGs. Learners should be empowered to make responsible decisions and take into account sustainability aspects, regardless of their specific role or function within an organisation (Weybrecht 2015).

(2) Learning for the Future of Work

Teaching professional "future skills" to learners that are becoming increasingly important in times of growing uncertainty and rapid technological development, such as creativity, adaptability and flexibility, as well as the ability to collaborate (LinkedIn Talent Solutions 2019; World Economic Forum 2016).

The focus of the summer school is on the students' empowerment to develop their own points of view and to link taught content individually with their personal interests: 'It is about moderating dynamic discussions and providing a space in which each student can develop his or her own understanding of sustainable management' (Kolb and Bungard 2018, p. 207). This is based on the understanding of sustainable management theory at the CBS. It should also go hand in hand with a necessary general cultural change in teaching, which does not focus exclusively on hard skills but also includes the personal development and creative competence of the learners (Schmidpeter and Kolb 2018). Learning should thus take place in a meaningful interplay between conventional and innovative formats (Lozano et al. 2019; Molthan-Hill 2017; Schmidpeter and Kolb 2018) as well as across locations, disciplines and results. During the seven-day course, students will be encouraged to learn independently. For this purpose, the role of the teaching staff is more oriented towards that of a moderator or a learning companion (MacVaugh and Norton 2012; Warburton 2003) who injects knowledge and impulses into the discussions and working formats. The learning events will follow a "learn to do" approach and distinguish themselves from the more traditional "learn to know" style (Hermens and Clarke 2009). Course content should be based on real examples and collaboration projects (Leal Filho et al. 2019) and clearly go beyond a *classroom as usual* experience.

The programme schedule is divided into the components "Knowledge, Mindset & Action," based on the "Head, Heart and Hands" approach by Sipos et al. (2008). Ultimately, it aims to create a transformative *mindset* conveyed through interactive, creative and practical teaching and learning formats (*action*). The applied pedagogical approaches thus promote the development of sustainability skills, for example interdisciplinarity, uncertainty tolerance, critical analysis, systemic thinking, empathy and communication skills (Lozano et al. 2019; Rieckmann 2018). Some suitable methods to promote these skills, which are incorporated into the summer school, include project- and problem-based learning, jigsaw teams, group work, critical discussions, case studies in combination with site visits and networking, as well as role play and reflection (Alden Rivers et al. 2015; Bass 2012; Leal Filho et al. 2018; Lozano et al. 2019; Moratis et al. 2006; Remington-Doucette and Musgrove 2015; Sipos et al. 2008).

A framework programme is moreover included which aims to strengthen reflective practices and the participants' personal development. An important element here, for example, is the introduction of learners to mindfulness practices, which can benefit both mental and physical health as well as creativity and the increased exploitation of learning potential (Yeganeh and Kolb 2009). This practice will be supported, for instance, by daily morning meditation and regular reflection on what has been learned and experienced in a personal journal as well as in open feedback rounds. Through activities such as a workshop in improvisational theatre or the inclusion of places of creative creation, the creativity of the participants will be stimulated (Ferrari et al. 2009). The summer school will also include short interactive exercises, coaching elements and learning games (Bass 2012; Sweeney et al. 2011). Students will further be able to participate actively in the programme design. After the participant selection process, the students can indicate whether they would like to participate through individual interests or skills (e.g. by writing blog posts during the summer school, offering workshops for their fellow students). It will be interesting to see how far the concepts used in the "Future Sustainability Manager" Summer School can be established as such, as well as be integrated more widely at the university; the format can be extended through continuous thematic development and new learning and teaching concepts in this setting, and international cooperation can be used to align the format in alternating cycles with one or more partner universities. Additionally, individual session formats can be integrated into existing lectures or programmes at the university, and the summer school could be developed into a new study programme or study focus at the university. Individual session formats can also be integrated into theme days or project weeks (e.g. sustainability days). The topics incorporated into the framework sessions of the summer school, such as meditation, mindfulness and reflection diaries, can be disseminated through the university's local curriculum or extra-curriculum offers.

Cultural Transformation Aspect within the Curriculum: Through the one-week programme, the "Future Sustainability Manager" Summer School creates an international, interdisciplinary set-up that most of all allows learners to break free from their usual study environment. The group experience and the focus on a framework encouraging participants to explore their purpose and impact through self-reflection, mindfulness and creative techniques aims to promotes a shift in mindset for the participants. Furthermore, it seeks to create an environment that pushes learners beyond their daily comfort zone and enables a personal transformation process.

3.3 21-Day Challenge—Pilot of Institutional Transformation Aspects Within the CBS Curriculum (Output Lead: Lappeenranta University of Technology)

With the "21-Day Challenge," partners work on an online learning platform that uses playful elements to convey knowledge and awareness of the topics covered by the

17 Sustainable Development Goals (SDGs). Building on the Online Collaborative Learning (OCL) methodology according to Harasim (2012), the programme aims to stimulate the sustainability discourse among learners and to initiate an effective change of awareness and behaviour through the continuous processing of the various tasks ("challenges") associated with the 17 SDGs.[1] According to Willenbacher et al. (2017), the integration of digital learning games can enable the successful transfer of the game's ideas into reality. Especially in the "Digital Natives" generation, such formats are very much accepted by participants. At the end of each challenge, participants receive points, which they can share publicly. Through gamification, i.e. the combination of pleasure and education, in the form of individual challenges, the motivation and commitment of the learners should be increased (Willenbacher et al. 2017). They additionally offer 'a significantly higher potential for appropriation through digital games […] than passive forms of reception' (Willenbacher et al. 2017, p. 2019). The platform can be used by anyone interested in the topic of sustainability and is motivated to deal more interactively with the topics of sustainable development. Although the platform is suitable for bringing students closer to learning content in a varied and self-responsible way, it also enables universities to encourage staff from all areas to engage with sustainability collectively. The "21-Day Challenge" ultimately aims to promote more sustainable habits throughout the entire university community.

The following three components are provided for each topic (SDG 1-17): first, a short video that introduces the respective topic and provides knowledge about the related challenges for society and the environment. Second, the "challenge" for the participant and third, feedback through the exchange of experience and discussion. Additional points are rewarded for interactions such as postings, uploads and comments on the platform, on top of the mere fulfilment of tasks, in order to encourage exchange and discussion among participants.

The "21-Day Challenge" format offers versatile possibilities for integration into university teaching. For example, it can be embedded in a regular course with a "learning diary" in which the students reflect on what they have learned. Further, it can be used for a Campus Challenge with prizes, in which all interested parties (students and staff) can participate. Lecturers can use it to precede the start of the seminar as a thematic introduction and preparation for the course. Moreover, it can be part of a student project group (e.g. "CSR Student Team" at the CBS), which reports on successes and experiences during the challenge and afterwards on social media, thus generating attention. Schools could also debate the introduction of an "elective" event with a concluding workshop day themed, for example, "Business in the context of SDGs" or "Responsible Business in Times of Global Transformation."

[1]Examples for challenges are: SDG4 - "Think about what sustainable development initiatives you believe are important to educate future generations about and what platforms you could use to promote your ideas. Consider why they are important, what do they aim to achieve, what are the consequences of not acting. Explore these ideas further, network with key people in those areas, follow them on social media and learn from them." SDG 6 – "Use a bowl in the sink when washing fruit, vegetables of dishes. Then use the wastewater to water your plants. Repeat at least 3 times.".

Institutional Transformation Aspect within the Curriculum: The 21-Day-Challenge supports a holistic implementation of SDGs within the curriculum by touching on all 17 subject matters, from responsible consumption, to life below water or sustainable cities and communities. Including experts not only from the business sector, but also from other scientific fields and institutions, such as NGOs or the political sector, allows for the inclusion of inter- and transdisciplinary perspectives in the business school curriculum.

3.4 Sustainability Escape Room—Pilot of Technological Transformation Aspects Within the CBS Curriculum (Output Lead: Budapest Business School)

Living in a time that is characterised by readily available entertainment and constant stimulation everywhere, the learner is to be "picked up" by the learning format of a "Sustainability Escape Room," based on the idea of gamified teaching. The Escape Room is also based on the idea of anchoring sustainability in an entertaining way (Willenbacher et al. 2017). Learning happens based on experiencing, so through the combination of entertainment, participation and focus, the learner should reach a state of "flow." First conceptualised by Csikszentmihalyi (1990), "flow" refers to a mental mode in which the person concerned pursues an intrinsically motivated activity. This happens in line with extreme focus, joy and full participation (Csikszentmihalyi 2014), which can have a positive effect on the learning experience (Hamari et al. 2016). Other than with the online-based 21-Day Challenge, learning takes place in the physical space, namely the "Escape Room." Digital elements, however, are integrated in the form of different tasks and riddles.

The Escape Room aims to test and expand players' knowledge in sustainable development by integrating real-world problems. Furthermore, participants should gain an increased awareness of global connections as well as for systemic transformation processes. Additionally, the format integrates positive learning effects in areas such as communication and cooperation, testing new conditions, dealing with change or reaching consensus (Moratis et al. 2006). The advantage of simulation formats compared to more traditional forms of teaching is that the learner is immediately confronted with the possible consequences of his or her action (Hermens and Clarke 2009; Kolb et al. 1984).

It will be possible to adapt the Escape Room for a broad range of applications. Students are thought to be the main target group, but depending on the setting, teachers, university staff and the interested public can also experience it. Integration within the university can happen through inclusion in a regular course. The experiences should be reflected in the following seminar lesson and worked on by the students in further tasks. The Escape Room is also a great platform to integrate into the context of theme days or a sustainability theme week ("Sustainability Days"). In the same sense, it can be used at open campus days or other events, conferences and fairs, and by faculty and administrative staff at the university, e.g. as part of a company outing or faculty days. When designing new events or programmes at the university, the Escape Room can always be considered in the planning process.

Technological Transformation Aspect within the Curriculum: With the Sustainability Escape Room, the ISSUE project is realising an innovative new format that transcends the usual classroom or site visit experience. By leveraging new technological possibilities and creating a gamified physical emersion experience for the learner, it perfectly displays the exploration of new learning and teaching potentials.

4 Conclusion

The main challenges that come with the innovation and reformation of university teaching will hardly be met by individual universities alone. Even though they can act as pilots and best-practice examples, the development of new teaching and learning methods discussed above is time-consuming and resource-intensive. This is why cooperation projects such as ISSUE, or networks such as Principles for Responsible Management Education (PRME) or the European School of Sustainability Science and Research (ESSSR), which serve as a platform and jointly test and implement new formats, are necessary. It is necessary for universities not only to cooperate, but also to build bridges to companies and politics, in order to successfully implement the United Nation's Agenda 2030. Second, in order to respond to the call for a "True University Responsibility" (Schneidewind 2014), the adoption of a more holistic perspective that goes beyond the triple bottom line logic of sustainability is of vital importance (Dyllick and Muff 2016; Muff et al. 2017; Schneidewind and Rehm 2019). According to the "Future Art" model approach, however, this sustainability can only be achieved if the cultural aspects of our current economic system (and all its respective players) change as well (Schneidewind 2019). Additionally, continuous monitoring of the alignment of course content and SDGs needs to be ensured. A key factor in achieving this would be a bottom-up approach to take care of the inclusion and willingness of lecturers to tackle and address proactively the SDGs in their course content and set-up.

Other universities can easily make use of this SDG Teaching Map and identify respective strengths and weaknesses throughout their curriculum. However, there needs to be more exchange regarding best practices in the field of higher education teaching that are targeted at improving RME. In regards to the way forward, using the SDGs for restructuring and improving the curriculum at CBS International Business School has provided a starting point to contribute to the Agenda 2030 through advancing higher education as the ultimate goal. Furthermore, the missing targets not covered in the curricula need to be addressed by CBS alone or in collaboration with other institutions. Additionally, it would be beneficial to use quantifiable indicators and goals that measure our advancements in addressing the SDGs and the change we can provoke in students' mindsets.

References

Alden Rivers B, Armellini A, Maxwell R, Allen S, Durkin C (2015) Social innovation education: towards a framework for learning design. High Educ Skills Work-Based Learn 5(4):383–400
Bass R (2012) Disrupting ourselves: the problem of learning in higher education. Educause Rev 47(2):23–33
Book S, Fischer K, Guldner J (2019) WirtschaftsWoche Hochschulranking 2019: Die besten Unis für Ihre Karriere. https://www.wiwo.de/my/erfolg/hochschule/hochschulranking-2019-die-besten-unis-fuer-ihre-karriere/24274822.html?ticket=ST-1851654-hWuGYKyUV5sEJctURHKs-ap2. Accessed 29 July 2019
CBS International Business School (2019) WiWo-Hochschulranking 2019: CBS an der Spitze der privaten Fachhochschulen Deutschlands im Bereich BWL. https://cbs.de/hochschule/news/cologne-business-school-verteidigt-spitzenrang-beim-wiwo-uni-ranking/. Accessed 23 July 2019
Cornelius N, Wallace J, Tassabehji R (2007) An analysis of corporate social responsibility, corporate identity and ethics teaching in business schools. J Bus Ethics 76(1):117–135
Csikszentmihalyi M (1990) Flow. Harper and Row, New York
Csikszentmihalyi M (ed) (2014) Flow and the foundations of positive psychology. the collected works of mihaly csikszentmihalyi. Springer, Dordrecht
Dyllick T, Muff K (2016) Clarifying the meaning of sustainable business: Introducing a typology from business-as-usual to true business sustainability. Organ Environ 29(2):156–174
Ferrari A, Cachia R, Punie Y (2009) Innovation and creativity in education and training in the eu member states: fostering creative learning and supporting innovative teaching. literature review on innovation and creativity in E&T in the EU member states (ICEAC). Luxembourg: Office for Official Publications of the European Communities
Giacalone RA, Thompson KR (2006) Business ethics and social responsibility education: Shifting the worldview. Acad Manag Learn Educ 5(3):266–277
Hamari J, Shernoff DJ, Rowe E, Coller B, Asbell-Clarke J, Edwards T (2016) Challenging games help students learn: an empirical study on engagement, flow and immersion in game-based learning. Comput Hum Behav 54:170–179
Harasim L (2012) Learning theory and online technologies. Routledge, New York
Hermens A, Clarke E (2009) Integrating blended teaching and learning to enhance graduate attributes. Educ Training 51(5/6):476–490
ISSUE (2019) Innovative solutions for sustainability in education. http://www.issue-project.eu
Jones Christensen L, Mackey A, Whetten D (2014) Taking responsibility for corporate social responsibility: the role of leaders in creating, implementing, sustaining, or avoiding socially responsible firm behaviors. Acad Manag Perspect 28(2):164–178
Kolb DA, Rubin IM, McIntyre JM (1984) Organizational psychology: an experiential approach to organizational behavior. Prentice-Hall, Englewood Cliffs, NJ
Kolb M, Bungard P (2018) Nachhaltiges Management lehren und lernen: Ein praktischer Ansatz zur Transformation. In: Raueiser M, Kolb M (eds) CSR und Hochschulmanagement. Sustainable Education als neues Paradigma in Forschung und Lehre, pp 199–211. Springer, Berlin, Heidelberg
Kolb M, Fröhlich L, Schmidpeter R (2017) Implementing sustainability as the new normal: Responsible management education—from a private business school's perspective. Int J Manag Educ 15(2):280–292
Kul B (2020) Responsible management education. In: Implementing the sustainable development goals in business school Curricula. Master Thesis in International Business: CBS International Business School, 108 pages
Leal Filho W, Raath S, Lazzarini B, Vargas VR, de Souza L, Anholon R, Quelhas OLG, Haddad R, Klavins M, Orlovic VL (2018) The role of transformation in learning and education for sustainability. J Clean Prod 199:286–295
Leal Filho W, Shiel C, Paço A, Mifsud M, Ávila LV, Brandli LL, Molthan-Hill P, Pace P, Azeiteiro UM, Vargas VR, Caeiro S (2019) Sustainable development goals and sustainability teaching at universities: falling behind or getting ahead of the pack? J Clean Prod 232:285–294

LinkedIn Talent Solutions (2019) 2019 Global Talent Trends. The 4 trends transforming your workplace. LinkedIn, Sunnyvale, CA

Lozano R, Barreiro-Gen M, Lozano F, Sammalisto K (2019) Teaching sustainability in european higher education institutions: assessing the connections between competences and pedagogical approaches. Sustainability 11(6):1602

MacVaugh J, Norton M (2012) Introducing sustainability into business education contexts using active learning. Int J Sustain High Educ 13(1):72–87

Macy J (2009). The great turning. joanna macy: the great turning is a shift from the industrial growth society to a life-sustaining civilization. https://www.ecoliteracy.org/article/great-turning. Accessed 10 May 2020

Molthan-Hill P (ed) (2017) The business student’s guide to sustainable management: principles and practice. Routledge, New York

Moratis L, Hoff J, Reul B (2006) A dual challenge facing management education. J Manag Dev 25(3):213–231

Muff K, Dyllick T, Drewell M, North J, Shrivastava P, Haertle J (2013) Management education for the world: a vision for business schools serving people and the planet. Edward Elgar Publishing, Cheltenham, UK, Northampton, MA, USA

Muff K, Kapalka A, Dyllick T (2017) The gap frame-translating the SDGs into relevant national grand challenges for strategic business opportunities. Int J Manag Educ 15(2):363–383

Raueiser M, Kolb M (eds) (2018) CSR und Hochschulmanagement. Sustainable Education als neues Paradigma in Forschung und Lehre. Springer, Berlin, Heidelberg

Remington-Doucette S, Musgrove S (2015) Variation in sustainability competency development according to age, gender, and disciplinary affiliation. Int J Sustain High Educ 16(4):537–575

Rieckmann M (2018) Learning to transform the world: key competencies in education for sustainable development. In: Leicht A, Heiss J, Byun WJ (eds) Issues and trends in education for sustainable development. United Nations Educational, Scientific and Cultural Organization, Paris, pp 39–60

Schmidpeter R, Kolb M (2018) Wirtschaft im Wandel—Neue Anforderungen an die Managementausbildung. In: Raueiser M, Kolb M (eds) CSR und Hochschulmanagement. Sustainable Education als neues Paradigma in Forschung und Lehre, pp 11–17. Springer, Berlin, Heidelberg

Schneidewind U (2014) Von der nachhaltigen zur transformativen Hochschule. Perspektiven einer „True University Sustainability “. Umweltwirtschaftsforum 22(4):221–225

Schneidewind U (2018) Die große Transformation. Eine Einführung in die Kunst gesellschaftlichen Wandels. S. Fischer, Frankfurt am Main

Schneidewind U (2019) Taktgeber der Großen Transformation: die Rolle der Zivilgesellschaft im 21. Jahrhundert Politische Ökologie 156:112–117

Schneidewind U, Rehm A (2019) Vom inside-out zum outside-in: Perspektivwechsel bei der Impact-Messung von transformativer Forschung. GAIA Ecol Perspect Sci Soc 28(2):168–170

Schneidewind U, Singer-Brodowski M (2013) Transformative Wissenschaft: Klimawandel im deutschen Wissenschafts-und Hochschulsystem. Metropolis Verlag, Marburg

Sipos Y, Battisti B, Grimm K (2008) Achieving transformative sustainability learning: engaging head, hands and heart. Int J Sustain High Educ 9(1):68–86

Sweeney LB, Meadows D, Mehers GM (2011) The systems thinking playbook for climate change: a toolkit for interactive learning. Eschborn: Deutsche Gesellschaft für Internationale Zusammenarbeit (GIZ) GmbH

UNESCO (2017) Education for sustainable development goals. Learning objectives. UNESCO, Paris

United Nations (2019a) Sustainable development knowledge platform. Sustainable development goal 10. Reduce inequality within and among countries. https://sustainabledevelopment.un.org/sdg10. Accessed 6 May 2020

United Nations (2019b) Sustainable development knowledge platform. Sustainable development goal 2. End hunger, achieve food security and improved nutrition and promote sustainable agriculture. https://sustainabledevelopment.un.org/sdg2. Accessed 6 May 2020

United Nations (2020). Sustainable development goals knowledge platform. Sustainable development goals. https://sustainabledevelopment.un.org/sdgs. Accessed 7 May 2020

Visser W (2020). Be an optimist. http://www.waynevisser.com/poetry/be-an-optimist. Accessed 6 May 2020

Warburton K (2003) Deep learning and education for sustainability. Int J Sustain High Educ 4(1):44–56

Weybrecht G (2015) 5 Key messages from businesses to business schools around sustainability. https://primetime.unprme.org/2015/09/07/5-key-messages-from-businesses-to-business-schools-around-sustainability/. Accessed 15 July 2019

Willenbacher M, Lepiorz R, Wohlgemuth V (2017) Serious Games, Umweltbewusstsein und Nachhaltigkeit. In: Eibl M, Gaedke M (eds) Informatik 2017. Gesellschaft für Informatik, Bonn, pp 2017–2026

World Economic Forum (2016) Global challenge insight report. The future of jobs. Employment, skills and workforce strategy for the fourth industrial revolution

Yeganeh B, Kolb D (2009) Mindfulness and experiential learning. OD Pract 41(3):8–14

Prof. Dr. habil. Elisabeth Fröhlich studied business administration at the Ludwig-Maximilians-University of Munich and at the University of Cologne. After her dissertation on the subject of "Supplier Evaluation," she continued her professional career at the seminar of Prof. Dr. U. Koppelmann at the University of Cologne and completed her habilitation on the subject of "Modelling Requirement Profiles in Procurement." Since the end of 2007, Elisabeth Fröhlich has been a professor at the CBS International Business School, responsible for the field of strategic procurement management. Since 1 May 2013, she has been managing the CBS International Business School in her function as President. She is a board member of several scientific organisations, e.g. the PRME Interim Management Board, PRME DACH Chapter Speaker, Rotary Club Köln Römerturm and BME e.V. Region Köln. She is the spokesperson for the "Kölner Wissenschaftsrunde" (Cologne Scientific Circle) and business ambassador for the City of Cologne. Her current research focuses on sustainable supply chain management and green procurement, qualification in purchasing and strategic supplier management. Procurement 4.0, as well as organisational, personal challenges of an "agile purchasing" and new, innovative teaching formats of Responsible Management Education, are further main areas of research. She has published several books and articles on the above-mentioned topics and supports several journals as an external reviewer.

Marina A. Schmitz works as a research associate and lecturer at the Center for Advanced Sustainable Management (CASM) at the CBS International Business School in Cologne, Germany. She lectures in Sustainable Management, Asian Business and Intercultural Management. Her research focus is in CSR in Emerging Economies, CSR and HRM and sustainability in higher education. After graduating at the University of Trier in Business Administration and Sinology, she worked as a research associate at the Chair of Human Resources Management and Asian Business at the University of Goettingen. Her dissertation deals with the voluntary labour turnover phenomenon of Chinese factory workers. She is involved in various research projects and engaged in networks like PRME, ESSSR and NBS to boost responsible management. She has also been involved in the organisation of sustainability conferences and has experience in consulting projects initiating the implementation of sustainable business models.

Silvia Damme works as a project manager and research associate at the Center for Advanced Sustainable Management at CBS International Business School in Cologne, Germany. She lectures in Sustainable Management and Intercultural Management. Her work and research focus is on sustainable management and innovative pedagogy in the field of sustainability education, where she is currently involved in various international research projects. Silvia Damme received her MA in International Business and Sustainability from the University of Hamburg, Germany, and holds a BA in Culture and Business Studies from Mannheim University. She is passionate about connecting sustainability change-makers from academia, business and society and has been involved in organising large-scale events and conferences in the fields of sustainable business and climate action in Germany, as well as in the United States. She is a Climate KIC Journey alumna and involved in networks like PRME, ESSSR and NBS to promote responsible management. She holds additional professional experience in sustainability consulting and the implementation of change processes.

The Implementation of the SD Programs for Management Curriculum: The Experience of Plekhanov Russian University of Economics

Siarhei Zenchanka, Nikolai Gorbatchev, Fernanda Frankenberger, and Amanda Lange Salvia

1 Introduction

Universities have a significant role in the achievement of the Sustainable Development goals (SDGs). Young people who start their education at university today will play critical role in sustainable development in a future. "A Guide for universities, higher educational institutions, and the academic sector" pointed, that "The SDGs cover a wide range of complex social, economic, and environmental challenges and addressing them will require transformations in how societies and economies function and how we interact with our planet. Education, research, innovation and leadership will be essential in helping society address these challenges. Universities, with their broad remit around the creation and dissemination of knowledge and their unique position within society, have a critical role to play in the achievement of the SDGs. Arguably none of the SDGs will be achieved without this sector" (SDSN Australia/Pacifc (2017)).

S. Zenchanka (✉)
Minsk Branch, Plekhanov Russian University of Economics, 40, Radialnaya Str, Minsk 220070, Belarus
e-mail: zench@tut.by

N. Gorbatchev
School of Business, Belarus State University, 4, Nezavisimosti Avenue, Minsk 220030, Belarus
e-mail: ngorbachev@sbmt.by

F. Frankenberger
Pontifical Catholic University of Paraná—PUCPR, R. Imac. Conceição,, 1155 - Prado Velho, Curitiba, PR 80215-901, Brazil
e-mail: ferfrank1@hotmail.com

A. L. Salvia
University of Passo Fundo, Campus I-Km 171-BR 285, Bairro São José, Passo Fundo, RS 99001-970, Brazil
e-mail: amandasalvia@gmail.com

W. Leal Filho et al. (eds.), *Universities, Sustainability and Society: Supporting the Implementation of the Sustainable Development Goals*, World Sustainability Series,
https://doi.org/10.1007/978-3-030-63399-8_32

One of the objectives identified for the 2030 Agenda and the Sustainable Development Goals is the acquisition of knowledge and skills to promote sustainable development and sustainable lifestyles (United Nations 2015a). According to the concept of Sustainable Development, Environment is one of its main areas and largely determines the economic and social development.

The need to include sustainable development issues in educational programs for business and management was considered, for example, by Minderman (2015). Study programs in Environmental management are usually implemented at the graduate level (Master's degree), while Bachelor programs (B.A. and B.Sc.) are connected with studying broader areas of environment management, such as resource management, environmental management of enterprises, waste management, among others. Bachelor programs are usually oriented to practical approaches, whilst master programs focus on the relations between science, management, and policy, providing scientific understanding of ecological, economic and social systems. It is rather difficult to analyse all programs of different areas and educational levels of curricula for environmental management. For example, there are 321 of such programs for bachelor level on BachelorsPortal (2019) and 492 master programs on MasterPortal (2019). All these programs refer to Earth Science and Environment Protection. Considering some of these programs, examples of the following universities can be used: Uppsala, Harvard, and Cambridge.

The objective of the syllabus for the course "Environmental Management in Practice" in Uppsala University is "to provide skills and improved understanding of how firms and organizations work with sustainability issues such as environmental and natural resource management and sustainability issues" (Uppsala University 2019). In Harvard, the course "Environmental Management" aims at studying scientific principles of environmental issues and environmental management practices, focusing on emerging topics such as air and water pollution, water use and management, aquatic ecosystems, energy and climate change, biodiversity, toxic substances in the environment, solid waste management, among others (Harvard University 2015). Another example is the Cambridge University syllabus on "Environmental Management Study" which is connected to environmental management and economics and pursues alternative technologies, allowing students to have local and global perspectives (Cambridge University 2019).

The examples mentioned above illustrate that different programs on Environmental Management give students the ability to analyse and understand the current state and past development, creating future scenarios and developing global and regional thinking.

Considering the implementation problem of Sustainable Development and Environmental Management courses and parts of these courses in Economics programs, Leal Filho and Manolas (2012, p. 55) allocated three barriers for this problem and possible ways overcome them:

1. Economics lecturers are not convinced of the relevance of embedding sustainable development in the curriculum mainly because it seems to conflict with the

employability of students. Such an obstacle could be overcome by developing sustainable products and markets, and/or through social marketing.
2. Those willing to embed sustainable development are uncertain as to what the key concepts are as well as how these concepts should be embedded in the curriculum. The way to overcome this barrier would be firstly to determine the concepts to be included, then survey the students to see what they know, followed by a survey at the end of the term to see what they were taught. Concepts not taught will be embedded during the following academic year.
3. If the problem of sustainable development involves multiple disciplines, the solution should be multi-disciplinary if not interdisciplinary. Barriers to the interdisciplinary work for economists could be that economics is not welcoming to other disciplines, the belief that economics has no limitations when answering questions of sustainable development and the feeling that some economists are insufficiently "trained" to communicate with other disciplines.

In some aspects, these barriers refer to management programs. In particular, this is connected with the need of using a multi-disciplinary and interdisciplinary approach.

In Belarusian universities, in a narrow sense the course "Environmental Management" refers mainly to environmental management of enterprises in accordance with ISO 14001. In a wide sense, environmental management is considered in the course "Environmental protection and rational use of natural resources" aimed at specialized training in the field of preserving and restoring the natural environment, rational use and reproduction of natural resources, preventing pollution, degradation, damage, depletion, destruction and other harmful effects and liquidation of its consequences. Basically, such courses are developed based on national legislation and the use of ISO standard only. It should also be noted that the international agreements of the Republic of Belarus emphasize the need to bring the norms of national legislation to an international one.

The aim of this paper is to present the result of the project "Environmental management in European Union course and MOOC" which was developed for management curriculum of Minsk Branch of the Plekhanov Russian University of Economics. This course was developed in the framework of Jean Monnet activity in the Erasmus + program and was directed at overcoming the gap between European and Belarus approaches to environmental management. Erasmus+ is the EU's programme to support education, training, youth, and sport in Europe. Jean Monnet Activity proposes the development of a teaching module connected with European Policy (Erasmus+ n.d.).

The analysis of sustainable development and its problems, as well as European legislation related to this field are the basis of the course, which includes 10 themes: Sustainable Development and Environmental Management, Global environmental issues, Indicators of Sustainable Development, International laws, International standards, Environmental monitoring, Development of environmental management systems, Natural resource accounting, Waste management and Environmental assessment. These topics use practical studies as means to verify cause-effect relations among problems and analyse statistical data related to the Sustainable Development

Goals. The knowledge acquired is used by students at a writing course and diploma projects and this is the main contribution of this experience to society: disseminating more knowledge on sustainable development and preparing and training future professionals to deal with sustainability challenges.

2 Methodology

To develop the course "Environmental management in the EU" five steps were taken, as described in Table 1.

These steps aimed at understanding different aspects of Environmental management in the EU. During these steps, it became clear that EU legislation and practice in Environment management are well developed and can be used as an example of good practice. Therefore, issues of European environmental legislation were considered in all developed courses.

The course includes lectures on the legislative basis of environmental management and lectures on the main phases of environmental management, including monitoring, implementation at the level of state and enterprises, and decision making.

Table 1 Steps for the development of the course

Step 1	Study the concept of sustainable development, environmental problems, and indicator of sustainable development, according to information available at the sustainable development goals—Knowledge Platform of United Nations (UN-Knowledge Platform n.d.). This step helped to define the scope of syllabus. Although the main topics of the syllabus being developed were proposed in the application to the Jean Monnet Foundation, it was necessary to develop a plan for each lecture and topics for practical seminars
Step 2.	Study the European environmental legislation taking as an example the European environment agency (EEA n.d.). This stage allowed determining the main milestones in the development of EU environmental legislation
Step 3	Study the European environmental action programs (European commission for environment n/d/). During the implementation of the environmental action programs in Europe, the basic principles of the environmental policy were determined and the main regulatory documents were adopted
Step 4	Study documents from different conferences, scientific and political reports such as Club of Rome, EU, JRC, UN, UNFCCC, among others. Such documents helped estimate last achievements in investigating sustainable development and environmental management
Step 5	Development of the course material. The course named "Environmental Management in the EU" consists of lectures, seminars and computer labs and is a part of the syllabus for managers and economists in the Minsk Branch of PRUE. Circular model of environmental management suggested by Zenchanka and Batyrev (2011) was used. This model includes Environmental monitoring, modelling, decision making and impact on environment. All stages of this model are supported by Legislative basis

During seminars, students report their findings of the legislative and practical aspects of environmental management. To study these aspects, the well-known Problem-Based Learning (PBL) approach was used. This approach helps students find causal relations between different phenomena, think critically and solve complex problems they will encounter in their lives (Boud and Feletti 2013).

Classes using computers are directed on analysis of statistical data for different Sustainable Development Goals in Belarus, Russia and some other European countries. Students must find data not from statistical compilations, but from different reports and articles, among others sources.

3 Results: Structure of "Environmental Management in EU Course and MOOC"

Based on the steps indicated above, the developed course includes the following topics: Sustainable Development and Environmental Management, Global environmental issues, Indicator of Sustainable Development, International laws, International standards, Environmental monitoring, Development of Environmental management systems, Natural resource accounting, Waste management, and Environmental assessment. All topics are detailed in the following sections, indicating the theoretical relevance of their consideration in the course.

This course structure was suggested for development in application form of Jean Monnet Module and was approved for implementation. Course topics correspond to a cyclical model of environmental management (Zenchanka and Batyrev 2011)—from problems definition to decision making—and are aimed at developing students' systemic and critical thinking.

3.1 Sustainable Development and Environmental Management

The main idea of this section is to present an introduction to Environmental management (EM) which is an integral part of the sustainable development (SD) concept, which arose in the middle of the twentieth century and was presented in reports of the Club of Rome and different World Summits. At the beginning stage, the reports of Forrester (1971) "World Dynamics" and Meadows et al. (1972) "The Limit of Growth" were the most remarkable. These reports showed the existence of limits for planetary development and launched numerous publications on sustainable development. Then Meadows et al. (2004) presented the results of a new research in which they sadly concluded that humanity continues to move along the undesirable trajectory of development. Models used in these books consider the planet as a whole, without taking into account the economic development of regions.

In 2012, Randers J., one of the co-authors of "The Limit to Growth" and "Limit to Growth: the 30-years update", published the results of his research with forecasts of humanity development for the following 40 years (Randers 2012). This report differs from the previous ones for having considered five regions of the world with different levels of economic development: (1) USA, (2) EU, Japan, Canada and the rest of the industrialized world; (3) China; (4) Brazil, Russia, India, South Africa, and 10 other big emerging economies; (5) and finally those at the bottom of the income ladder.

The main findings of this study are:

- The impact will differ among the five regions analysed in the book. The world in 2052 will certainly not be uniform or flat—conditions in the five regions will differ dramatically;
- The increase in global population and economy (GDP) will not be as fast as many expect;
- Growth will be slow in OECD countries and also in the poorest parts of the world. As a consequence, the global demand for resources will remain lower than it would have been if the population and the economy were bigger.

The report "A finer future is possible. How humanity can avoid system collapse and craft a better economic system" from Lovins et al. (2016) presents ways to overcome collapse:

- The challenge facing humanity today is to create a convincing new story and a powerful movement for a new economic system based on different values, such as strengthening human dignity, the common good and proper management of nature.
- Humanity must work within the planetary boundaries and ensure human well-being and dignity. We need a "safe and working space for humanity" (Lovins et al. 2016). The growing population—combined with the extravagant way of life among the rich—will lead to difficulties in relation to planetary boundaries. It is necessary to limit the wealthy way of life and make efforts to limit population growth.
- In a circular economy, products are designed to facilitate processing, reuse, disassembly and rework. It replaces the traditional linear model "Take, Make and Dispose", which have dominated the economy so far and will promote more efficient use of energy and resources.

Environmental management systems are components of the sustainable development of any state and its national security. These systems are developed at the state and local levels. At the state level, sustainable development and environmental management are considered in sustainable development strategies and regulations. At the local level, in documents of environmental management systems, such as policies, guidelines, standards of enterprises and organisations. A circular model of environmental management is considered, which includes the following steps: environmental monitoring, environmental modelling, decision making and environmental impact.

To evaluate the learning, students should answer the following topics on the seminar:

1. The main development stages of the sustainable development concept.
2. Club of Rome–its creation, development of ideas, main results.
3. "Limit to growth" and "Limit to growth: The 30-years Update": similarities and differences
4. Main findings of the investigation "2052: A Global Forecast for the Next Forty Years"
5. "A finer future is possible. How humanity can avoid system collapse and craft a better economic system"—possible ways to overcome collapse.

3.2 Global Environmental Issues

Global environmental issues are considered from the point of view of cause-effect relationships. Forrester (1971) considered that birth control should be introduced in order to counteract the processes of exponential growth, which would support decreasing both the population size and production and consumption. In turn, Meadows et al. (1972) considered Climate Change (CC) and Ozone Layer Depletion (OLD) as the most important modern problems of mankind. In this section, the cause-effect relation chain starts from the problem of overpopulation (Malthus Law) and ends with the problems of CC and the OLD. It is clear that all problems in this chain are interconnected. Climate Change, for example, influences Biodiversity Change, and so on.

Earth's population growth is an initial problem. Its consequences are the problems of production and consumption, poverty and changes in the structure of nature management. Next elements of this chain are resource depletion, air, water, and soil pollutions, and, at last, ozone layer depletion and climate change (Fig. 1).

To manage these problems, different international agreements were adopted such as the UN Framework Convention on Climate Change (UNFCCC 1992), the Vienna Convention for the protection of the Ozone Layer (United Nations Environment Programme 1985), Montreal Protocol (1987), Kyoto Protocol (United Nations 1998), the Convention on Biological Diversity (United Nations 1992), and the Paris Agreement (United Nations 2015b), among others.

To investigate the cause-effect relation in these problems, students should take classes on computers and present their results in the form of a table. The example of such investigation is presented in Table 2.

Each group of students was divided into subgroups with one global problem for studying. Then they present their findings for further discussion with other colleagues. Such discussions help clarify data and make recommendations. Column "Notes" should contain relevant references.

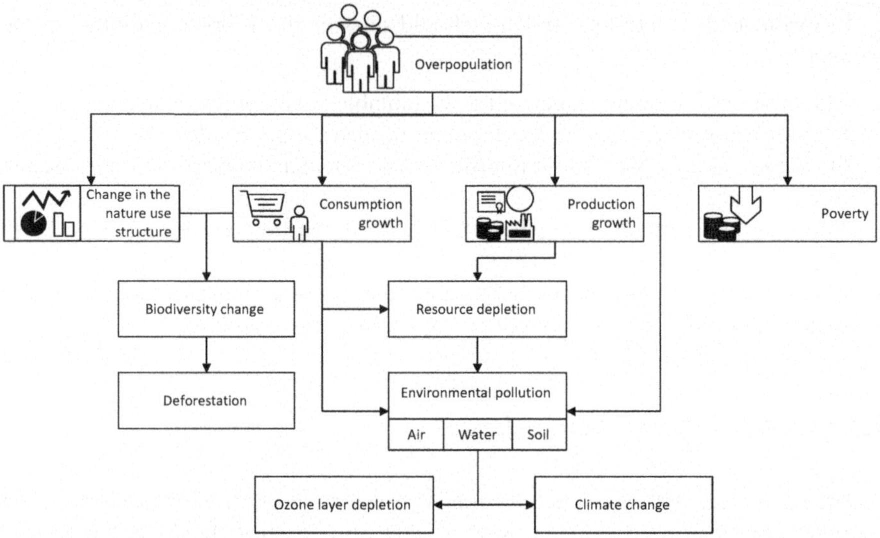

Fig. 1 Global environmental issues

Table 2 Causal relationship for climate change issue

Problem	Climate change			
Aspects	*Economy aspects*	*Environment*	*Society*	Notes
	Agriculture productivity	GHG, OLD	Human health	
Main cause	Change climatic zones	Resource depletion	High UV	
Main effect	Crop change	Biodiversity	Skin cancer, eyes disease	

3.3 Indicator of Sustainable Development

The indicators support several issues in management. First of all, they can help managers classify and aggregate information for better decision making. They can also help introduce the knowledge of natural and social science in decision making, as well as assess progress in achieving the goals of sustainable development.

Several types of indicators are considered:

- Indicators of the Commission on Sustainable Development (United Nations 2007);
- Indicators for Millennium Development Goals (United Nations 2000);
- Sustainable Development Goals 2030 (United Nations 2015a).

The main focus was on Sustainable Development Goals 2030 which contains 169 targets and 230 indicators. In order to facilitate the implementation of the system of

Table 3 Sustainable development goal 7

Goal 7. Ensure access to affordable, reliable, sustainable and modern energy for all			
Target 7.1. By 2030, ensure universal access to affordable, reliable and modern energy services			
Global Indicators	National Indicators	Russia	Belarus
7.1.1 Percentage of population with access to electricity	7.1.1 Percentage of population with access to electricity	96%	100%
7.1.2 Percentage of population with primary reliance on clean fuels and technology	7.1.2. The proportion of the population using clean fuels for cooking (natural gas, electricity) in the total population	Natural gas—88.1% Electricity—4.2%	Natural gas—80,8% Electricity—5.8%

global indicators, all of them were sorted according to the degree of methodological study and the availability of data at global level to the following three levels:

- Level I: there is an established methodology and standards, and data is provided by countries on a regular basis;
- Level II: there is an established methodology and standards, but data is not presented regularly by countries;
- Level III: there is no established methodology or standards for the indicator or the methodology/standard relevant to the indicator is under development or verification.

According to these levels, countries can include different indicators for estimation of sustainability and develop their own indicators.

During the classes on computers, students were invited to consider the sustainable development indicators achievement in Russia and in Belarus. As sources, it is recommended to use the data of the statistical committee and different ministries of the Republic of Belarus and Eurostat. Some results for SDG 7 are presented in Table 3.

Each subgroup of the class gets its own goal for studying and students offer their results for discussion. In case official data is not available and statistical reports cannot be found, use of data from mass media is allowed. During the discussion, there is an exchange of views on the various goals of sustainable development, which helps in developing systems thinking.

3.4 International Laws

Analysis of environmental legislation begins with the definition of aspects arising from the environmental legislation: scientific, economic and social—from the Malthus law to the concept of sustainable development. EU legislation addressed four main types of documents: the Treaty on the functioning of the Union, which is

Table 4 Sample questions of the test applied as part of the topic International laws

Test 1. Specify the period of validity of the 6th program	Test 2. Integration of the environmental dimension into all major policy areas was suggested in
(a) 1993–2000 (b) 2002–2012 (c) 2014–2020	(a) 4th EAP (b) 5th EAP (c) 6th EAP (d) 7th EAP

constitutional basis; the Declarations, which are political documents and determines the direction of development; programs such as Environmental Action Programs (EAP), which are non-binding political documents but define the principles and direction of development; and Regulations, Directives and Decisions.

Seven Environmental Action Programs have been considered. The first four programs were directed to the development of different environmental directives and decisions. Fifth program "Towards Sustainability" (1993–2000) was directed to the search for solutions to the following environmental issues: climate change, nature and biodiversity, air quality, city environment, coastal zone, waste management, water management, (http://ec.europa.eu/environment/archives/actionpr.htms). The sixth program "Our future. Our choice" (2002–2012) included four key directions (European Commission 2011a): climate change, nature and biodiversity (nature net completing), environment and health, and resources and waste. The Seventh Programme (2014–2020) has three key objectives (Official Journal of the European Union 2013), as presented below:

- To protect, conserve and enhance the Union's natural capital;
- To turn the Union into a resource-efficient, green, and competitive low-carbon economy;
- To safeguard the Union's citizens from environment-related pressures and risks to health and wellbeing.

At the seminar, students had to perform a short test on these objectives and results of all environmental action programs. Examples of questions are presented in Table 4.

3.5 *International Standards of the Environment Management*

Starting from the Rio Conference in 1992, Environmental Management Standards have been developed by the International Standards Organization for different applications (ISO 2017). This section of the course gives an overview of international environmental standards and it is presented in accordance with thematic issues (Environmental management systems, Environmental auditing, and related environmental investigations, Environmental performance evaluation, Environmental labelling, Life cycle assessment, Environmental communication, Environmental aspects of product design and development, Environmental aspects in product standards, Terms and definitions, Greenhouse gas management and related activities, and Measuring the carbon footprint of products). The role of standards in organizing and conducting environmental management and standards interconnection is considered.

Organisations around the world, as well as their stakeholders, are increasingly aware of the need for environmental management, socially responsible behaviour and sustainable growth and development. International standards are becoming increasingly important for organisations working in the direction of general and environmental management, to support the sustainability of the organization, products, and services. These standards focus on the preservation and development of the environment as well as on the development of methods for its assessment, such as water and carbon footprint, ecological footprint, assessment of greenhouse gases, and so on.

This section comprises different tasks for students. One of these tasks was the definition of environmental footprint using online platforms. Students should collect data according to the experience with their families and evaluate water and energy consumption, waste generation, transportation services, among other issues, and calculate the environmental footprint.

3.6 Environmental Monitoring

Environmental monitoring describes the processes and activities that need to take place to characterize and monitor the quality of the environment. Environmental monitoring is used in the preparation of environmental impact assessments, as well as in many circumstances in which human activities carry a risk of harmful effects on the natural environment. All monitoring strategies and programs have reasons which are often designed to establish the current status of an environment or to establish trends in environmental parameters.

Two main European monitoring programs are considered in this part (European Commission 2015; JRC 2017). These programs closely interconnect and use the results of each other. Main areas of monitoring are: Atmosphere, Marine Environment, Land, Climate Change, Emergency Management and Security. The monitoring results are used in agriculture, forestry, city management, biodiversity, climate change, air and water quality, clean energy, among others.

One of the tasks in the creation of monitoring systems is their compatibility. Students must compare national environmental monitoring systems with the European ones. Results of the comparison should be entered in Table 5.

Table 5 Comparison of monitoring programs

Item	Europe		Russia	Belarus
System of environmental monitoring	Copernicus	JRS	State system of environmental monitoring	National system of environmental monitoring
Subsystems	Satellite observations; ground, sea and air observations	Different information systems—GHG, water monitoring, etc.	Monitoring systems belong to different government bodies	Different monitoring systems are joined to national system of environmental monitoring
Tasks	Collecting data on the state of the environment	Collecting data on the state of the environment	Collecting data on the state of the environment	Collecting data on the state of the environment

3.7 Development of Environmental Management Systems

One of the most remarkable results of the conference in Rio de Janeiro in 1992 was the recognition of environmental problems that have arisen to humanity. There was a need for an instrument that would be aimed at reducing the impact of enterprises and organizations on the environment.

In 1992, the British standard BS 7750 was published; in 1996 the International Organization for Standardization (ISO) published the first ISO 14001:1996 related to environmental management standard. In 2015, the third version of ISO 14001 was published. This version corresponds to high-level structure common for all ISO standards. This structure helps the creation of integrated management systems (ISO 2015).

ISO 14004 (ISO 2016) is an additional standard to ISO 14001:2015. This standard provides guidance on the principles, systems, and methods for ensuring the operation of environmental management systems.

In 1993, the European Commission developed the Eco-Management and Audit Scheme (EMAS n.d.), which also allows enterprises and organizations to develop environmental management systems. In 2010, the third version of the EMAS was adopted, requiring pre-certification in accordance with ISO 14001. EMAS for small and medium-sized enterprises is considered too.

Students must complete the following tasks for different types of enterprises and organizations:

- Prepare Environmental Policy;
- Prepare questionnaire for internal audit.
- By using sites such as https://ec.europa.eu/environment/efe/themes/environmental-law_en and http://minpriroda.gov.by/ru/ students must compare environmental legislation of EU, Russia, and Belarus, and define similarities and differences (Table 6).

Table 6 Comparison of the environmental management standards

Europe	Russia	Belarus
Eco-management and audit scheme (EMAS) ISO 14001:2015	GOST R ISO 14001:2016—environmental management systems	STB ISO 14001:2017—environmental management systems
Similarities		
All standards are aimed at the implementation of the environmental management systems of enterprises (organizations) and use identical stages		
Differences		
1. To get an EMAS certification, the organization must develop the environmental management system in accordance with ISO 14001		
2. The organization must present an environmental statement to get EMAS certification		

3.8 Natural Resource Accounting

Sustainable economic development requires consideration of economic, social and environmental factors. Since the last decade of the twentieth century, there is a need for statistical data reflecting the state of the natural environment and its impact on society. At the same time, statistics are required to reflect the impact of ecosystems in the economy.

In the document of Agenda 21 (1992), adopted in Rio de Janeiro, the topics "Integrating Environment and Development in Decision Making" (Chap. 8) and "Conservation and Rational Use of Resources for Development" (Sect. 2) were defined as main development priorities.

Ecosystems relate to renewable natural capital. They can process ecosystem products, such as wood, as well as reverse ecosystem services, such as carbon dioxide absorption. Examples of non-renewable capital are fossil fuels and mineral resources.

The ecosystem is defined as a dynamic complex of communities of plants, animals and microorganisms and an inanimate environment interacting as a functional unit (Biodiversity Information System for Europe n.d.). People are an integral part of ecosystems.

The EU addresses the definition of ecosystem services, as well as approaches for their assessment and accounting. In accordance with these approaches to the definition, assessment, and accounting of ecosystem service, students must give an answer to the following questions:

1. What ecosystem services are presented in Russia and Belarus?
2. What legislation governs ecosystem services in Russia and Belarus?
3. What biodiversity measures have been taken?

Table 7 Waste management in EU, Russia and Belarus

Type of waste	Percent of recycling		
	EU	Russia	Belarus
Plastic	60	30	30
Municipal	60	5–7	10–15

3.9 Waste Management

Waste is an issue that affects all of us. We all produce waste: on average, each of the 500 million people living in the EU throws away around half a ton of household rubbish every year. This is on top of huge amounts of waste generated from activities such as manufacturing (360 million tons) and construction (900 million tons), while water supply and energy production generate other 95 million tons. Altogether, the European Union produces up to 3 billion tons of waste every year.

All this waste has a huge impact on the environment, causing pollution and greenhouse gas emissions that contribute to climate change, as well as significant losses of materials—a particular problem for the EU which is highly dependent on imported raw material.

The amount of waste we generate is increasing and the nature of waste itself is changing, partly due to the dramatic rise in the use of hi-tech products. This means waste now contains an increasingly complex mix of materials, including plastics, precious metals and hazardous materials that are difficult to deal with safely. EU waste management policies aim to reduce the environmental and health impacts of waste and improve Europe's resource efficiency. The long-term goal is to turn Europe into a recycling society, avoiding waste and using unavoidable waste as a resource wherever possible. The aim is to achieve much higher levels of recycling and to minimize the extraction of additional natural resources. Proper waste management is a key element in ensuring resource efficiency and the sustainable growth of European economies (European Commission 2011b).

Based on the European experience, students should consider waste management problems and suggest ways to improve waste management in Russia and Belarus. As an activity, they should complete Table 7, presented in this paper as an example.

3.10 Environmental Assessment

According to the Directive 2011/92/EU (Official Journal of the European Union 2011) "on the assessment of the effects of certain public and private projects on the environment", environmental assessment is a procedure to ensure an impact assessment before making a decision which may cause an impact on the environment. This is carried out both for individual objects (e.g. the construction of dumbs, roads, factories, etc.), as well as for strategic plans and programs in accordance with Directive 2001/42/EU.

The Environmental Assessment Directive ensures a high level of protection and integration of measures to reduce impact at the stage of preparation of programs, plans, and projects. A key element of the assessment is a public consultation, which is designed to improve the quality of decision making and sustainable development of territories. Moreover, the widespread use of the ISO standards of 14000 series help organizations to minimize the impact on the environment. In addition, many large companies have internal documents regulating the decision making that may affect the environment.

During seminars, students must discuss projects such as "Construction of a battery plant" and "Construction of nuclear power plant" and present their environmental assessment.

4 Discussions

The "Environmental management in EU course and on-line resources (MOOC)" project was developed and was included in various courses.

Lecture materials are based on published materials which give the students a view of the problem. At the same time, this course covers many topics, which can be presented as a separate course. Hence, this course should be considered as an introduction to the environmental management problem and process.

Seminars, practical tasks, and computer labs allow students to better understand the lecture material. For a better understanding of the European approach to environmental management, practical tasks were mainly aimed at comparing the regulatory and practical approaches to the organization of environmental management in Europe and in Russia and Belarus. This helps students to identify differences in approaches and outline possible ways of corresponding changes. Analysis of publications in the media allows clarifying statistics and a deeper understanding of existing problems associated with the implementation of sustainable development goals. Many tasks were based on problem-based learning that allowed students to establish cause-effect relations between problems.

Final testing on all topics of the course helps identify problematic issues of individual lectures and make appropriate changes to the program.

By comparing this approach with other options of courses, the closest in structure is the Environmental management program of Open learning College, Brentwood (2016). This program contains topics that largely coincide with the topics covered in this project: Introduction to Environment, Global Environmental Issues, Environmental Management, Environmental Management Systems, Environmental Law, Environmental Policy, Planning for Environmental Improvement, Implementation-Improving Environmental Performance, Checking Environmental Performance, Reviewing and Reporting Environmental Performance. The duration of this course is one year, which allows for a more detailed look at all the topics and additionally includes issues related to pollution of all ecosystems, including air, water and the soil. It also provides an opportunity to establish cause-effect relationships and develop systems thinking. As a result of having this course, students receive a diploma in environmental management, while the course "Environmental Management in the EU" is part of a training program for managers.

Pallant et al. (2020) from Allegheny College (USA), consider five different courses in Environmental Science and Sustainability (ESS): Introduction to Sustainability, "Think Sustainability"; Environmental Problem Analysis - "Analyze Sustainability"; Environmental Research Methods—"Research Sustainability"; Junior Seminar in Sustainable Development (Semester long)—"Apply Sustainability" and Senior Project (Year Long)—"Becoming an Agent of Change". In many aspects, the content of these courses coincides with the themes of the "Environmental Management in EU" program. Graduates of this 4-years curriculum get degree in ESS whereas the course "Environmental management in EU" is an integral part of programs for managers and economists.

Oxford University syllabus (Pallister n.d) on Environmental Management also contents 10 themes referring to its different aspects but as separate lectures and more focused on environment and natural resources rather than on global/general sustainability aspects. Cambridge and Harvard programmes on Environmental Management (Cambridge University 2019; Harvard University 2015) consider synthesis from particular (separate phenomena, such as Agriculture and the Environment, Water Management, among others) to general (Global Environmental Issues, such as Resource depletion or Atmosphere pollution) approaches whereas the presented programme considers analysis from general (Global Environmental Issues such as Overpopulation) to particular (Waste management, Environmental assessment) approach. This programme permits to establish causal relationships between problems and phenomena. Uppsala University Master programme "Environmental Management in Practice" (Uppsala University 2019), in turn, has a more practical orientation and it is directed to provide skills and an improved understanding of how organisations work with sustainability issues.

5 Conclusion

The course development was made in accordance with Jean Monnet Action module "Environmental management in EU course and on-line resources (MOOC)" (2014–2017). During the development of the course, it became clear this opportunity could serve as an introduction to courses such as Sustainable Development, Environmental Management of Enterprises, Waste Management and so on. In all lectures, special attention was paid to the regulatory framework. It should be noted that the regulatory framework of environmental management in the European Union is well developed and is constantly updated, which required constant monitoring of these changes and their inclusion in lecture materials.

The need to introduce the Environmental management and Sustainable Development courses in the training of managers and economists was confirmed by the results of the Davos Economic Forum in January 2018, at which economic development issues were considered in direct connection with the state of the environment and climate change. The WEF (2018) considered top 5 global risks in 2018 in terms of impact: (1) Weapons of mass destruction, (2) Extreme weather events, (3) Natural disasters, (4) Failure of climate-change mitigation and adaptation, and (5) Water crises. Four of these risks are linked with environmental issues and this confirms the need in studying environmental management and sustainable development.

This course provides students with the knowledge, skills and motivation in different areas of sustainable development and is addressing on implementation SDGS solutions in economics, society and environment.

The advantage of the proposed approach is the systematic consideration and use of a causal analysis. The course "Environmental Management in EU" was developed in Russian and is presented on http://www.reu.by/index.php/2014-05-13-19-05-45/moos. This paper represents then an opportunity to share to a wider audience the approach used by the course and how the ten chosen themes are worked. It may useful to other universities which are in the process of developing courses in the context of environmental management or considering ways to add the SDGs to the curricula.

Some limitations of this study include the fact that the course was developed taking into account only a Russian university and was based only on documents from Russia and the European Union, with the exception of ISO. For future research in the topic, it is suggested the application of this course in other countries, so different data bases and documents about environmental sustainability can be used. With this, the course might be adjusted to integrate other views, and it could be possible to reach a module-based course to teach Environmental Management to students.

Acknowledgements The authors would like to thank Jean Monnet Fund for its financial support to the project in accordance with Agreement n/2014-2385/001-001 PROJECT NUMBER–553553-EPP-1-2014-1-BY-EPPJMO-MODULE.

References

Agenda 21 (1992) United Nations conference on environment and development, Rio de Janeiro, Brazil, 3 to 14 June 1992, available at: https://sustainabledevelopment.un.org/content/documents/Agenda21.pdf. Accessed 20 Jan 2019

BachelorsPortal (2019) Bachelors in environmental studies and earth sciences. Environmental Management, available at: https://www.bachelorsportal.com/disciplines/287/environmental-management.html. Accessed 20 Jan 2019

Biodiversity Information System for Europe (n.d). Ecosystems and habitats, available at: https://biodiversity.europa.eu/topics/ecosystems-and-habitats. Accessed 20 July 2019

Boud D, Feletti G (2013) The challenge of problem-based learning. Routledge, London

Brentwood (2016) BrentwoodOpen learning colledge. Diploma in environmental management, available at: www.bolc.co.uk. Accessed 12 June 2019

Cambridge University (2019) Environmental Management, available at: https://www.cambridgeinternational.org/Images/329760-2019-2021-syllabus.pdf. Accessed 20 Jan 2019

EEA (n.d.) Celebrating Europe and its environment, available at: https://www.eea.europa.eu/environmental-time-line/1970s. Accessed 4 June 2019
EMAS (n.d.) Environment. Eco-Management and audit scheme, available at: http://ec.europa.eu/environment/emas/. Accessed 4 June 2019
Erasmus+ (n.d.) Jean monnet modules. Erasmus+, available at: https://ec.europa.eu/programmes/erasmus-plus/node_en. Accessed 4 June 2019
European Commission (2011a) Communication from the commission to the European parliament, the council, the European economic and social committee and the committee of the regions. Final Assessment. Brussels, 31.8.2011, available at: https://eur-lex.europa.eu/LexUriServ/LexUriServ.do?uri=COM:2011:0531:FIN:EN:PDF. Accessed 19 July 2019
European Commission (2011b) Report from the comission to the European rarliament, the council, the european economic and social committee and the committee of the regions on the thematic strategy on the prevention and recycling of waste. Brussels, 19.1.2011, available at: https://eur-lex.europa.eu/LexUriServ/LexUriServ.do?uri=COM:2011:0013:FIN:EN:PDF. Accessed 19 July 2019
European Commission (2015) Europe's eyes on Earth. Copernicus Brochure. European Commission. Directorate-General for communication publications. Brussels, Belgium, available at: https://www.copernicus.eu/sites/default/files/documents/Copernicus_brochure_EN_web_Oct 2017.pdf. Accessed 19 July 2019
European Commission for Environment (n.d.) Environment action programme to 2020, available at: http://ec.europa.eu/environment/action-programme/. Accessed 4 June 2019
Forrester JW (1971) World dynamics. Right-Allen Press, Cambridge, Mass
Harvard University (2015) Environmental management, available at: https://canvas.harvard.edu/courses/4345/assignments/syllabus. Accessed 20 Jan 2018
ISO (2015) 14001-2015 Environmental management systems. Requirements with guidance for use, available at: https://www.iso.org/ru/iso-14001-environmental-management.html. Accessed 20 Jan 2019
ISO (2016) 14004 2016, Environmental management systems general guidelines on implementation, available at: https://www.iso.org/ru/iso-14001-environmental-management.html. Accessed 20 Feb 2019
ISO International Organization for Standardization (2017) ISO/TC 2017, available at: https://www.iso.org/committee/54808/x/catalogue/p/0/u/1/w/0/d/0. Accessed 4 June 2019
JRC (2017) JRC annual report 2016. Publications Office of the European Union, Luxemburg
Leal Filho W, Manolas E (2012) Implementing sustainable development in higher education. In: Gonçalves F, Pereira R, Leal Filho W, Peter Lang GmbH (eds) Contribution to the UN Decade of education for sustainable development. Frankfurt am Main
Lovins H, Wijkman A, Fullerton J, Wallis S, Maxton G (2016) A finer future is possible. How humanity can avoid system collapse and craft a better economic system, available at: https://www.clubofrome.org/wp-content/uploads/2016/08/A-finer-future.pdf. Accessed 25 Mar 2019
MasterPortal (2019) Masters in environmental studies and earth sciences. Environmental management, available at: htps://www.mastersportal.com/disciplines/287/environmental-management.html. Accessed 25 Mar 2019
Meadows DH, Meadows DL, Randers J, Behrens WW III (1972) The limit to growth. Universe book, New York
Meadows DH, Randers J, Meadows DL (2004) Limit to growth: the 30-yeas update. ChelseaGreen, White River Junction, VT
Minderman M (2015) Business and management education for sustainable development: a process for creating sustainable public value. In: Filho L et al (eds) Integrative approaches to sustainable development at university level, pp 467–478. World Sustainability Series. Springer International Publishing, Swizerland
Montreal Protocol (1987) Montreal protocol on substances that deplete the ozone layer. Montreal, 16 September 1987, available at: https://treaties.un.org/doc/Treaties/1989/01/19890101%2003-25%20AM/Ch_XXVII_02_ap.pdf. Accessed 15 Jan 2019

Official Journal of the European Union (2011) Directive 2011/92/EU 201) of the European parliament and of the council of 13 December 2011 on the assessment of the effects of certain public and private projects on the environment, available at: https://eur-lex.europa.eu/legal-content/EN/TXT/?uri=CELEX%3A32011L0092. Accessed 4 June 2019

Official Journal of the European Union (2013) Decision 2013 No 1386/2013/EU of the European parliament and of the council of 20 November 2013 on a General Union Environment Action Programme to 2020 'Living well, within the limits of our planet', available at: http://eur-lex.europa.eu/legal-content/EN/TXT/PDF/?uri=CELEX:32013D1386&from=EN. Accessed 20 Jan 2019

Pallister J (n.d.) Environmental management. A core text for O level and IGCSE. Teaching guide, 2nd edn. OXFORD University Press

Pallant E, Choate B, Haywood B (2020) How do you teach undergraduate university students to contribute to UN SDGs 2030*?* In: Filho L et al (eds) Universities as living labs for sustainable development. World Sustainability Series, Springer Nature. Switzerland. https://doi.org/10.1007/978-3-030-15604-6_5

Randers J (2012) 2052: A global forecast for the next forty years. The future in practice. The state of sustainability leadership. University of Cambridge

SDSN Australia/Pacifc (2017) Getting started with the SDGs in universities: a guide for universities, higher education institutions, and the academic sector. Australia, New Zealand and Pacifc Edition. Sustainable development solutions network—Australia/Pacifc, Melbourne

UNFCCC (1992) United Nations framework convention on climate change, available at: http://unfccc.int/resource/docs/convkp/conveng.pdf. Accessed 15 January 2019

United Nations (1992) Convention on biological diversity, available at: https://www.cbd.int/doc/legal/cbd-en.pdf. Accessed 20 Jan 2019

United Nations (1998) The Kyoto protocol to the United Nation framework convention on climate change, available at: http://unfccc.int/resource/docs/convkp/kpeng.pdf. Accessed 20 Jan 2019

United Nations (2000) United Nations millennium declaration. 55/2. United Nations general assembly, available at: https://www.un.org/millennium/declaration/ares552e.htm. Accessed 15 Jan 2019

United Nations (2007) Indicators of sustainable development: guidelines and methodologies, 3rd edn. United Nations, New York, available at: https://sustainabledevelopment.un.org/content/documents/guidelines.pdf (accessed 19 July 2019)

United Nations (2015a) Transforming our world: the 2030 Agenda for sustainable development. A/RES/70/1, available at: https://www.un.org/en/development/desa/population/migration/generalassembly/docs/globalcompact/A_RES_70_1_E.pdf. Accessed 15 Jan 2019

United Nations (2015b) Paris Agreement, available at: https://unfccc.int/files/essential_background/convention/application/pdf/english_paris_agreement.pdf. Accessed 25 Jan 2019

United Nations Environment Programme (1985) Handbook for the Vienna convention for the protection of the ozone layer. 12th edn (2019), available at: https://ozone.unep.org/sites/default/files/Handbooks/VCHandbook_W_2019.pdf. Accessed 10 July 2019

UN-Knowledge Platform (n.d.) Helping governments and stakeholders make the SDGs a reality, available at: https://sustainabledevelopment.un.org. Accessed 04 June 2019

Uppsala University (2019) Syllabus for environmental management in practice, available at: http://www.uu.se/en/admissions/master/selma/kursplan/?kKod=1TV027. Accessed 15 Jan 2019

WEF (2018) Global risk report. World Economic Forum. Geneva, available at: https://www.weforum.org/reports/the-global-risks-report-2018. Accessed 20 Feb 2019

Zenchanka S, Batyrev V (2011) Environmental management: from environmental monitoring to decision making. In: Sustainable Development: society, education, technology, economy, ecology. Abstracts of European seminar, pp 98–99. Minsk. Belarus (in Russian)

Zenchanka Siarhei is a professor of the chair of Management, Accounting and Finance of the Minsk Branch of the Plekhanov Russian University of Economics. He has a Ph.D. in Applied Science. His research focus is in the areas of the sustainable development and education for sustainable development.

Gorbatchev Nikolai is a Vice-Dean of Faculty of Retraining and Advanced Studies, School of Business of Belarus State University, He has a Ph.D. in Economics ("Mathematical and Instrumental Methods of Economics"). He is the author of more than 50 scientific works on modeling of information systems, sustainable development, e-learning technologies.

Fernanda Frankenberger is a Ph.D. Student at the Business School at PUCPR and did her Master of Science in Strategic Management (PUCPR 2016). She is Bachelor of Administration PUCPR (2001). She is currently lecturer at Universidade Positivo in Brazil. Her research focus is in the areas such as sustainable development, sustainable universities and sustainable businesses.

Amanda Lange Salvia has a degree in Environmental Engineering (2014), has a PhD in Civil and Environmental Engineering at Passo Fundo University, Brazil. She has experience in research since 2011, and current interests include energy efficiency, sustainable cities, environment management and the Sustainable Development Goals.

Sustainable Development in Higher Education Institutions: Accounting for Sustainability in Higher Education Institutions by a Data Mining Analysis of Publications

Samara da Silva Neiva, Wellyngton Silva de Amorim, André Borchardt Deggau, Rafael Gustavo de Lima, Gabriélli do Livramento Gonçalves, Flávio Ceci, Robert Samuel Birch, and José Baltazar Salgueirinho Osório de Andrade Guerra

1 Introduction

As evidence grows of the environmental impact of human activity on our planet, the role of sustainable development (SD) is increasingly important in the behavior of organizations. The Sustainable Development Goals (SDGs) of the United Nations

S. da Silva Neiva (✉) · W. S. de Amorim · A. B. Deggau · R. G. de Lima ·
G. do Livramento Gonçalves · F. Ceci · J. B. S. O. de Andrade Guerra
Center for Sustainable Development (Greens), University of Southern Santa Catarina (UNISUL), Santa Catarina, Brazil
e-mail: samara.neiva@unisul.br

W. S. de Amorim
e-mail: wellyngton.amorim@gmail.com

A. B. Deggau
e-mail: andre.deggau@gmail.com

R. G. de Lima
e-mail: rafael.lima@ufsc.br

G. do Livramento Gonçalves
e-mail: gabylivramento@gmail.com

F. Ceci
e-mail: flavio.ceci@unisul.br

J. B. S. O. de Andrade Guerra
e-mail: baltazar.guerra@unisul.br

R. S. Birch
School of Engineering, University of Liverpool, Liverpool, UK
e-mail: r.s.birch@liverpool.ac.uk

W. Leal Filho et al. (eds.), *Universities, Sustainability and Society: Supporting the Implementation of the Sustainable Development Goals*, World Sustainability Series,
https://doi.org/10.1007/978-3-030-63399-8_33

General Assembly Resolution 70/1, "Transforming our world: the 2030 Agenda", offers a broad framework of interdependent organizational goals covering issues of social and economic development (UN General Assembly 2015). Thus, sustainability science has emerged as a fresh line of academic research to create new knowledge in tandem with the advancing policies and practices of sustainability. Consequently, a growing body of work is now available in all disciplines which investigates how sustainable development can be progressed. However, it is important to note that issues such as water, energy, health, agriculture and biodiversity (i.e., the so-called WEHAB targets) are not fully described by any single individual discipline but, instead, require an inter-disciplinary approach.

Much of the literature on the subject is produced by higher education institutions (HEIs) through non-profit making research but funding agencies still seek justification and an eventual market value that reflects the value of investment for people and knowledge; Both are difficult to measure in a traditional sense of financial management. A sustainability assessment of the higher education sector should in some way reflect the contribution that HEIs make to the implementation of SDGs. In turn, it is recognized that applying such a processes aids in the integration of good sustainability practices into their organizational structure, helps their decision-making and external communications with stakeholders (Hopwood et al. 2010). Accounting for sustainability in terms of physical data and presenting this information in a clear and meaningful way goes some way to addressing the major challenges and risks posed by sustainability-related issues. But, deciding what key performance indicators (KPIs) should be gathered, measured, analyzed and reported on requires careful consideration. A structured approach is necessary in order to assess their significance to SDGs, the implications for individual HEIs and the higher education sector as a whole.

There are many KPIs in the higher education sector that could be looked at in terms of sustainability. The work presented in this article examines how an accounting for sustainable analysis process could be applied to the publication output appearing in the well-known International Journal of Sustainability in Higher Education (IJSHE). The process investigates the frequency of terms appearing in these publications and their significance to the implementation of sustainability in the operational management, research and teaching activities in HEIs. Accounting for sustainability in the higher education sector in this way offers a readily understandable performance metric and potential influencer on local communities and wider society in general. This investigation is achieved through a data mining method based on a semi automatic approach for the recognition, correlation and extraction of named entities and relations from textual documents as reported by Ceci et al. (2012) which are used to create and classify Web-based data.

2 Sustainability in HEIs: A Brief History

In the last few decades, several initiatives to promote SD in HEIs have emerged, demonstrating how relevant universities are in driving this initiative forward (Aleixo et al. 2018). Since the 1970s, there is an increasing international awareness about the inherent connections there are between HEIs, environmental education and SD education (Mckeown and Hopkins 2010; Pereira et al. 2014). In the drive towards understanding these connections, some conferences, declarations and initiatives are highlighted as key milestones. In particular, the Tbilisi Declaration, the Talloires Declaration and the United Nations Decade of Education for Sustainable Development and the Sustainable Development Goals (Berchin et al. 2020).

Generally, environmental education is a term used to address relations between education and sustainability awareness that should also be implemented in HEIs operational activities as much as it is in their teaching curricula (i.e., practice what you teach). This example helps the wider society to become more aware and willing to participate actively in actions that consider sustainable behavior and thinking (Zsóka et al. 2013; Dmochowski et al. 2016). To be effective, sustainability must be implemented in all sectors of HEI operations, i.e., education, research, management, community engagement, culture and governance (Aleixo et al. 2018). These activities should be monitored so that greater effort is made to implement them, stimulating dialogue and participation among all members of the university (Calitz et al. 2018). Thus, the implementation of initiatives that stimulate environmental awareness and practical actions contributes to the creation of greening campus processes, improving academic sustainability and the integration between HEIs' users and the society (Ribeiro et al. 2017).

3 Sustainable Development in HEIs: An Overview

HEIs are the places where future leaders are educated (Freidenfelds et al. 2018). As they have a critical role in promoting research and innovation, they promote experiences, shape behaviors, create and disseminate knowledge—making them drivers of social changes (Berchin et al. 2018). HEIs are also major employers and economic actors. For example, if all British HEIs were put together, they would create the fifth biggest city in the UK. As leaders in sustainability challenges, HEIs can develop technology and raise environmental awareness between students and the wider community (Wadud et al. 2019). HEIs focus on teaching, extension/outreach projects and research which contributes to the development of greening campus processes and, consequently, setting a standard for the reduction of waste and energy.

Traditionally HEIs are seen are as guardians of reason based on the pillars of creating knowledge through research and disseminating it through their teaching and publications. However, it can be observed that their purpose has changed somewhat to a focus on social mobility allowing more people to transform their lives. But they still they aim to hold a position at the frontier of knowledge always pushing at the boundaries in pursuit of more and looking for way to disseminate it effectively. Their capacity to shape opinion, society and create awareness is a powerful force to achieving sustainable development, especially when emphasizing interdisciplinary and multidisciplinary approaches to research (Berchin et al. 2018). This integration between society and science helps to break the sense of mistrust that sometimes permeates between them which, as yet, has not been fully overcome by current science structure (Čada and Ptáčková 2013).

Scientific discoveries and technological decisions affect human society and the natural environment while people make decisions for social and environmental reasons that ultimately guide the work of scientists and engineers. Therefore, the idea of HEIs as places to implement sustainable development is exploitable and is recognized by the United Nations from the early 1970s (Casarejos et al. 2017). Since the UN Conference on Human Environment, in 1972, HEIs have taken increasingly active measures to support sustainability in their campuses such as declarations, redesign of curricula and regional and global partnerships. The UN has continued to promote events for stimulating sustainability in HEIs, such as the UN Decade of Education for Sustainable Development (2005–2014) (Findler et al. 2019).

Casarejos et al (2017) identified 14 international initiatives to foster sustainability in HEIs, the first of them being the Talloires Declaration (1990). This declaration recognized the role universities play in meeting sustainability challenges and its broader themes are the basis of several sustainability plans in HEIs (Berchin et al. 2018). The Talloires Declaration aimed to increase awareness of the population, create a culture of sustainability within HEIs, educate for sustainable development and support primary and secondary schools to address issues such as population and sustainability in an interdisciplinary context (Association of University Leaders for Sustainable Development 1990). The concept of SD has emerged out of concerns about the future of Earth and the humanity in face of limited resources, considering social, environmental and economic spheres—the so-called triple bottom line of sustainability (Zemigala 2019). The goal is to promote a SD that meets the needs of today without compromising the ability of future generations to meet their own needs of tomorrow (WCED 1987). Therefore, SD aims to create and maintain social, economic and environmental wealthy systems in order to achieve a better and more sustainable future for all (Folke et al. 2002).

Since its formal introduction in literature in 1987, when the Brundtland Report was published, SD has been a key argument that guides and shapes political decision-making around the world (Gazzola et al. 2019). Although some barriers seem immovable and the institutions seem ill-equipped at times for life in the Anthropocene (Gaffney 2014), countries have committed in the Rio+20 United Nations conference in 2012 to then create SDGs adding to the Millennium Goals (Griggs et al. 2013). These SDGs were inserted in the 2030 Sustainable Development Agenda as blueprint for peace and prosperity for people and the planet and are separated in 17 categories that work as an urgent call for action by all countries in a global partnership (UN General Assembly 2015). Building on the Millennium Development Goals (MDGs), SDGs continue to prioritize the fight against hunger and poverty, while embracing other fundamental concerns for development and welfare of humanity and its environments (World Health Organization 2015).

Despite being designed to be universal in the sense of embodying a universally shared common global vision of progress towards a safe, just and sustainable space for all human beings to thrive on the planet, all goals and targets present different levels of challenges and ambition, considering current development states and other conditions that are particular to each nation (Osborn et al. 2015). Due to the need for global, integrated and scientifically based information on sustainable development (Hák et al. 2016), SDGs represent a more efficient agenda when compared to MDGs—since they represent an integrated system, facilitating interaction between policy-makers (Le Blanc 2015) and being comprehensive in meeting the needs of developed and developing countries.

4 Methodology

The previous section illuminated the context and importance of the role that HEIs can play in promoting SD. But what are the issues that fall under the ambit of sustainable development and how do we translate these concerns into an analytical accounting context that distinguishes social and/or environmental accounting more clearly from any putative accounting for SD? Clearly, the literature output from the HEIs offers a measurable parameter but what analysis do we apply considering the rapidly growing volume of Web-based free-text documents containing information on science and technology (Chu et al. 2005)? To answer these questions an analysis of SD research in HEIs was conducted using a computational ontology approach based on Ceci et al. (2012) This data mining methodology represents a significant improvement over conventional manual methods in representing large amounts of information and is a process applied five distinct phases as shown in Fig. 1.

The first phases consists of a literature audit of papers in a target journal IJSHE, a reputable and typical forum for the publication of SD activities in the HEI sector and is published by Emerald Publishing. Emerald itself was founded in 1967 with its mission to defend new ideas that promote research and practice in business, management, health and social care, education and engineering and features an extensive portfolio of more than 300 journals, 2500 books and 1500 teaching cases (Emerald 2019b). It is abstracted and indexed in: A + Education, Australian Education Index, BFI (Denmark), Cabell's Directory of Publishing Opportunities in Educational Psychology and Administration, Current Contents/Social and Behavioral Sciences, EBSCO, Environmental Abstracts, e-psyche, ERIC, HEDBIB, Journal Citation Reports/Social Sciences Edition (ISI), LexisNexis Environmental, Norwegian Journal for Scientific Journals, QUALIS (A2), ReadCube Discover, The Publication Forum (Finland) (Emerald 2019a). It has a 2020 impact factor of 1.73 and a 5 year impact factor of 2.377 (Emerald 2019a). Additionally it is ranked by Scopus since 2000 with the information: Cite Score 2018: 2.29; SCImago Journal Rank (SJR) 2018: 0.542 and Source Normalized Impact per Paper (SNIP) 2018:

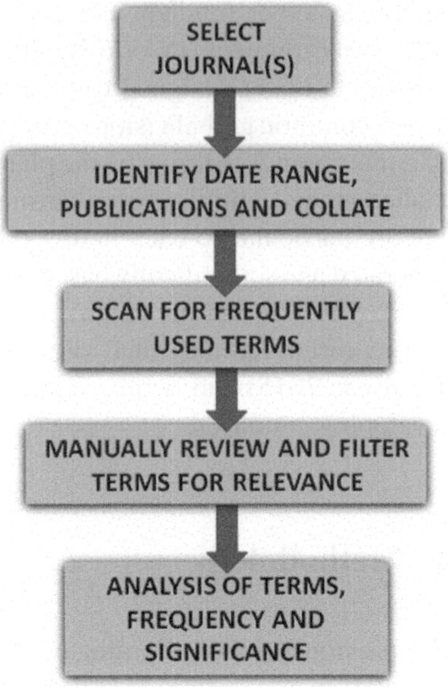

Fig. 1 Methodology flow chart used to analyze the literature

1.061 (Scopus 2019). The journal was chosen because it aims to provide up-to-date information on new developments and trends in sustainability focused on the context of higher education. It also seeks to catalyze networks and information exchange on sustainable development as a whole and on SD globally within the following areas of interest:

- Environmental management systems
- Sustainable development
- Sustainable Development Goals
- Curricular innovation
- Campus greening
- Operational aspects of universities
- Energy, water, recycling, waste management
- Planning and design of campuses
- Environmental reports
- Environmental policies and action plans
- Staff and student initiatives
  (Emerald 2019a).

Using these areas of interest, the second phase identified relevant articles published in the IJSHE from its inception in 2000 until the end of 2018. The third phase analyzed the abstracts of these articles to select the most important terms. The fourth phase organized and submitted the articles to a process of placement mining (Ceci et al. 2012) that created a list of the most important terms being used in these articles based on the frequency of citation. The fifth and final phase of the process carried out a validation of these terms where the authors reviewed the terms manually to select those most relevant to answer the research questions.

5 Results and Discussion

Ultimately, the process outlined in the methodology analyzed 618 articles and, considering the depth to which the areas of interest could be explored, care was taken to focus only on the main connections between terms related to the implementation of sustainability in universities. In this way, 100 key terms were validated by the authors manually and summarized as being used in the IJSHE publication in relation to the implementation of sustainable development in activities of HEIs (e.g., terms such as "full text" and "case study' were disregarded as irrelevant). A full list of these 100 validated terms appears in the Appendix together with their frequency of use ranking.

Each phase of the methodology yields a level of analysis that can be employed towards a strategy in accounting for sustainability. A first analysis of the IJSHE considers the growth in the number of publications per year since its first year of publication in 2000. It can be seen in Fig. 2 that, beginning with 18 articles (published in 3 issues), there is a small but steady rate of publication until 2014 when a rapid experiential growth begins.

In 2005, the IJSHE had its first special issue regarding a specific theme (Issue 3) of that year addressed Engineering Education for Sustainable Development. Other special issues followed, such as Issue 4 in 2008 regarding "Education for Sustainability" and Issue 2 of 2010, focused in the "Asia-Pacific Region". The publication of these special issues increased the overall number of articles published in the IJSHE. In 2015, following resolution 70/1 by the United Nations, that created the concept of

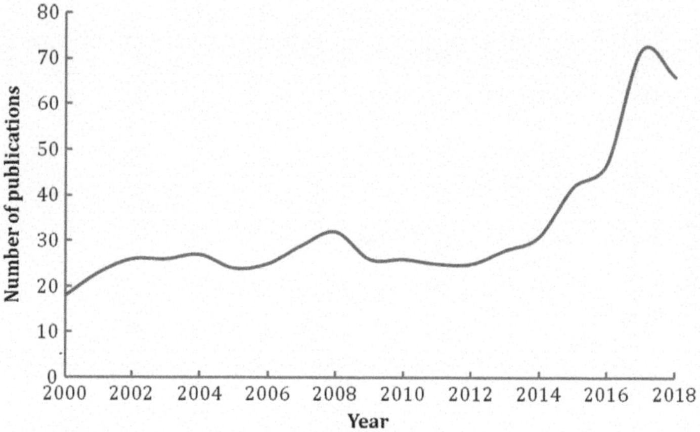

Fig. 2 IJSHE publications per year

SDGs, awareness on the study of sustainable development grew internationally and this is reflected clearly in Fig. 2. In the following years, the IJSHE had the biggest number of publications in a single year in 2016, publishing 47 articles including two special issues, with seven articles each: the first regarding sustainable development in African HEIs and the second addressing Campus Sustainability and Social Sciences.

In 2017, the IJSHE saw again the largest number of articles in a single year: 72 articles published, including 24 from special issues. In Fig. 3 it is possible to observe the number of citations that the journal presented per year (between 2011 and 2018) based on Scopus ranking (Scopus 2019). It can be seen that the number of citations per year of the journal is increasing, which confirms the relevance of IJSHE for the international scientific community working in sustainability. Data presented in Table 1 shows that most publications in the IJSHE come from developed economies. In total, 26 countries have four or more articles published in the IJSHE; 19 of them are developed countries, and only seven are classed as developing countries.

Generally, countries with higher levels of educational development and human development index will be more likely to publish articles on sustainable in HEIs, since they are more likely to experience such practices as a reality. It is also noteworthy that the four countries with more publications have English as an official language, which might facilitate easier access of researchers in these countries to publishing. Table 1 also shows that developed economies are more deeply involved with aspects of sustainable development in HEIs and possibly have more case studies in their universities to contribute to academic debates. Consequently, there is a smaller number of publications coming from developing countries. It might be concluded here that there is a need to create internal and external incentives to increase engagement in these nation's HEIs to implement sustainable actions.

Figure 4 provides are visual representation of IJSHE publication attributed to counties. Clearly visible is the large number of publications coming from northern countries, especially in Europe and North America. An analysis on the number of

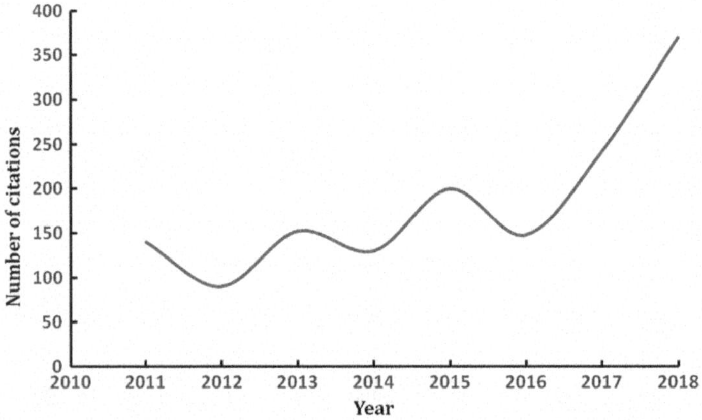

Fig. 3 IJSHE citations per year

Table 1 Number of publications in IJSHE by country

Country	Frequency	Country	Frequency
USA	196	New Zealand	11
Australia	61	Portugal	11
United Kingdom	59	Switzerland	11
Canada	54	Mexico	9
Germany	30	Brazil	7
Spain	26	China	7
Sweden	22	Taiwan	6
South Africa	IS	Denmark	5
Netherlands	13	Hong Kong	5
Italy	14	Finland	4
Malaysia	12	Ireland	4
Austria	11	Latvia	4
Greece	11	Others	72
Japan	11		

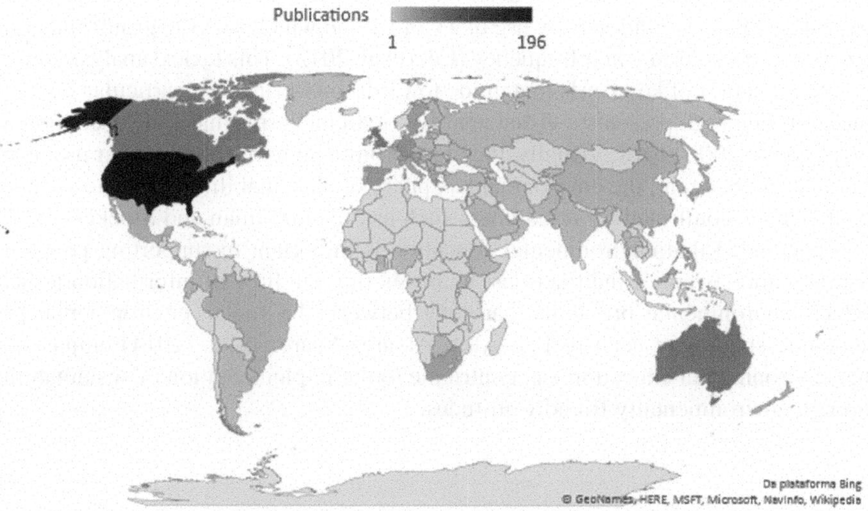

Fig. 4 World map of countries with four or more articles published in IJSHE

publications by country not only allows us to understand academic engagement in these nations, but also helps understanding how the world develops and shares their experiences with other nations. This set of data can also be helpful to identify where good practices are being applied, facilitating international cooperation among researchers and universities, boosting academic debates and incentivizing new techniques, tools, experiences and knowledge. The absence or low presence of some nations in these articles and in the academic debate regarding sustainability in HEIs suggests that there is a great gap that can work as incentive for these regions to be represented and share their own studies and experience with the academia.

The next phase of analysis process considers the validated frequency of use of the terms appearing in articles. These results, shown in the Appendix, rank "environmental education"; "environmental management", "environmental sustainability" as the top three most used terms with "environmental studies" appearing in the tenth position. By definition these focus on environmental concerns and from this we might infer that authors in the IJSHE prioritize these in the programs and curricula of HEIs. It may also reflect the influence of the 2030 Agenda and SDGs. Although these high ranking terms are have obvious connections with SD, lower ranked terms can still highlight important information about the HEI sector with regard to their strategies for promoting sustainability in activities and programs. For example, other frequently quoted terms that were validated as being relevant but not necessarily related directly to the environmental sphere, are "engineering education", "social responsibility", "engineering students", "learning outcomes", "teacher education", "social sciences" and "services learning".

The final phase of the process is to present the results of the analysis in a concise meaningful way that is readily understood by a target audience. A useful method of presenting this information is by use of a word cloud which seeks to group and organize words considering their frequency (Ceci et al. 2012). This lexical analysis allows the identification of keywords that guide towards answering to a particular research question. Figure 5, presents a visual analysis of all the terms that were used more in the period of 2000–2018. Visually, the more prominent terms have greater use in the literature with, again, the environmental sphere of sustainability being most significant. This is confirmed by researchers such as Salequzzaman and Stocker (2001), who concluded that environmental education is important in supporting people to become more aware of their actions, by promoting the flow of information to help reduce environmental problems, harmony between the social, environmental and economic spheres of each reality. Also, Álvarez-Suárez et al. (2014) emphasizes that environmental education can contribute to the implementation of sustainability through environmentally friendly attitudes.

Fig. 5 Word cloud for the IJSHE publication period 2000–2018

In relation to the term "environmental issues", higher education institutions can be of great help to educate on environmental issues through a variety of practical activities that stimulate universities HEIs to participate in their programs (Figueredo and Tsarenko 2013). Jarchow et al. (2018), presented in their study the importance of "learning outcomes" for sustainability, from the evaluation of curricula of educational courses focused on sustainability and the measurement of student performance throughout the duration of the course. Additionally, Hill and Wang (2018), consider the learning outcomes serve to create, evaluate, share and apply knowledge so that students can be prepared to be accountable leaders.

Christensen et al. (2009), present an analysis on how the HEIs can contribute to sustainable development, through "environmental management", from the implementation of new policies and plans and with the development of specific programs for sustainability. Knowledge for environmental management can be carried out in a number of ways, such as the creation of research groups, which include students and researchers at various levels from undergraduate to doctoral and post-doctoral to professorial. It is this extension from teaching and diffusion through to communities that attracts interdisciplinary relationships. Its adoption from environmental management practices such as reduction, reuse and recycling, or awareness involves all members of the HEI community, namely, students, academics and staff (Jabbour 2010). These observations from the data support the work of James and Card (2012), who cited the main factors that contribute to environmental sustainability as being green campuses, operational measures, campus administration, organization and leadership, teaching and research.

Although the word cloud shown in Fig. 5 offers an immediate summary of the results it only offers narrow view of the data and it would unwise to draw too many conclusions from this. A further dimension to the analysis is added by considering the frequency of use of the terms appearing in the IJSHE. It can be seen from Fig. 6 that

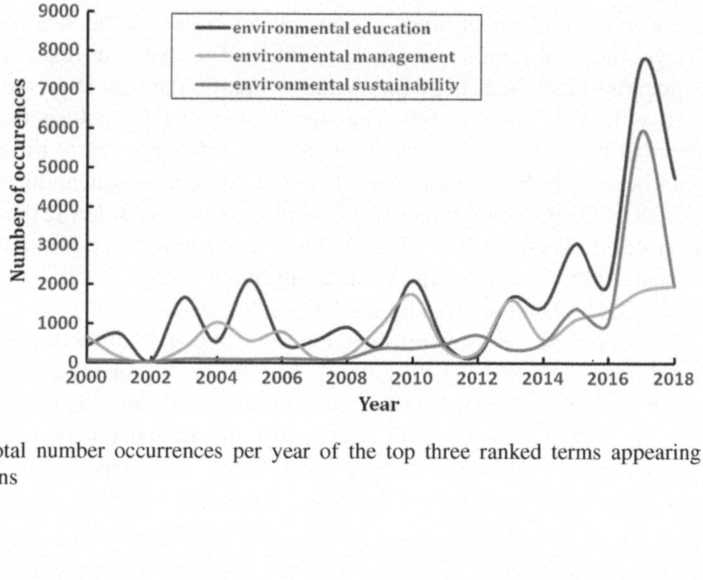

Fig. 6 Total number occurrences per year of the top three ranked terms appearing in IJSHE publications

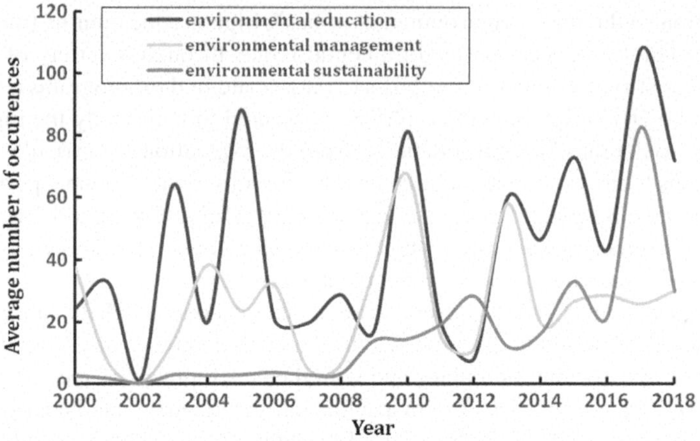

Fig. 7 Annual average occurrence of the top three ranked terms appearing in IJSHE publications

the total number of times the top three most relevant terms have appeared in IJSHE publications per year. Again, with the exception of a few years, 2000, 2004, 2006 and 2009 "environmental education" appears as the most used term with the exception of years 2000 and 2018, when "environmental management" overtook it briefly. Further observation shows that "environmental sustainability" had a large growth in its use in 2017 when it appeared nearly 6000 times (the second most used term). We might draw the conclusion here that the relationship between environmental education and sustainability strengthened at this stage.

Consideration must also be given to the fact that the rapid growth in the use of terms "environmental education" and "environmental sustainability" from 2015 in Fig. 6 is related to a large change in the total number of publications in that same year, as shown in Fig. 2. In order to investigate this comparability further, Fig. 7 considers an average number of times a term appears based on the total number of uses divided by the total number of articles published in that year (e.g., suppose a terms has appeared 1200 times in 20 publications a particular year then, the average for this terms will be 1200/20 = 60). The significance of this analysis means that further comparisons are possible considering all the eighteen years of the analyzed period. It can be seen in Fig. 7 that again that "environmental education" is nearly always the most used term in this analysis of averages. Even considering the increase in the number of articles from 2015 (Fig. 2), both "environmental education" is still the predominant term. One may argue that again that this change can be related to the Agenda 2030, launched in 2015 by the United Nations.

So far, the analysis of the terms presented in Figs. 5, 6 and 7 provides some indication about the main concerns of the authors, and their institutions, who published work in IJSHE during the 2000–2018 period. Although their use fluctuates over the years, there is an overall sense of stability of growth meaning that they have not lost their importance as time progresses. To examine this further, we need to examine clusters of terms and associations beyond those with the highest ranking. By excluding

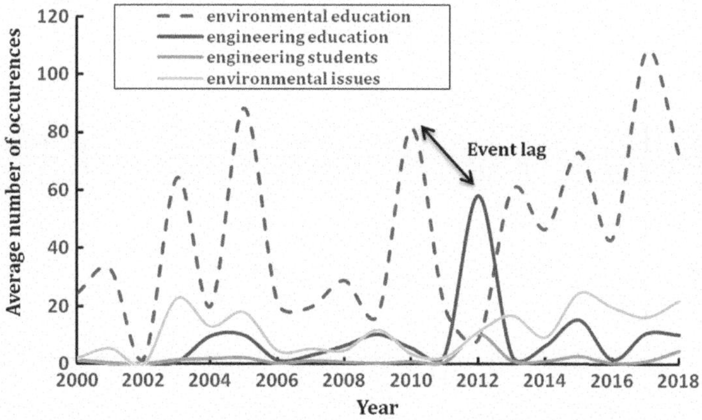

Fig. 8 Average of annual occurrence of term "environmental education" with reference to "engineering education", "engineering students" and "environmental issues"

terms "environmental sustainability" and "environmental management" and considering only "environmental education" as a reference, we can examine its association with the lower ranked terms, "engineering education", "engineering students" and "environmental issues". The results of this analysis are shown in Fig. 8 using the same time related form where key features can be observed and interpreted more easily. In this case, the bias of the cluster is towards education. The clearly defined peak of use in the term "engineering education" in 2012 may be linked to a fall-off of a peak in the use of "environmental education" two years earlier in 2010. If we consider the two events are linked in some way by historical events of the time then we can define it as an event lag triggered by some change in environmental education. In this way offers an indication about the connective inertia that may exist between the two terms. Further evidence of a link appears as a similar significant peak in use of the term "engineering students" that also occurs in 2012 together with a rise in the use of "environmental issues".

Similar to the previous analysis, Fig. 9 examines the clustered relationship between the third highest ranked reference term, "environmental sustainability", and the lesser ranked terms "social sciences", "social responsibility" and "environmental studies". Here, the bias is more towards the outward going social aspects of sustainability and by deepening the analysis we see that these unconnected terms are linked at simultaneous peak usage in 2011. These precede a marked peak in the use of "environmental sustainability" in 2012. In this case we can say these events lead and possibly indicate the presence of influencing links. It is noteworthy that "Social responsibility" also grew in importance despite finding little use up until 2007 before growing to a use of 15 times per article in average in 2018.

Following on from this line of analysis, Fig. 10 examines the relationship between the second highest ranked term, "environmental management" and the lesser terms of "learning outcomes", "service learning" and "teacher education" where the bias of this cluster may be considered to be towards the management aspects of sustainability

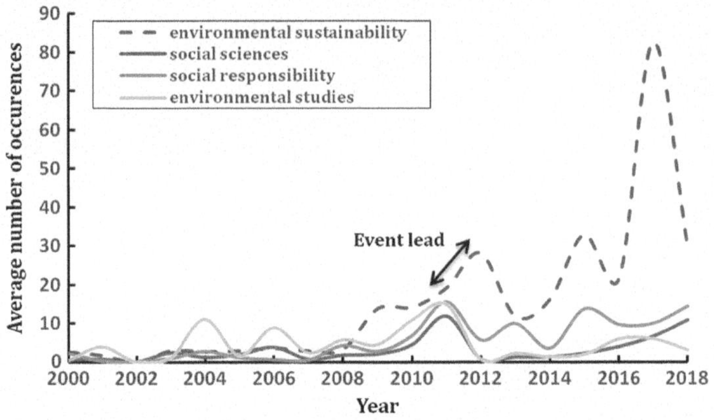

Fig. 9 Average of annual occurrence of term "environmental sustainability", with reference to "social sciences", "social responsibility" and "environmental studies"

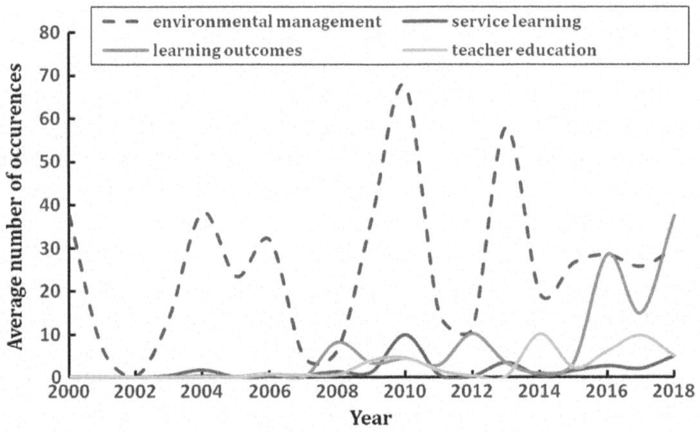

Fig. 10 Average of annual occurrence of term "environmental management" and the lesser terms of "learning outcomes", "service learning" and "teacher education

in HEIs. It can be seen that no use of the term "learning outcomes" was made in the IJSHE up until 2008 when it appeared for the first time as a peak after which it fluctuated but increased in use nearly 40 times per article. Again possible event leads and lags can be identified in the data and used to determine what the linkages there are between them.

Finally, an important consideration is that many more terms do not show high and consistent levels of relevance in IJSHE articles, such as "interdisciplinary research", "problem based learning", "socio economic", "transformative learning", "green buildings" (see Appendix) despite being relevant in the context of all terms, and being very relevant to this research. In this sense, it is also important to highlight that research paths and themes inside universities are heavily influenced by what publishing guidelines may consider to be acceptable. Under these circumstances, it might be argued that journals and publications have some influence in shaping scientific production and thinking.

6 Conclusions

The work presented in this article examines how an accounting for sustainable analysis could be applied to the publication output appearing in the IJSHE. The process investigated the frequency of terms appearing in these publications and their significance to the implementation of sustainability in the operational management, research and teaching activities in HEIs. Accounting for sustainability in the higher education sector in this way offers a readily understandable performance metric and potential influencer on local communities and wider society in general. The investigation was achieved through a data mining methods based on previous research reported by Ceci et al. (2012) which were used to create and classify Web-based data.

A basic analysis of the of the publication data shows that United States, Australia, United Kingdom and Canada stand out as significant contributors to sustainability. This could imply that these countries and their researchers are more interested in the issue of sustainability in HEIs and/or that authors from these countries have better access to publishing and funding for their research. Conversely, this analysis also shows global regions that perhaps should be encouraged to be more engaged in achieving sustainability in their own HEIs.

A more advanced analysis produced a list of the top 100 most used terms appearing the IJSHE publications and is presented in the Appendix in order of their frequency of use ranking. The most frequently used terms "environmental education", "environmental management" and "environmental sustainability" indicate that authors prioritize their focus on the environmental aspect of sustainability in their research and curricula within HEIs.

A further analysis investigated clusters of terms in order to examine potential relationships between highly ranked terms and secondary ones with a lower ranking. This analysis offers an additional tool to identify possible contributions and shortcomings of the current analysis on sustainability and allows a greater understanding of how authors and scientists interact and share their studies with the academic environment.

Such analysis must be put into context considering the triple bottom-line of sustainability: environmental, social and economic spheres. While author's perceptions might be influenced by the 2030 Agenda, it is important that SDGs are regarded in a holistic approach. In this sense, it would be beneficial for the IJSHE and other publications to incentivize content that allows the social and economic dimensions of sustainability to gain more space in the academia. This movement is equally built by authors, editors and researches, and is part of a broader effort of the scientific community and the world to better understand the concept of sustainable development in all levels.

7 Suggestions for Further Work

Future studies in the area of research presented in this article should aim to investigate how can/do universities implement sustainable development. A deeper analysis of the highest ranking frequently used terms would allow better comprehension in the role of each for making activities and programs more sustainable. Not only the most important terms, such as "environmental education", would benefit from deeper analysis but also some of the lesser terms ranked terms that have grown in the past few years. Finally, the same methodology should be applied to other relevant journals, in order to understand whether the trends observed here are global or exclusive to the IJSHE.

Acknowledgements This study was conducted by the Centre for Sustainable Development (Greens), from the University of Southern Santa Catarina (Unisul) in the context of the project BRIDGE—Building Resilience in a Dynamic Global Economy: Complexity across scales in the Brazilian Food-Water-Energy Nexus; funded by the Newton Fund, Fundação de Amparo à Pesquisa e Inovação do Estado de Santa Catarina (FAPESC), Coordenação de Aperfeiçoamento de Pessoal de Nível superior (CAPES), National Council for Scientific and Technological Development (CNPq) and the Research Councils United Kingdom (RCUK).

Appendix

Frequency of use ranking	Term	Frequency of use ranking	Term
1	Environmental education	51	Organizational change
2	Environmental management	52	Learning activities
3	Environmental sustainability	53	High school
4	Environmental issues	54	Curriculum design
5	Teaching learning	55	Interdisciplinary research
6	Engineering education	56	Academic institutions
7	Social responsibility	57	Project based learning
8	Learning outcomes	58	Triple bottom line
9	Environmental science	59	Strategic planning
10	Environmental studies	60	Qualitative data
11	Professional development	61	Human resources

(continued)

(continued)

Frequency of use ranking	Term	Frequency of use ranking	Term
12	University students	62	Interdisciplinary approach
13	Future generations	63	Environment development
14	Social sciences	64	Academic programs
15	Environmental policy	65	Policy makers
16	Experiential learning	66	Socio cultural
17	Teacher education	67	Political science
18	Ecological footprint	68	Human development
19	Learning process	69	Educational programs
20	Social science	70	Learning objectives
21	Decision makers	71	Stakeholder engagement
22	University campus	72	Teaching learning
23	Science education	73	Learning environments
24	Environmental sciences	74	Strategic plan
25	Engineering students	75	Integrated approach
26	Economic social	76	Social cultural
27	Development education	77	Environmental planning
28	Sustainable design	78	Institutional change
29	Service learning	79	Inter disciplinary
30	Curriculum development	80	Development issues
31	Environmental behaviour	81	Engineering curriculum
32	Active learning	82	Information technology
33	Social economic	83	Program development
34	Socio economic	84	Environmental research
35	Problem based learning	85	Green buildings
36	Learning environment	86	Organisational change
37	Environmental management system	87	Comparative study
38	University community	88	Education program
39	Environmental attitudes	89	Human environment
40	Education policy	90	General public
41	Learning processes	91	Different approaches

(continued)

(continued)

Frequency of use ranking	Term	Frequency of use ranking	Term
42	Sustainability management	92	Sustainability indicators
43	Transformative learning	93	University teachers
44	Learning experiences	94	Higher educational institutions
45	Education programs	95	Education training
46	Teaching methods	96	Data collected
47	Teacher training	97	Empirical study
48	Course content	98	Staff development
49	Formal education	99	Public sector
50	Social learning	100	Information systems

Top one hundred most used terms appearing in IJSHE publications years 2000–2018

References

Aleixo AM, Azeiteiro U, Leal S (2018) The implementation of sustainability practices in Portuguese higher education institutions. Int J Sustain High Educ 19(1):146–178. https://doi.org/10.1108/ijshe-02-2017-0016

Álvarez-Suárez P, Vega-Marcote P, Mira RG (2014) Sustainable consumption: a teaching intervention in higher education. Int J Sustain High Educ 15(1):3–15. https://doi.org/10.1108/IJSHE-06-2011-0044

Association of university leaders for a sustainable future (1990) Association of university leaders for a sustainable future the talloires declaration. The talloires declaration, 1. Retrieved from http://www.ulsf.org/programs_talloires.html

Berchin II, Sima M, de Lima MA, Biesel S, dos Santos LP, Ferreira RV, de Andrade JBSO, Ceci F (2018) The importance of international conferences on sustainable development as higher education institutions' strategies to promote sustainability: a case study in Brazil. J Clean Prod 171:756–772. https://doi.org/10.1016/j.jclepro.2017.10.042

Berchin II, de Amorim WS, Valduga IB, Heerdt ML, de Andrade Guerra JBSO (2020) Sustainable Campuses as Living Labs for Sustainable Development: An Overview of a Brazilian Community University. In: Leal Filho W et al (eds) Universities as living labs for sustainable development. World Sustainability Series, Springer, Cham

Čada K, Ptáčková K (2013) Possibilities and limits of collaboration between science and NGOs in the Czech Republic. J Clean Prod 49:25–34. https://doi.org/10.1016/j.jclepro.2012.07.039

Calitz A, Bosire S, Cullen M (2018) The role of business intelligence in sustainability reporting for South African higher education institutions. Int J Sustain High Educ 19(7):1185–1203. https://doi.org/10.1108/IJSHE-10-2016-0186

Casarejos F, Frota MN, Gustavson LM (2017) Higher education institutions: a strategy towards sustainability. Int J Sustain High Educ 18(7):995–1017. https://doi.org/10.1108/IJSHE-08-2016-0159

Ceci F, Pietrobon R, Gonçalves AL (2012) Turning text into research networks: information retrieval and computational ontologies in the creation of scientific databases. PLoS ONE 7(1):e27499

Christensen P, Thrane M, Herreborg Jørgensen T, Lehmann M (2009) Sustainable development: assessing the gap between preaching and practice at Aalborg University. Int J Sustain High Educ 10(1):4–20. https://doi.org/10.1108/14676370910925217

Chu WW, Liu Z, Mao W, Zou Q (2005) A knowledge-based approach for retrieving scenario-specific medical text documents. Control Eng Pract 13:1105–1121

Dmochowski JE, Garofalo D, Fisher S, Greene A, Gambogi D (2016) Integrating sustainability across the university curriculum. Int J Sustain High Educ 17(5):652–670. https://doi.org/10.1108/IJSHE-10-2014-0154

Emerald (2019a) Int J Sustain High Educ Inf. Available at: http://www.emeraldgrouppublishing.com/products/journals/journals.htm?id=ijshe#news

Emerald (2019b) About Emerald. Available at: http://www.emeraldgrouppublishing.com/about/index.htm

Figueredo FR, Tsarenko Y (2013) Is "being green" a determinant of participation in university sustainability initiatives? Int J Sustain High Educ 14(3):242–253. https://doi.org/10.1108/IJSHE-02-2011-0017

Findler F, Schönherr N, Lozano R, Reider D, Martinuzzi A (2019) The impacts of higher education institutions on sustainable development: a review and conceptualization. Int J Sustain High Educ 20(1):23–38. https://doi.org/10.1108/IJSHE-07-2017-0114

Folke C, Carpenter S, Elmqvist T, Gunderson L, Holling CS, Walker B (2002) Resilience and sustainable development: building adaptive capacity in a world of transformations. AMBIO: A J Hum Environ 31(5):437–440. https://doi.org/10.1579/0044-7447-31.5.437

Freidenfelds D, Kalnins SN, Gusca J (2018) What does environmentally sustainable higher education institution mean? Energy Procedia 147:42–47. https://doi.org/10.1016/j.egypro.2018.07.031

Gaffney O (2014) Sustainable development goals: improving human and planetary wellbeing. Glob Change 82:20–23

Gazzola P, Del Campo AG, Onyango V (2019) Going green vs going smart for sustainable development: Quo vadis? J Clean Prod. https://doi.org/10.1016/j.jclepro.2018.12.234

Griggs D, Stafford-Smith M, Gaffney O, Rockström J, Öhman MC, Shyamsundar P, Steffen W, Glaser G, Kanie N, Noble I (2013) Sustainable development goals for people and planet. Nature 495(7441):305–307. https://doi.org/10.1038/495305a

Hák T, Janoušková S, Moldan B (2016) Sustainable development goals: a need for relevant indicators. Ecol Ind 60:565–573. https://doi.org/10.1016/j.ecolind.2015.08.003

Hill LM, Wang D (2018) Integrating sustainability learning outcomes into a university curriculum: a case study of institutional dynamics. Int J Sustain High Educ 19(4):699–720. https://doi.org/10.1108/IJSHE-06-2017-0087

Hopwood A, Unerman J, Fries J (2010) Accounting for sustainability—practical insights. Routledge, London

Jabbour JC (2010) Greening of business schools: a systemic view. Int J Sustain High Educ 11(1):49–60. https://doi.org/10.1108/14676371011010048

James M, Card K (2012) Factors contributing to institutions achieving environmental sustainability. Int J Sustain High Educ 13(2):166–176. https://doi.org/10.1108/14676371211211845

Jarchow ME, Formisano P, Nordyke S, Sayre M (2018) Measuring longitudinal student performance on student learning outcomes in sustainability education. Int J Sustain High Educ 19(3):547–565. https://doi.org/10.1108/IJSHE-11-2016-0200

Le Blanc D (2015) Towards integration at last? the sustainable development goals as a network of targets. Sustain Dev 23(3):176–187. https://doi.org/10.1002/sd.1582

Mckeown R, Hopkins C (2010) EE p ESD: defusing the worry. Environ Educ Res 9(1):117–128. https://doi.org/10.1080/13504620303469

Osborn D, Cutter A, Ullah F (2015) Universal sustainable development goals. Understanding the transformational challenge for developed countries

Pereira GDSM, Jabbour C, de Oliveira SVB, Teixeira AA (2014) Greening the campus of a Brazilian university: cultural challenges. Int J Sustain High Educ 15(1):34–47. https://doi.org/10.1108/IJSHE-10-2011-0067

Ribeiro JMP, Barbosa SB, Casagrande JL, Sehnem S, Berchin II, da Silva CG, da Silveira, ACM, Zimmer, GAA, Faraco RÁ, de Andrade JBSO (2017) Promotion of sustainable development at universities: the adoption of green campus strategies at the University of Southern Santa Catarina, Brazil. In: Handbook of theory and practice of sustainable development in higher education, pp 471–486. Springer, Cham

Salequzzaman M, Stocker L (2001) The context and prospects for environmental education and environmental careers in Bangladesh. Int J Sustain High Educ 2(2):104–127. https://doi.org/10.1108/14676370110388309

Scopus (2019) Int J Sustain High Educ. Available at: https://www-scopus.ez222.periodicos.capes.gov.br/sourceid/144819#

UN General Assembly (2015) Transforming our world: the 2030 Agenda for sustainable development. Available at: https://sustainabledevelopment.un.org/content/documents/21252030%20Agenda%20for%20Sustainable%20Development%20web.pdf

Wadud Z, Royston S, Selby J (2019) Modelling energy demand from higher education institutions: a case study of the UK. Appl Energy 233–234(April 2018):816–826. https://doi.org/10.1016/j.apenergy.2018.09.203

WCED (World Commission on Environmental and Development) (1987) Our common future. Oxford University Press, New York. Available at https://sustainabledevelopment.un.org/content/documents/5987our-common-future.pdf

World Health Organization (2015) Health in 2015: from MDGs, millennium development goals to SDGs, sustainable development goals

Zemigala M (2019) Tendencies in research on sustainable development in management sciences. J Clean Prod 218:796–809. https://doi.org/10.1016/j.jclepro.2019.02.009

Zsóka Á, Szerényi ZM, Széchy A, Kocsis T (2013) Greening due to environmental education? Environmental knowledge, attitudes, consumer behavior and everyday pro-environmental activities of Hungarian high school and university students. J Cleaner Prod 48:126–138. https://doi.org/10.1016/j.jclepro.2012.11.030

Introduction of the Research Project EcoING—Development and Implementation of an Ecodesign Learning Factory for the University Engineering Education

Daniela Kattwinkel and Beate Bender

1 Introduction and Motivation

The topic of sustainability has been in the public for some time and still enjoys an increasing attention. Especially for companies, a sustainable orientation offers the chance for technological innovation and increases the economic competitiveness (Die Bundesregierung 2016). This requires the development of new or improved products and services. However, the development of environmentally compatible and sustainable products represents a complex challenge, since a multitude of partially competing technical, economic and ecological requirements have to be considered (Abele et al. 2008, p. 2). Ecodesign (or environmentally compatible product development) is a developmental approach that integrates these requirements throughout the entire product lifecycle. "The objective is to create sustainable solutions that satisfy human needs and desires" (Karlsson and Luttropp 2006, p. 1291).

Ecodesign is understood to be a proactive, systematic design approach, in which a product's environmental impacts are already considered during the product development to minimize the use of resources throughout the entire product life cycle without impairing other key targets such as the product's functionality, quality or cost (Johansson 2002, p. 98). Within the product development, the choice of functions, functional principles and materials determine the ecological, technical and economic product properties (Feldhusen and Grote 2013, p. 253). According to this, an early testing and improvement of the ecological properties of a product is the starting point for a successful product development (Oberender 2006, p. 31). Over

D. Kattwinkel (✉) · B. Bender
Chair for Product Development, Faculty of Mechanical Engineering, Ruhr-University Bochum, Universitaetsstraße 150, 44801 Bochum, Germany
e-mail: kattwinkel@lpe.ruhr-uni-bochum.de

B. Bender
e-mail: bender@lpe.ruhr-uni-bochum.de

W. Leal Filho et al. (eds.), *Universities, Sustainability and Society: Supporting the Implementation of the Sustainable Development Goals*, World Sustainability Series,
https://doi.org/10.1007/978-3-030-63399-8_34

the past decades, the role of the product developer has altered "from meeting needs to stimulating desires" (Cooper, 1999, p. 8). Ecodesign also seems to be promising approach to encourage the development of sustainable desires and thus to foster a more sustainable consumption among consumers for example through an increased product durability (Karlsson and Luttropp 2006, p. 1293). Altogether, a widespread application of Ecodesign will also assist in reaching the Sustainable Development Goals adopted by the United Nations, such as ensuring "access to affordable, reliable, sustainable and modern energy for all" (Goal 7) through an improvement in energy efficiency, promoting "sustained, inclusive and sustainable economic growth, full and productive employment and decent work for all" (Goal 8) through technological upgrading and innovation and ensuring "sustainable consumption and production patterns" (Goal 12) by encouraging companies to implement more sustainable practices (United Nations 2015).

Over the past decades, many international and European Ecodesign laws, standards and guidelines have been released, that support a sustainable product design such as the ISO 14040 for life cycle assessment. In addition, numerous methods and tools for Ecodesign, such as Design for Energy Efficiency or the CAD Plugin "SolidWorks Sustainability" have been developed, mainly in the academic sector (Abele et al. 2005). Their benefit, significance and application within companies has however been very limited so far (Rossi et al. 2016). Reasons for the insufficient integration of Ecodesign into the product development practice are among other things the lack of knowhow and non-existent competences in the field of Ecodesign as well as not enough time to acquire information or to receive further training (Vallet et al. 2009). A central approach for the integration of Ecodesign into the product development in companies is the development of competences among employees by means of training and education (Johansson 2002, S. 105). In order to realize this, it is necessary to include Ecodesign into the education of future employees, as there will hardly be any additional time for extensive trainings in their daily working routine. Consequently, Ecodesign should be integrated into the curricula of prospective product developers. This strategy is also emphasized by the universities themselves. A major survey of all German universities for product and industrial design showed that 79% of lecturers see the university as the most important driver for implementing sustainable design in the industry (Bader 2013, p. 5). Despite this significance, Ecodesign has hardly found its way into the curriculum of young engineers and prospective product developers in Germany. A recently conducted study accentuates this fact. We analyzed the degree course schemes, curricula and module manuals of 118 German universities (and universities of applied sciences) with a technical focus regarding the inclusion of environment-related topics (see Fig. 1) (Kattwinkel et al. 2018). In conclusion, only eleven universities (of applied sciences) offer courses that explicitly address the topic of Ecodesign in a specially designed course. These include, for example, the courses "Ergonomics and Ecodesign" (Aalen University) or the "Blue Engineering Seminar" (Technical University Berlin). A total of 98 universities offer courses on topics that are only implicitly associated with or verged on the concept of environmentally compatible product development. These include courses on topics such as lightweight engineering, environmental protection, renewable energies and

Topic of the course / PLC Categorisation	Ethics, Sustainability, Environment	Sustainable product development, Ecodesign	Lightweight engineering	Alternative drive technology	Sustainable materials	Sustainable manufacturing	Reycling and Disposal	Energy efficiency	Environm. engineering and protection	Sustainable energies	Energy and climate management	Energy economy
Manufacturing	○	◑	◑	○	●	●	○	◑	◑	○	○	○
Usage	○	◑	◑	◑	○	○	○	◑	○	○	◑	○
Recycling / Disposal	○	◑	○	○	○	○	●	◑	●	○	○	○
Greater context	●	◑	○	○	◑	○	○	◑	●	●	●	●
Total number of German courses	**6**	**11**	**60**	**15**	**1**	**22**	**21**	**29**	**54**	**76**	**27**	**34**
	5%	*9%*	*51%*	*13%*	*1%*	*19%*	*18%*	*25%*	*46%*	*64%*	*23%*	*29%*

● Full coverage ◑ Partial coverage ○ No coverage

Fig. 1 Dissemination of environment-related and sustainable courses at German universities and correlation of the course contents with the phases of the product life cycle (PLC) (based on Kattwinkel et al. 2018)

recycling. If the topics of the courses (first row) are assigned to the individual phases of the product life cycle (PLC, first column), it becomes apparent that the Ecodesign courses presently offered by German universities address all life cycle phases. No course, however, focuses on the usage phase of a product.

Thereby, the usage phase offers a considerable potential for ecological improvements, which is not yet fully exploited. For many technical products, the most significant environmental impacts are caused during a product's usage (up to 95% in the case of household appliances such as coffee machines or vacuum cleaners) (Oberender 2006, p. 18). In addition, the sub-phases of the usage phase (e.g. initial operation, use, maintenance, repair) also influence the upstream and downstream life cycle phases (Dannheim 1999, p. 46). For example, an improper use can lead to an accelerated product disposal.

The analysis of the university curricula and module manuals we carried out, also revealed, that no course systematically evaluates the possibilities for product developers to influence the environmental impacts of a product in its usage phase and the correlation between these impacts and other product properties (such as the relationship between the energy consumption and the weight of a movable object) and other life cycle phases (e.g. manufacturing or recycling) (Kattwinkel et al. 2018). Furthermore, the specific possibilities for reducing the environmental impacts of technical products in the usage phase are not sufficiently addressed in the analyzed courses. In most cases, the key elements of a sustainable or environmentally compatible product development as well as related methods and tools along the life cycle are taught without determining the causes, context and interrelations of the environmental impacts caused in the usage phase. A course on sustainable product development should not only impart knowledge but enable the students "to reflect on further effects and the complexity of behavior and decisions in a future oriented and global perspective of responsibility" (Barth et al. 2007, p. 2).

Within most engineering degree programs, traditional teaching formats (e.g. teacher-centered lectures) still dominate, although they do not comply with the current findings of academic didactics (Preißler et al. 2010, p. 1). Because of the reorientation in the course of the Bologna Process almost 20 years ago, the so-called "shift from teaching to learning" is supposed to be carried out at German universities. Especially complex and multifaceted subjects such as Ecodesign and sustainable design, in which products are developed regarding diverse and sometimes conflicting requirements call for didactically suitable teaching formats, in order to achieve a long-lasting learning success.

Against this background, the present publication addresses the following research question: How should a university course be designed that addresses the complexity of environmentally compatible product development and sensitizes engineering students effectively and lastingly to the issue of sustainability? In order to answer this question, the research project "EcoIng" as well as its challenges are summarized in Sect. 2. In Sect. 3, exemplary content and planned learning scenarios are described to create a vision of the future learning factory. Afterwards, the approach to develop this specific university course is outlined in Sect. 4. This publication ends with a conclusion and a short discussion (Sect. 5).

2 Targets and Challenges of the Ecodesign Learning Factory

The research project *"EcoING – Development and implementation of an Ecodesign Learning Factory for the university engineering education"* (sponsored by the Deutsche Bundesstiftung Umwelt—DBU) pursues the goal of enabling engineering students, with the help of an appropriate educational concept, to ecologically improve products without impairing their technical and economical properties. The innovative essence is here the unique transfer of the concept of a learning factory to the field of environmentally compatible product development. In the learning environment of the learning factory, students can acquire necessary environmentally relevant skills and competencies in a problem-oriented manner within realistic work settings. In this context, operational strategies for reducing the environmental impacts of products in the usage phase are implemented with corresponding product examples in practical learning scenarios, which represent the various activities of product development. Within the project, a new university course for environmentally compatible product development will be conceived around the Ecodesign learning factory. This course as well as the physical elements of the learning factory will be implemented at the Ruhr-University Bochum in Germany.

Due to its already described large ecological significance and the identified educational gap, the project will focus on the environmental impacts caused in the usage phase. A high practical relevance within the project is ensured by an interdisciplinary project consortium. Experienced partner companies from the manufacturing industry represent the engineering perspective (e.g. Vorwerk Elektrowerke or Ingpuls GmbH). One partner, a resource efficiency consulting agency (Effizienz-Agentur NRW), for example, has already initiated and accompanied more than 3500 projects to increase the resource efficiency in companies and can thus provide a good insight into the challenges of organizational constraints in practice. Other professional perspectives are represented by project partners from product design (Elbe Eichhorn/Duesseldorf), consumer protection, environmental authorities, associations and other academic chairs and faculties. The transferability of the concept to other teaching institutions is ensured by a practical guide that describes the conception and implementation of the Ecodesign learning factory.

2.1 Challenges of Environmentally Compatible Product Development

The terms *Sustainable Design* or *Design for Sustainability* are linked closely to the concept of a sustainable development and provide a framework for Ecodesign. *Ecodesign* (usually used synonymously with *environmentally compatible product development*) adds an additional requirement, the environment, to the classic product development objectives such as safety, reliability, technical feasibility or ergonomics (Tischner and Moser 2015) without losing sight of social aspects (Verhulst and van Doorsselaer 2015, p. 774). In the industry however, products are mostly optimized based on financial or technical criteria. Ecological product properties are often not prioritized. This is primarily due to the fact that the environmental requirements compete with numerous other requirements in product development and these conflicting objectives cannot be balanced in a trivial way (Luttropp and Lagerstedt 2006, p. 1397). As a result of these complex requirements and boundary conditions, the decision for a best possible compromise to solve this problem has to be reached in product development based upon a multitude of possible, sometimes contradictory combination of product characteristics. Thereby, it is not always possible to derive the actual properties of a product (e.g. the weight) from the previously chosen characteristics (e.g. the choice of material and the geometry) (see Fig. 2) (Weber et al. 2004, p. 2). In reality, most of the times the actual product properties (P_{Actual}) differ (ΔP) from the targeted properties (P_{Target}), because of external interferences (e.g. the user and the surroundings) that influence the system itself as well as a change of the system over time. For example, a mishandling or incorrect operation of a product can increase the immediate resource consumption and lead to a premature wear in the future.

Figure 2 illustrates the complex relations within the environmentally compatible product development. Here the product developer can principally choose between two central strategies. On the one hand, it can be evaluated which actual product

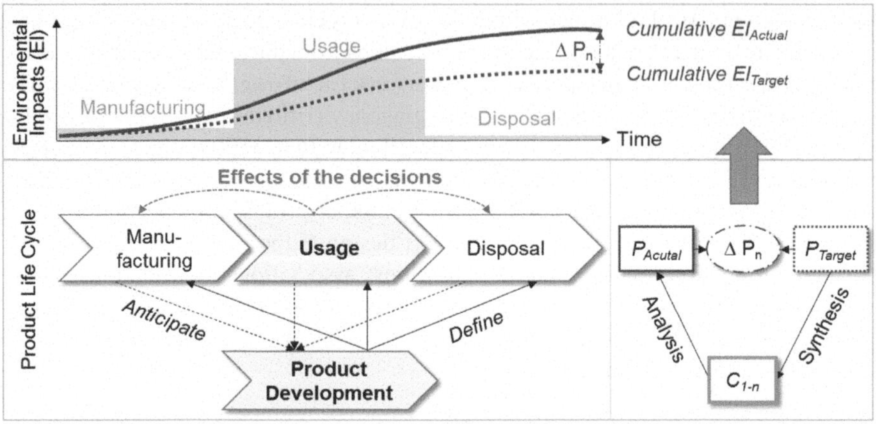

Fig. 2 Understanding of the product life cycle (left) and the relationship between properties (P) and characteristics (C) (right)

properties (P_{Actual}) will result from existing characteristics (C_{1-n}) (analysis). On the other hand, characteristics (C_{1-n}) can be derived based on anticipated properties (P_{Target}) (synthesis). In the product development practice, these two strategies do not solely occur separately but rather as a combination within iterative problem-solving cycles (Feldhusen and Grote 2013, p. 285).

2.2 *Challenges of a Learning Factory for Environmentally Compatible Product Development*

The approach of a *learning factory* provides a special contribution towards a competence orientation in the engineering education, since it is characterized by deliberate actions in the form of structured self-learning processes of the learners in authentic surroundings with a high degree of reality (Abel et al. 2013, p. 240). Classically, a learning factory is defined as a realistic working (or factory) environment with direct access to production processes to foster problem-oriented and action-oriented learning (Abele et al. 2010). In the terms of learning success, this offers many advantages over traditional teaching methods that have dominated the engineering education for so long. For various production-related sectors (e.g. process optimization or resource efficiency) not only numerous university learning factories exist so far, but also companies build learning environments tailored specifically to their needs and use this method for the training of their employees (Tisch and Metternich 2017, p. 90). While learning factories are an established teaching method in the production community, the approach to use a learning factory for the environmentally compatible product development focusing the usage phase can be declared as unique. A central challenge in the development of the Ecodesign learning factory is the exemplary realization of the authentic working environment, because a representative depiction of the actual working processes within product development has hardly been realized so far (Song et al. 2016, p. 2103) and appears much more complicated compared to the production sector. The complexity primarily results from the fact that product developers in this context must anticipate the product usage as well as all other life cycle phases and the expectations and possibilities of all involved parties (e.g. from the management, the manufacturing or from suppliers, customers or legislatures), in order to define a reasonable and appropriate combination of product properties. Each product development is thereby associated with individual goals, activities and constraints, resulting in a unique sequence of processes.

3 Outlook on the Content of the Learning Factory

Due to the challenges of teaching Ecodesign (Sect. 2.1), the concept of the new Ecodesign course will be based on authentic learning scenarios, which enable students to experience, discover and understand the processes involved in the development of environmentally compatible products. Within the learning scenarios, the students acquire experiences working with practical examples. Suitable, multi-layered product examples were selected on which strategies for improving the environmental impacts during the usage phase could be applied and which would enable a transfer of the gained knowledge to other more complex product groups. The following products are intended as demonstrational objects within the course:

- A **vacuum cleaner** is used as a representative example for energy-dominated systems, on which strategies for reducing the energy consumption during the usage will primarily be illustrated.
- A **washing machine** will act as an example of a material-converting system, which will be used to demonstrate strategies for reducing the material consumptions (in form of consumables and auxiliaries) during the usage.
- An **automatic coffee machine** will represent an example for an improvement of the service life. To identify strategies to improve the life time of the product, all usage, maintenance and service processes have to analyzed and the maintainability and reparability of the entire product as well as of single components that may lead to an early and unscheduled decommissioning have to be investigated.

The infrastructure of the learning factory will consist of both physical and virtual components, which are closely interlinked and developed simultaneously to form an optimal didactic symbiosis. The Ecodesign learning factory is divided into two connected rooms—the Design Space and the EcoLab—which represent the two views: product development (defining product characteristics) und product usage (observing product properties) (see Fig. 3). The *EcoLab* depicts a realistic living environment that can be deliberately and systematically manipulated like a laboratory set-up. In this way, different phenomena of the product usage can be observed separately (e.g. the influence of the room preparation on the usage time of the vacuum

Fig. 3 Layout of the learning factory: the design space (above) and the EcoLab (below)

cleaner and thus on its energy consumption). Not only the periphery (e.g. furniture, flooring, dust quantity) can be varied, but also the product examples themselves can be manipulated. A targeted variation of the filling level of the dust bag, its size and material or the material and degree of soiling of the dust filter or the surface finish of the suction hose are possible examples. In the specific case of the vacuum cleaner, these factors influence the suction capacity individually and collectively as a result of their complex interrelations and with that they also affect the energy requirement of the appliance. The effects of a targeted variation of the product characteristics on the product properties such as the energy consumption can be revealed in the EcoLab.

The *Design Space* will be installed in the immediate vicinity of the EcoLab. It will act as a kind of mechanics workshop in which students learn to understand and to improve the environmental impacts of the exemplary products and to discover the potential for their improvement through an analytical and systematic approach. Here students can dismantle the products for a more detailed inspection and test subsystems and components detached from the entire product system. For example, to analyze the influence of the motor on the energy consumption and the suction power of the vacuum cleaner, there will be a motor testing facility for the students.

4 Procedure for the Planning and Design of the Course

Due to the novelty and unique character of the Ecodesign learning factory, it is not possible to simply rely on standardized learning environments. Rather, a new concept must include the specific boundary conditions of the project, such as the target group and learning outcomes, requirements of the stakeholders, time and financial resources. In addition to the planning and design, the implementation and the evaluation regarding the defined objectives has to be a part of the approach applied in the project. The scientific foundation for the conception of the learning environment of the Ecodesign learning factory is laid down by the established concept of instructional design according to Seel (1999). *Instructional design* (or didactic design) was created for the systematic development of efficient (multimedia) learning environments that are suitable for promoting the acquisition of competencies. These are situation-specific models that focus on independent learning activities, an active role of the learners and support cooperative or collaborative learning (Niegemann et al. 2008, p. 23). Figure 4 illustrates the elements of instructional design and the division of the work packages of the project. The process for developing the learning factory is divided into five central work packages (see Fig. 4): planning (WP1), design of the learning environment (WP2), prototypical implementation (WP3), evaluation and further development (WP4) and documentation (WP5).

First of all, the initial planning (WP1) involves the analysis of the available resources, the assessment of the needs and general objectives as well as the analysis of the tasks (in environmentally compatible product development) and the target group. Within the resource analysis, for example, it is examined which technical, economic and legal factors influence of even restrict the design of the learning environment. Some of the results of this work packages have already been described in Sect. 2.1 (Challenges of environmentally compatible product development) and Sect. 2.2 (Challenges of a learning factory for environmentally compatible product development).

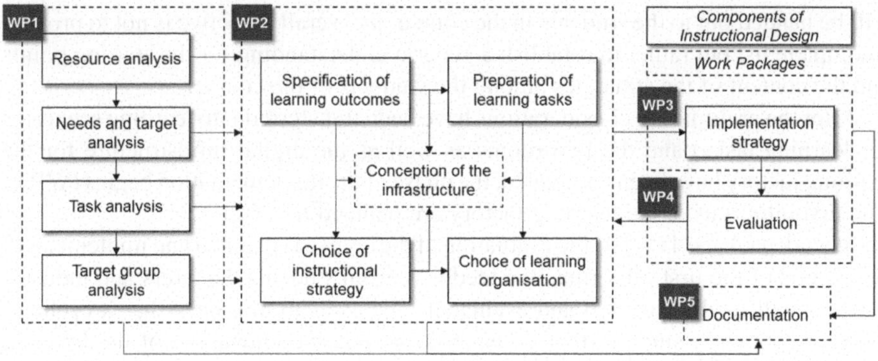

Fig. 4 Elements of instructional design according to Seel (1999) and planned work packages of the project EcoING

The design phase (WP2) includes the specification of the learning outcomes, the choice of instructional strategy, the preparation of the learning tasks, the choice of the learning organization and the conception of the digital and physical infrastructure. First results of this work packages can already be presented in Sect. 3 (Outlook on the content of the learning factory). The described learning scenarios are only examples. A systematic definition of the actual learning scenarios (in the context of instructional design, we speak of learning tasks) and their elaboration is currently in progress. The didactic conception of the learning scenarios is based on a theoretical model, which combines the systemic correlations between product characteristics and properties with various Ecodesign strategies to reduce a product's environmental impacts. For this purpose, within a literature study, numerous Ecodesign strategies could be identified and systemized regarding their relation to the usage phase. These strategies do not only focus a reduction of the energy and resource consumption but also address the issue of improving a product's service life (in accordance with the three different product examples). Specific strategies for improving the service life, which are to be dealt with in the learning scenarios are, for example (based on Pigosso 2012):

- Indicate the wear for the user (e.g. indication of required maintenance intervals).
- Simplify the access and the disassembly of components requiring maintenance.
- Use recycled or refurbished components and spare parts.
- Design the product with a modular structure so that obsolescence occurs in the components and not in the entire product.
- Enable and simplify a hardware upgrading.
- Design products for multifunctional purposes.
- Design products for a collective use (e.g. collective usage concepts such as car sharing).

This collection of strategies will be further extended within work package 2 and a systematization of the strategies according to the presented main strategies the reduction of the energy and material consumption as well as the improvement of the service life will be finalized. This catalogue of ecological improvement strategies will be handed out to the students in the course. Its overall objective is not to provide a complete list but rather to establish a generic understanding for the starting points and the content of the strategies among the students of the course.

After the preliminary considerations have been transferred into learning scenarios (or learning tasks) and the two rooms as well as the digital infrastructure for the learning factory have been completed, the phase of the implementation begins (WP3). The first pilot runs of the learning factory are planned for 2021.

In work package 4 (WP4) the evaluation of the course is planned and implemented. At the end of the first pilot runs, the feedback from students, lecturers and industry partners will be documented and evaluated. The evaluation process has a cyclical-iterative character, since further revision steps (e.g. an adjustment of the learning scenarios or a redefinition of the learning outcomes) can follow the evaluations, depending on their results.

The experiences and knowledge gained in all work packages will be documented and serve as a basis for the practical implementation guide (WP5). This should enable the dissemination of the concept to other universities. In addition to a collection of learning tasks and teaching materials, the guide should also include the experiences gained during the didactic conception of the learning environment of the learning factory for environmentally compatible product development. In particular, it will address factors that are considered to be critical for the implementation at different locations. These factors include, for example, the number of participants and suitable learning methods (e.g. teamwork or individual work), required prior knowledge or even the focus of specific learning outcomes due to the strategic orientation of a university or a course of study.

5 Summary and Discussion

In order to promote sustainability and thus to increase the development of sustainable products, it is essential to include sustainability into the university education of future engineers presented within this publication. The research project EcoING, the development and implementation of an Ecodesign learning factory, pursues the target to sensitize students for an environmentally compatible product development at an early stage of their professional life. The contents and central challenges in the development and implementation of such a learning factory were explained based on didactic concepts and first results were discussed. Currently, the specific skills and necessary competences for environmentally compatible product development are analyzed and set in the context of household appliances (due to the product examples for the learning scenarios). This knowledge is needed to derive the specific learning outcomes. Based on a newly developed model for the reducing the environmental impacts of technical products during their usage phase, the contents of the practical learning scenarios are being developed, suitable teaching and testing formats are selected, and the infrastructure is conceived. The first prototypical implementation of learning scenarios in the learning factory is expected in 2021.

In scientific literature and teaching practice, there is currently a discussion whether individual courses for sustainability are required or whether the topic of sustainability should be integrated more comprehensively into existing engineering modules and become a part of the curriculum (Olsen et al. 2015, p. 161). The future Ecodesign learning factory addresses both targets. On one hand, a new course for environmentally compatible product development will be designed that takes places in the learning factory and on the other hand, the infrastructure of the learning factory and individual learning scenarios can be integrated into existing courses even from other faculties. Furthermore, it is essential to continue to include the topic of sustainability into the curricula of degree courses not only from the field of engineering sciences and more importantly to establish it as an elementary component of a university education.

This research is sponsored by DBU (2018-2021).

sponsored by

DBU

Deutsche
Bundesstiftung Umwelt

www.dbu.de

References

Abel M, Czajkowski S, Faatz L, Metternich J, Tenberg R (2013) Kompetenzorientiertes Curriculum für Lernfabriken: Ein didaktisches hinterlegtes Konzept für Lernfabriken. wt Werkstatttechnik online 103(3)

Abele E, Anderl R, Birkhofer H (eds) (2005) Environmentally-friendly product development: methods and tools. Springer

Abele E, Anderl R, Birkhofer H, Rüttinger B (2008) EcoDesign: Von der Theorie in die Praxis. Springer, Berlin Heidelberg

Abele E, Tenberg R, Wennemer J, Cachay J (2010) Kompetenzentwicklung in Lernfabriken für die Produktion. ZWF 105(10):909–913

Bader N (2013) Nachhaltige Produktentwicklung an deutschen Hochschulen: Auswertung - Keyfindings. White Lobster

Barth M, Godemann J, Rieckmann M, Stoltenberg U (2007) Developing key competencies for sustainable development in higher education. Int J Sustain High Edu 8(4):416–430. https://doi.org/10.1108/14676370710823582

Cooper T (1999) Creating an economic infrastructure for sustainable product design. J Sustain Prod Des (8):7–11

Dannheim F (1999) Die Entwicklung umweltgerechter Produkte im Spannungsfeld von Ökologie und Ökonomie: Eine Analyse unter besonderer Berücksichtigung der Nutzungsphase. Zugl.: Darmstadt, Techn. Univ., Diss., 1999. Fortschritt-Berichte VDI Reihe 1, Konstruktionstechnik, Maschinenelemente: Bd. 320. VDI-Verl

Die Bundesregierung (ed) (2016) Deutsche Nachhaltigkeitsstrategie - Neuauflage 2016. Berlin

Feldhusen J, Grote K-H (eds) (2013) Pahl/Beitz Konstruktionslehre. Springer, Berlin Heidelberg. https://doi.org/10.1007/978-3-642-29569-0

Johansson G (2002) Success factors for integration of ecodesign in product development. Environ Manag Health 13(1):98–107. https://doi.org/10.1108/09566160210417868

Karlsson R, Luttropp C (2006) EcoDesign: what's happening? An overview of the subject area of EcoDesign and of the papers in this special issue. J Clean Prod 14(15–16):1291–1298. https://doi.org/10.1016/j.jclepro.2005.11.010

Kattwinkel D, Song Y-W, Bender B (2018) Analysis of ecodesign and sustainable design in higher education. In: Design conference proceedings, proceedings of the design 2018 15th international design conference. Faculty of Mechanical Engineering and Naval Architecture, University of Zagreb, Croatia. The Design Society, Glasgow, UK, pp 2451–2460. https://doi.org/10.21278/idc.2018.0305

Luttropp C, Lagerstedt J (2006) Ecodesign and the ten golden rules: generic advice for merging environmental aspects into product development. J Clean Prod 14(15–16):1396–1408. https://doi.org/10.1016/j.jclepro.2005.11.022
Niegemann H, Domagk S, Hessel S, Hein A, Hupfer M, Zobel A (eds) (2008) Kompendium multimediales Lernen. Springer, Berlin Heidelberg. https://doi.org/10.1007/978-3-540-37226-4
Oberender C (2006) Die Nutzungsphase und ihre Bedeutung für die Entwicklung umweltgerechter Produkte. Fortschritt-Berichte VDI. Reihe 1, Konstruktionstechnik/Maschinenelemente: nr. 385. VDI Verlag
Olsen SI, Nielsen SB, Ejlertsen ML, McAloone T (2015) Teaching sustainable solutions in engineering. Int J Innov Sustain Dev 9(2):157. https://doi.org/10.1504/IJISD.2015.068788
Pigosso D (2012) Ecodesign maturity model: a framework to support companies in the selection and implementation of ecodesign practices. University of Sao Paulo, Sao Carlos
Preißler I, Hammerschmidt J, Müller R, Scholl S (2010) Hochschuldidaktik trifft Ingenieursausbildung: Segen oder Fluch? Didaktik der Physik. Frühjahrstagung – Hannover
Rossi M, Germani M, Zamagni A (2016) Review of ecodesign methods and tools. Barriers and strategies for an effective implementation in industrial companies. J Clean Prod 129:361–373. https://doi.org/10.1016/j.jclepro.2016.04.051
Seel NM (1999) Instruktionsdesign: Modelle und Anwendungsgebiete. Unterrichtswissenschaft 1(27):2–11
Song Y-W, Herzog M, Kreimeier D, Bender B (2016) Prototype of a new learning factory—an educational approach to integrate production and product development. In: International design conference—design 2016 Dubrovnik, Croatia, 16–19 May, pp 2101–2110
Tisch M, Metternich J (2017) Potentials and limits of learning factories in research, innovation transfer, education, and training. Procedia Manuf 9:89–96. https://doi.org/10.1016/j.promfg.2017.04.027
Tischner U, Moser H (eds) (2015) Was ist Ecodesign: Praxishandbuch für Ecodesign inklusive Toolbox (1. Aufl.). Umweltbundesamt
United Nations (eds) (2015) Transforming our world: the 2030 agenda for sustainable development: A/RES/70/1
Vallet F, Millet D, Eynard B (2009) Investigation the use of eco-design guides: presentation of two case studies. In: International conference on engineering design, ICED'09, 24–27 August 2009, Stanford University, Stanford, CA, USA, pp 441–452
Verhulst E, van Doorsselaer K (2015) Development of a hands-on toolkit to support integration of ecodesign in engineering programmes. J Clean Prod 108:772–783. https://doi.org/10.1016/j.jclepro.2015.06.083
Weber C, Steinbach M, Botta C, Deubel T (2004) Modelling of product-service systems (PSS) based on the PDD approach. In: International design conference, Dubrovnik, 18–21 May 2004

Daniela Kattwinkel born in 1986, has studied Sales Engineering and Product Management as well as Mechanical Engineering at Ruhr-University Bochum. After her studies, she completed an international trainee program and began working in the position of sales coordination in a global engineering and consulting company. In 2016, she became a research assistant at the chair for product development in the research field methodical product development and has since worked with Ecodesign and sustainable product development. As part of her doctoral thesis, Daniela Kattwinkel is working on the development and implementation of a university course about environmentally compatible product development for engineering students.

Beate Bender Professor Beate Bender has many years of experience in teaching, researching and working with product development. She heads the chair for product development at the faculty of Mechanical Engineering since 2013. In addition, she was the managing director of the Institute Product and Service Engineering at the Ruhr University Bochum from 2015 until 2019. Since

2015 Prof. Bender is a member of the Board of Directors of the Institute of Corporate Governance (ifu) at Ruhr University. As a member of the Scientific Community for product development (WiGeP) and the Design Society, she has a national and international network. Prof. Bender's research interests address Engineering Design and relate to many years of experience in the industry. Since 2020 she is Co-Editor of the international standard work on Engineering Design, first edited in 1977, which has been translated into several languages: Pahl/Beitz Konstruktionslehre; Methoden und Anwendung erfolgreicher Produktentwicklung (Springer Verlag).

Universities, Sustainability and Society: A SDGs Perspective

Walter Leal Filho

1 Introduction

As seen throughout this book, sustainable development and sustainability have been considered and implemented in several sectors over the past decades (Ragazzi and Ghidini 2017). Also, universities and higher education institutions have been highlighted as major contributors to sustainable development initiatives. This is mainly attributed to the educational institutions having the power to shape the education of future generations and-inter alia-of society, by fostering principles and values among learners, that promote sustainability (Ragazzi and Ghidini 2017; Mohamed et al. 2020).

This contribution has become even more important if ones takes into account the process of implementation of the Sustainable Development Goals (SDGs), which were unanimously adopted by all member states of the United Nations in September 2015 (UN 2015). These goals encompass a broad range of economic, social, and environmental dimensions of sustainable development and entail a set of specific targets for their implementation. In order that the timeframe set by the UN to achieve these ambitious goals by 2030 is met, many mechanisms need to be put in place and deployed at the same time. One of them is the active contribution of higher education.

There are in this case some reasons for optimism. Universities are playing a key role in the SDGs implementation process, at the same time that they fulfil their societal obligations. In more recent times, sustainable practices associated with the SDGs have also been introduced in higher education institutions and universities, via the implementation of special education and research programmes, community projects

W. Leal Filho (✉)
Faculty of Life Sciences, Research and Transfer Centre "Sustainable Development and Climate Change Management", Hamburg University of Applied Sciences, Ulmenliet 20, 21033 Hamburg, Germany
e-mail: walter.leal2@haw-hamburg.de

W. Leal Filho et al. (eds.), *Universities, Sustainability and Society: Supporting the Implementation of the Sustainable Development Goals*, World Sustainability Series,
https://doi.org/10.1007/978-3-030-63399-8_35

and campus based initiatives (Mohamed et al. 2020). This is particularly helpful, since universities can indeed be considered to be "small cities" due to the amount of people and movement occurring within them (Alshuwaikhat and Abubakar 2008). Therefore, the environmental impact of higher education remains high, unless sustainable practices are timely implemented (Mohamed et al. 2020).

Traditionally, universities have been able to promote sustainability both internally and externally. Internally, they are able to function as organisations that implement practices and policies compatible with their educational mission. Externally, universities may act as facilitators in their own regions, thus aiding society in the implementation of sustainable practices (Dagiliūtė and Liobikienė 2015).

Universities also have the power to shape a student's training and societal identity via the course plans and curricula that they develop. In doing so, universities may indeed contribute towards transforming the future (Beynaghi et al. 2016). The adjustment of education courses and operations has ensured that there is a reduction in carbon emissions, and in the carbon footprints of individuals. A systematic approach enables students to take responsibility for their actions towards the environment, and further motivating them to engage in more sustainable practices (Garcia et al. 2017).

Furthermore, students make up a large population at any university, and therefore, play an important role in implementing sustainable practices within the campus. Moreover, students often express their enthusiasm to engage in more sustainable practices, and have displayed pro-environmental behaviour which is advantageous to the implantation of the sustainable development goals (Ting and Cheng 2017; Dagiliūtė et al. 2018).

From a point of view of operations, sustainability at universities has been predominantly pursued in four main areas. This includes **green buildings** which ensures that the construction and maintenance of buildings occur in an environmentally friendly sustainable way that does not use toxic, hazardous materials. The **management of waste**, which is carried out in a manner that protects the environment and takes sustainability into account. Additionally, **sustainable procurement** of goods and services occurs at certain universities ensuring that green policy guidelines are met. Lastly, **sustainable mobility** is also taken into account to ensure that staff and students travel in more sustainable and cost-effective ways. This can include the use of non CO_2 emitting vehicles such as bicycles, electric vehicles, or by encouraging the use of car-sharing services to minimise environmental damages (Leal Filho et al. 2019; Fissi et al. 2020).

2 A SDGs Perspective

The general slogan of the SDGs, "No one shall be left behind", gives a hint of the deep societal needs associated with their implementation. This slogan can also be interpreted as a promise, connected with the goals for sustainable development that all UN member states want to achieve by 2030. The total of 17 goals and 169 targets

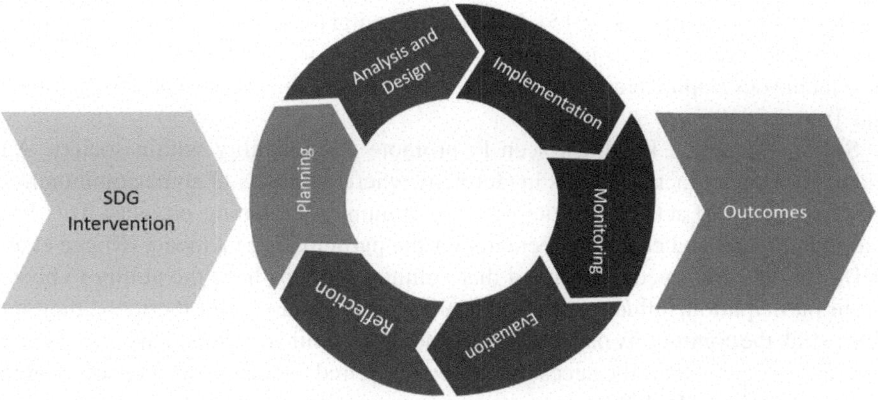

Fig. 1 Elements to be deployed to enhance the impact of sustainability interventions

also contain many concerns and demands, all of which also require political support if they are to be fully met. It should be noted that the formulation of the goals for sustainable development has taken into account many of the experiences gained in the years 2000 to 2015 in implementing the Millennium Development Goals. But what is rather unclear, is how best can universities and societies may work, hand in hand, succeeding in involving everyone in the process of comprehensive sustainable development?

The answer to this questions is not simple. But there are some factors which may be deployed, so as to increase the effectiveness and impacts of sustainability-related interventions. These are summarised in Fig. 1.

Operationally, the process of creation and support to sustainable communities is a complex task. These are communities, where members feel a strong need to ensure that they are able to live in an area and in the future, whilst ensuring that the resources for future generations are not depleted. Such communities have strong values and principles, and members actively participate in governance to sustain the wellbeing and health of individuals while protecting the environment. Sustainable communities are often developed—or supported—through university-community partnerships (Kagan et al. 2017). These partnerships are based on the principle of values, addressing the concerns of a given community, highlighting capacity building, assessing power relations and promoting the participation and engagement of members of society, sometimes adopting multi-layered approaches to achieve change (Kagan et al. 2017).

Aside from the role played by educational institutions, society itself has a major role to play in guaranteeing that the sustainable development goals are achieved. Learning institutions are also able to impact the general public. However, one needs to admit that a large percentage of individuals do not attend higher education institutions. In these instances individuals rely of the knowledge passed through the general community (Noguchi et al. 2015). It is therefore vital to ensure that suitable community projects are set up to help members of society to understand the concept of sustainable development and their role in achieving it (Chaichana et al. 2017; Franco and Tracey 2019).

Social interaction has been seen to promote sustainability within society. An example of such practice occurs in Germany where members of some communities are encouraged to actively participate in community gardening projects. This has many ecological and economic benefits for the participating members (Rogge et al. 2018). In other instances it is noted that voluntary groups have the ability to boost public participation, influence opinions, and channel resources. Furthermore, they are able to link the community participation to the local administrations. This ensures that the community receives the necessary support required to engage in more sustainable practices (Fu and Ma 2020).

A critical analysis of sustainable development across societies shows that indigenous knowledge is, unfortunately, often disregarded. However, the dissemination of knowledge between the various generations of a society allows for indigenous knowledge to be preserved and incorporated into modern practices. This often enhances sustainability, and is known to be cost effective in the long run. This is often achieved through lifelong learning, where members of the community are educated on sustainability issues related to everyday living such as farming and fishing (Noguchi et al. 2015), outside of institutional settings.

The practice of lifelong learning is especially beneficial to societies where formal educational training is unavailable and the bulk of knowledge is attained through people sharing experiences and social interactions (Wilson et al. 2018).

In all these occasions, the influence of higher education institutions may occur at various levels, and can be direct or indirect. Although we now know well what needs to be done and changed in many areas to foster sustainable development across society, the changes needed are not progressing fast enough and in some cases are not yet being implemented on a broad scale. That is why we need to look more intensively at the paths to successful change, where the role of society is better considered. This is an important prerequisite for further developing an effective model of sustainability, which is more firmly anchored in political and social practice. Apart from academia, this process of change includes the contributions of individuals and social groups, states, companies and the international community.

3 Conclusions

The current ecological, economic, health and political challenges, many of which have been exacerbated by the COVID-19 pandemic, make a comprehensive social change towards sustainability necessary. It is also now clear that setting goals, scenarios, or instruments for the implementation of the SDGs is not enough: we need concrete measures and indicators, which may allow progress to be measured.

In order to ensure that societally relevant pathways to sustainable development are followed, also (but not only) in order to implement the SDGs, a set of key components is needed in teaching, extension and in outreach initiatives as part of the so-called "third mission". Some of these components are:

(a) due emphasis to environmental boundaries and the carrying capacity of ecosystems when planning and implementing activities and projects;
(b) due considerations to demographic elements and profiles;
(c) the correct design and implementation of technological innovation to support each effort;
(d) the creation of opportunities for social and economic development;
(e) the fostering of the capacity of people and institutions to implement the SDGs, i.e. the promotion of education for sustainable development in an integrated way.

Finally, the pursuit of a sustainable societal development also needs to take into account matters related to governance. Here, decision-makers, societal actors and the scientific community should work together, in order to overcome the many challenges associated with the implementation of the SDGs. Taking all these items into account, it is clear that the many challenges associated with the nexus sustainability, university and society can only be met, if tackled in an integrated way. In this context, all members of the university community can play a key role, and should be encouraged to become engaged, and to interact with other societal actors.

References

Alshuwaikhat HM, Abubakar I (2008) An integrated approach to achieving campus sustainability: assessment of the current campus environmental management practices. J Clean Prod 16(16):1777–1785. https://doi.org/10.1016/j.jclepro.2007.12.002

Beynaghi A, Trencher G, Moztarzadeh F, Mozafari M, Maknoon R, Leal Filho W (2016) Future sustainability scenarios for universities: moving beyond the United Nations decade of education for sustainable development. J Clean Prod 112:3464–3478. https://doi.org/10.1016/j.jclepro.2015.10.117

Chaichana C, Wongsapai W, Damrongsak D, Ishihara KN, Luangchosiri N (2017) Promoting community renewable energy as a tool for sustainable development in rural areas of Thailand. Energ Procedia 141:114–118. https://doi.org/10.1016/j.jclepro.2020.123655

Dagiliūtė R, Liobikienė G (2015) University contributions to environmental sustainability: challenges and opportunities from the Lithuanian case. J Clean Prod 108:891–899. https://doi.org/10.1016/j.jclepro.2015.07.015

Dagiliūtė R, Liobikienė G, Minelgaitė A (2018) Sustainability at universities: Students' perceptions from Green and Non-Green universities. J Clean Prod 181:473–482. https://doi.org/10.1016/j.jclepro.2018.01.213

Fissi S, Romolini A, Gori E, Contri M (2020) The path toward a sustainable green university: the case of the University of Florence. J Clean Prod 279:123655. https://doi.org/10.1016/j.jclepro.2020.123655

Franco IB, Tracey J (2019) Community capacity-building for sustainable development. Int J Sustain High Edu 20(4):691–725. https://doi.org/10.1108/IJSHE-02-2019-0052

Fu Y, Ma W (2020) Sustainable urban community development: a case study from the perspective of self-governance and public participation. Sustainability 12(2):617. https://doi.org/10.3390/su12020617

Garcia J, da Silva SA, Carvalho AS, de Andrade Guerra JBSO (2017) Education for sustainable development and its role in the promotion of the sustainable development goals. In: Davim J (eds) Curricula for sustainability in higher education. Management and industrial engineering. Springer, Cham. https://doi.org/10.1007/978-3-319-56505-7_1

Kagan C, Lawthom R, Clennon O, Fisher J, Diamond J, Goldstraw K (2017) Sustainable Communities: University-Community Partnership Research on Social Dimensions of Sustainable Development. In: Leal Filho W (ed) Sustainable Development Research at Universities in the United Kingdom. World Sustainability Series. Springer, Cham. https://doi.org/10.1007/978-3-319-47883-8_14

Leal Filho W, Shiel C, Paço A, Mifsud M, Ávila LV, Brandli LL, Vargas WR (2019) Sustainable Development Goals and SUSTAINABILITY TEACHING at universities: falling behind or getting ahead of the pack? J Clean Prod 232:285–294. https://doi.org/10.1016/j.jclepro.2019.05.309

Mohamed NH, Noor ZZ, Sing CLI (2020) Environmental sustainability of universities: critical review of best initiatives and operational practices. In: Yaser A (eds) Green engineering for campus sustainability. Springer, Singapore. https://doi.org/10.1007/978-981-13-7260-5_2

Noguchi F, Guevara JR, Yorozu R (2015) Communities in Action—lifelong learning for sustainable development. UNSCO, Hamburg

Ragazzi M, Ghidini F (2017) Environmental sustainability of universities: critical analysis of a green ranking. Energ Procedia 119:111–120. https://doi.org/10.1016/j.egypro.2017.07.054

Rogge N, Theesfeld I, Strassner C (2018) Social sustainability through social interaction—a national survey on community gardens in Germany. Sustainability 10(4):1085. https://doi.org/10.3390/su10041085

Ting DH, Cheng CFC (2017) Measuring the marginal effect of pro-environmental behaviour: guided learning and behavioural enhancement. J Hospitality Leisure Sport Tourism Edu 20:16–26. https://doi.org/10.1016/j.jhlste.2016.12.001

Wilson B, Osborne M, Guevara R (2018) Lifelong learning and sustainable development from the guest editors' desk. Aus J Adult Learn 58(3):299. https://search.informit.com.au/documentSummary;dn=069152661155767;res=IELAPA;type=pdf

The manufacturer's authorised representative in the EU is Springer Nature Customer Service Centre GmbH, Europaplatz 3, 69115 Heidelberg, Germany. If you have any concerns regarding our products, please contact ProductSafety@springernature.com

Printed and bound by CPI Group (UK) Ltd, Croydon, CR0 4YY

30/09/2026

02225690-0019